AIR POLLUTION RESEARCH REPORT 43

DIMETHYLSULPHIDE: OCEANS, ATMOSPHERE AND CLIMATE

Commission of the European Communities

DIMETHYLSULPHIDE: OCEANS, ATMOSPHERE AND CLIMATE

Proceedings of the International Symposium
held in Belgirate, Italy, 13–15 October 1992

Organized by the Commission of the European Communities

Directorate-General for Science, Research and Development
DG XII/D, Environment Programme

Joint Research Centre
Environment Institute

Cosponsored by the
Danish Centre for Atmospheric Research

Edited by

GIAMBATTISTA RESTELLI
Joint Research Centre, Environment Institute, Ispra, Italy

and

GIOVANNI ANGELETTI
DG XII/D, Environment Programme, Brussels, Belgium

SPRINGER SCIENCE+BUSINESS MEDIA, LLC

Library of Congress Cataloging-in-Publication Data

Dimethylsulphide : oceans, atmosphere, and climate : proceedings of the international symposium held in Belgirate, Italy, 13-15 October 1992 / organized by the Commission of the European Communities, Directorate-General for Science, Research, and Development, DG XII/D, Environment Programme, Joint Research Centre, Environment Institute ; cosponsored by the Danish Centre for Atmospheric Research ; edited by Giambattista Restelli, Giovanni Angeletti.

p. cm.

Includes index.

ISBN 978-90-481-4325-2 ISBN 978-94-017-1261-3 (eBook)

DOI 10.1007/978-94-017-1261-3

1. Atmospheric chemistry--Congresses. 2. Dimethyl sulphate--Environmental aspects--Congresses. 3. Climatic changes--Environmental aspects--Congresses. 4. Sulphur cycle--Environmental aspects--Congresses. 5. Biochemichemical cycles--Environmental aspects--Congresses. I. Restelli, G. II. Angeletti, G., 1943- III. Commission of the European Communities. IV. Danish Centre for Atmospheric Research.

QC879.6.D56 1993

551.5'11--dc20 93-30136

ISBN 978-90-481-4325-2

Publication arrangements by
Commission of the European Communities
Directorate-General Telecommunications, Information Industries and Innovation,
Scientific and Technical Communications Unit, Luxembourg

EUR 14796

Printed on acid-free paper

CONTENTS

FIELD MEASUREMENTS

ATMOSPHERIC CHEMISTRY

GAS-TO-PARTICLE CONVERSION AND CCN PRODUCTION

GLOBAL MODELLING AND CLIMATIC IMPLICATIONS

PREFACE

Since the discovery by J.E.Lovelock, R.J.Maggs and R.A.Rasmussen, in 1972, of its ubiquity in sea water, dimethylsulphide (DMS), a biologically produced sulphur compound, has been the subject of continuously increasing interest by the scientific community. DMS was immediately recognized as an important component of the biogeochemical sulphur cycle, and is now indicated as the second most important source of sulphur in the atmosphere, after anthropogenic SO_2 emission from fossil fuel combustion and industry.

DMS reacts rapidly in the atmosphere where it is oxidized to condensable acidic sulphur products; in fact, rainwater acidification, observed in remote areas, is attributed to DMS emissions. The hypothesis of a climatic role of DMS was made already in 1983 by B.Shaw, and by B.C.Nguyen, B.Bonsang and A.Gaudry. In 1987, a study appeared in Nature, in which R.J.Charlson, J.E.Lovelock, M.O.Andreae and S.G.Warren suggested the possibility of a partial control of the climate by the biosphere through a chain of processes, linking production of DMS by marine phytoplankton with changes in clouds albedo. The publication of this paper triggered a strong debate and stimulated new efforts to describe the various aspects of the DMS cycle in the environment. The paper was timely and added to the discussion on the relative roles of atmospheric sulphur and greenhouse gases in the Earth's radiative budget. In the following years, understanding the cycle of DMS in the marine troposphere and its interaction with aerosol budget and cloud properties, became a key research target. This knowledge could improve our ability to evaluate the radiative forcing of the anthropogenic component of atmospheric sulphur and to decrease uncertainties in our estimate of the climate sensitivity, both of which are badly needed for the application of enlightened decision policies on energy use in our society.

The symposium, "Dimethylsulphide: Oceans, Atmosphere, Climate" was organized by the Commission of the European Communities in collaboration with the Danish Centre for Atmospheric Research (DCAR) with the primary objective of updating the state of our knowledge of this subject. The focus put on the marine environment is coherent with the belief that, if a climatic role of DMS exists, it should certainly be played over the oceans. The symposium was held in Belgirate (Italy), 13-15 October 1992. It followed the conference "Biogenic Sulphur in the Environment" held, with a wider, more general scope, in 1987 in New Orleans. As in that meeting, the Belgirate Symposium involved a wide range of disciplines, from biology, to chemistry and physics, and scientists active in DMS studies pertinent to these disciplines were gathering together. The second objective of the meeting was, in fact, to encourage cooperative investigations of the climatic impact of DMS, which certainly requires multidisciplinary approaches for its comprehension.

Over 70 participants coming from 17 countries, contributed the 41 papers, which are now presented in these proceedings. The

matter, following the same scheme applied to the symposium, is divided into five sections: 1. DMS production by marine phytoplankton; 2. Field measurements; 3 Atmospheric chemistry; 4. Gas-to-particle conversion and CCN production; 5. Modelling and climatic implications.
The animated discussion at the round table closing the symposium, while recording the significant progress made in various areas, emphasized that our understanding of the DMS cycle is still fragmentary. This is true equally for the development of a theoretical model and for the field studies which mostly are affected by incomplete coverage of the wide range of relevant parameters. The symposium was successful in reinforcing the consciousness that a congruence of efforts in the important cooperative projects in progress, mainly in the framework of the IGBP-IGAC (International Geosphere Biosphere Programme-International Global Atmospheric Chemistry) Core Project, is strongly needed. The development of a comprehensive validated model of the processes involved in the environmental cycle of DMS appears a more and more exciting and feasible target. We hope that the Symposium and the publication of the proceedings can in some way contribute to this effort.

Acknowledgements. Support in various phases of the preparation and during the meeting has been received by our colleagues, H. Barth, W. Eifler, J. Hjorth, N. Murray, F. Raes and, from the Danish Centre for Atmospheric Research (DCAR), by N.R. Jensen.
The conduction of the sessions by the Chairmen and in particular the preparation and the conduction by R.J.Charlson and H.Rodhe of the lively round table discussion are gratefully acknowledged.
The valuable help of the Public Relations Unit of the Joint Research Centre-Ispra, before and during the symposium, is appreciated.

G.Angeletti G.Restelli

PRODUCTION OF DMS BY MARINE PHYTOPLANKTON

P.S. LISS, G. MALIN and S.M. TURNER

School of Environmental Sciences,
University of East Anglia,
Norwich,
NR4 7TJ,
United Kingdom

SUMMARY

Dimethyl sulphide (DMS) is the dominant sulphur gas found in surface marine waters, and there is compelling evidence that it is formed biologically in these environments. In all areas so far investigated, the oceans are found to be highly supersaturated with respect to atmospheric levels of DMS, which indicates a net flux of the gas out of the oceans. In the air DMS is oxidised to a variety of products which play important roles in the chemistry and physics of the atmosphere. It is clearly important to understand the controlling mechanisms for DMS formation in seawater in order to quantify its present, past and possible future roles in the environment.

In the paper what is currently known about how and why organisms produce DMS is summarised in terms of: i) Basic biology and biochemistry; ii) The role of nutrients; iii) Zooplankton grazing; iv) Cycling of DMSP (dimethylsulphoniopropionate, the precursor of DMS) and DMS in the upper oceans; v) Relationship between DMS/DMSP and chlorophyll 'a'; vii) Inter-organism differences in ability to form DMSP/DMS. Finally, the air-sea exchange of DMS is compared with the equivalent exchange of nitrogen as ammonia and carbon as CO_2.

1. INTRODUCTION

If a sample of surface seawater is analysed for its content of volatile sulphur compounds, the dominant gas is almost always dimethyl sulphide (DMS, $(CH_3)_2S$). This has been known for more than two decades, since the pioneering measurements of Lovelock et al.[35]. Relative to concentrations of DMS in the atmosphere, the surface oceans are typically two orders of magnitude supersaturated[3], implying a net flux of the gas from the oceans to the atmosphere. A recent estimate of the global sea-to-air flux gives a value of 35×10^{12} g S yr^{-1} (11), with an uncertainty of about a factor of two. The methods used to calculate air-sea fluxes for gases in general are reviewed by Liss and Merlivat[33], with their particular application to DMS being discussed in Turner et al.[46] .

Once in the atmosphere DMS molecules are subject to oxidation by radical attack (mainly OH and NO_3) to yield a

G. Restelli and G. Angeletti (eds.), Dimethylsulphide: Oceans, Atmosphere, and Climate, 1–14.

variety of products, the most important of which are sulphur dioxide, sulphate and methanesulphonate[38]. These oxidation products play at least three important roles in atmospheric chemistry:

1) It has been known for more than 30 years that the sulphur cycle at the Earth's surface cannot be balanced without a substantial flux of volatile sulphur from the oceans to the atmosphere[18]. The carrier molecule was initially assumed to be hydrogen sulphide, but we now know that a large fraction of the volatile sulphur is DMS[31].

2) Since most of the oxidation products of DMS in air are acidic they contribute to the acidity of atmospheric particles and rain. This is particularly so in areas remote from anthropogenic activities where the major part of the total acidity is estimated to be derived from DMS oxidation[41,50].

3) Away from the continents, sub-micron size sulphate particles in the atmosphere are mainly derived from the oxidation of marine DMS. These particles affect the radiation balance of the atmosphere directly by backscattering solar radiation. They also affect it indirectly, since they are the major source of cloud condensation nuclei (CCN) which are a major determinant of the albedo of clouds[13,42]. The latter authors hypothesise that a negative feedback loop may operate by which changes in atmospheric temperature lead to compensating alterations in the flux of DMS from the oceans.

Many of the ideas outlined above are shown diagrammatically in Figure 1 (which is adapted from[4]). In view of the importance that DMS has assumed in atmospheric chemistry in recent years, several of the chapters in this volume give detailed accounts of the roles of DMS oxidation products in the atmosphere. The present chapter is largely devoted to summarising what is known about the production of the gas in seawater. This is clearly a prerequisite to understanding the importance of DMS in the environment now, as well as to predicting how that role differed in the past and might alter in the future.

2. FACTORS CONTROLLING THE FORMATION OF DMS IN SEAWATER

Biology and biochemistry

There is clear evidence that DMS is formed by biological processes. This contrasts with the situation for several other trace gases studied in seawater where the mode of formation is not so apparent (for example, various low molecular weight hydrocarbons and organo-halogens), and for which both biological and photochemical mechanisms have been proposed.

DMS formation by both macro- and micro-algae (seaweeds and phytoplankton) is well established. In all of these the biochemical precursor of DMS is known to be dimethylsulphoniopropionate (DMSP, $(CH_3)_2S^+(CH_2)_2COO^-$), as first proposed by Challenger and Simpson[12] from studies on the seaweed *Polysiphonia fastigata*. The DMSP molecule comprises DMS and acrylic acid in a 1:1 molar ratio. The function of DMSP in the algae is thought to be as an osmolyte; for example, with increase in salinity DMSP is found to increase in several species of phytoplankton[16,17,48]. Other suggested roles are

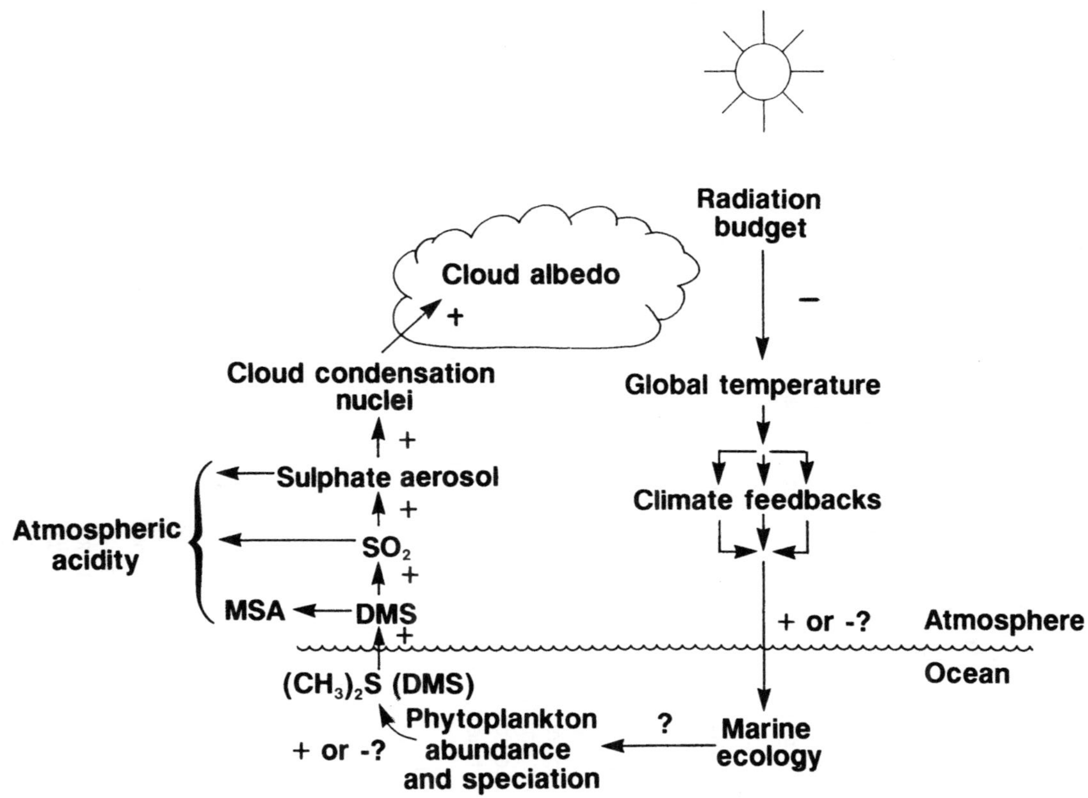

Figure 1. The air-sea cycle of DMS showing its atmospheric oxidation products and the proposed feedback loop between climate and DMS production in seawater. The pluses and minuses indicate if an increase in the value of the preceding parameter in the cycle is expected to lead to an increase (+) or decrease (-) in the value of the subsequent parameter.

through the bactericidal properties of the acrylic acid formed when DMSP is cleaved[43] and as a cryoprotectant for organisms living in very low temperature marine environments, such as polar waters and ice[22,28].

In view of its importance at the cellular level, it isn't surprising to find that DMSP (and the resulting DMS) are linked in a general way to the level of biological activity in the seawater. This is manifested by strong seasonality in concentrations of DMSP and DMS in temperate waters, which is particularly marked in shelf and coastal areas[34,47]. Similarly vertical profiles of both compounds show highest concentrations in surface or near-surface waters, with very low or undetectable amounts at depth[2,45]. However, the relationship between DMSP, DMS and biological activity cannot be taken too far, as we shall see later, when the tightness of the association between these compounds and general indicators of algal activity such as levels of chlorophyll 'a' are examined.

Nutrients

As mentioned above, the major role of DMSP in algal cells is thought to be as an osmolyte. However, DMSP is not the only compound with the necessary structural properties (e.g. it is a zwitter ion) to fulfil this function. Well established alternative osmolytes are the compounds proline and glycine betaine (GBT, $(CH_3)_3N^+CH_2COO^-$)(9). The structure of the latter molecule is similar to that of DMSP but with nitrogen as the central atom rather than sulphur. It has been suggested that since the energy algae need to expend to synthesise DMSP is greater than that required to form GBT and proline[3], it could be advantageous for them to make DMSP rather than GBT or proline under conditions where nitrogen is in short supply and needed for other purposes.

Two lines of experimental evidence in support of this hypothesis are given in Turner et al.[45]. Firstly, they report that the coccolithophore *Emiliania huxleyi* grown in cultures containing high concentrations of nitrate produces less DMSP per cell than under low nitrate conditions. Secondly, for geographically adjacent samples from the North Sea having no clear differences in phytoplankton species composition, the levels of intracellular DMSP are significantly higher for samples where nitrate in the seawater is below the analytical detection limit (<0.2 μmol) than for samples containing detectable amounts of NO_3^-. Other evidence comes from the recent work of Gröne and Kirst[20] who find that intracellular DMSP levels in cultures of *Tetraselmis subcordiformis* increased by 75% within a day of being subjected to nitrogen deficiency. Further, Kiene and Service[27] measured increased concentrations of DMS in seawater containing phytoplankton after the water had been spiked with GBT. The explanation for this unexpected finding is suggested to be that the phytoplankton may have been taking up the added GBT for use as an osmolyte in preference to internally formed DMSP, which was released to produce the increased amounts of DMS observed in the water.

Zooplankton grazing

It seems to be generally agreed that there is little release of DMSP or DMS from actively growing phytoplankton cells in culture[23,45]. However, Dacey and Wakeham[15] showed from a laboratory study that addition of macrozooplankton (*Labidocera aestiva*) to a phytoplankton culture (*Gymnodinium nelsoni*) led to a substantial increase in the concentration of DMS in the water. Similar results using a microzooplanktonic predator (*Oxyrrhis marina*) and prey (*E. huxleyi*) have recently been reported[37].

Grazing by zooplankton also seems to be an important process for DMSP/DMS release from phytoplankton in the marine environment. For example, Leck et al.[29] found a significant correlation between DMS levels measured in the Baltic Sea and copepod and total zooplankton biomass. Further, recent research indicates the effect is not confined to macrozooplankton species. Microzooplankton are also important, particularly for some of the smaller phytoplankton species (such as *E. Huxleyi* and *Chrysochromulina* sp.) which are inefficiently grazed by larger zooplankton forms. From a detailed study in which the organisms were size fractionated, Belviso et al.[8] have shown that microzooplankton grazing is important for release of DMSP/DMS from samples from the NW Mediterranean.

Transformations of DMSP and DMS

Once released to the seawater the DMSP and DMS are subject to a variety of biological and photochemical transformations. Those so far identified are included in Figure 2 (from Malin et al.[36]), which summarises our current knowledge of the cycling of DMS in marine surface waters.

The right hand side of the diagram shows temperature, light and nutrients as the probable major factors controlling DMSP formation in algae. The algal DMSP is either cleaved intracellularly and the resulting DMS emitted to the water or, as seems more probable, it is released from the cells as either DMS or free DMSP (i.e. dissolved in the water) when the organisms die or are grazed by zooplankton. On the left of Figure 2 is shown a whole web of bacterially mediated processes to which DMSP and DMS are thought to be subject. These are discussed in Malin et al.[36,37] and only some highlights are presented here.

It has been reported[49,52] that phototrophic bacteria can oxidise DMS to dimethylsulphoxide (DMSO) and cometabolic oxidation of DMS has also been observed[53]. In addition, in laboratory studies it has been found that DMS can be oxidised to DMSO photochemically[10]. These various processes of DMS oxidation to DMSO probably account for the relatively high concentrations of the latter in ocean waters first reported by Andreae[1] and since confirmed by Gibson et al.[19] and Kiene (personal communication, 1992). In the reverse direction, it has been shown that DMSO can be utilised by bacteria both anaerobically[14] and aerobically[44].

The importance of internal cycling within the water, relative to sea-to-air exchange, has only been generally realised in the last few years[25,36,51]. This is because of the difficulties of trying to quantify the various production,

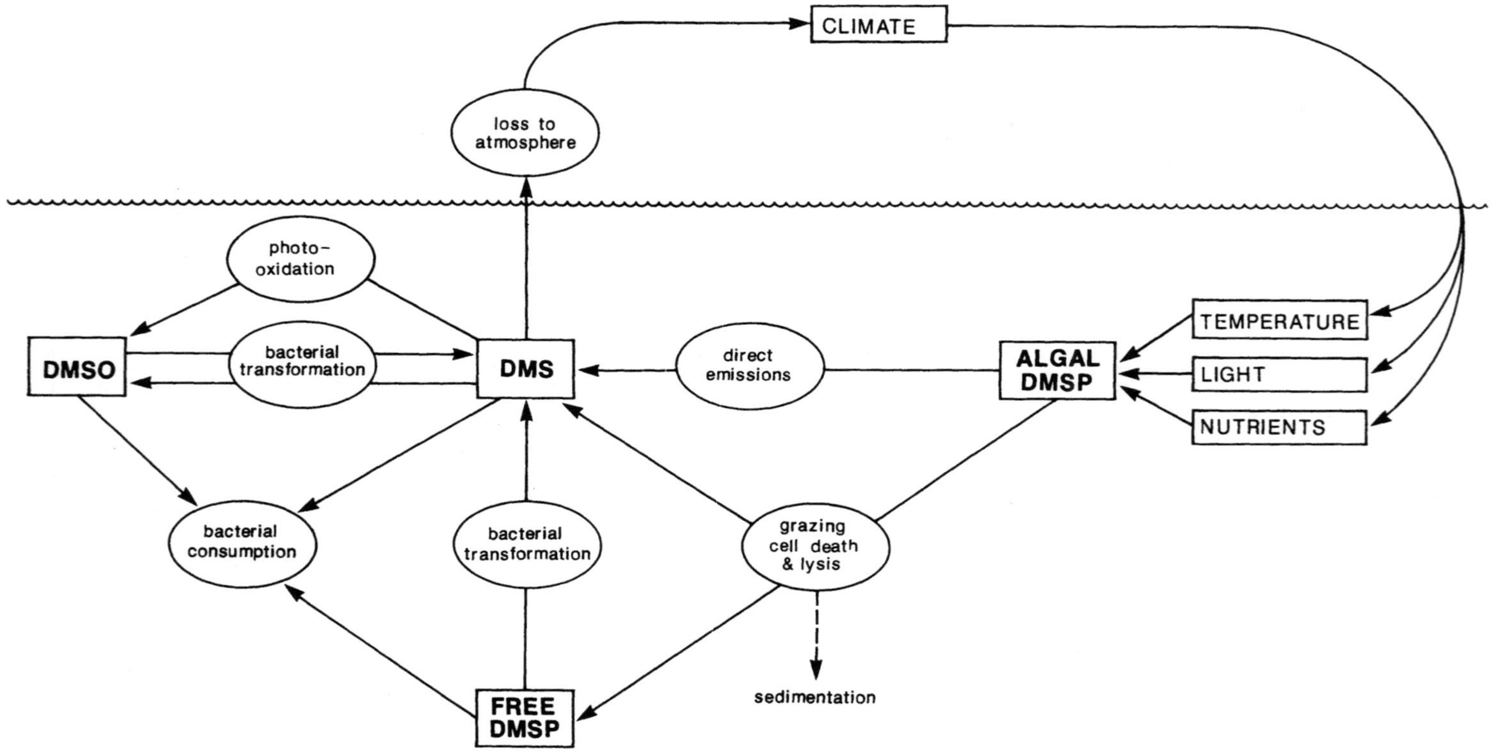

Figure 2. The marine biogeochemical cycle of DMS: production, transformation, and utilisation pathways that may ultimately influence the quantity of DMS lost to the atmosphere.

transformation and consumption processes. Some attempts have been made, for example Brimblecombe and Shooter[10] estimated that the loss of DMS by photooxidation to DMSO was at about the same rate as that by gas exchange across the sea surface. For bacterial transformation/consumption of DMS Kiene and Bates[26] measured the rate to be between 3 and 430 times faster than loss to the atmosphere in the Equatorial Pacific. Clearly these estimates are rather preliminary and only a limited number of the possible processes in the water column have been examined. However, such studies are vital if we are ever to be able to predict DMS emissions to the atmosphere under circumstances when they cannot be measured directly (i.e. in the past and under a variety of possible future scenarios). What has become clear from the studies carried out to date is that there is very active cycling of biogenic sulphur in the surface oceans, with the flux of DMS from sea to air as just a small bleed. However, a small perturbation of water column processes could have a very significant effect on the size of the atmospherically important DMS flux.

Relationship to chlorophyll

Since DMSP is clearly synthesised by marine algae, with little suggestion that it can be formed in any other way, there is the possibility that concentrations of DMSP and DMS in seawater could be related to some general indicator of plant production, such as chlorophyll. Were it possible to establish such a relationship then satellite data for ocean colour (which was formerly available from the CZCS satellite and should be provided in the future by SeaWIFS) would offer the possibility of global estimates of surface DMS/DMSP concentration fields without the need for direct measurement. However, most attempts to show a relationship between DMS and chlorophyll have led to the conclusion that the correlation is rather poor[5,7,21,45]. It tends to improve when the area examined is dominated by one type of phytoplankton and chlorophyll is more likely to be significantly correlated with DMSP than DMS[37]. This latter result is to be expected since DMSP is a constituent of algal cells, whereas there are many processes external to the cell which can affect the aqueous concentration of DMS (as illustrated in Figure 2).

Another very important reason for these generally poor relationships is highlighted in a paper by Turner et al.[45]. Here it is shown that for samples from around the U.K. continental shelf the overall correlation between DMS and chlorophyll is poor. However, when the samples are stratified into groups within which one type of organism is dominant then the correlations improve, but with different DMS/chlorphyll ratios for each of the groups. This clearly explains why samples containing mixed populations show such poor correlations.

From the above discussion of field data it is apparent that various marine phytoplankton show different abilities to form DMSP and thence DMS. In laboratory culture studies using a wide range of organisms Keller[23] has confirmed that this is indeed the case. Further, from our studies[34] around the U.K. we have found that field samples also show large inter-species variations in DMSP/DMS production (expressed in terms of ratio of S to chlorophyll 'a'). These results are shown in Table 1. There is general agreement between our field results and the laboratory culture findings of Keller et al.[24]. From Table 1 we conclude that the ability of phytoplankton to form DMSP increases in the order:

diatoms < *Gyrodinium* < *Phaeocystis* < coccolithophores

The top and bottom of the series span a factor of about 20. Such knowledge starts to lay the foundations for the way ahead in the use of satellites for global mapping of DMSP (and maybe DMS) in surface oceans waters. The SeaWIFS satellite has significantly enhanced spectral resolution compared to CZCS and may therefore be able to measure particular plant pigments. Since there is evidence that different groups of microorganisms produce specific pigments (for example fucoxanthin by diatoms, peridinin by dinoflagellates, and 19'-hexanoyl-oxy-fucoxanthin by prymnesiophytes)[6], coupled with quantitative understanding of the ability of different organisms to synthesise DMSP (such as in Table 1), this approach holds the potential for remote sensing of DMSP/DMS distributions in the future.

3. AIR-SEA EXCHANGE OF DMS, NH_3 AND CO_2

Dimethyl sulphide and ammonia

For many years there has been uncertainty about the flux of ammonia across the sea surface, with even the net direction of exchange being unclear[30]. However, the recent studies by Quinn et al.[39,40] in the Pacific Ocean definitely indicate a net flux from sea to air. If this net upward flux proves to be general then it has important consequences for the chemistry of marine aerosols from remote locations.

Chemical analysis of aerosol particles, collected over the oceans far from land, indicates that they generally have a composition between ammonium bisulphate (NH_4HSO_4) and ammonium sulphate ($(NH_4)_2SO_4$), i.e.with N and S in a molar ratio between 1:1 and 2:1 (40). In their study along 170°W and from 50°N to 11°S in the Pacific, Quinn et al.[40], as well as measuring the composition of the aerosols given above, also estimated the flux of reduced nitrogen (as NH_3) and sulphur (as DMS) all along the cruise track. Their calculations indicate that the two fluxes are within a factor of two of each other. The quantitative agreement between the emission fluxes of both nitrogen and sulphur and what is found in the aerosol is a remarkable finding. The most obvious interpretation is that over the remote oceans the sea is apparently the sole supplier of the two elements, so that there is no need to invoke terrestrial sources. A fuller discussion of this interesting coupling between the sulphur and nitrogen cycles, both of which are largely driven by biological processes in the water, is given in Liss and and Galloway[32].

Table 1. DMS and $DMSP_t$[ƒ] to chlorophyll a ratios for different species of phytoplankton in European coastal and shelf waters (from Liss *et al.*, 1993).

Species	% Total Carbon	S/chl.a Ratio* DMS	S/chl.a Ratio* $DMSP_t$	n	Area
Gyrodinium aureolum	> 80	16 (18)	162 (146)	18	A
Ceratium lineatum	> 77	115 (73)	- -	6	B
Cyclococcolithus leptoporus	> 42	61 (38)	- -	6	C
Coccolithophores (inc. *E. huxleyi*, *C. pelagicus*)	> 50	170 (76)	1292 (540)	16	D
P. pouchetii	> 80	24 (17)	476 (345)	11	E
P. pouchetii	> 50	172 (92)	381 (198)	4	F
P. pouchetii	a	26 (8)	442 (366)	4	G
P. pouchetii	a	35 (10)	379 (101)	6	H
P. pouchetii	a	83 (34)	429 (141)	16	I
Diatoms	b	4 (2)	64 (26)	9	J

ƒ $DMSP_t$ = Intracellular DMSP + Dissolved DMSP

* 10^{-7} mol S m^{-3}/mg Chl.a m^{-3}, (σn-1)

Explanatory notes for DMS and $DMSP_t$/Chl.a Ratios:

a: Visual observations of colonies, Chl.a > 5 mg m^{-3}
b: Visual observations of bloom, Chl.a > 3mg m^{-3}
A: July/August 1985, English Channel
B: July/August 1985, Orkney Islands
C: July/August 1985, West of Scotland
D: July/August 1985, N.E. Atlantic
E: May 1986, S. North Sea
F: June/July 1987, Faeroe Islands
G & H: April 1989, S. North Sea
I: May 1989, S. North Sea
J: April 1989, off East Anglia

Dimethyl sulphide and carbon dioxide

Another cycle in the oceans where biology plays a major role is that of carbon, with CO_2 as the gaseous form which exchanges with the atmosphere. Carbon dioxide is well acknowledged as a major determinant of the earth's radiation balance. As discussed earlier, DMS through its atmospheric oxidation to sulphate CCN also plays an important role in climate, with higher production of DMS leading to increased cloud albedo and hence a cooling of the atmosphere. Further, we showed that different types of phytoplankton have considerably differing abilities to form DMSP/DMS (Table 1). Thus, it is possible to assess the likely change in climate if phytoplankton speciation was different in the past or alters in the future.

To do this it is necessary to note that in terms of CO_2 balance there are two main types of phytoplankton - those which fix carbon in carbohydrate and other complex organic molecules (e.g. diatoms and *Phaeocystis*), as opposed to those which additionally form their structure by producing calcium carbonate from the seawater (i.e. coccolithophores). In the former case increase in plant growth leads to a decrease in dissolved CO_2, whereas formation of calcium carbonate produces a net input of the gas to the water, as the following equations illustrate:

$$CO_2 + H_2O \Rightarrow CH_2O + O_2 \quad \text{(Carbon Fixers)}$$

$$Ca^{2+} + 2HCO_3^- \Rightarrow CaCO_3 + CO_2 \quad \text{(Calcifiers)}$$

Since different phytoplankton have the potential to influence the air-sea exchange of CO_2 and DMS to differing degrees, it is of interest to speculate on the consequences for climate should shifts in phytoplankton assemblages result from changes in environmental conditions. For example, if diatoms were to increase, the tendency would be for a reduction in both CO_2 and DMS in the atmosphere, suggesting cooling (from decrease in CO_2) and warming (from decrease in DMS). In the case of coccolithophores becoming more significant, CO_2 and DMS would both tend to increase, again suggesting opposing effects on temperature, but in the opposite sense, i.e. increase in DMS leading to cooling and increase in CO_2 producing warming. A further scenario is that for increase in species such as *Phaeocystis*, where atmospheric CO_2 and DMS would tend to decrease and increase, respectively, the coupled effects of which would both lead to cooling.

The above description is clearly a considerable simplification of the real situation. However, it does show that in trying to understand past and future climatic scenarios it is necessary to consider the air-sea exchange of DMS as well as carbon dioxide, and also to take into account any changes in phytoplankton speciation.

ACKNOWLEDGEMENTS

This work was supported by research grants GR3 8050 and GST/02/653 from the U.K. Natural Environment Research Council.

REFERENCES

(1) Andreae, M.O. (1980). Dimethylsulfoxide in marine and fresh waters. Limnol. Oceanogr., 24: 440-452.
(2) Andreae, M.O. (1985). Dimethylsulfide in the water column and the sediment pore waters of the Peru upwelling area. Limnol. Oceanogr., 30: 1208-1218.
(3) Andreae, M.O. (1986). The ocean as a source of atmospheric sulfur compounds. In: The Role of Air-Sea Exchange in Geochemical Cycling (P. Buat-Menard, ed.). Reidel, Dordrecht, pp.331-362.
(4) Andreae, M.O. (1990). Ocean-atmosphere interactions in the global biogeochemical sulfur cycle. Mar. Chem., 30: 1-29.
(5) Andreae, M.O. and Barnard, W.R. (1984). The marine chemistry of dimethylsulfide. Mar. Chem., 14: 267-279.
(6) Barlow, R.G., Mantoura, R.F.C., Gough, M.A. and Fileman, T.W. (1993). Pigment signatures of the phytoplankton composition in the northeastern Atlantic during the 1990 spring bloom. Deep-Sea Res., 40: 459-477.
(7) Barnard, W.R., Andreae, M.O., Watkins, W.E., Bingemer, H. and Georgii, H.W. (1982). The flux of dimethylsulfide from the oceans to the atmosphere. J. Geophys. Res., 87: 8787-8793.
(8) Belviso, S., Kim, S.-K., Rassoulzadegan, F., Krajka, B., Nguyen, B.C., Milhalopoulos, N. and Buat-Menard, P. (1990). Production of dimethylsulfonium propionate (DMSP) by a microbial food web. Limnol. Oceanogr., 35: 1810-1821.
(9) Blunden, G. and Gordon, S.M. (1986). Betaines and their sulphonio analogues in marine algae. In: Progress in Phycological Research (F.E. Round and D. Chapman, eds.). Biopress Ltd., Bristol, pp.39-80.
(10) Brimblecombe, P. and Shooter, D. (1986). Photo-oxidation of dimethylsulphide in aqueous solution. Mar. Chem., 19: 343-353.
(11) Brimblecombe, P., Hammer, C., Rodhe, H., Ryaboshapko, A. and Boutron, C.F. (1989). Human influences on the sulphur cycle. In: Evolution of the Global Biogeochemical Sulphur Cycle (P. Brimblecombe and A Lein, eds.). Wiley, Chichester, pp. 77-121.
(12) Challenger, F. and Simpson, M.I. (1948). Studies on biological methylation XII: A precursor of the dimethyl sulphide evolved by *Polysiphonia fastigata* dimethyl-2-carboxyethyl-sulphonium hydroxide and its salts. J. Chem. Soc., 3: 1591-1597.
(13) Charlson, R.J., Lovelock, J.E., Andreae, M.O. and Warren, S.G. (1987). Oceanic phytoplankton, atmospheric sulphur, cloud albedo and climate. Nature, 326: 655-661.
(14) Clarke, G.J. and Ward, F.B. (1988). Purification and properties of trimethylamine-n-oxide reductase from *Shewenella* sp. NCMB 400. J. Gen. Microbiol., 134: 379-386.

(15) Dacey, J.W.H. and Wakeham, S.G. (1986). Oceanic dimethylsulfide: Production during zooplankton grazing on phytoplankton. Science, 233: 1314-1316.
(16) Dickson, D.M. and Kirst, G.O. (1987). Osmotic adjustment in marine eukaryotic algae: The role of inorganic ions, quaternary ammonium, tertiary sulphonium and carbohydrate solutes. I. Diatoms and a rhodophyte. New Phytol., 106: 645-655.
(17) Dickson, D.M. and Kirst, G.O. (1987). Osmotic adjustment in marine eukaryotic algae: The role of inorganic ions, quaternary ammonium, tertiary sulphonium and carbohydrate solutes. II. Prasinophytes and haptophytes. New Phytol., 106: 657-666.
(18) Eriksson, E. (1960). The yearly circulation of chloride and sulphur in nature: Meteorological, geochemical and pedological implications. Part 2. Tellus, 12: 63-109.
(19) Gibson, J.E.A., Garrick, R.C., Burton, H.R. and McTaggart, A.R. (1990). Dimethylsulphide and the alga *Phaeocystis pouchetii* in Antarctic coastal waters. Mar. Biol., 104: 339-346.
(20) Gröne, T. and Kirst, G.O. (1992). The effect of nitrogen deficiency, methionine and inhibitors of methionine metabolism on the DMSP contents of *Tetraselmis subcordiformis* (Stein). Mar. Biol., 112: 497-503.
(21) Holligan, P.M., Turner, S.M. and Liss, P.S. (1987). Measurements of dimethyl sulphide in frontal regions. Continental Shelf Res., 7: 213-224.
(22) Karsten, U., Wienke, C. and Kirst, G.O. (1990). The β-dimethylsulphoniopropionate (DMSP) content of macroalgae from Antarctica and southern Chile. Botanica Marina, 33: 143-146.
(23) Keller, M.D. (1989). Dimethyl sulfide production and marine phytoplankton: The importance of species composition and cell size. Biol. Oceanogr. 6: 375-382.
(24) Keller, M.D., Bellows, W.K. and Guillard, R.R.L. (1989). Dimethyl sulfide production in marine phytoplankton. In: Biogenic Sulfur in the Environment (E.S. Saltzman and W.J. Cooper, eds.). American Chemical Society, Washington, D.C., pp. 183-200.
(25) Kiene, R.P. (1993). Microbial sources and sinks for methylated sulfur compounds in the marine environment. In: Microbial Growth on C_1 Compounds (J.C. Murrell and D.P. Kelly, eds.). Intercept, Andover, pp. 15-33.
(26) Kiene, R.P. and Bates, T.S. (1990). Biological removal of dimethyl sulphide from sea water. Nature, 345: 702-705.
(27) Kiene, R.P. and Service, S.K. (1993). The influence of glycine betaine on dimethyl sulfide and dimethylsulfoniopropionate concentrations in seawater. In: The Biogeochemistry of Global Change: Radiatively Active Trace Gases (R.S. Oremland, ed.). Chapman and Hall, New York, in press.
(28) Kirst, G.O., Thiel, C., Wolff, H., Nothangel, J., Wanzek, H and Ulmke, R. (1991). Dimethylsulphoniopropionate (DMSP) in ice-algae and its possible biological role. Mar. Chem., 35: 381-388.

(29) Leck, C., Larsson, U., Bagander, L.E., Johansson, S. and Hadju, S. (1990). DMS in the Baltic Sea - annual variability in relation to biological activity. J. Geophys. Res., 95: 3353-3363.
(30) Liss, P.S. (1983). Gas transfer: Experiments and geochemical implications. In: Air-Sea Exchange of Gases and Particles (P.S. Liss and W.G.N. Slinn, eds.). Reidel, Dordrecht, pp. 241-298.
(31) Liss, P.S. (1991). The sulphur cycle. In: Climate and Global Change (J.C. Duplessy, A. Pons and R. Fantechi, eds.). CEC, Brussels, pp. 75-89.
(32) Liss, P.S. and Galloway, J.N. (1993). Air-sea exchange of sulphur and nitrogen and their interaction in the marine atmosphere. In: Interations of C, N, P and S Biogeochemical Cycles and Global Change (R. Wollast, F.T. Mackenzie and L. Chou, eds.). Springer, Berlin, pp. 259-281.
(33) Liss, P.S. and Merlivat, L. (1986). Air-sea exchange rates: Introduction and synthesis. In: The Role of Air-Sea Exchange in Geochemical Cycling (P. Buat-Menard, ed.). Reidel, Dordrecht, pp. 113-127.
(34) Liss, P.S., Malin, G., Turner, S.M. and Holligan, P.M. (1993). Dimethyl sulphide and *Phaeocystis*: A review. J. Mar. Systems, in press.
(35) Lovelock, J.E., Maggs, R.J. and Rasmussen, R.A. (1972). Atmospheric sulphur and the natural sulphur cycle. Nature, 237: 452-453.
(36) Malin, G., Turner, S.M. and Liss, P.S. (1992). Sulfur: The plankton /climate connection. J. Phycol., 28: 590-597.
(37) Malin, G., Liss, P.S. and Turner, S.M. (1993). Dimethyl sulphide:Production by prymnesiophytes and atmospheric consequences. In: The Biology of the Prymnesiophyta (J.C. Green and B.S.C. Leadbeater, eds.). Oxford University Press, in press.
(38) Plane, J.M.C. (1989). Gas-phase atmospheric oxidation of biogenic sulphur compounds: A review. In: Biogenic Sulfur in the Environment (E.S. Saltzman and W.J. Cooper, eds.). American Chemical Society, Washington, D.C., pp. 404-423.
(39) Quinn, P.K., Charlson, R.J. and Bates, T.S. (1988). Simultaneous observations of ammonia in the atmosphere and ocean. Nature, 335: 336-338.
(40) Quinn, P.K., Bates, T.S., Johnson, J.E., Covert, D.S. and Charlson, R.J. (1990). Interactions between the sulfur and reduced nitrogen cycles over the central Pacific ocean. J. Geophys. Res., 95: 16,405-16,416.
(41) Savoie, D.L. and Prospero, J.M. (1989). Comparison of oceanic and continental sources of non-sea-salt sulphate over the Pacific Ocean. Nature, 339: 685-687.
(42) Shaw, G.E. (1987). Aerosols as climate regulators: A climate - biosphere linkage ? Atmos. Environ., 21: 985-986.
(43) Sieburth, J. McN. (1961). Antibiotic properties of acrylic acid, a factor in the gastrointestinal antibiosis of polar marine animals. J. Bacteriol., 82: 72-79.

(44) Suylen, G.M.H. (1988). Microbial Metabolism of Dimethyl Sulphide and Related Compounds. Ph.D. Thesis, University of Technology, Delft, The Netherlands, 162 pp.
(45) Turner, S.M., Malin, G., Liss, P.S., Harbour, D.S. and Holligan, P.M. (1988). The seasonal variation of dimethyl sulfide and dimethylsulfonio-propionate concentrations in near-shore waters. Limnol. Oceanogr., 33: 364-375.
(46) Turner, S.M., Malin, G., Nightingale, P.D. and Liss, P.S. (1993). Seasonal variation of dimethyl sulphide in the North Sea and an assessment of fluxes to the atmosphere. Manuscript in preparation.
(47) Uchida, A., Ooguri, T., Ishida, T. and Ishida, Y. (1992). Seasonal variations in dimethylsulfide in the water of Maizuru Bay. Nippon Suisan Gakkaishi, 58: 255-259.
(48) Vairavamurthy, A., Andreae, M.O. and Iverson, R.L. (1985). Biosynthesis of dimethylsulfide and dimethylpropiothetin by *Hymenomonas carterae* in relation to sulfur source and salinity variations. Limnol. Oceanogr., 30: 59-70.
(49) Visscher, P.T. and van Gemerden, H. (1991). Photoautrophic growth of *Thiocapsa roseopersicina* on dimethyl sulphide. FEMS Microbiol. Lett., 81: 247-250.
(50) Wagenbach, D., Gorlach, U., Moser, K. and Munnich, K.O. (1988). Coastal Antarctic aerosol: The seasonal pattern of its chemical composition and radionuclide content. Tellus, 40B: 426-436.
(51) Wolfe, G.V., Bates, T.M. and Charlson, R.J. (1991). Climatic and environmental implications of biogas exchange at the sea surface: Modeling DMS and the marine biologic sulfur cycle. In: Ocean Margin Processes in Global Change (R.F.C. Mantoura, J.-M. Martin and R. Wollast, eds.). Wiley, Chichester, pp. 383-400.
(52) Zeyer, J., Eicher, P., Wakeham, S.G. and Schwarzenbach, R.P. (1987). Oxidation of dimethylsulphide (DMS) to dimethyl sulphoxide (DMSO) by phototrophic purple bacteria. Applied and Environmental Microbiol., 53: 2026-2032.
(53) Zhang, L., Kunyoshi, I., Hirai, M. and Shoda, M. (1991). Oxidation of dimethylsulphide by *Pseudomonas acidovorans* DMR-11 isolated from peat biofilter. Biotechnology Lett., 13: 223-228.

ASSESSMENT OF THE ROLE OF ZOOPLANKTON IN THE CYCLING OF DMSP AND DMS IN THE WATER COLUMN DURING EUMELI-4 (FRANCE-JGOFS)

S. BELVISO, M. CORN and P. BUAT-MENARD

Centre des Faibles Radioactivités,
Laboratoire Mixte CNRS-CEA,
Avenue de la Terrasse,
91198 Gif-sur-Yvette Cedex,
France

SUMMARY

DMS (dimethyl sulfide) plays an important role in the marine environment because it contributes to the acidity of precipitations and may be to marine cloud formation and climate regulation. Since DMS appears to arise primarily from dimethylsulfonio propionate (DMSP), it is important to understand the processes controlling the cycling of DMSP in the oceanic upper water column. During the fourth EUMELI (France-JGOFS) cruise (May-June 1992) in the tropical northeastern Atlantic Ocean, DMSP analyses on sediment trap material were performed to assess the sink of DMSP by particulate sedimentation and the involvement of zooplankton. In open ocean waters 1500 km off the Mauritanian coast, under conditions of low autotrophic plankton biomass (chlorophyll a <0.2 µg/l), low nitrates (<0.01 µM) and low particulate DMSP (20 nM) concentrations, the particulate DMSP flux at 200 m depth varied on a daily cycle. Similar diel cycles were observed under conditions of high chlorophyll a (6 µg/l), high nitrates (10 nM) and high particulate DMSP (70-150 nM) in the active upwelling band off Mauritania. DMSP fluxes appeared to be markedly lower during the night than during the day. From these data it is evident that zooplankton nycthemeral migration controls the vertical flux of particulate DMSP independently of the intensity of the source of particulate matter, of the particulate DMSP content of upper water layers and of the zooplankton communities.

1. INTRODUCTION

On the basis of present knowledge, DMS is thought to arise primarily from dimethylsulfoniopropionate (DMSP). Since DMSP in particulate and dissolved form is ubiquitous in seawater (1), it is thus important to understand the processes controlling the cycling of DMSP in the upper water column. Dacey and Wakeham (2) demonstrated experimentally that zooplankton grazing results in the release of DMS from algal cells containing DMSP and gave reasonable indication for the presence of DMSP in fecal material. The loss of particulate DMSP from surface seawater via sinking particles and zooplankton fecal pellets, however, has not yet been addressed experimentally. Project EUMELI gave us the opportunity to investigate the fate of particulate DMSP in different trophic regimes in the tropical northeastern Atlantic Ocean.

EUMELI is based on the study of three sites, respectively eutrophic, mesotrophic and oligotrophic, off Mauritania, at latitude 20°N, along an ecological continuum where the intensity of the "source" of particulate matter varies at least within a 1 to 100 ratio. This ratio range is representative of the global ocean. All three sites are sufficiently close that the weather conditions, the properties of the water bodies, the fauna and the food webs do not differ too much between sites. Moreover, at such a latitude, the seasonal effects are minimised.

The oligotrophic site (21°N-31°W) is located 1500 km off the Mauritanian coast, on the abyssal plain (Figure 1).

The mesotrophic site (18°30'N-21°W), above a terrace at 3000 m, is subjected to the effects of the Mauritanian upwelling that can still be felt in spite of the distance from the coast (>300 km).

G. Restelli and G. Angeletti (eds.), Dimethylsulphide: Oceans, Atmosphere, and Climate, 15–19.

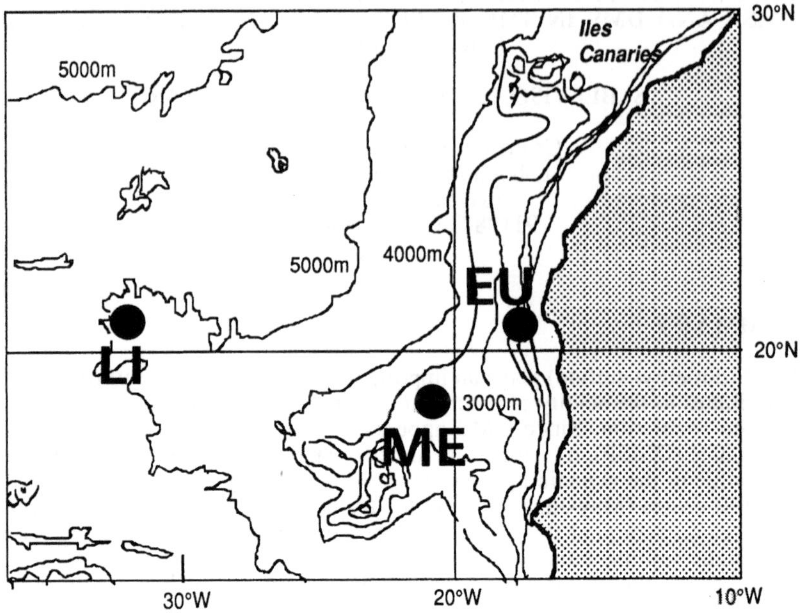

Figure 1: Positions of the EUtrophic, MEsotrophic, oLIgotrophic sites of the EUMELI project.

The eutrophic site (20°30'N-18°30'W), of about 500 m water depth, is as close as possible to the active upwelling band. The coast is arid and riverless.

Field measurements were carried out in May-June 1992 during the fourth EUMELI cruise of R.V. L'ATALANTE. Free-drifting sediment trap experiments were conducted to assess the downward flux of particulate DMSP in the upper 200 m of the water column. Briefly, the trap manufactured by Technicap is of the type PPS5/2, has a 1 m^2 collection area and can sample sequentially up to 24 samples of 250 ml each. A free-drifting system was deployed for 36 hours on 13-14 June at the eutrophic station, on 16-17 June at the mesotrophic station and on 22-23 June at the oligotrophic station, at a depth of 200 m. Measured current profiles show that the water masses and the sediment trap system drifted in parallel. Sequential samples were collected automatically for 4h each. Trap sample bottles contained DMSP-free deep seawater (250 m). No fixative solution was added in order to prevent interferences with the gas chromatographic analytical method. Living or dead mesozooplankton were removed systematically so that only passively sinking inert particles and organisms too small to be removed were analysed for DMSP.

DMSP levels were determined after treatment with cold alkali followed by GC/FPD analysis of DMS evolved (3) and converted to fluxes (µg DMS.m^{-2}. $4h^{-1}$).

2. RESULTS AND DISCUSSION

At the eutrophic station under conditions of low mixed layer water temperature (18°C), high chlorophyll a, nitrate and particulate DMSP concentrations (6 µg/l, 10 µM, 70-150 nM respectively), the particulate DMSP flux at 200 m depth varied on a daily cycle (Figure 2a). DMSP fluxes appeared to be at least one order of magnitude lower during the night (20-04 local time) than during the day (04-20 local time). At the mesotrophic site under conditions of higher mixed layer water temperature (21.5°C), lower chlorophyll a and nitrate concentrations (1.5 µg/l and 1 µM respectively), but comparable particulate DMSP concentrations (50-200 nM), a similar diel cycle was observed (Figure 2b). Finally, under conditions of mixed layer water temperature of 23.5°C, chlorophyll a, nitrate and particulate DMSP concentrations of <0.2 µg/l, <0.01 µM and 20 nM respectively, the same diel cycle was observed (Figure 2c).

It is well known that trapped zooplankton in sediment traps are a source of contamination and alteration of collected materials. Moreover, some bacterial decomposition of DMSP is likely to have occurred since the samples were not poisoned. Consequently, the absolute DMSP fluxes may be somewhat different from the reported values.

Nevertheless, it is apparent from these data that the same mechanism is controlling the vertical flux of particulate DMSP at 200 m depth independently of the intensity of the source of particulate matter and of the particulate DMSP content of upper water layers. The strong diel changes in the flux of DMSP in sinking material show most certainly the influence of nycthemeral vertical migrations of zooplankton.

Preliminary observations carried out on the organisms removed from the traps reveal a marked difference in the zooplankton composition. At the eutrophic and mesotrophic stations numerous pteropods, small and large >3 mm individuals, together with copepods, were observed in the day time samples. At the oligotrophic station no pteropods were found in the samples and copepods were less numerous and smaller. This suggests that diel changes in DMSP downward flux are also independent of the type of migrating zooplankton or that small copepods are the major carriers of DMSP between the upper waters and 200 m depth.

The role of copepods in the oceanic DMS cycle was highlighted by Dacey and Wakeham (2). These authors suggested that DMS production associated with grazing of phytoplankton by zooplankton may be the major mechanism of DMS production in many marine settings. Leck et al. (4) indeed found significant correlations of DMS concentration with copepods and total zooplankton biomass for samples collected in the Baltic Sea. Our sediment trap data obtained in contrasting trophic regimes characterises the general involvement of zooplankton in the cycle of the major DMS precursor, DMSP.

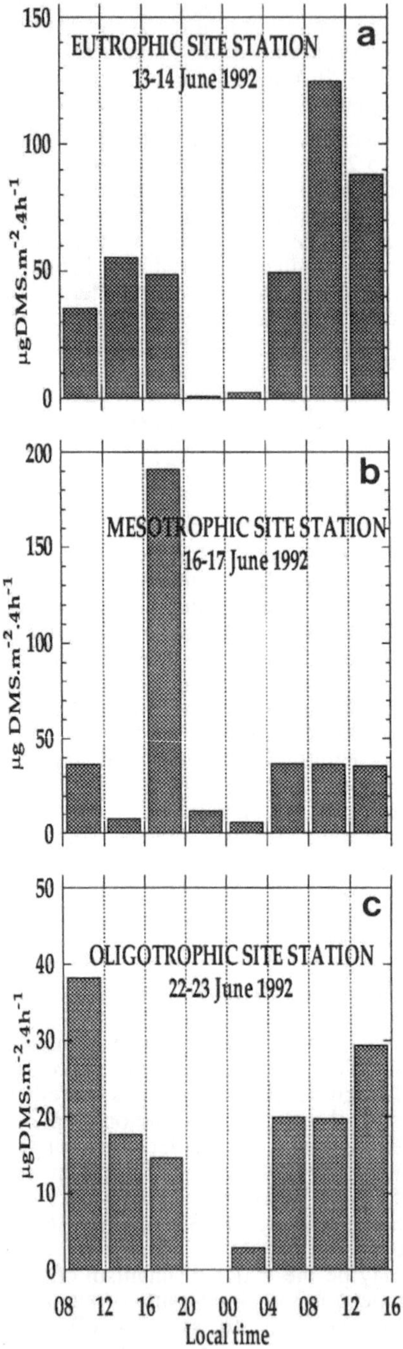

Figure 2: Time series of downward particulate DMSP flux at the three EUMELI sites during June 1992.

3. REFERENCES

(1) MALIN, G., TURNER, S.M. and LISS, P.S. (1992). Sulfur: the plankton/ climate connection. J. Phycol., 28, 590-597.
(2) DACEY, J.W.H. and WAKEHAM, S.G. (1986). Oceanic dimethylsulfide: production during zooplankton grazing on phytoplankton. Science, 233, 1314-1316.
(3) BELVISO, S., KIM, S.-K., RASSOULZADEGAN, F., KRAJKA, B., NGUYEN, B.C., MIHALOPOULOS, N. and BUAT-MENARD, P. (1990). Production of dimethylsulfoniumpropionate (DMSP) and dimethylsulfide (DMS) by a microbial food web. Limnol. Oceanogr., 35, 1810-1821.
(4) LECK, C., LARSSON, U., BAGANDER, L.E., JOHANSSON, S. and HAJDU, S. (1990). Dimethylsulfide in the Baltic Sea: annual variability in relation to biological activity. J. Geophys. Res., 95, 3353-3363.

4. ACKNOWLEDGMENTS

The authors wish to thank the Captain and Crew of R.V. L'Atalante, A. Vigot, D. Tailliez and C. Bournot in charge of the sediment traps and hydrographic work at sea, and J. Neveux, P. Raimbault and P. Nival for chlorophyll, nitrate and zooplankton data respectively. This work was supported by CNRS, INSU, IFREMER and CEA.

THE CYCLING OF SULFUR IN SURFACE SEAWATER DURING PSI-3(*)

TIMOTHY S. BATES[1], BYRON W. BLOMQUIST[2], KURT R. BUCK[3], FRANCISCO P. CHAVEZ[3], RUSSELL L. CUHEL[4], RONALD P. KIENE[5], PATY A. MATRAI[6] and GORDON V. WOLFE[7]

(1) NOAA/Pacific Marine Environmental Laboratory, 7600 Sandpoint Way N.E., Seattle, WA 98115; 206-526-6248
(2) Drexel University, Philadelphia, PA
(3) Monterey Bay Aquarium Research Inst., Pacific Grove, CA
(4) Center for Great Lakes Studies, Univ. of WI, Milwaukee, WI
(5) University of Georgia Marine Institute, Sapelo Island, GA
(6) RSMAS, Univ. of Miami, Miami, FL
(7) JISAO, Univ. of Washington, Seattle, WA

ABSTRACT

Oceanic dimethylsulfide (DMS) emissions to the atmosphere are potentially important to the Earth's radiative balance. Since these emissions are driven by the surface seawater concentration of DMS, it is important to understand the processes controlling the cycling of sulfur in surface seawater. During the third Pacific Sulfur/Stratus Investigation (PSI-3, April 1991) we measured the major sulfur reservoirs (total organic sulfur, total low molecular weight organic sulfur, ester sulfate, protein sulfur, DMSP, DMS, DMSO) and quantified many of processes that cycle sulfur through the upper water column (sulfate assimilation, DMSP consumption, DMS production and consumption, air-sea exchange of DMS, loss of organic sulfur by particulate sinking). Under conditions of low plankton (< 0.4 g/L chlorophyll a) and high nutrient concentrations (5 M nitrate), 250 km off the Washington State coast, DMSP and DMS were 24% and 1%, respectively of the total particulate organic sulfur pool. DMS production from the cleavage of DMSP accounted for 25% of the total sulfate assimilation. However, only 0.3% of sulfate-S assimilated was released to the atmosphere. From these data it is evident that air-sea exchange is currently only a minor sink in the seawater sulfur cycle and thus there is the potential for much higher DMS emissions under different climatic conditions.

(*) The full paper has been submitted to: J. Limnology and Oceanography.

G. Restelli and G. Angeletti (eds.), Dimethylsulphide: Oceans, Atmosphere, and Climate, 21.

ECOPHYSIOLOGY OF ICE ALGAE (ANTARCTICA): DIMETHYLSULFONIOPROPIONATE CONTENT AND RELEASE OF DIMETHYLSULFIDE DURING ICE MELT.

G. O. KIRST, M. WANZEK, R. HAASE
Marine Botany, FB2, University of Bremen, P.O.Box 330440, D-2800 Bremen (FRG)
S. RAPSOMANIKIS, S. DEMORA, G. SCHEBESKE, M. O. ANDREAE
Biogeochemistry, MPI Chemie, P.O.Box 3060, D-6500 Mainz (FRG)

SUMMARY

During a cruise in the area of the northern Antarctic Peninsula (December 1991; RV "Polarstern" ANT X1b) several phytoplankton blooms of varying species composition were encountered. The DMSP concentration correlated with the chlorophyll values, however, it changed indicating variation due to species specific production rates of DMSP. After crossing the Antarctic Convergence the species composition of the phytoplankton resembled that of recently released ice algal assemblages. High chlorophyll and DMSP concentrations were observed in the Antarctic Sound and in the adjacent part of the Weddell Sea due to a mass development of *Phaeocystis*. The correlation between chlorophyll, DMSP, DMS and concentration of condensation nuclei (CN) was clearly evident. Maximum values measured were: DMSP 75 nM, DMS(water) 250 nM, DMS(air) > 2000 ng/m^3, while CN concentration ranged between 1000 and 1800 /mL. This indicates that during the ice melt Antarctica is an area of especially high DMSP and DMS production.

1. INTRODUCTION

During sea ice formation a highly concentrated brine remains unfrozen in small channels which interlace the ice. These brine-pockets are inhabited by numerous ice algae, mostly diatoms and dinoflagellates which accumulate organic osmolytes, such as proline or dimethylsulfoniopropionate (DMSP) (1). In early summer when released into the water column during ice melt the ice algae initiate blooms which may result in increasing amounts of DMSP and DMS in the water and in the atmosphere (2; 3). These conditions provide a good opportunity to further test the DMS - cloud condensation nuclei (CCN) - albedo hypothesis (4). We therefore sumultaneously measured organismic DMSP in phytoplankton samples, DMS concentration in the sea water and in the atmosphere and the number of condensation nuclei (CN) in the early austral summer of 1991/92 during a cruise in the Bransfield Strait , the Antarctic Sound and the adjacent Weddell Sea.

2. MATERIALS AND METHODS

The samples were collected during the RV "Polarstern" cruise ANT X/1b (Puerto Madryn - Antarctica - Punta Arenas; 10 December 1991 - 2 January 1992) in an area extending from the South Shetland Islands (Bransfield Strait) to the northern tip of the Antarctic Peninsula (Antarctic Sound; Weddell Sea) - Elephant Island. Surface sea water samples were taken with a bucket and prefiltered through a 200 µm net. In the water column samples were obtained with a rosette sampler equipped with 12 Niskin bottles (12

G. Restelli and G. Angeletti (eds.), Dimethylsulphide: Oceans, Atmosphere, and Climate, 23–36.

L). Temperature and salinity were recorded with a Neil Brown CTD sonde attached to the rosette.

Phytoplankton algae were collected on glass fibre filters (Whatman GF/C) for DMSP and chlorophyll measurement as described by (5) and (1). DMSP was analysed shortly after collection employing gas chromatography and chlorophyll was determined by introducing the sample (in vivo measurement) and additionally the acetone extracts (90%) to a Kleinfeld Fluorometer.

Photosynthetic capacity was also determined with the Kleinfeld Fluorometer and calculated as the ratio of in vivo fluorescence before and after addition of DCMU (3-(3,4-dichlorophenyl)-1,1-dimethylurea), a potent inhibitor of photosynthesis which resulted in maximum chlorophyll fluorescence (6).

For phytoplankton cell counting and species identification 200 ml samples were preserved with neutralized formaldehyde solution (5).

Determination of DMS in the sea water and in the atmosphere was carried out according to literature methods (7).

Condensation nuclei were measured as number per mL in air using a TSI 3020 CN counter. Sampling was midships on the port side. Each data point in Fig. 2 represents a mean value between 4 or up to 12 measurements close to the time of plankton sampling.

3. RESULTS AND DISCUSSION

Fig. 1 a - c summarise the measurements of chlorophyll A, organismic DMSP (based on organisms filtered from 1 L sea water and expressed as nM), DMS in sea water from the same sample as DMSP, and DMS in the atmosphere along the cruise course. For clarity the range lap S 60° to S 65° is shown seperately for chlorophyll, DMSP and DMS(water), DMS(air) in Fig. 1 b and c. There is a general correlation between marine biological activity and aqueous particulate DMSP, aqueous DMS and atmospheric DMS.

The first bloom close to the Falkland current (Fig. 1a: S 46° 30; W 61° 45) was low in chlorophyll (ca. 2 µg/L) but very high in DMSP (150 nM). The dominant species were dinoflagellates, principally *Exuviella antarctica* (73 to 91 % of the species counted; Table 1), and coccolithophores. The low photosynthetic capacity may indicate a decaying bloom which was confirmed on microscopic examination. Especially high rates of DMSP and DMS production have been observed in old blooms (8). Before entering the Antarctic Convergence (S 57° 33 to S 57° 45; W 64° 23) a rapid change from high (> 7,5 µg/L) to low (0,5 µg/L) and back to high chlorophyll contents indicated the presence of two adjacent blooms (Table 1, sample 9 and 11). The first bloom was dominated by *Chaetoceros* sp. while the latter contained a substantial amount of *Phaeocystis* sp.. The DMSP content correlated with the chlorophyll values, however, the amount was less than observed in the bloom close to the Falkland current although chlorophyll A and cell numbers were higher.

Table 1: Summary of species composition and photosynthetic capacity

Latitude (Sample No.)	Dominant Species (% of cell numbers)	Photosynthetic capacity in % (100% = max. capacity)
S 46° 30 (5)	*Exuviella antarctica* (91%) small Coccolithophores (not counted)	35 - 40
S 53° 13 (9)	*Chaetoceros sp* (86%) *Phaeocystis sp.* (9%)	60 - 70
S 54° 47 (11)	*Chaetoceros sp.* (44%) *Phaeocystis sp.* (25%) *Odontella sp.* (14%) *Exuviella antarctica* (7%)	90
S 60° 3 / S 63° 43 (17 - 38) Southern part of the Drake Passage; Bransfield Strait	*Phaeocystis sp.* (48%) *Chaetoceros sp.* (23%) *Nitzschia sp.* (14%) Dinoflagellates (10%)	53±12 (n=22) range: 37 -78
S 63° 34 / S 61° 22 (40 - 60) Antarctic Sound; Weddell Sea (marginal ice zone); Elephant Island	*Phaeocystis sp.* (20-84%) *Chaetoceros sp.* (2-63%) *Nitzschia sp.* (1-18%) *Thalassiosira sp.* (2-8%) *Rhizosolenia sp.* (0,4-7%) *Odontella sp.* (0-3%) Dinoflagellates (0,5-10%)	75±13 (n=21) range: 59 -97
S 63° 34; W 54° 14 (a) TIME COURSE (Fig. 2) 28 h (Chl. A 6,2 µg/L)	*Phaeocystis sp.* (66%) *Chaetoceros sp.* (12%) *Rhizosolenia sp.* (7%) Dinoflagellates (8,5%)	84
S 63° 6; W 54° 25 (b) TIME COURSE (Fig. 2) 36 h (Chl. A 14,4 µg/L)	*Chaetoceros sp.* (63%) *Phaeocystis sp.* (20,5%) *Rhizosolenia sp.* (5%) *Odontella sp.* (3%) Dinoflagellates (3%)	95

After crossing the Antarctic Convergence the species composition of the phytoplankton resembled that of recently released ice algal assemblages containing high numbers of diatom species, always accompanied by varying amounts of *Phaeocystis* (Table 1: samples 17 - 38; Fig. 1b). Here, on average, chlorophyll as well as DMSP and DMS were on a minimum (Fig. 1b, c). The mean photosynthetic capacity also indicated a lower biological activity compared to the samples obtained in the marginal ice zone (Table 1: samples 40 - 60).

The chlorophyll data given in Fig. 1 are in good agreement with other reports from this area (9; 10; 11).

In the marginal ice zone (Antarctic Sound and in the adjacent part of the Weddell Sea) high chlorophyll content was measured, ranging from 6 to more than 14 µg/L. This was mainly due to a mass development of *Phaeocystis,* known to be a major DMSP producer. The correlation between chlorophyll, DMSP and DMS is clear. Maximum values measured were: DMSP 75 nM, DMS(water) 250 nM, DMS(air) > 2000 ng/m^3. The spatial patchiness of areas of high relative abundance of *Phaeocystis* is well documented in the literature and explained by the release of these algae from the ice edge in pulses probably related to localized melting events (12).

The concentration of the condensation nuclei (CN) as a function of latitude is presented in Fig. 2. Contaminated data (i.e. stack gases) have been eliminated. This explains the lack of data for the Drake Passage. Going southwards, the winds were persistently from the starboard. The Weddell Sea CN data ties in nicely with the DMSP and DMS data.

Values from the sampling period of over 36 hours (Fig. 3), which was conducted during the passage through the Antarctic Sound into the Weddell Sea, confirms the relationship between the phytoplankton species composition and production of DMSP and DMS (Table 1). The first maximum in chlorophyll A content represents a bloom with a high proportion of *Phaeocystis*. As to be expected, the DMSP and the DMS(water) concentrations were also high. The second bloom was dominated by *Chaetoceros* and other diatoms and exhibited even more chlorophyll A and a five times higher cell number per liter. *Phaeocystis*, however, was still present but comprised of to only 20% of the total cell number, which is about the same absolute amount as in the first bloom. The DMSP content in the second bloom is low if related to the chlorophyll value but similar to that found in the first bloom when based on the *Phaeocystis* portion. Hence, in this example the DMSP production may be due almost exclusively to *Phaeocystis*, as has been reported from other parts of Antarctica (3; 12). The remarkable increase in DMS(air) reflects the increase of DMSP and DMS in the water column (Fig. 3; note the logarithmic scale).

Depth profiles: Again a close relationship between chlorophyll A, DMSP, DMS(water) was observed. DMS was also present in deep zones due to mixing of the water column down to the sea floor on the shelf of the Weddell Sea (depth: ca. 260 m; Fig. 4a,b). This contrasts with the abundance restricted to the uppermost 10 - 80 m in temperate regions reported by (8; 13), which was also found at the slope of the Weddell Sea shelf (depth: more than 2500 m; Fig. 4c,d). The deep mixing of the water column may have an influence on the biodegradation of DMSP and DMS in the waterbody by microbes, e.g. in the Eastern Pacific DMS consumption rates were higher in the surface mixed layer than at 100 m depth (14). In addition we have to assume that a large portion of the ice algae released from the sea ice are subject to rapid sedimentation (15). Hence, a substantial amount of DMSP / DMS will be removed at least transiently from the global cycle.

Conclusions

1. During the season of ice melt, Antarctica is an area of particularly high production of DMSP and DMS. This is confirmed by long time observation (1983 to 1991) of atmospheric sulphur at the Georg von Neumayer Station: Sulphate and MSA concentrations increase dramatically in the months December to February (Minikin and Wagenbach, personal communication, 16). A clear seasonal cycle with a minimum of atmospheric sulfur dioxide and DMS in winter and a maximum in summer was also measured in the Southern Indian Ocean (17).

2. *Phaeocystis* appears to be the major source of DMSP. However, some ice diatoms also produce DMSP (e.g. *Nitzschia lecointeii*; Sasse and Kirst, unpublished results).

3. There is a direct correlation between DMSP (organismic), DMS(water), DMS(air) and CN formation.

4. DMS is also found in water as deep as 200 m under deepwater mixing conditions which occur frequently during ice melt close to the marginal ice zone (ice edge).

REFERENCES

(1) KIRST, G.O., THIEL, C., WOLFF, H., NOTHNAGEL, J., WANZEK, M., and ULMKE, R. (1991). Dimethylsulfoniopropionate (DMSP) in ice-algae and its possible biological role. Mar. Chem. 35, 381-388.

(2) FOGELQVIST, E. (1991). Dimethylsulfide (DMS) in the Weddell Sea surface and bottom water. Mar. Chem. 35, 169-177.

(3) GIBSON, J.A.E., GARRICK, R.C., BURTON, H.R. AND MCTAGGART, A.R. (1990). Dimethylsulfide and the alga *Phaeocystis pouchetii* in Antarctic coastal waters. Mar. Biol. 104, 339-346.

(4) CHARLSON, R.J., LOVELOCK, J.E., ANDREAE, M.O. AND WARREN, S.G. (1987). Oceanic phytoplankton, atmospheric sulphur, cloud albeda and climate. Nature 326, 655-661.

(5) BÜRGERMEISTER, S., BINGEMER, H., KIRST, G.O., JANSSEN, M. AND ERNST, W. (1990). On the biogenic origin of dimethyl sulfide: relation between chlorophyll, ATP, organismic dimethylsulfoniopropionate and dimethyl sulfide distribution in Atlantic surface waters and atmosphere. J. Geophys. Res. 95, 20 607-20 615.

(6) ROY, S. and LEGENDRE, L. (1979). DCMU-enhanced fluorescence as an index of photosynthetic activity in phytoplankton. Mar. Biol. 55, 93-101.

(7) ANDREAE, M.O. and BARNARD, W.R. (1983). Determination of trace quantities of dimethyl sulfide in aqueous solutions. Anal. Chem. 55, 608-612.

(8) MATRAI, P.A. and KELLER, M.D. (1992). Dimethylsulfide in a large-scale coccolithophore bloom in the Gulf of Maine. Cont. Shelf Res. (in press).

(9) HOLM-HANSEN, O. and MITCHELL, B.G. (1991). Spatial and temporal distribution of phytoplankton and primary production in the western Bransfield Strait region. Deep-Sea Res. 38, 961-980.

(10) BIANCHI, F., BOLDRIN, A., CIOCE; F., DIECKMANN, G., KUOSA, H., LARSSON, A.-M., NÖTHIG, E.-M., SEHLSTEDT, P.-I., SOCAL, G. and SYVERTSEN, E.E. (1992). Phytoplankton distribution in relation to sea ice, hydrography and nutrients in the northwestern Weddell Sea in early spring 1988 during EPOS. Polar Biol. 12, 225-235.

(11) JACQUES, G. and PANOUSE, M. (1991). Biomass and composition of size fractioned phytoplankton in the Weddell-Scotia Confluence area. Polar Biol. 11, 315-328.

(12) FRYXELL, G.A. and KENDRICK, G.A. (1988). Austral spring microalgae across the Weddell Sea ice edge: spatial relationships found along a northward transect during AMERIEZ 83. Deep-Sea Res. 35, 1-20.

(13) TURNER, S.M., MALIN, G. and LISS, P. (1989). Dimethyl sulfide and Dimethylsulfoniopropionate in European coastal and shelf waters. In: Biogenic sulfur in the environment (ed. SALTZMAN, E.S. and COOPER, W.J.), American Chemical Society 393, 183-200.

(14) KIENE, R.P. (1992). Dynamics of dimethyl sulfide and dimethylsulfoniopropionate in oceanic water samples. Mar. Chem. 37, 29-52.

(15) RIEBESELL, U., SCHLOSS, L. and SMETACEK, V. (1991). Aggregation of algae released from melting sea ice: implications for seeding and sedimentation. Polar Biol. 11, 239-248.

(16) MINIKIN, A. and WAGENBACH, D. (1990). The biogenic sulfur fraction of the coastal Antarctic aerosol body. Presented at the VIIth Symposium of CACGP, Chamrousse, 5-11 September.

(17) PUTAUD, J.P., MIHALOPOULOS, N., NGUYEN, B.C., CAMPIN, J.M. and BELVISO, S. (1992). Seasonal variations of atmospheric sulfur dioxide and dimethylsulfide concentrations at Amsterdam Island in the Southern Indian Ocean. J. Atmos. Chem. 15, 117-131.

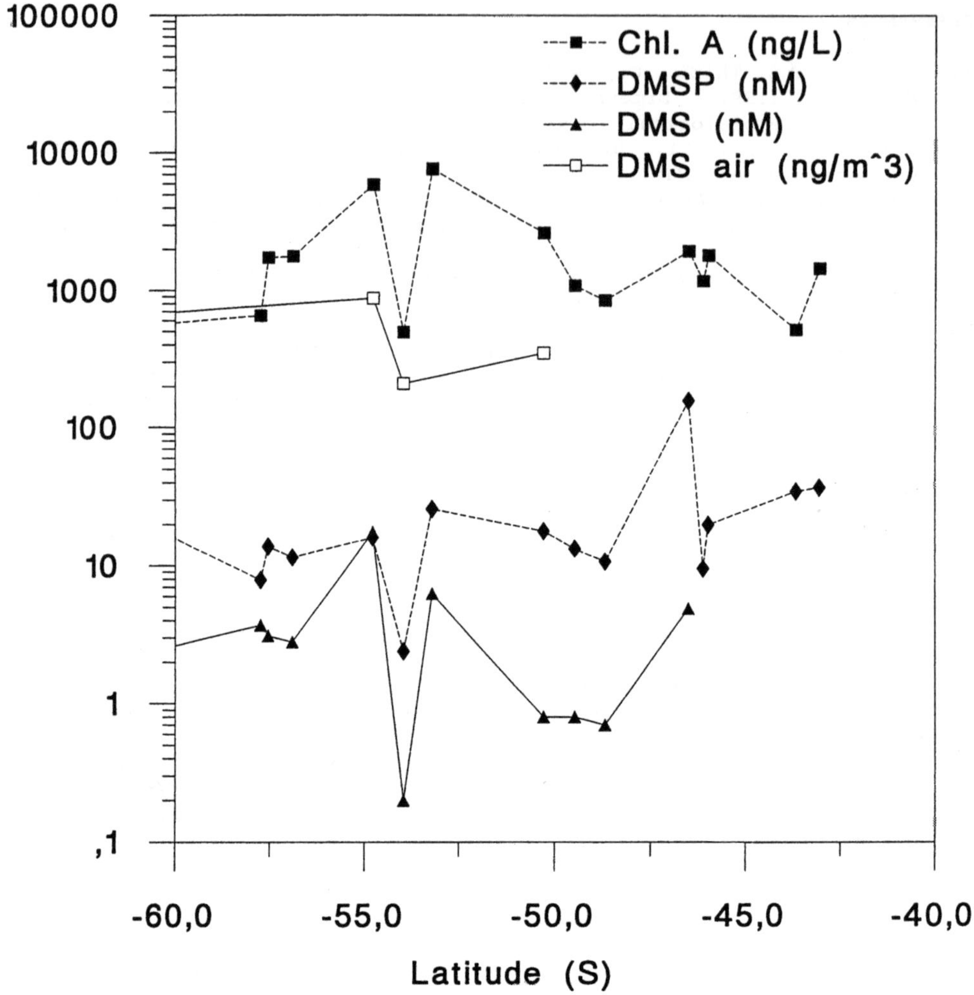

Fig. 1a: Chlorophyll A and DMSP content of plankton samples; DMS concentration in the water and atmosphere (air): Falkland current; Drake Passage; Antarctic Convergence.

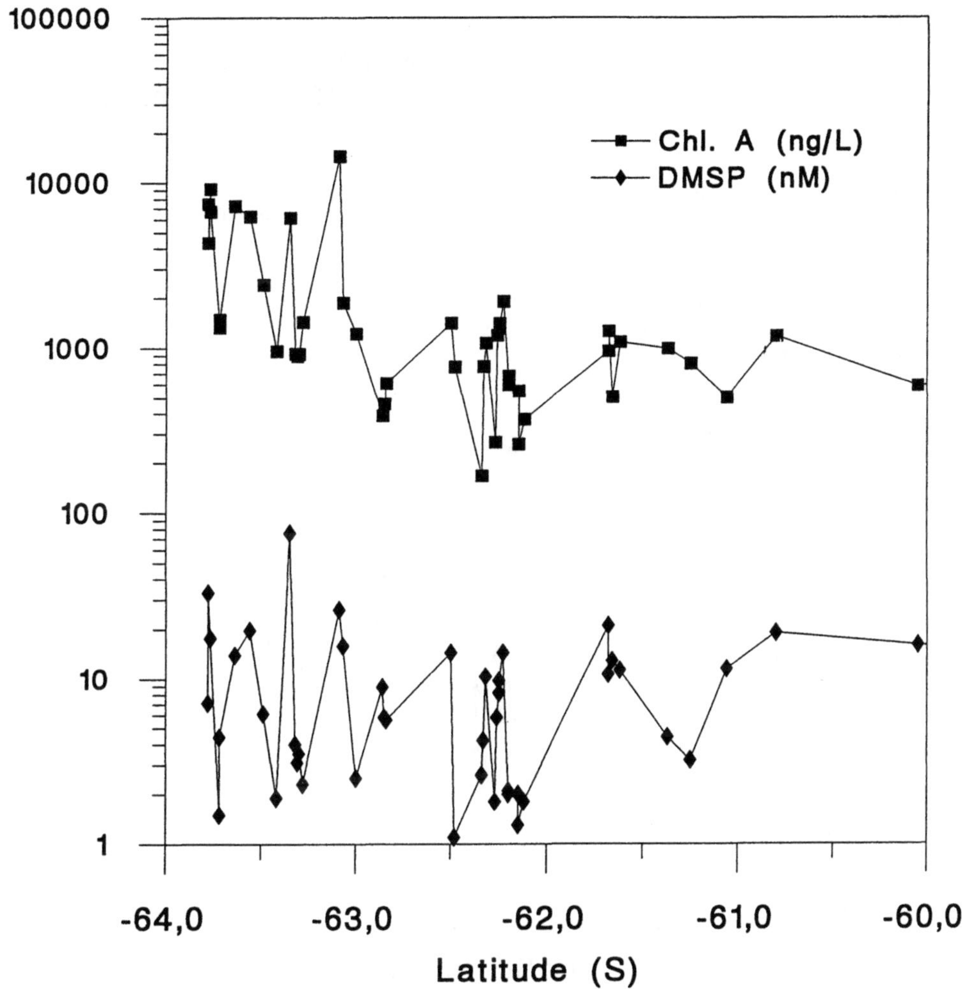

Fig. 1b: Chlorophyll A and DMSP content of plankton samples: Bransfield Strait, Antarctic Sound, Weddell Sea

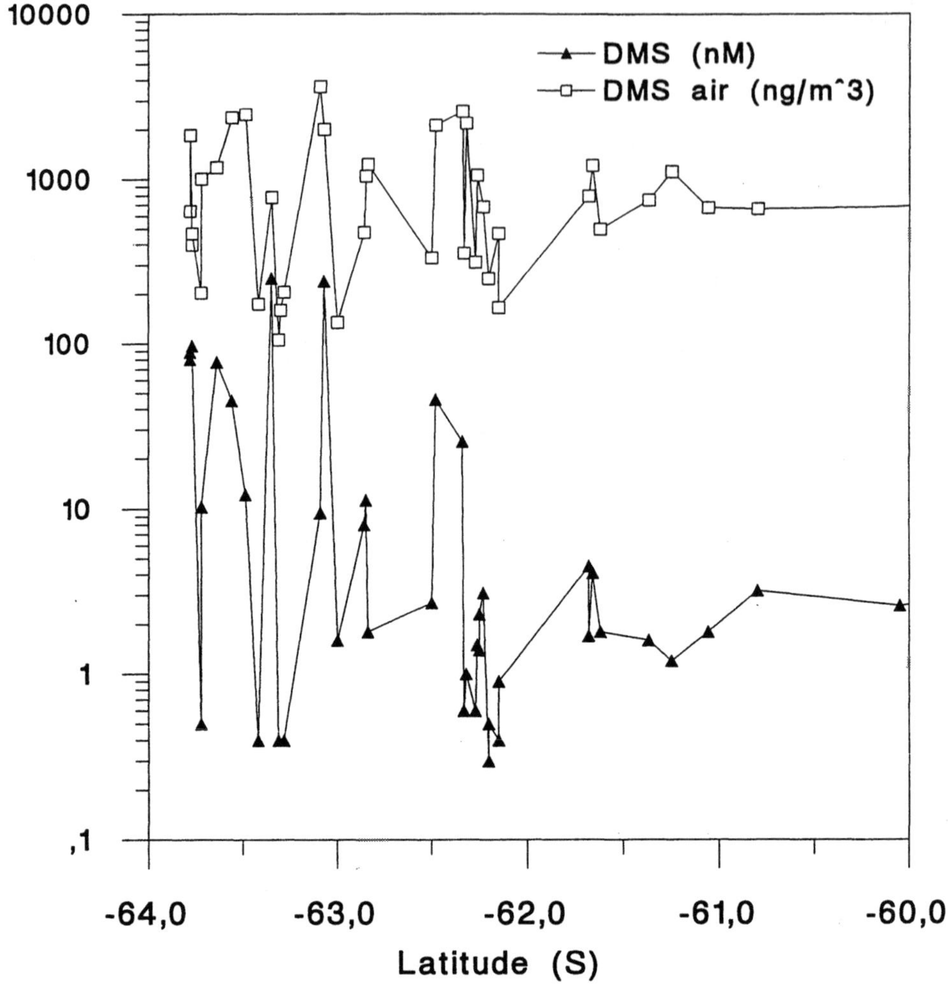

Fig. 1c: DMS concentration in the water and atmosphere (air)

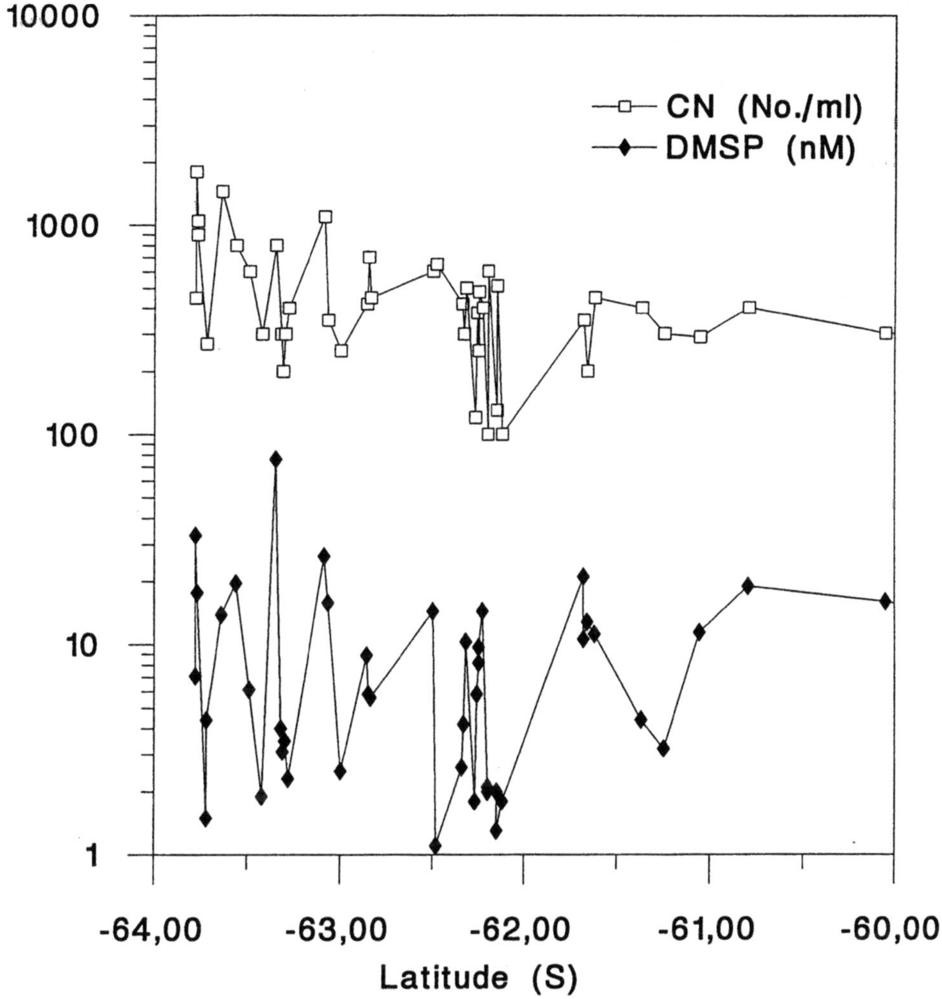

Fig. 2: Condensation nuclei (CN) and DMSP data, shown for comparison. Samples taken in the Bransfield Strait, Antarctic Sound and Weddell Sea (see Fig. 1b,c).

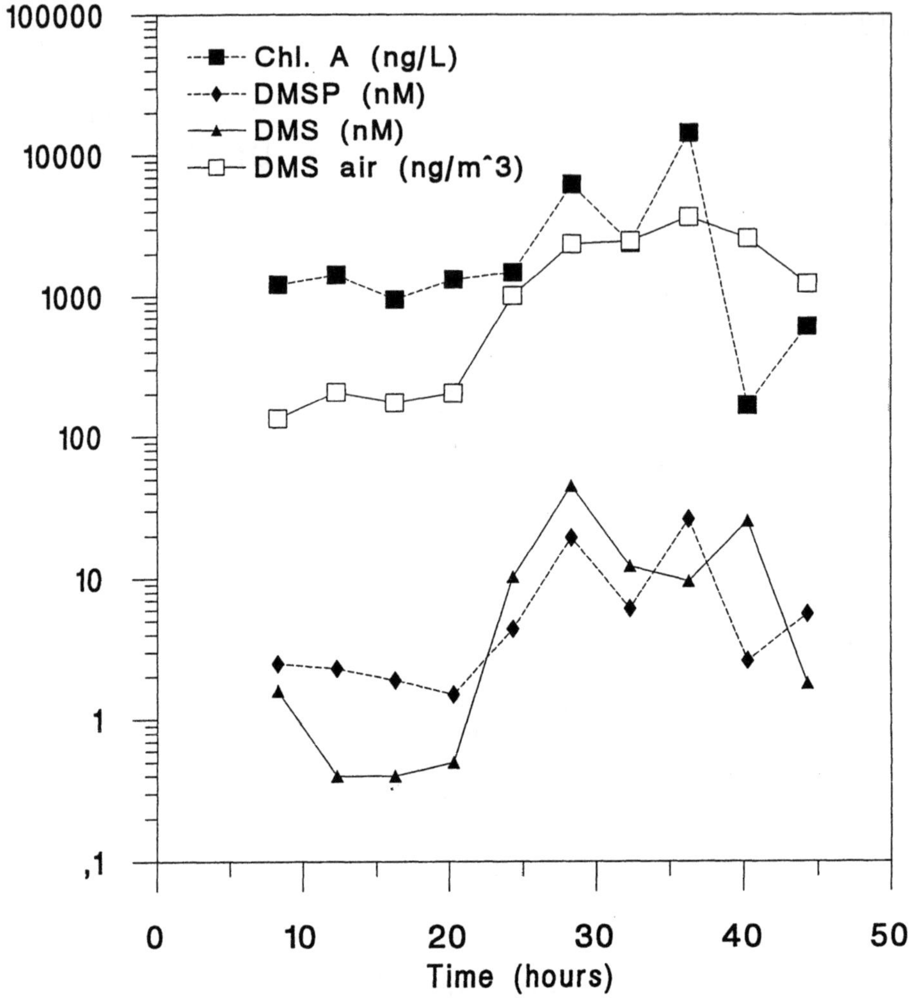

Fig. 3: Time course during the passage through the Antarctic Sound into the Weddell Sea. Chlorophyll A and DMSP content of plankton samples; DMS concentrations in the water and atmosphere (air).

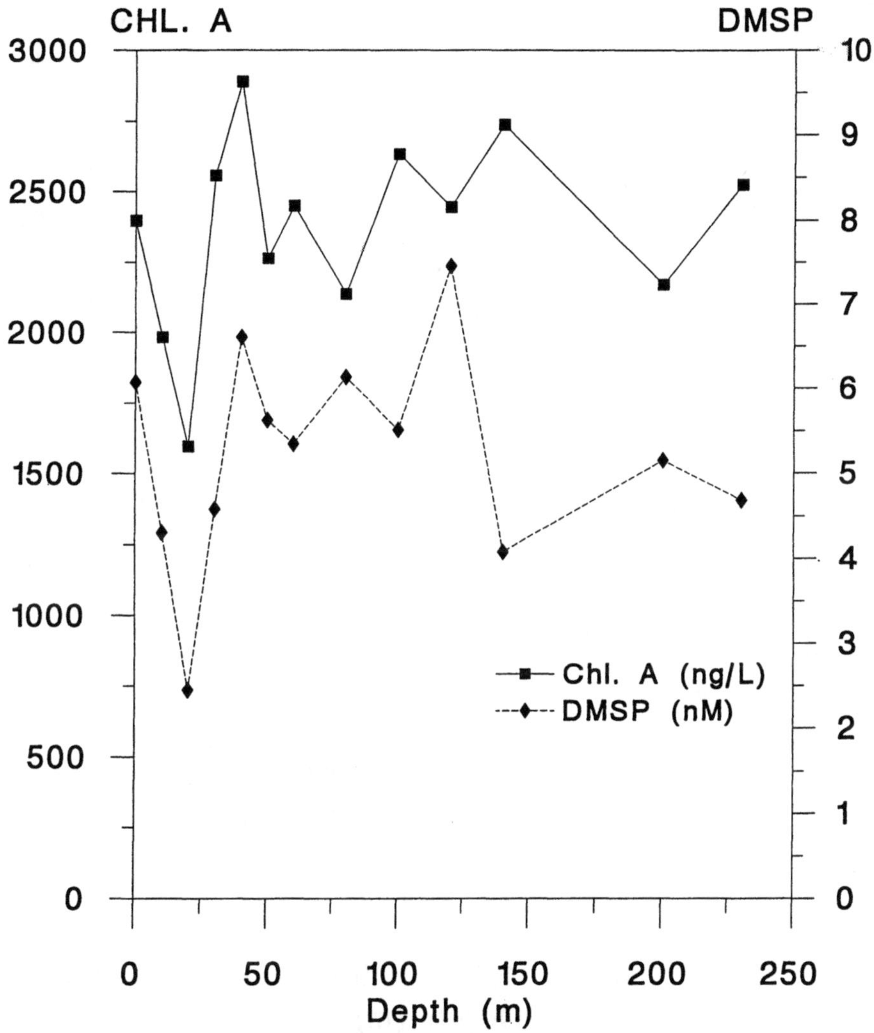

Fig. 4a: Depth profile on the shelf of the Weddell Sea: Chlorophyll A and DMSP content of plankton samples.

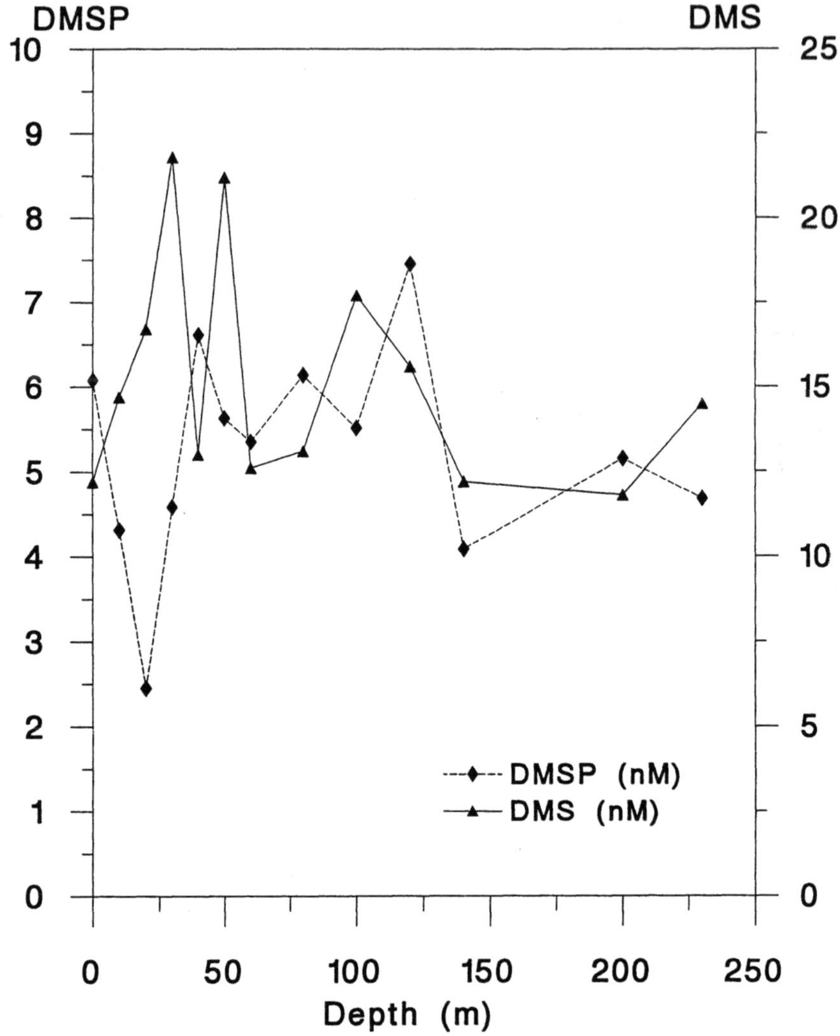

Fig. 4b: Depth profile: DMS concentration in the water; DMSP data as in Fig. 4a, shown for comparison.

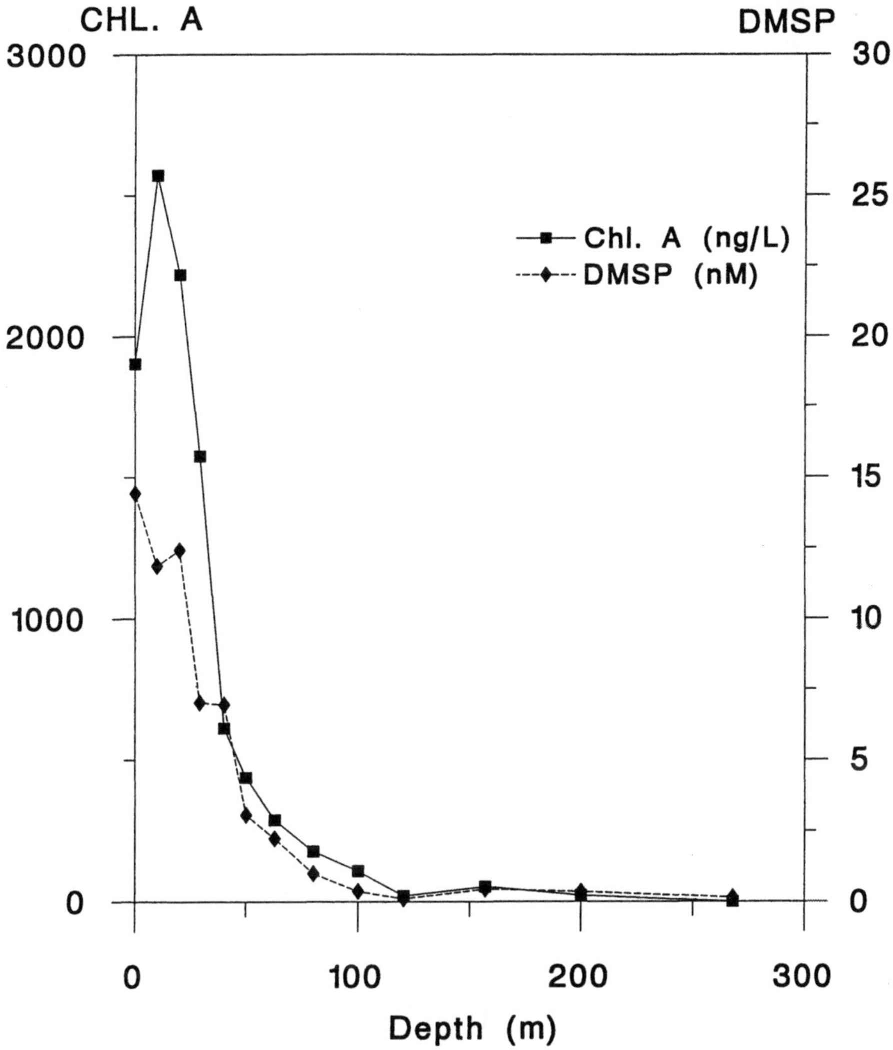

Fig. 4c: Depth profile: Chlorophyll A and DMSP content of plankton samples

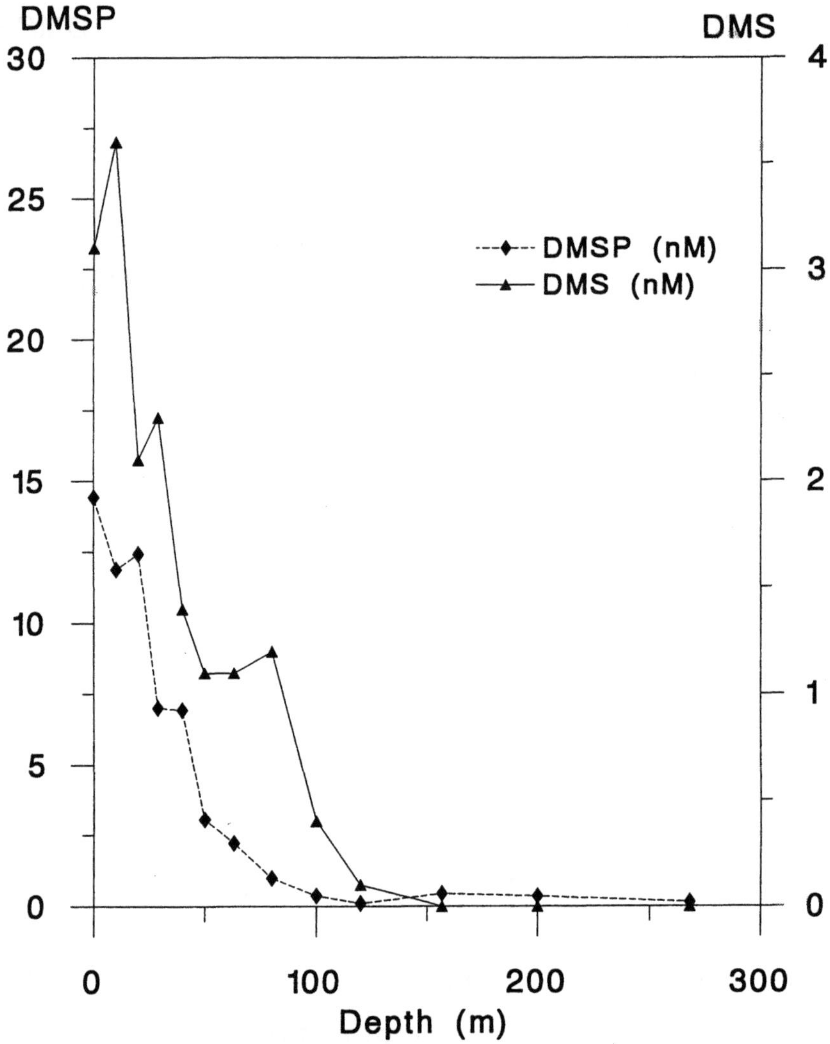

Fig. 4d: Depth profile: DMS concentration of the water, DMSP data as in Fig. 4c, shown for comparison.

ISOLATION OF MARINE DIMETHYLSULFIDE-OXIDIZING BACTERIA

T.A. HANSEN, P. QUIST, M.J.E.C. VAN DER MAAREL and L. DIJKHUIZEN

Department of Microbiology,
University of Groningen,
Kerklaan 30
9751 NN Haren, The Netherlands

SUMMARY

The low dimethylsulfide (DMS) concentrations found in sea water are the result of a dynamic balance between its production by algae and bacteria (mainly from dimethylsulfoniopropionate present in algae), consumption by bacteria, chemical oxidation and escape to the atmosphere.
Quite a variety of DMS-utilizing bacteria have been isolated from freshwater, waste water purification plants and peat filters but few data are available on such bacteria from marine environments.
The objective of our current study is the isolation and characterization of important DMS-utilizing marine bacteria. Using samples from various environments (North Sea water, intertidal sediments, non-axenic algal cultures) and employing both batch and continuous culture techniques we succeeded in isolating several pure cultures of aerobic DMS-utilizing marine bacteria. All isolated bacteria are short rods. A full characterization of these strains is underway.

1. INTRODUCTION

The quantitatively most important source of DMS in nature is dimethylsulfoniopropionate (DMSP), a compound which occurs in certain marine algae in high concentrations and which is thought to function as an osmolyte (18). Other sources of DMS include methoxylated aromatic compounds which in the presence of sulfide can be demethoxylated by the action of certain anaerobic bacteria (1).
The DMSP from algae can be cleaved to yield DMS and acrylate, a reaction which occurs chemically at high pH values (values that do not normally occur in the sea) or through the action of the enzyme DMSP lyase. Bacteria that split DMSP into DMS and acrylate use the acrylate formed but do not oxidize the DMS (7). Not all of the DMSP is degraded via the lyase pathway. Taylor and Gilchrist (19) demonstrated the existence of marine bacteria that degrade DMSP via an initial demethylation to methiolpropionate. These bacterial strains did not grow with DMS or methanethiol as a carbon and energy source but they did produce methanethiol from DMSP and methiolpropionate. Anaerobically the breakdown of DMSP may also proceed via a lyase reaction to DMS and acrylate or via demethylation (10). Pure bacterial cultures that carry out the anaerobic demethylation are not available. Because of the importance of the flux from DMSP to DMS and the demethylation pathway there is an increased interest in the bacterial breakdown of DMSP by marine bacteria. Bacteria may not be the only organisms that can break DMSP down to DMS. In the Biological Centre of the University of Groningen Stefels and van Boekel recently demonstrated that senescent axenic (i.e. without contaminating bacteria) cultures of the alga *Phaeocystis* degraded part of their DMSP to DMS. This clearly shows that the microalgae themselves may also be involved in the production of DMS.

The flux of DMS to bacteria that metabolize the compound is most probably virtually always far more important than the flux of DMS to the atmosphere (8). Despite the evident importance of bacterial metabolism in determining the DMS concentration in sea water and thereby the flux to the atmosphere up to recently little was known about the bacteria that carry out the degradation of DMS. In their 1991 paper on aerobic marine DMSP-degrading bacteria Taylor and Gilchrist (19) concluded that no pure cultures of aerobic DMS-metabolizing marine bacteria were available. Knowledge of the nature and the growth characteristics of DMS-degrading bacteria is essential in understanding the DMS concentrations in sea water and the occurrence and levels of DMS peaks in relation to season-dependent algal bloooms. With this in mind we are working on the isolation and characterization of marine DMS-degrading bacteria.

G. Restelli and G. Angeletti (eds.), Dimethylsulphide: Oceans, Atmosphere, and Climate, 37–41.

2. DMS DEGRADATION BY BACTERIA

Bacterial degradation of DMS by a pure culture of bacteria was discovered by Sivelä and Sundman in 1975 (15). From a waste water treatment system of a wood pulp industry in Finland these authors isolated a *Thiobacillus* culture that did not only grow with reduced inorganic sulfur compounds such as thiosulfate as an energy source but also with DMS. Unfortunately, the DMS-oxidizing culture was lost before extensive work had been carried out. Since this pioneering work several pure cultures of DMS-degrading bacteria have been described. These bacteria fall into three categories, namely aerobes, non-phototrophic anaerobes and anoxygenic phototrophs (Table 1).

The aerobic DMS-utilizing bacteria that oxidize DMS to CO_2 and sulfate belong either to the genus *Hyphomicrobium* or to *Thiobacillus*. These genera do not only differ in the shape of the cells and the mechanism of cell division but also in the pathway used for the conversion of DMS to cell material. *Hyphomicrobium* is a so-called budding bacterium. Unlike most other bacteria the cells do not divide via binary fission, a process in which two daughter cells of equal size arise from the mother cell. For the synthesis of cell material from DMS *Hyphomicrobium* employs the so-called serine pathway. In this route formaldehyde (formed from DMS) is used at the level of methylenetetrahydrofolate to convert glycine to serine. *Thiobacillus*, a genus of rod-shaped bacteria that divide by binary fission, can use the oxidation of inorganic sulfur compounds such as sulfide, sulfur and thiosulfate to meet its energy requirements and it can use CO_2 as its sole carbon source. When growing with DMS *Thiobacillus* does not use the formaldehyde as its main source for the synthesis of cell material but it fixes CO_2.

Table 1. Pure cultures of bacteria that have been shown to metabolize DMS

ORGANISM	SOURCE	PRODUCTS	REFERENCE
Aerobes			
Thiobacillus MS1	biofilter	CO_2, sulfate	(15)
Thiobacillus thioparus TK-m	sewage	CO_2, sulfate	(6)
Thiobacillus thioparus E6	pond water	CO_2, sulfate	(16)
Thiobacillus thioparus T5	marine microbial mat	CO_2, sulfate	(23)
Hyphomicrobium S	soil	CO_2, sulfate	(3)
Hyphomicrobium EG	biofilter	CO_2, sulfate	(17)
Hyphomicrobium 155	peat biofilter	CO_2, sulfate	(26)
Pseudomonas acidovorans DMR	peat biofilter	DMSO	(27)
Anoxygenic phototrophs			
Thiocystis sp.	intertidal mud	DMSO	(25)
Thiocapsa roseopersicina M1	marine microbial mat	DMSO	(21)
Anaerobes			
Methanohalophilus zhilinae	alkaline/saline lake	CH_4, CO_2, HS^-	(12)
Methanohalophilus oregonense	alkaline/saline aquifer	CH_4, CO_2, HS^-	(11)
Methanolobus siciliae	oil well	CH_4, CO_2, HS^-	(13)
methanogen GS-16	estuarine mud	CH_4, CO_2, HS^-	(14)
methanogen MTP4	marine mud	CH_4, CO_2, HS^-	(4)

The *Thiobacillus* strain isolated by Visscher et al. (23; see Table 1) of the University of Groningen was in fact the first aerobic marine DMS-oxidizing bacterium studied. This strain was obtained as an isolate that grew with thiosulfate as an energy source and only later it was found out that the culture utilized DMS as well. The microbial mats these workers were studying contain rather high concentrations of DMS (up to 0.22 mM) most probably because of the dominant presence in the mats of a DMSP containing cyanobacterium, *Microcoleus chthonoplastes* (22). In the mats there is also an active production of sulfide and thiosulfate. Oxidation of DMS to CO_2 and sulfate by *Hyphomicrobium* and *Thiobacillus* proceeds via methanethiol, formaldehyde and formate (and sulfide) as intermediates. The initial step is the conversion of DMS in a monooxygenase reaction with a concomitant NADH oxidation; the enzyme is rather unstable and has not been characterized in detail (3). Methanethiol is oxidized in a reaction catalyzed by methanethiol oxidase; the oxygen is reduced to H_2O_2. The methanethiol oxidase has been purified to homogeneity from *Hyphomicrobium* EG and *Thiobacillus thioparus* TK-m (5,18). Methanethiol concentrations above 14 μM were inhibitory to the *Hyphomicrobium* enzyme.

Of special interest is the recent discovery of the oxidation of DMS to dimethyl sulfoxide (DMSO) by a *Pseudomonas (Comamonas) acidovorans* strain (27). This organism required an additional carbon substrate for growth. Possibly, this type of bacterial oxidation of DMS may also occur in seas. The *Hyphomicrobium* strains discussed above and the *Thiobacillus* strain of (23) were also able to metabolize DMSO; the initial step of its metabolism was the reduction to DMS.

Anaerobic oxidation of DMS to DMSO can be carried out by certain anoxygenic phototrophic bacteria. A marine *Thiocystis* and a strain of *Thiocapsa roseopersicina* were shown to be able to use DMS as a photosynthetic electron donor (see Table 1); these organisms normally use hydrogen sulfide for that purpose. Under anaerobic conditions several bacteria can reduce DMSO to DMS. They use DMSO as an alternative electeron acceptor (24). For that reason the anaerobic oxidation of DMS to DMSO does not really lead to the degradation of DMS.

Anaerobic breakdown of DMS in the absence of potential electron acceptors leads to the formation of methane. Pure cultures of methanogenic bacteria have been isolated, for instance from estuaries and saline lakes; the conversion of DMS proceeds according to the following equation:

$$(CH_3)_2S + H_2O \dashrightarrow 1.5\,CH_4 + 0.5\,CO_2 + H_2S$$

Studies with anaerobic estuarine sediments on the fate of added labelled DMS and the effects of inhibitors of methanogenesis and sulfate reduction suggested that at the low DMS concentrations present in the sediments DMS is mainly broken down to CO_2 by sulfate-reducing bacteria (9). Attempts to obtain pure cultures of sulfate-reducing bacteria that degrade DMS including those in our laboratory have been unsuccessful thus far.

3. ISOLATION OF NEW TYPES OF AEROBIC, MARINE, DMS-OXIDIZING BACTERIA

Cultivation of DMS-oxidizing bacteria is difficult. DMS is a volatile, malodorous compound; in earlier studies (17) it appeared to be tolerated in only low concentrations by the aerobic bacteria that could degrade it. One of the ways to circumvent these problems is the use of DMSO as a growth substrate. This approach was used by de Bont et al. (3) and also by Suylen et al. (17) in their work with *Hyphomicrobium* strains. Only a few growth experiments were carried out with DMS as a substrate. It is however unlikely that DMS-oxidizing bacteria in general are also able to oxidize DMSO. For this reason we carried out all our enrichment and isolation experiments with DMS as a substrate. To avoid toxicity problems two techniques were applied: a) DMS-limited chemostat cultures and b) cultivation in flasks incubated in closed containers with a relatively low DMS concentration in the atmosphere. In both cases a mineral medium (culture solution) according to (2) was used or slight modifications of it; it contained 150 nM vitamin B12. All experiments were carried out 20°C.

The chemostat enrichment was carried out at a dilution rate of 0.03 h^{-1}. In order to limit the loss of DMS from the liquid to the gas phase, the DMS-containing solution was fed directly into the liquid. To prevent backgrowth of bacteria into the feed line the medium was divided into two solutions, one solution containing all of the ingredients with the exception of DMS at a pH 4.0, and a second solution of 10 mM DMS at pH 12. In the culture vessel the pH was kept at pH 8.0 by automatic titration with 1.0 N HCl. In the initial phase of the enrichment a DMS concentration of 300 μM was used. As an inoculum we used a suspension prepared with *Ulva lactua* collected from the Dutch Wadden Sea; the sample was filtered through a 5 μm-filter to remove potential predators such as protozoa. After 5 volume changes the chemostat culture was approaching a steady state: microscopically the culture appeared stable and the DMS level remained below 5 μM. A pure culture was obtained via repeated plating on similar media solidified with agar and incubated under an atmosphere with no more than 0.3 % (v/v) DMS. It took up to 3 weeks to obtain colonies on plates.

Cultures of DMS-oxidizing bacteria were also obtained via batch enrichments under an atmosphere with up to 0.3 % DMS and via the highest dilution that scored positive in counts according to the most probable number technique. As inocula we used mud/water mixtures from the Dutch Wadden Sea, *Ulva lactua* collected at the Dutch coast, North Sea water, non-axenic algal cultures obtained from the Department of Marine Biology of the University of Groningen and samples from the mesocosms studied by Kwint (see paper in this volume). For a more constant DMS level in the gas phase of the containers in which our cultures were incubated we made use of the property that a DMS solution in silicon oil is in equilibrium with a far lower DMS concentration in the gas phase than an aqueous DMS solution. Alternatively, DMS could be supplied by injection of DMS through a septum into the container.

By these techniques we obtained 7 pure bacterial cultures that could grow slowly with DMS as sole energy source and additional strains that metabolized DMS in the presence of acrylate. None of these cultures was able to degrade DMSP or DMSO. Although several strains grew with methanol as a carbon and energy source, none of the strains was specialized in using C1-substrates. Furthermore, the cultures did not grow in media for *Thiobacillus*. All of the strains were short rods that divide by binary fission and did not have the characteristic *Hyphomicrobium* morphology. Thus it appears that we have obtained several strains of new type(s) of DMS-oxidizing bacteria. Currently the biochemical pathway(s) of DMS oxidation and cell carbon assimilation are being studied and the taxonomic affiliation of these bacteria is object of further work. In addition, the growth parameters (such as affinity for DMS) and their occurrence will be determined and used in attempts to model DMS fluxes and concentrations in (semi)-natural systems.

REFERENCES

(1) BAK, F., FINSTER, K. and ROTHFUSS, F. (1992). Formation of dimethylsulfide and methanethiol from methoxylated aromatic compounds and inorganic sulfide by newly isolated anaerobic bacteria. Arch. Microbiol. 157:529-534.

(2) BAUMANN, P., BAUMANN, L., and MANDEL, M. (1971). Taxonomy of marine bacteria: the genus *Beneckea*. J. Bacteriol. 107:268-294.

(3) DE BONT, J.A.M., VAN DIJKEN, J.P. and HARDER, W. (1981). Dimethyl sulphoxide and dimethyl sulphide as a carbon, sulphur and energy source for growth of *Hyphomicrobium* S. J. Gen. Microbiol. 127:315-323.

(4) FINSTER, K., TANIMOTO, Y. and BAK, F. (1992). Fermentation of methanethiol and dimethylsulfide by a newly isolated methanogenic bacterium. Arch. Microbiol. 157:425-430.

(5) GOULD, W.D. and KANAGAWA, T. (1992). Purification and properties of methyl mercaptan oxidase from *Thiobacillus thioparus* TK-m. J. Gen. Microbiol. 138:217-221.

(6) KANAGAWA, T. and KELLY, D.P. (1986). Breakdown of dimethyl sulphide by mixed cultures and by *Thiobacillus thioparus*. FEMS Microbiol. Lett. 34:13-19.

(7) KIENE, R.P. (1990). Dimethyl sulfide production from dimethylsulfoniopropionate in coastal seawater samples and bacterial cultures. Appl. Environ. Microbiol. 56:3292-3297.

(8) KIENE, R.P. and BATES, T.S. (1990). Biological removal of dimethyl sulphide from sea water. Nature 345:702-705.

(9) KIENE, R.P., OREMLAND, R.S., CATENA, A., MILLER, L.G. and CAPONE, D. (1986). Metabolism of reduced methylated sulfur compounds in anaerobic sediments and by a pure culture of an estuarine methanogen. Appl. Environ. Microbiol. 52:1037-1045.

(10) KIENE, R.P. and TAYLOR, B.F. (1988). Demethylation of dimethylsulfoniopropionate and production of thiols in anoxic marine sediments. Appl. Environ. Microbiol. 54:2208-2212.

(11) LIU, Y., BOONE, D.R. and CHOY, C. (1990). *Methanohalophilus oregonense* sp. nov., a methylotrophic methanogen from an alkaline, saline aquifer. Int. J. Syst. Bacteriol. 40:111-116.

(12) MATHRANI, I.M., BOONE, D.R., MAH, R.A., FOX, G.E. and LAU, P.P. (1988). *Methanohal ophilus zhilinae* sp. nov., an alkaliphilic, halophilic, methylotrophic methanogen. Int. J. Syst. Bacteriol. 38:139-142.

(13) NI, S. and BOONE, D.R. (1991). Isolation and characterization of a dimethyl sulphide-degrading methanogen, *Methanolobus siciliae* HI350, from an oil well, characterization of *M. siciliae* T4/M', and emendation of *M. siciliae*. Int. J. Syst. Bacteriol. 41:410-416.

(14) OREMLAND, R.S., KIENE, R.P., MATHRANI, I., WHITICAR, M.J. and BOONE, D.R. (1989). Description of an esturaine methylotrophic methanogen which grows on dimethyl sulfide. Appl. Environ. Microbiol. 55:994-1002.

(15) SIVELÄ, S. and SUNDMAN, V. (1975). Demonstration of Thiobacillus-type bacteria which utilize methyl sulphides. Arch. Microbiol. 103:303-304.

(16) SMITH, N.A. and KELLY, D.P. (1988). Isolation and physiological characterization of autotrophic sulphur bacteria oxidizing dimethyl disulphide as sole source of energy. J. Gen. Microbiol. 134:1407-1417.

(17) SUYLEN, G.M.H. and KUENEN, J.G. (1986). Chemostat enrichment and isolation of *Hyphomi crobium* EG, a dimethyl sulphide oxidizing methylotroph, and reevaluation of *Thiobacillus* MS1. Antonie van Leeuwenhoek 52:281-293.

(18) SUYLEN, G.M.H., LARGE, P.J., VAN DIJKEN, J.P. and KUENEN, J.G. (1987). Methyl mercaptan oxidase, a key enzyme in the metabolism of methylated sulphur compounds by *Hyphomi crobium* EG. J. Gen. Microbiol. 133:2989-2997.

(19) TAYLOR, B.F. and GILCHRIST, D.C. (1991). New routes for aerobic biodegradation of dimethylsul foniopropionate. Appl. Environ. Microbiol. 57:3581-3584.

(20) TAYLOR, B.F. and KIENE, R.P. (1989). Microbial metabolism of dimethylsulfide. In: Saltzman, E.S. (ed.), Biogenic sulfur in the environment. American Chemical Society, Washington, pp. 202-221.

(21) VISSCHER, P.T. and VAN GEMERDEN, H. (1991). Photoautotrophic growth of *Thiocapsa roseopersicina* on dimethyl sulfide. FEMS Microbiol. Lett. 81:247-250.

(22) VISSCHER, P.T. and VAN GEMERDEN, H. (1991). Production and consumption of dimethylsul foniopropionate in marine microbial mats. Appl. Environ. Microbiol. 57:3237-3242.

(23) VISSCHER, P.T., QUIST, P. and VAN GEMERDEN, H. (1991). Methylated sulfur compounds in microbial mats: *in situ* concentrations and metabolism by a colorless sulfur bacterium. Appl. Environ. Microbiol. 57:1758-1763.

(24) WIDDEL, F. and HANSEN, T.A. (1991). The dissimilatory sulfate- and sulfur-reducing bacteria. In: Balows, A., Trüper, H.G., Dworkin, M., Harder, W. and Schleifer, K-H (eds.) The Prokaryotes. Springer-Verlag, New York, pp. 583-624.

(25) ZEYER, J., EICHER, P., WAKEHAM, S.G. and SCHWARZENBACH, R.P. (1987). Oxidation of dimethyl sulfide to dimethyl sulfoxide by phototrophic purple bacteria. Appl. Environ. Microbiol. 53:2026-2032.

(26) ZHANG, L. HIRAI, M. and SHODA, M. (1991). Removal characteristics of dimethyl sulphide, methanethiol an dhydrogen sulphide by *Hyphomicrobium* sp. 155 isolated from peat biofilter. J. Ferm. Bioeng. 72:392-396.

(27) ZHANG, L., KUNIYOSHI, I., HIRAI, M. and SHODA, M. (1991). Oxidation of dimethyl sulfide by *Pseudomonas acidovorans* DMR-11 isolated from peat biofilter. Biotechnol. Lett. 13:223-228.

ASPECTS OF THE BIOGEOCHEMISTRY OF DIMETHYLSULFIDE (DMS) AND DIMETHYLSULFONIUMPROPIONATE (DMSP) AT AN ANTARCTIC COASTAL SITE

A. McTAGGART and H. BURTON

Australian Antarctic Division, Channel Highway, Kingston 7150, Tasmania, Australia

Summary

The influence of the summer stratification and increased primary productivity in the Antarctic coastal water column on the production of dimethylsulfide (DMS) and dimethylsulfoniumpropionate (DMSP) was determined. Differences in sample handling procedure resulted in DMS concentration changing by a factor of 2.75. The ratio of dissolved DMSP and DMS was maintained at *ca.* 2.0 during winter and summer. The December summer thermocline divided the production of DMSP into two zones. DMSP was closely related to the increase in *Phaeocystis pouchetii* pigments in both zones whereas DMS was only closely related below the base of the thermocline.

1. Introduction

In Antarctic coastal waters, sea-ice coverage and the season dictate, (a) the level of light available for photosynthesis, (b) the temperature of the water and (c) the phytoplankton assemblage in the water column. The concentration of DMS in seawater depends strongly upon whether there are DMSP producing phytoplankton. There has been only three published studies investigating seasonal change in DMS concentration in the water column. Each study has attributed the control of the concentration of DMS and dissolved DMSP to either (a) the net effects of biological production in the euphotic zone, (b) consumption by bacterioplankton, (c) grazing of specific DMSP producing phytoplankton or (d) ventilation to the atmosphere (1, 2). The third study took place at an inshore site near Davis Station, Antarctica (3). The increase in DMS was closely correlated to the number of colonial cells of *Phaeocystis pouchetii*. The aim of this work was to investigate the biogeochemistry of DMS and dissolved DMSP relative to indicators of primary productivity in an Antarctic coastal water column.

2. Sample location and handling procedures

Seawater samples were collected at 68° 31.8' S 77° 47.9' E using a 2.5 litre Kemmerer water sampler on 21 August, 22 September, 24 October, 24 November, 22 December 1988 and on 17 January 1989. The sampling site was 5 km magnetic north of Magnetic Island in a water

G. Restelli and G. Angeletti (eds.), Dimethylsulphide: Oceans, Atmosphere, and Climate, 43–52.

depth of *ca.* 85 metres. Samples were taken from the surface (or the ice/seawater interface), 2, 5, 7, 10, 15, 20, 30, 40, 60, and 80 m. Seawater for DMS and dissolved DMSP analysis was transferred directly into 250 ml high density polyethylene (Duronal) bottles containing 0.1% w/v mercuric chloride. The bottles were stored in the dark at *ca.* 0 °C until analysis which occurred within 48 hours. The influence of the length of the storage period on the concentration of DMS in Antarctic seawater samples has been previously described (McTaggart and Burton, *in press*). The release of DMS due to the sample handling procedure was assessed during the height of the Antarctic summer phytoplankton bloom. A comparison was made between the concentration of DMS in seawater poured directly into 250 ml high density polyethylene bottles and a sub-sample that had been reverse filtered at the sample site. Reverse filtration gently concentrates the phytoplankton syphoned from the water sampler into the bottom of a wide shallow perspex cylinder allowing a phytoplankton free filtrate to emerge under gravity through a Whatman 541 filter (4). This experiment was repeated without $HgCl_2$.

Separate casts were made for nutrients, dissolved oxygen and phytoplankton pigments. Samples were taken at the same depths as described above. The seawater taken in the field for nutrient analyses was stored in seawater aged high density polyethylene bottles and kept at 0 °C. These samples were subsequently filtered under vacuum in the laboratory through 0.45 µm type HA Millipore membrane filter then frozen in the dark to -20 °C. Nutrients were analysed upon return to Australia. A 2.0 l seawater sample for phytoplankton pigments was taken at a depth of 15 m each month. In December, phytoplankton pigments were taken from each sample depth. The seawater was later filtered through a Whatman GF/C to retain particulate material. The filters were wrapped in aluminium foil and stored at -20 °C. Upon return to Australia, the filters were transferred into liquid nitrogen and were stored for a year prior to analysis. Seawater samples taken for oxygen determination were gently syphoned from the bottom of the Kemmerer bottle into 500 ml glass stoppered oxygen bottles before fixing reagents were added. Analyses were carried out in the laboratory after the bottles had reached laboratory temperature (*ca.* 19 °C). A further 250 ml of seawater was taken for salinity measurements. These samples were stored in 250 ml high density polyethylene bottles which contain a polyethylene insert inside the caps to prevent leakage or evaporation. The samples were stored at 0 °C in the dark until analysed back in Australia.

3. Methods

Seawater temperature was measured by a mercury thermometer housed inside the Kemmerer water sample marked to 0.05 °C intervals. Salinity was measured using a YEOKAL inductive salinometer calibrated with International Standard seawater (Plymouth Marine Laboratory). Oxygen was determined in fixed seawater samples at the Davis laboratory using the Winkler method (5). Nitrate, phosphate and silicate were analysed by a 3 channel automated nutrient analyser (6). Phytoplankton pigments were analysed by HPLC (7).

DMS was analysed using a purge and cryotrap method (8). In brief, a 40 ml (August to November) or 10 ml (December and January) sample was gently filtered under hand pressure through a 0.45 μm type HA Millipore filter into a silanised 100 ml glass purge vessel (Dreschel bottle). The sample was purged with 50 ml/min helium for 20 minutes and the volatile constituents of the seawater sample were trapped at *ca.* -90 °C on glass beads packed in a Teflon coated stainless steal tube. After heating the trap with boiling water, the trapped volatiles were separated using GC/FPD. The relative standard deviation was better than 10% based on DMS seawater sub-standards and a DMS permeation tube (G.C. Industries, USA). Dissolved DMSP was determined as DMS after the addition 2M NaOH. DMSP breaks down to stoichiometric equivalent amounts of DMS and acrylic acid under mild alkaline conditions at room temperature (9). The reaction was complete after 20 minutes of sparging (10). The DMS released from the dissolved DMSP was determined as described above.

4. Results

Salinity and temperature throughout the water column showed very little variation between August and November. Sea-ice covered the sample site until the December sample occasion. The seawater temperature of the water column was -1.8 °C except for October when the surface and 2 m samples warmed slightly to -1.5 and -1.6 °C respectively. The mean salinity between August and November was 34.412 psu (S.D. 0.197, n=44). There was a decrease in salinity throughout the water column in October (mean : 34.263, S.D. : 0.330, n=11). In December a strong thermocline developed between the surface and 15 m which decreased in depth in January to 40 m. The sea-ice cover had broken by December such that the edge of the pack ice was at the sample site. By January, the pack ice had cleared and the sampling took place on an ice flow situated near the sample site. Seawater temperature in the December and January ranged from 0 °C to -0.8 °C at the base of the thermocline. Below the thermocline the temperature ranged from -1.2 to -1.5 °C. Oxygen showed little variation between August and November. The mean concentration was 352.9 μM (S.D. : 3.1, n=37). There was a marked increase in oxygen concentration in the summer thermocline. Oxygen concentrations ranged from 425 μM at 2 m to 395 μM at 15 m in December while in January the oxygen concentration ranged from 427 μM at the surface to 409 μM at the base of the thermocline. Below the thermocline the oxygen concentration ranged from 358 to 382 μM, decreasing with depth.

Nutrient concentrations were corrected for dilution if the salinity of the sample showed a decrease from the mean winter salinity (34.412) of 1%, otherwise the original concentration was taken (11). The mean concentration of nitrate and dissolved phosphate throughout the water column from August to November was 21.9 μM (S. D.: 1.8, n=33) and 1.93 μM (S.D. : 0.2, n=33) respectively. There was little variation from month to month. The concentration of nitrate and phosphate decreased rapidly in the summer thermocline by up to *ca.* 58 and 75 % respectively of their mean winter concentrations.

Phytoplankton pigment concentrations measured at 15 m depth each month are shown in Table 1. The concentration of chlorophyll *a* increased by 2 orders of magnitude in mid- summer relative to the winter chlorophyll *a* concentrations. The other two pigments, fucoxanthin and 19'-hexanoyloxyfucoxanthin, may provide useful markers of phytoplankton populations in Antarctic water samples (12). Both pigments showed concomitant increases in concentration with chlorophyll *a* concentration. There were two maxima in concentrations for all three pigments. The first was confined to the warmer fresher waters above the thermocline and the second was situated below the base of the summer thermocline. This would suggest that the position of the summer thermocline in December separates the water column into two biogeochemical zones.

Table 1 Phytoplankton pigment concentrations at a depth of 15 m between August 1988 and January 1989

Month	Chlorophyll *a* (mg m^{-3})	Fucoxanthin (mg m^{-3})	19'-H-O-Fuco[a] (mg m^{-3})
August	0.010	0.006	0.004
September	0.029	0.030	0.010
October	0.024	0.010	0.004
November	0.053	0.140	0.020
December	2.316	1.080	0.620
January	2.130	1.640	0.520

[a]19'- Hexanoyloxyfucoxanthin

The affect of sample handling on the concentration of DMS and dissolved DMSP in seawater was assessed by comparing samples that had been poured directly into the sample bottles to samples that were first gently reverse filtered before syphoned into the sample bottles. A comparison was also made between samples containing 0.1% $HgCl_2$ and samples that did not (see Table 2). The samples were taken on two separate occasion in early and mid January and at different depths in the water column. The addition of $HgCl_2$ did not significantly affect the concentration of DMS compared to the effect of reverse filtration. DMS concentration decreased by a factor of *ca.* 2.75 relative to the direct addition method on both sample dates no matter whether $HgCl_2$ was added or not. As a result, all previous measurements of DMS made using the direct addition method were decreased by a factor of 2.75. The source of the increase in DMS is most likely to be dissolved DMSP. The shock of pouring the sample by direct addition into the sample bottle instead of gently separating the organic material from the dissolved constituents was apparently sufficient to break down dissolved DMSP to DMS. The difference between the former DMS concentrations and the recalculated DMS therefore represented dissolved DMSP. This has been added to the dissolved DMSP previously determined after the analysis of DMS by direct addition.

Table 2 Effect of reverse filtering on mean DMS concentration (nM) in seawater.

Sample	Direct Addition		Reverse filtration	
	$HgCl_2$ added	No $HgCl_2$	$HgCl_2$ added	No $HgCl_2$
3/1/89 (n=2)	123	127	43	49
17/1/89 (n=3)	119	117	46	39

A justification of this reappraisal of DMS and dissolved DMSP concentrations is taken up in the Discussion section of this paper. The mean and standard deviation of the concentration DMS and dissolved DMSP recalculated for the effect of direct addition is shown in Table 3. The concentration of DMS and DMSP in the water column from August to November showed a relatively uniform monthly concentration throughout the water column. The concentration of DMS in August and September were 2-3 nM on average greater than in October and November. The concentration of DMSP in September was 4 nM greater on average relative to August and October. The concentration of DMS and DMSP increased rapidly throughout the mid-summer with a maxima in concentrations always present in the December and January thermocline.

Table 3 Mean and standard deviation of the concentration of DMS (nM) and dissolved DMSP during 1988 - 89.

Month	DMS		DMSP	
	Mean	S.D	Mean	S.D.
August	2.9	0.9	5.9	1.7
September	3.9	1.4	9.4	3.6
October	1.0	0.2	5.9	1.7
November	1.9	1.4	NA	NA
December	28.2	24.0	95.5	71.7
January	91.0	30.5	214.2	69.3

NA : Not analysed

4. Discussion

The difference in DMS concentration between the two methods of sample handling has been ascribed to the rapid cleavage of DMSP due to the shock of pouring seawater into a sample bottle rather than gently syphoning and reverse filtering the sample. The source of the DMS is most likely to be DMSP present in a dissolved form or part of phytoplankton cells. Prior to the 1987/88 summer phytoplankton bloom (3), the ratio of dissolved DMSP to DMS from May to November of 1987 was 1.8 as compared to 4.2 for samples collected in the English channel and the North Sea (13). The analysis of DMSP in seawater collected during the summer phytoplankton bloom of 1987/88 proved impossible as DMSP concentrations were found to increase during the storage period. The initial determination of DMS and DMSP by direct addition during winter showed that DMS was in excess, ranging from a factor of 11.3 in August, 4.7 in September and 0.6 in October, based on the mean concentrations in the water column. This would appear contrary to the conception that DMS has a high turnover in the water column (14) and that DMSP is relatively stable in seawater (15). The concentrations of DMSP in seawater would tend to be greater than DMS concentration. After the DMS concentrations determined between August and November of 1988 were decreased by a factor of 2.75 and the difference added to the DMSP concentration, the ratio of dissolved DMSP to DMS was 1.9. As the 1988 winter ratio is about the same as the 1987 winter ratio, the difference in DMS concentration between seawater samples that were directly added to the sample bottles and those that had been syphoned and reverse filtered was due to the break down of dissolved DMSP. The method of transferring seawater into sample bottles in 1987 involved the gentle release of seawater by a sprung valve from the base of the Kemmerer water sampler. The method of transferring the sample during 1988 involved releasing the entire base of the Kemmerer bottle such that *ca.* 250 ml was dumped rapidly via a funnel into the sample bottle. This difference in sample transfer was sufficient to cleave any dissolved DMSP to DMS. The concentration of DMS and dissolved DMSP recalculated for the summer months of November to January 1988/89 had a dissolved DMSP to DMS ratio of 2.1. Both the summer and winter ratios were very similar. In Antarctic coastal waters throughout winter and summer, it would appear that the ratio of dissolved DMSP and DMS were kept in equilibrium by other biological processes, probably bacterial or enzymatic, rather than from the seasonally variable phytoplankton growth.

The production of DMS in Antarctic coastal waters during 1987/88 was closely associated with the increase in the numbers of *Phaeocystis pouchetii* (3). High concentrations on DMS has also been found in samples dominated by *Phaeocytis pouchetii* taken from the Bering Sea (16). It remains to be seen whether the increase in DMS and dissolved DMSP during the Antarctic austral summer is dependent upon the increase in primary productivity. The change in the concentration of DMS and dissolved DMSP determined throughout 1988/89 was related to the change temperature and salinity, to assess the influence of the thermocline and to nutrients and oxygen, to assess indirectly the effect of primary productivity (see Table 4). There was no strong correlation of DMS and dissolved DMSP with temperature, salinity, nutrients or dissolved oxygen during the winter which reflects the adverse conditions for biological growth under the sea-ice. In the summer months from November to January, an increase in temperature was strongly related to an increase in the concentration of DMS and dissolved DMSP. This was not coupled with changes to salinity, suggesting increasing temperatures favoured DMS produced from biological activity rather than from osmoregulation by the phytoplankton. This was supported by the good correlation of DMS and dissolved DMSP with phosphate and oxygen. As phosphate concentrations decreased, and oxygen production increased, DMS and dissolved DMSP increased. This would support the arguement that the increase in DMS and dissolved DMSP was dependent upon primary production in Antarctic coastal waters. An increase in the concentration of phytoplankton pigments is a direct measurement of the increase in primary productivity in the water column. The concentration of chlorophyll *a* rose rapidly from late spring to a level of 2.3 mg m^{-3} at a depth of 15 m by 20 December. Two other phytoplankton pigments measured at the same time, fucoxanthin and 19'-hexanoyloxyfucoxanthin, also showed maxima during the mid-summer.

Table 4 Linear regression of DMS and dissolved DMSP v temperature, salinity, nutrients and dissolved oxygen

	Summer		Winter	
	DMS r^2 (n=33)	DMSP r^2 (n=22)	DMS r^2 (n=33)	DMSP r^2 (n=22)
Temperature	0.936	0.922	0.112	0.017
Salinity	0.473	0.237	0.148	0.009
Nitrate	0.591	0.379	0.377	0.077
Phosphate	0.734	0.605	0.514	0.003
Silicate	0.115	0.382	0.481	0.011
Oxygen	0.752	0.708	0.172	0.000

The largest concentration of fucoxanthin was measured on the 17 January at 1.64 mg m^{-3} while the maximum concentration of 19'-hexanoyloxyfucoxanthin was measured earlier on 20 December at 0.62 mg m^{-3}. One of the sources of 19'-hexanoyloxyfucoxanthin is likely to be the colonial algae *Phaeocystis pouchetii*. The algae was observed to be in bloom proportions in a seawater sample taken from 15 m on the 20 December (17), on the same date and location that the DMS and dissolved DMSP were measured. High concentrations of this fucoxanthin derivative had also been measured in deeper waters adjacent to Davis throughout which *Phaeocystis* was also wide spread (12). The change in the concentration of phytoplankton pigments throughout the December water column coupled with the presence of a thermocline from the surface down to 15 m suggested that there were two biogeochemically different zones for primary productivity. The thermocline was warmer, fresher and would limit the depth to which lower density phytoplankton could sink. Beneath the base of the thermocline, the waters were colder and saltier, with less available surface light for photosynthesis. Linear regressions of DMS and DMSP with the concentrations of chlorophyll *a*, fucoxanthin and 19'-hexanoyloxyfucoxanthin have been divided according to whether the samples were taken above or below the thermocline (Table 5). The production rate of dissolved DMSP and chlorophyll *a* were very similar above and below the thermocline. The production of DMS showed a similar relation below the thermocline for chlorophyll *a*, fucoxanthin and 19'-hexanoyloxyfucoxanthin but that relation uncoupled above the thermocline. The mean ratio of 19'-hexanoyloxyfucoxanthin to fucoxanthin throughout the water column was 0.564 above the thermocline and 0.342 below the thermocline on the 20 December. The ratio of 19'-hexanoyloxyfucoxanthin to fucoxanthin at a depth of 15 m was 0.572 and the sample contained about 2 orders of magnitude more *Phaeocystis pouchetii* colonial algae cells than diatoms (17).

Table 5 Linear regression of phytoplankton pigments v DMS and dissolved DMSP

	Chlorophyll *a*		Fucoxanthin		19'-H-O-Fucoxanthin	
	Above r^2	Below r^2	Above r^2	Below r^2	Above r^2	Below r^2
DMS	0.100	0.928	0.199	0.900	0.080	0.906
DMSP	0.636	0.951	0.704	0.904	0.610	0.930

The warmer fresher waters of the thermocline were therefore dominated by *Phaeocystis pouchetii* colonial cells. Below the thermocline, the phytoplankton population contained *Phaeocystis pouchetii* colonial cells that were most probably at an earlier stage of development as significant differences in the proportion of these pigments had been observed between strains and age of *Phaeocystis pouchetii* cultures (12). This was also supported by similar production rates of dissolved DMSP and chlorophyll *a* above and below the thermocline. The above data provides strong evidence that the production rate of DMSP was dependent upon the growth of the colonial algae *Phaeocystis pouchetii* during its growth in the December Antarctic coastal water column.

REFERENCES

1. Wakeham, S. G., Howes, B. L. and Dacey, J. W. H. (1987). Biogeochemistry of dimethylsulfide in a seasonally stratified coastal salt pond. Geochim. Cosmochim. Acta 51, 1675-1684.
2. Leck, C., Larsson, U., Bågander, L. E., Johansson, S. and Hadju, S. (1990). Dimethylsulfide in the Baltic Sea: Annual variability in relation to biological activity. Geophys. Res., 95 (C3), 3353-3363.
3. Gibson, J. A. E., Garrick, R. C., Burton, H. R. and McTaggart, A. R. (1990). Dimethylsulfide and the algae *Phaeocystis pouchetii* in Antarctic coastal waters. Mar. Biol., 104, 339-246.

4. Dodson, A. N. and Thomas, W. H. (1978). Reverse filtration. In Sournia, A. (Ed.) UNESCO Phytoplankton manual, Monographs on oceanographic methodology, 104-107.
5. Strickland, J. D. H. and Parsons, T. R. (1968) A practical handbook of seawater analysis. Fish. Res. Board Can. Bull., 167, 1-311.
6. Airey, A and Saunders, G. (1987). Automated analysis of nutrients in seawater. CSIRO Marine Laboratories, Report 166. Hobart Tasmania
7. Wright, S. W. and Shearer, J. D. (1984). Rapid extraction and high performance liquid chromatography of chlorophylls and carotenoids from marine phytoplankton. J. Chromatog, 294, 281-295.
8. McTaggart, A. R. and Burton, H. R. (1992, *in press*). Dimethylsulfide concentrations in the surface waters of the Australasian Antarctic and Subantarctic Oceans during an austral summer. J. Geophys. Res.
9. White, R. H. (1982). Analysis of dimethyl sulfonium compounds in marine algae. J. Mar. Res., 40, 529-536.
10. Turner, S. M., Malin, G., Bågander, L. E. and Leck, C. (1990). Inter laboratory calibration and analysis of dimethyl sulfide in water. Mar. Chem., 29, 47-62.
11. Jennings, Jr, J. C., Gordon, L. I. and Nelson, D. M. (1984). Nutrient depletion indicates high primary productivity in the Weddell Sea. Nature Lond., 309, 51-54.
12. Wright, S. W. and Jeffrey, S. W. (1987). Fucoxanthin pigment markers of marine phytoplankton analysed by HPLC and HPTLC. Mar. Ecol. Prog. Ser., 38, 259-266.
13. Turner, S. M., Malin, G., Liss, P. S., Harbour, D. S. and Holligan, P. M. (1988). The seasonal variation of dimethyl sulfide and dimethylsulfoniopropionate concentrations in near shore waters. Limnol. Oceanogr., 33, 364-375.
14. Kiene, R. P. and Bates, T. S. (1990) Biological removal of dimethyl sulfide from seawater. Nature, 345, 702-704.
15. Dacey, J. W. H. and Blough, N. V. (1987). Hydroxide decomposition of dimethylsulfoniopropionate to form dimethylsulfide. Geophys. Res. Lett., 14, 1246-1249.
16. Barnard, W. R., Andreae, M. O. and Iverson, R. L. (1984). Dimethylsulfide and *Phaeocystis pouchetii* in the southeast Bering Sea. Contin. Shelf. Res., 3, 103-113.
17. Davidson, A. T. and Marchant, H. J. (1992, *in press*). Protist interactions and carbon dynamics of a *Phaeocystis*-dominated bloom at an Antarctic coastal site. J. Mar. Biol.

THE PRODUCTION OF DMS BY A PLANKTON COMMUNITY: A MESOCOSM EXPERIMENT

RIK L.J. KWINT[1,2] KEES J.M. KRAMER[2]
ARTHUR C. BAART[3] & HENK L.M. VERHAGEN[3]

[1] University of Groningen, Dept. of Microbiology, P.O. Box 14, 9750 AA Haren, The Netherlands
[2] Laboratory for Applied Marine Research, IMW-TNO, P.O. Box 57, 1780 AB Den Helder, The Netherlands.
[3] Department of Environmental Chemistry, IMW-TNO, P.O. Box 6011, 2600 JA Delft, The Netherlands

SUMMARY.

Dimethylsulfide (DMS) is one of the most important precursors of biogenic non-sea-salt-sulfate which is a major source of cloud condensation nuclei. Despite this important role, factors controlling the release of DMS from algae into the water and from the marine environment into the atmosphere are not yet clearly understood. Since laboratory experiments are not easily extrapolated to the field and difficulties will arise when trophic interactions are studied over longer timescales, experimentation at an intermediate level between laboratory and field seems desirable. For the present paper pelagic mesocosms were available to study plankton communities under semi-natural conditions. The results show clearly that in all systems the maximum production of DMS was linked to algal senescence. As changes in DMS concentrations in water were relatively fast, care must be taken when extrapolating results of simple field surveys to global DMS budgets.

1. INTRODUCTION:

Although dimethylsulfide (DMS) is considered to be the largest source of biogenic volatile sulfur compounds in the atmosphere and may have a considerable effect on climate due to the formation of cloud condensating nuclei (2), mechanisms controlling its release into water are still poorly understood.

No simple correlation between chlorophyll-a and DMS concentration exists in oceanic surface waters (3, 16, 17) However, Belviso et al. (1) found that dimethylsulfoniopropionate (DMSP), a precursor for DMS, was predominantly associated with small dinoflagellates. These authors suggest that the release of DMS in ocean surface waters is initiated by microzooplankton grazing. Other suggestions are among others: metabolic excretion accompanying phytoplankton growth (1, 19); phytoplankton grazing by zooplankton (4, 12, 18) and the senescence of microalgae (14). Gibson et al. (5) found a significant correlation between DMS concentration and biomass of the algae *Phaeocystis pouchetii*.

It is difficult to relate laboratory experiments to the field, as experiments with several trophical levels are not easily realised in the laboratory and are often executed on a scale that is too small to reveal trophic interactions on longer timescales.

This problem may be overcome by using pelagic mesocosms, or plankton enclosures (plastic bags) with about 1500 litres content. In this way the plankton dynamics can be studied under semi-field conditions. An important advantage is that the same plankton community can be followed under various experimental conditions and that repetitive samples can be collected from the same waterbody without major disturbance during the experiment. A disadvantage could be that the cycle of growth senescence and mineralization of a phytoplankton bloom is compressed in time depending on the size of the system, when compared with the natural system. In our case, a

G. Restelli and G. Angeletti (eds.), Dimethylsulphide: Oceans, Atmosphere, and Climate, 53–62.

natural sequence from March to September can be observed in 35 days. However, former studies by Kuiper (10) showed that in these types of mesocosms replicable plankton successions were found in systems with similar treatments. It should therefore be possible to relate differences between the individual bags to the different circumstances under which the plankton is incubated. New developments allow the measurement of actual fluxes of DMS from the water column to the atmosphere in these systems (figure 1).

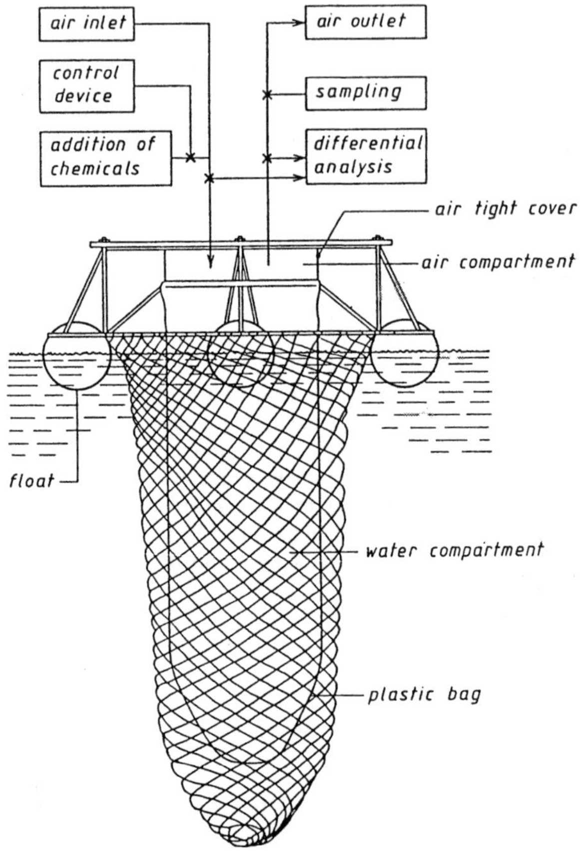

Figure 1. Schematical representation of the pelagic mesocosm, or model plankton system

The main objective of our experiment was to investigate whether phytoplankton will release DMS in the water during growth and/or senescence, and the role of grazing by copepod zooplankton in this process. In this paper we describe a full cycle of the formation and decay of a phytoplankton bloom and its DMS production, whether in the presence or in the absence of zooplankton. Both the DMS concentrations in the water column and the measured fluxes will be discussed. To compare the mesocosm findings to a field situation of North Sea coastal water, a full year cycle of the aquatic [DMS] and the chlorophyll-a concentration was monitored.

2. MATERIALS AND METHODS.

Experimental set-up

In the Den Helder harbour, four mesocom systems were installed alongside a raft. Each system consisted of a plastic bag (polythene/polyamide two-layered foil) with a depth of approximately 3 metres and 0.75 m diameter. The bags were suspended in the water, and covered with clear perspex shields. Mixing of the watercolumn resulted from the wave action, while ambient temperature and light remained close to the natural situation (see Kuiper (10); figure 1). One of the enclosures could temporarily be closed with an airtight lid, creating a mesocom with a water and an air compartment. Thus we were able to measure actual fluxes of DMS to the atmosphere under controlled meteorological conditions. To homogenize the air compartment, a small propellor provided a moderate windspeed of about 1.3 $m.s^{-1}$. The actual measurement was performed after equilibration of the system for a minimum of 45 min. The system was only closed during measurements, to allow for optimal comparability.

Seawater was collected in the Marsdiep channel (coastal North Sea) at high tide on 23 March 1992 by means of a floating watertank. The tank was left alongside the raft until the next day to allow most of the suspended particulate matter to be deposited. All enclosures were filled simultaneously via a branched pipe (10). The water was filtered over a 2 mm mesh to remove the larger predators. To create a difference in grazing pressure the water of two of the bags was additionally filtered over a 55 μm mesh in order to remove copepods.

The depth of all the bags was measured to calculate the volume. The water in the systems was left to rest for another day until 25 March 1992, to allow fine suspended material to settle.

Sampling from the systems

Water samples were collected through teflon tubing from a sampling depth of about 1.5 m, using a peristaltic pump. Samples for chlorophyll-a were collected in one litre polythene bottles and stored in the dark. Nutrient samples were deepfrozen until analysis. Phytoplankton samples were fixed with lugol. Water for the determination of $[DMS]_{water}$ was collected in glass stoppered dark bottles (100 ml) and stored in an isolated box at a temperature of about 10 °C until further treatment.

Zooplankton samples were collected using a 3 m pvc pipe, equipped with a ball valve at the end, according to Kuiper(9). The pipe was lowered 5 times on different spots in each enclosure. A total of 35 litres was filtered through a 55 μm net and the zooplankton was immediately fixed with glutaraldehyde.

The samples for the Marsdiep monitoring were collected several hundred meters from the coast. Most measurements were carried out at least in duplicate. We aimed to sample always at high tide and at about 10.00 AM in order to rule out tidal and/or diurnal effects. This automatically implied sampling intervals of about two weeks.

Samples that were obtained during the Marsdiep monitoring were treated in the same way as during the mesocosm experiment.

Analytical methods.

The water samples for the determination of DMS were placed in a refrigerator at 4 °C within two 2 hours. According to Kiene and Service(8) DMS solutions in seawater will remain stable for at least 16 hours at this temperature. A volume of 25 ml water was carefully filtered (a paper filter was used) in a glass purge vessel, equipped with a glass frit. Through this frit, high grade helium at 45 ml/min was purged through the sample via a teflon tube at room temperature. The purge gas, containing the volatile compounds, was dried through a 'Nafion' permeation drier (Dupont, model MD-125 P/F). Nitrogen was used as drying-gas. The dried helium was passed through a cold trap, consisting of a straight glass tube containing 200 mg of Tenax-ta 60/80 (Chrompack). This cold trap was placed horizontally over a 'Dewar-flask', containing liquid nitrogen. Cooling was established with aluminium strips, placed over the glass tube into the nitrogen. Thus the temperature at the tube could be maintained at about -120 °C. After 20 minutes purging, the collection tube was closed on both ends with 'Swagelock' caps (stainless steel, fitted with teflon ferrules) and stored in liquid nitrogen until analysis (adapted from Lindqvist (13); Lindqvist, pers. comm.). Tests on the storage procedure with calibration gas revealed that DMS samples can thus be stored for at least 8 weeks without obvious change. The DMS flux was measured based on the difference between the inflow and outflow of synthetic air (at a rate of 10 $l.min^{-1}$). Air samples, based on 15 min collection of synthetic air (3 l) for the flux measurements, were trapped cryogenically on 'Tenax' tubes as well, and analyzed on the day of collection.

The DMS was analyzed on a 'Varian 3700' gaschromatograph equipped with a linear plot column and a photo-ionization detector (10.2 eV), according to Lindqvist (12). At the end of the capillary column, sulfur compounds are catalytically converted to H_2S in a platinum containing pyrolyser at 1100 °C. Only sulfur species are thus detected. The analyses were compared with a dynamic dilution of the weight loss of permeation/diffusion tubes containing COS (carbonylsulfide), DMS, DMDS (dimethyldisulfide) and water. This calibration gas was sampled via a similar Nafion permeation drier and trapped on Tenax cryogenic traps. The detection limit for DMS was 100 pg.

Chlorophyll-a samples were immediately filtered on glass fibre filters (Whatmann GF/C), extracted with 90% acetone and analyzed on a spectrophotometer according to standard procedures (15). Nutrient samples were analyzed for orthophosphate, ammonia, nitrate, nitrite and reactive silicate using a 'Technicon' auto-analyzer (15). Phytoplankton samples, preserved with Lugol, were counted in a 5 ml inverted microscope chamber. At least 25 fields were counted (magnification 400x). Diatoms were identified to species level while flagellates were divided into size classes. Zooplankton organisms, preserved with 2.5% glutaraldehyde, were counted microscopically and identified in a 1 ml chamber. Species were identified and a division was made in adult copepods, copepodites and nauplii.

3. RESULTS AND DISCUSSION.

Figure 2 shows the development of the chlorophyll-a concentrations in the mesocosms. The results are typical for this kind of experiment. A complete spring and summer plankton bloom is compressed into 35 days. The first peak, around day 8, is mainly caused by the diatoms *Skeletonema costatum* and *Thalasiosira nordenskioldii*. The second peak, which is less pronounced (day 20-30), consists mainly of flagellates like *Phaeocystis pouchetii*. It will be clear that the development of phytoplankton in all systems (expressed as chl-a) is very similar not withstanding treatment by filtering. The presence of copepods (55 μm filtered v.s. no filtered) seemed to have little effect on the phytoplankton development.

In figure 3 the concentrations of soluble silicate, nitrate and phosphate are given. The silicate is depleted from the water in one week by the diatoms and reappears only in very low

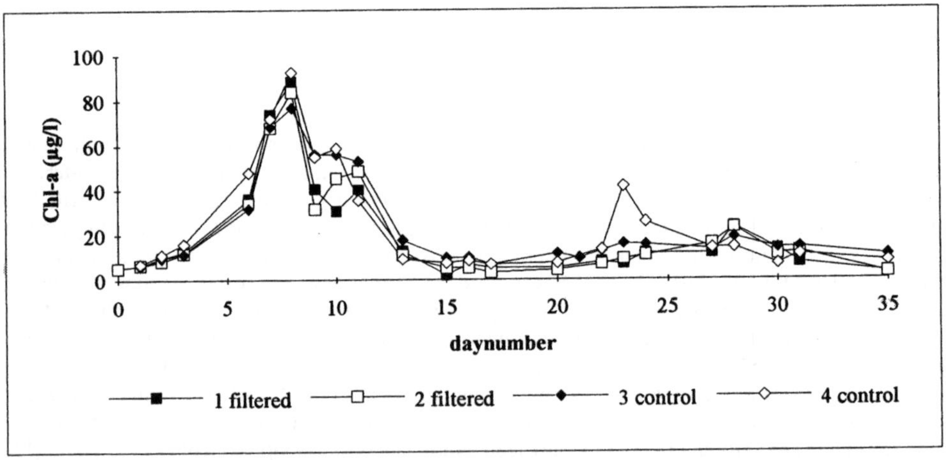

Figure 2. Chlorophyll-a concentrations during the mesocosm experiment of March/April 1992 Shown are results of 55 μm filtered (1,2) and non filtered (3,4) duplicates

concentrations after day 20, indicating that re-mineralization had started. The nitrate concentration becomes limiting after day 10 and stays low during the rest of the experiment. During this time the second phytoplankton bloom utilises the mineralized nitrate in this period. The phosphate concentration does become very low, but is never limiting during the experiment. No major

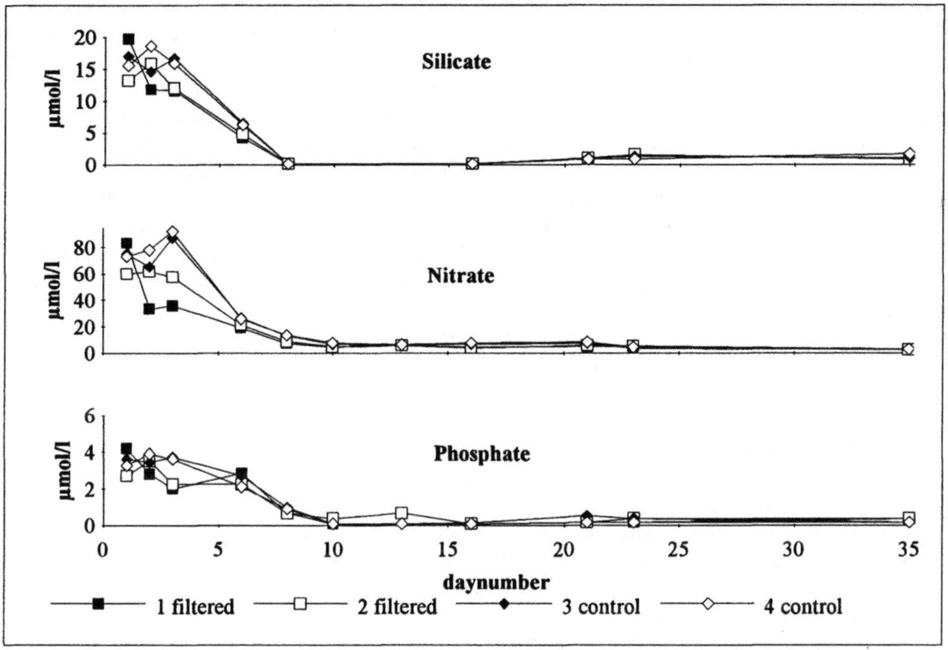

Figure 3. Nutrient concentrations in the mesocosm experiments of March/April 1992

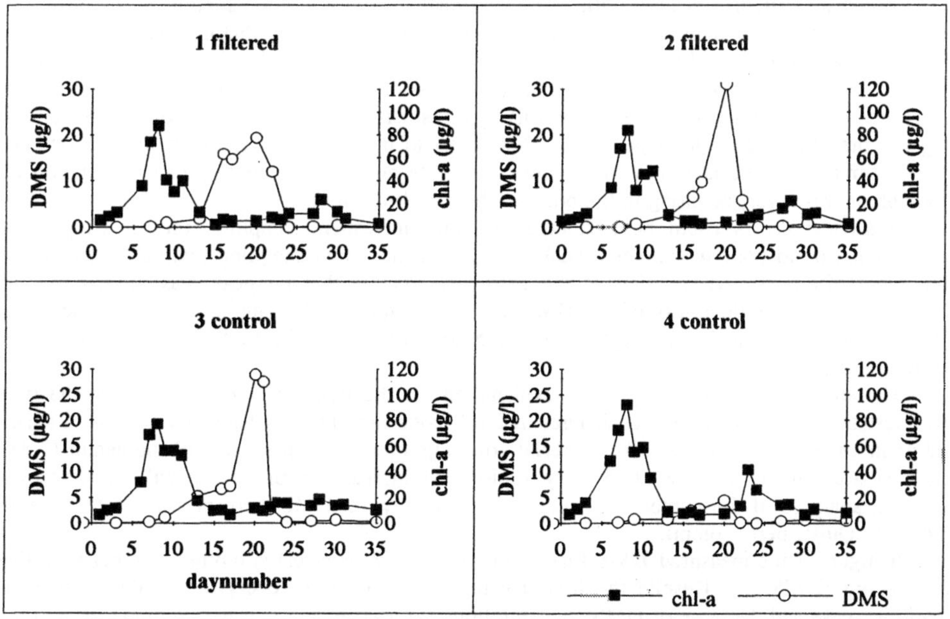

Figure 4. Chlorophyll-a and DMS concentrations in the four mesocosms of March/April 1992

differences in the nutrient behaviour were observed in the four mesocosm systems.

In figure 4, the results of the DMS measurements in the water phase are shown, together with the chlorophyll-a concentrations. The $[DMS]_{water}$ initially increases slowly, but only a few days after the collapse of the diatom peak the $[DMS]_{water}$ increases dramatically in most systems. One observation of separate peaks for chlorophyll and DMS evolutions has been reported for a field situation by Leck et al. (11). This strongly suggests that not algal growth but senescence is a most important mechanism of DMS release in seawater. This agrees well with the findings of Nguyen et al.(14). Also the collapse of the DMS peak is fast. In only one day the concentration decreases from about 30 $\mu g.l^{-1}$ to about 3 $\mu g.l^{-1}$. This suggests that under natural conditions also, the period of maximum emission of DMS will be relatively short, and that changes in DMS concentration in the water column may be quite rapid. This will have implications for surveys of natural waters. During a transect for instance not only different plankton species may be present, but different phases of a bloom development may also be present and that could result in highly variable DMS measurements. Large variations of $[DMS]_{water}$ exist, even between days, and short sampling intervals or synoptic sampling seems a prerequisite before proper isolines can be drawn in natural situations.

Unfortunately there is a difference in DMS behaviour between the duplicate control mesocosms, despite the good agreement in chl-a and nutrient behaviour (figure 2 and 3). There may be several causes for this discrepancy:

* a difference in phytoplankton species distribution despite chl-a similarity (5);
* a difference in zooplankton grazing pressure (11);
* a difference in bacterial consumption of DMS (7);
* sampling and/or analytical errors.

Figure 5 shows the phytoplankton composition during the experiment in the control mesocosms. The diatom species *Skeletonema costatum* and *Thalasiosira nordenskioldii* dominate the first bloom while the second bloom is dominated by flagellates. There is no obvious difference in algal composition between duplicate systems.

In figure 6 the zooplankton development in all four systems is shown. A large difference between the filtered systems (enclosures 1 and 2) and untreated systems (3 and 4) exists, but not between duplicate systems. At least until day 25 no major activity of zooplankton can be observed in the 55 μm filtered systems. In the other non-treated systems, zooplankton consisted of about 70% of the species *Temora longicornis*. Other species were: *Acartia clausii* (11%), *Centropages hamatus* (11%) and *Pseudocalanus elongatus* (2%). From figure 6 it is not expected that the grazing pressure by copepods in the (control) systems has a major influence on DMS release during the first 20 days of the experiment. This seems to be in contrast to other observations (4, 11, 19). Different time scales may affect the rate of the zooplankton development in relation to phytoplankton, however, resulting in the non-matching of life stages.

Analytical errors could be attributed to either infrequent sampling, thus missing peak concentrations, or to poor instrumental analysis. With a maximum sampling interval of two days during the DMS peak and the size of the peak, we assume that no peak was missed in our observations. The reproducibility of the DMS analyses is quite high. The coefficient of variation (CV) for independent analyses ranges between 5 and 17 %. In figure 8 results for replicate analyses are displayed.

The above results do not give an explanation for the large discrepancy between the two control systems in $[DMS]_{water}$ but as changes in DMS concentration may happen very fast, a small timeshift in an essential process, like bacterial consumption, may explain this phenomenon (7,8). If bacteria are able to recycle DMS very rapidly no increase in the water will be detected. Insufficient analyses of the number of bacteria able to grow on DMS were performed to support this idea (P. Quist, pers. comm).

In figure 7 the measured DMS flux from enclosure 3 (control) is plotted together with the corresponding $[DMS]_{water}$. Despite the high frequency of sampling (every 1 - 3 days), both a dramatic increase and decrease of DMS concentrations is observed in the water column, and in the

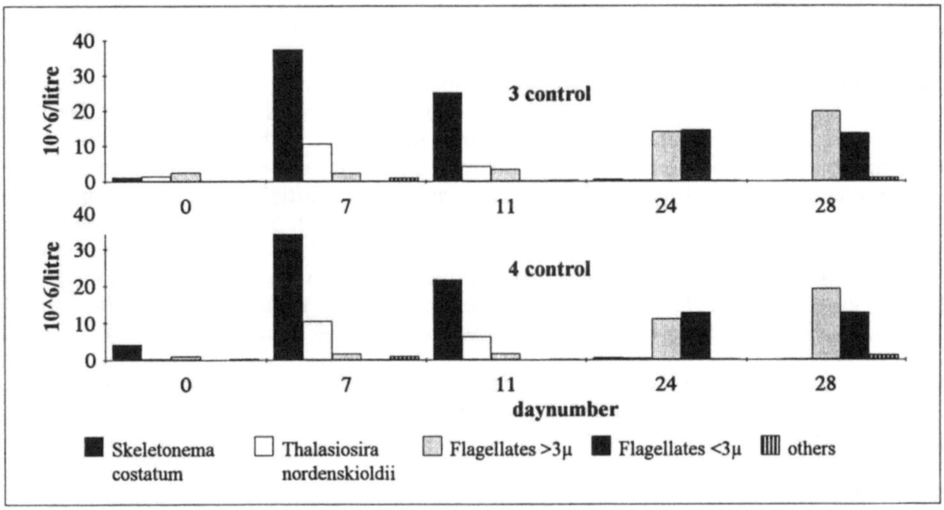

Figure 5. Phytoplankton composition in the control systems during the mesocosm experiment of March/April 1992

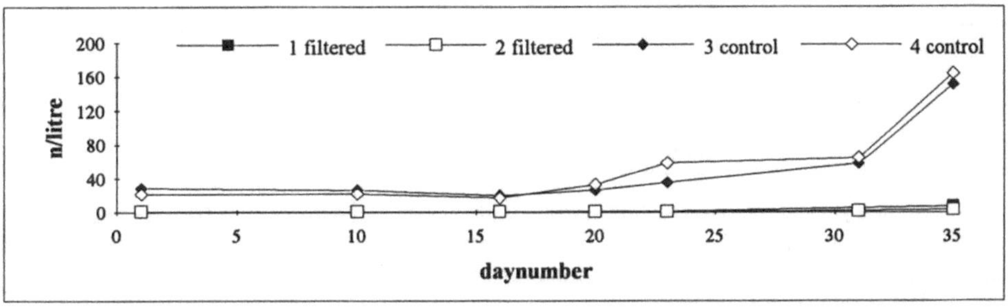

Figure 6. Zooplankton development in the control and the 55 μm filtered mesocosm systems

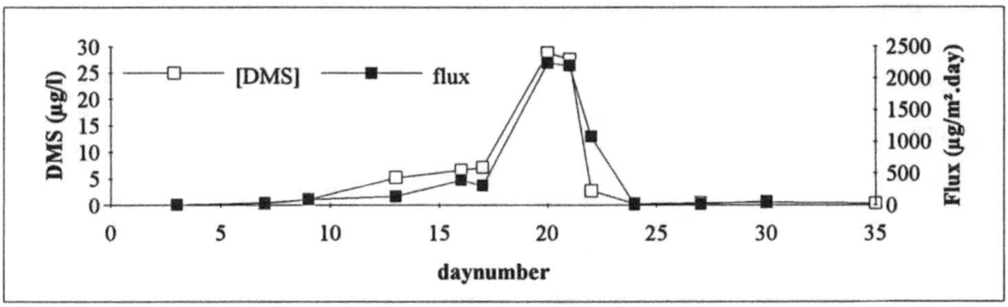

Figure 7. The variation of the DMS concentration in the water and the measured DMS flux to the atmosphere under standardized meteorological conditions in the control system 3

pattern of the measured fluxes. This implies a direct coupling of the fluxes to the gradient in DMS, no delay seems to be present. The measured fluxes are consistent with the 'Liss-Slater' model for the adjusted windspeed of about 1.3 $m.s^{-1}$ (6). When we calculate the total flux per day, assuming similar conditions as during the measurements, and compare this with the decrease of the total amount of DMS in the water column (table 1) it is observed that the amount of DMS lost from the aquatic system can only partly be explained by the flux to the atmosphere. We have to realize, however, that the normalized conditions in the air compartment (about 1.3 $m.s^{-1}$) are at the lower side of the natural wind speed spectrum, and that the airtight closure of the mesocosm only lasted during actual flux measurements (about 1 hour per day). Therefore, actual fluxes may be higer, possibly a factor 10. Furthermore, the kinetics of [DMS] increase and decrease observed are very rapid, and the present calculations are based only on relatively few measurements. Results of experiments on the coupling of the mesocosm studies regarding fluxes under variable experimental (wind) conditions will be reported elsewhere (Baart & Verhagen, in prep.).

Table 1: Calculated total DMS and decrease of DMS in the water column per day compared to the calculated total DMS flux to the atmosphere per day, extrapolated from the measurements and assuming constant conditions

day nr.	ΣDMS_{water} μg	ΔDMS_{water} $\mu g.d^{-1}$	DMS flux $\mu g.d^{-1}$ (windspeed 1.3 ms^{-1})
20	41,438		
21	39,650	-1,788	1,036
22	3,955	-35,695	765
23			
24	311	-1,820	253

Assuming that we underestimate the flux (perhaps by a factor of 5 - 10), we still are not able to explain the rapid disappearance of DMS at maximum rate. This could imply that an additional sink for the DMS in the water column exists. Possible sinks are:

* photochemical conversion;
* diffusion through the mesocosm material;
* bacterial/algal consumption of DMS.

Photochemical conversion seems unimportant as the chemical half-life of DMS in water is in the order of months. Additionally, the water was rich in humic material, limiting penetration of UV light. Diffusion through the plastic enclosure material could have played a role as the

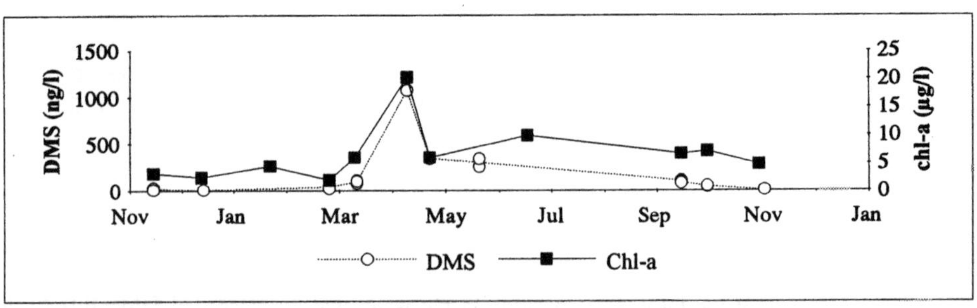

Figure 8. Field observations of a full year cycle in the Marsdiep tidal inlet of the concentration of DMS in the water and chlorophyll-a

concentration outside the bags was measured as being lower. The importance of this factor will have to be estimated in future experiments. Bacterial consumption is likely to have been important in removing DMS from the water (7,8) but as already mentioned we were not able to collect sufficient data to estimate bacterial DMS consumption.

Figure 8 shows the results of the monitoring of the one year cycle at the Marsdiep tidal inlet. The results show a relatively low background level in winter time of about 5 to 50 $ng.l^{-1}$. In spring $[DMS]_{water}$ starts to rise and a maximum of about 1000 $ng.l^{-1}$ was measured on 10 April 1992. At first sight, development of $[DMS]_{water}$ is parallel with the development of chlorophyll-a in the water column, whereas in the enclosure experiments the [DMS] peaked distinctly after the chlorophyll maximum. Because of the different timescale and large sampling intervals in the natural series these results do not necessarily have to be inconsistent with each other. In the natural situation the phyto- and zooplankton blooms develop more slowly than in a mesocosm experiment. Chlorophyll peaks tend to be broader indicating that senescence could take place during the chlorophyll maximum. Due to the relatively large sampling intervals and the rapid kinetics of DMS described above, the actual position of both chlorophyll a and of $[DMS]_{water}$ peaks may have been missed in the time series. In addition, although samples were collected at the same station, sampling was never from the same watermass.

4. CONCLUSIONS.

The results from the mesocosm (enclosure) experiment show that $[DMS]_{water}$ can be highly variable in time and under apparently identical circumstances. Increase and decrease of DMS in the water can be very rapid. In a matter of days $[DMS]_{water}$ can change by one order of magnitude. DMS release in the watercolumn seems to be highly correlated with phytoplankton activity, but the relation is due to senescence rather than to growth.

Mesocosms are very useful in research on the mechanisms controlling the release of DMS into the water. They form an intermediate between laboratory experiments and the field, although care must be taken to extrapolate results directly to the natural situation.

ACKNOWLEDGEMENTS.

The authors are grateful for the practical assistance of Gerrit Hoornsman, Marijke van der Meer, Ad van Mullem, the captains of the research vessel Nauplius II and Peter Quist of the Department of Microbiology of the University of Groningen. Cooperation and stimulating discussions with Hub Diederen is gratefully acknowledged. Without the expertise of Finn Lindqvist in the analysis of DMS and related compounds, and hospitality of the Department of Analytical Chemistry, this project could not have been carried out.

This project was funded by the Dutch Government (Ministry of VROM) under contract number NOLK/026/90.

REFERENCES.

(1) BELVISO, S., KIM, S.K., RASSOULZADEGAN, F., KRAJKA, B., NGUYEN, B.C., MIHALOPOULOS, N., BUAT-MENARD, P. (1990). Production of dimethylsulfonium propionate (DMSP) and dimethylsulfide (DMS) by a microbial food web. Limnol. Oceanogr. 35, 1810-1821

(2) CHARLSON, R.J., LOVELOCK, J.E., ANDREAE, M.O., WARREN, S.G. (1987). Oceanic phytoplankton, atmospheric sulfur, cloud albedo and climate. Nature, 326, 655-661

(3) COOPER, W.J. & MATRAI, P.A. (1989). Distribution of dimethylsulfide in the oceans. In: Biogenic sulfur in the environment. Saltzman, E.S. and Cooper W.J. (eds). ACS

symposium series 393. American Chemical Society, Washington D.C., pp. 140-151
(4) DACEY, J.W.H., KING, G.M. & WAKEHAM, S.G. (1987). Factors controlling emission of dimethylsulphide from salt marshes. Nature 330, 643-645
(5) GIBSON, J.A.E., GARRICK, R.C., BURTON, H.R. & MCTAGGART, A.R. (1990). Dimethylsulfide and the alga *Pheaocystis pouchetii* in Antartic coastal waters. Mar. Biol. 104, 339-346
(6) LISS, P.S. & SLATER, P.G. (1974). Flux of gases across the air-sea interface. Nature 247, 181-184
(7) KIENE, R.P. & BATES, T.S. (1990). Biological removal of dimethyl sulfide from sea water. Nature 345, 702-705
(8) KIENE, R.P. & SERVICE, S.K. (1991). Decomposition of dissolved DMSP and DMS in estuarine waters: dependence on temperature and substrate concentration. Mar. Ecol. Prog. Ser. 76, 1-11
(9) KUIPER, J. (1981). Fate and effects of cadimum in marine plankton communities in experimental enclosures. Mar. Ecol. Prog. Ser. 6, 161-174
(10) KUIPER. J. (1982). The use of enclosed plankton communities in aquatic ecotoxecology. Fate and effects of mercury, cadmium and selected aromatic organics in a marine model ecosystem. PhD. thesis, Univ. Wageningen, pp. 238
(11) LECK, C., LARSSON, U., BAGANDER, L.A., JOHANSSON, S. & HAJDU, S. (1990). Dimethyl sulfide in the Baltic Sea: Annual variability in relation to biological activity. J. Geophys. Res. 95/C3, 3353-3363
(12) LINDQVIST, F. (1989). Sulphur-specific detection in air by photoionization in a multiple detector gas chromatography system. J. High Res. Chrom. 12, 628-631
(13) LINDQVIST, F. (1987). Detectie van vluchtige zwavelhoudende verbindingen in lucht met reactiegaschromatografie/fotoionisatie detector. MT-TNO-report R87/392, Delft, pp. 18 (in Dutch)
(14) NGUYEN, B.C., BELVISO, S., MIHALOPOULOS, N., GOSTAN, J. & NIVAL, P. (1988). Dimethyl sulfide production during natural phytoplanktonic blooms. Mar. Chem. 24, 133-142
(15) PARSONS, R.T., MAITA, Y. & LALLI, C.M. (1984). A manual of chemical and biological methods for seawater analysis. Pergamon press, Oxford, pp. 173
(16) TURNER, S.M., MALIN, G. & LISS, P.S. (1989). Dimethylsulfide and (dimethylsulfonio)propionate in European coastal and shelf waters. In: Biogenic sulfur in the environment. Saltzman, E.S. and Cooper W.J. (eds). ACS symposium series 393. American Chemical Society, Washington D.C., pp. 183-200
(17) TURNER, S.M., MALIN, G., LISS, P.S., HARBOUR, D.S. & HOLLIGAN, P.M. (1988). The seasonal variation of dimethyl sulfide and dimethyl sulfonio-propionate concentrations in nearshore waters. Limnol. Oceanogr. 33, 364-75
(18) VAIRAVAMURTHY, A., ANDREAE, M.O. & IVERSON, R.L. (1985). Biosynthesis of dimethylsulfide and dimethylpropiothetin by *Hymenomonas cartarae* in relation to sulfur source and salinity variations. Limnol. Oceanogr. 30, 59-70
(19) WAKEHAM, S.G. & DACEY, J.W.H. (1989). Biogeochemical cycling of dimethyl sulfide in marine environments. In: Biogenic sulfur in the environment. Saltzman, E.S. and Cooper W.J. (eds). ACS symposium series 393. American Chemical Society, Washington D.C., pp. 152-166

A MODEL OF DIMETHYLSULPHIDE PRODUCTION DURING A PHYTOPLANKTON BLOOM

A.GABRIC,[1] C. N. MURRAY [2], L. STONE [3] and M. KOHL[4]

1. Faculty of Environmental Sciences, Griffith University, Nathan, Queensland, Australia, 4111.
2. Institute for Remote Sensing Applications, CEC Joint Research Centre, Ispra Site, I-21020 Ispra (Va) Italy
3. Department of Environmental Sciences, Weizmann Institute, Rehovot 76100, Israel.
4. Environment Institute, CEC Joint Research Centre, Ispra Site, I-21020 Ispra (Va) Italy

SUMMARY

The production of the dimethylsulphide (DMS) during a phytoplankton bloom episode has been simulated using an eight compartment network flow model. The model ecosystem includes phytoplankton, bacteria, zooflagellates, protozoa and zooplankton with both nitrogen and sulphur flows being considered. The microbial consumption of DMS has been included. Model predictions compare favourably with phytoplankton bloom experiments done in laboratory tanks and sensitivity analyses confirm the important role of the microbial loop in DMS cycling.

1. INTRODUCTION

Dimethylsulphide (DMS; CH_3SCH_3) is an important sulphur-containing trace gas produced by some classes of marine phytoplankton (8, 16,17, 35). It is present in oceanic surface waters at concentrations sufficient to sustain a considerable net flux to the atmosphere and is estimated to constitute 90% of the marine sulphur flux, accounting for nearly half of the global biogenic input of sulphur to the atmosphere (2).

Charlson *et al* . (10) have suggested that a major source of cloud-condensation nuclei (CCN) over the oceans is due to DMS which is produced by planktonic algae in seawater. DMS is oxidised in the atmosphere to form non-sea-salt sulphate (nss - SO_4^{2-}) and methane sulphonate (MSA) aerosols. Because the formation of clouds is sensitive to CCN density it has been postulated that biological regulation of the climate is possible by affecting cloud albedo and thus the overall radiation budget. This would be due to the effect of temperature and solar radiation on phytoplankton growth, however it is still unclear whether the change in albedo will cause a positive or negative feedback on climate.

G. Restelli and G. Angeletti (eds.), Dimethylsulphide: Oceans, Atmosphere, and Climate, 63–81.

Support for the hypothesis that a change in DMS flux can affect global temperature has come from analysis of Vostock ice-core sulphate data (20). The data suggest that significant correlation between CO_2 and nss - SO_4^{2-} also exists, implying a global significance of these changes and a possible link between marine productivity, CO_2 and DMS emissions. Further evidence of the connection between atmospheric DMS and aerosol sulphur species has been presented by Ayers *et al.* (4) who demonstrated a strong coherence between sea-level DMS concentration, MSA and nss -SO_4^{2-} at Cape Grim, Tasmania.

Regional studies in the North Sea (35) and the east coast of the USA (16) have been undertaken in order to estimate regional DMS fluxes and to attempt to correlate these with biological parameters such as phytoplankton biomass (chl *a*), however good correlations were found only for coccolithophores and dinoflagellates. Leck *et al.* (19), in a study of the Baltic Sea, found that there was a significant correlation of DMS concentration with zooplankton biomass, confirming the observation of Dacey and Wakeham (12) that grazing is a major factor in the liberation of DMS and its biogenic precursor DMSP (dimethylsulphoniopropionate or dimethylpropiothetin, DMPT) from phytoplankton cells.

The lack of a strong correlation between DMS atmospheric flux and gross biological parameters such as chlorophyll *a* limits the use of current ocean colour remote sensing algorithms (23) to specific sites and time periods. Recently Thompson *et al.* (34) have explored the feasibility of using remote sensed ocean colour data to map sources of DMS and note that the method is limited to regions in which phytoplankton pigment is well correlated with DMS production. These authors conclude that the increased spectral information provided by future ocean colour sensors such as the Sea-viewing Wide Field-of-View Sensor (Sea-WiFS) scheduled for launch in August 1993, may enable species-specific algorithms to be developed so that classes of DMS-producing phytoplankton can be mapped individually - some progress along these lines has already been reported (15, 1) for the coccolithophore *Emiliana huxleyi* . It has also been noted (34) that since the spatio-temporal distribution of DMS is not simply related to instantaneous measure of phytoplankton concentration or productivity, a time-dependent model of DMS formation in surface waters would be a necessary complement to future improved ocean colour algorithms.

Our model is designed to demonstrate the interdependence of the DMS cycle and the major ecological processes that occur over a range of scales in time and space. Using a modelling approach it becomes possible to examine DMS cycling within the framework of the recent conceptual picture of marine food webs (13, 5, 11), where micro organisms are recognised as significant components (for a general review see 14). It is only recently that bacteria and protozoa have been considered in models of aquatic ecosystems. Aquatic ecosystems were thought of as a linear trophic food chain in which major nutrients move vertically from primary producers to zooplankton to fish. However, over the last decade there has been something of a paradigm shift and it is now realised that bacteria and protozoa can control large nutrient fluxes through a myriad of pathways. The microbial loop has the potential to divert a large proportion (over 50%) of primary production and nutrients out of the classical food-chain. Further, because microorganisms have very high rates of turnover (~1/d) they have the ability to rapidly recycle and remineralise nutrients at the base of the food web.

The role of bacteria in the anaerobic removal of DMS has been known for some time (38, 39), and its aerobic catabolism has been reviewed in (33). Recently a number of studies have reported that there is a definite link between the microbial food web and the oceanic cycling of DMS. It has been shown (18) that DMS turnover in sea water is of the same order as microbial activity, ie 3-340 times faster than turnover by atmospheric ventilation. In a similar vein, the results of Belviso *et al.* (8) suggest that ciliates are largely responsible for DMS production. The production of DMS is apparently closely linked to the dynamics of the marine pelagic food web and it is clear that the microbial loop should be included when modelling the DMS cycle.

At the present time the complex mechanisms responsible for DMS cycling are not well understood, and the ecological ramifications even less so. In the approach presented here, we attempt to synthesise some of the diverse information reported in the literature and place it within the broader context of a (model) aquatic ecosystem. The aims of the present study are several :

- to understand the influence of the pelagic food web on DMS cycling
- to investigate the relative importance of the various physical, chemical and ecological processes involved in dissolved DMS production
- to provide a model framework that can be incorporated in future remote sensing ocean colour algorithms

2. MODEL DESCRIPTION

In order to analyse the dynamics of DMS production in the upper ocean we have chosen a flow network approach and adapted the ecological structure of the plankton community model of Moloney *et al* . (22) to include DMS production, its bacterial consumption, chemical transformation and venting to the atmosphere (see Fig. 1) . The microbial loop plays a crucial role in the model providing the link between the N and S cycles since we assume bacteria can consume (and grow on) dissolved DMS as well as nitrogen. Our model is time-dependent and applies for non-equilibrium conditions, constituting an initial attempt at quantifying the oceanic components of the overall DMS-climate hypothesis.

The model network consists of eight compartments, of which three are abiotic : dissolved N, DMS and DMSP. The biotic compartments comprise the primary producers (generic phytoplankton) and four consumer groups (bacteria, zooflagellates, non-photosynthetic protozoa and micro-meso zooplankton). The various fluxes between compartments are described in Table 1. The cycling of nitrogen and sulphur (as DMSP or DMS) occurs in a mixed layer whose bottom boundary coincides with the thermocline. Nitrogen (as nitrate) is injected into the euphotic zone at the start of the simulation mimicking the rapid input of new nutrients that occurs during wind mixing of intermediate waters or coastal upwelling. Phytoplankton growth is assumed not to be limited by light in the mixed layer although it is clear that vertical light attenuation over deep mixed layers or during very intense blooms may invalidate this assumption. Loss of particulate organic matter (phytodetritus) occurs by sedimentation below the thermocline. Nitrogen (as ammonium) is regenerated via the microbial loop however we have not distinguished between nitrate and ammonium in the nutrient uptake parameterisation. For simplicity the present model assumes spatial homogeneity although the inclusion of vertical variability is feasible.

Dissolved DMS and its biogenic precursor DMSP are both phytoplankton metabolites although the actual mechanism of release is unclear. DMSP is transformed to DMS by enzymatic cleavage both within the algal cell (at a slow rate) and in the water column. Belviso *et al.* (8) have suggested that a second pathway exists for DMSP decomposition to DMS, namely demethylation during anaerobic digestion by microzooplankton. We assume DMSP to be present in the excreta of zooplankton and large protozoa both of which consume DMSP-containing phytoplankton, while both DMSP and DMS are assumed to be released by phytoplankton. Nguyen *et al.* (25) found that the ratio of DMS production during algal cell lysis to that during the growth phase was 7 for a diatom species and 26 for a dinoflagellate species. A similar finding was reported by Dacey and Wakeham (12) who demonstrated that zooplankton grazing significantly increased the flux of DMSP from phytoplankton to the water column.

Loss mechanisms for dissolved DMS include photochemical oxidation in the euphotic zone (9) and ventilation to the atmosphere. DMS flux to the atmosphere is a function of the concentration gradient across the air-sea interface and the wind speed at the water surface. Given the large air-sea gradient in DMS concentration, the atmospheric concentration is usually neglected in determining the flux (7). Particulate DMS/DMSP in the phytoplankton and zooplankton compartments can also be lost from the system by sinking of fecal material or dead cells below the thermocline. In order to incorporate the cycling of DMS via the microbial loop, the model uses dual elemental currencies of nitrogen and sulphur. To estimate the growth of bacteria due to DMS uptake, we first calculate the proportion of carbon in DMS and then assume the ratio of carbon to nitrogen uptake is the same as the elemental composition of the biotic compartment ie. C:N equals 5.0 for bacteria and 6.0 for phytoplankton (32, 24).

The flow of N or S between the various compartments is described by a system of coupled ordinary differential equations (1-8) given in Table 2. Where both N and S fluxes occur in the same equation a conversion factor has been applied for dimensional consistency, for example in equations (2) and (8). Since the magnitude of nitrogen fluxes associated with the sulphur cycle is relatively small we neglect the fluxes of DMSP and DMS (viz. F17, F18, F47, F57) in equations (1),(4) and (5). The sole coupling between the N and S cycles is via bacterial uptake of S in (2). Equations (7) and (8) are in S(DMSP) and S(DMS) currencies respectively - and the flux conversion factor in (8) is given by $\beta = 2.16$ mg S(DMS)/mg S(DMSP).

The functional form of fluxes follow the prescription given in Moloney *et al*. (22). Michaelis-Menten kinetics are assumed for N uptake by phytoplankton and heterotrophic bacteria, thus the rate of nutrient uptake ν is given by,

$$\nu = V_{max} N/(K_S + N)$$

where V_{max} is maximum uptake of nutrient N, and K_S is the half-saturation concentration. Following the suggestion of Moloney *et al.* (22) a two-parameter asymptotic formulation,

$$Fmn = k_i \; Xn \exp[1 - \exp(k_j \, Xm)]$$

has been used to avoid computational difficulties in the integration. The correspondence between the Michaelis-Menten parameters, V_{max} and K_S and the parameters of the asymptotic formulation is, $V_{max} = k_i$ and $K_S = \ln 2 / k_j$.

Simpler single parameter expressions have also been used where possible. A Lotka-Volterra expression is employed to model consumption of prey compartment n by predator compartment m, namely, $Fmn = k_i \, Xn \, Xm$. The release of N or S from compartment n to compartment m is modelled by the expression $Fnm = k_i \, Xn$, that is, by assuming first-order kinetics. We have used the same S:N ratio γ for the large protozoa (F47) and zooplankton (F57) fluxes since it appears that DMSP does not accumulate in zooplankton body tissue (12).

The parameter values assumed for the mixed layer model are given in Table 3 and a range has been shown where data is available. Uncertainty exists in the biological parameters associated with phytoplankton and bacteria nutrient uptake (k23, k24, k25, k26) which vary according to cell size and physiological status as well as ambient conditions such as temperature and nutrient concentration (21, 27). The values used here are based on those chosen by Moloney *et al.* (22) and Moloney and Field (21) allowing where appropriate for the different mixed layer depth used by these authors, and may be considered relevant to coastal phytoplankton in temperate waters.

The DMS photo-oxidation rate constant (2.07 d^{-1}) quoted by Brimblecombe and Shooter (9) is appropriate only for the first metre or so of the water column where light intensity is maximal. For use in the model a depth-averaged estimate of the photo-oxidation rate may be derived: for example, the average light intensity over a 20m mixed layer, assuming a vertical extinction coefficient $k_e = 0.042\ m^{-1}$ appropriate for clear oceanic waters (27), is 0.67 times the surface value.

Keller *et al.* (17) report cellular concentration of DMSP for twelve classes of phytoplankton and note that there is a strong correlation between DMS production and taxonomic position with the most important producers being Dinophyceae and Prymnesiophyceae. The DMSP content per cell is shown for a selection of species in Table 4. The measured cell volumes are approximate because osmotic volume was not estimated by Keller *et al.* (17). The carbon content per cell volume has been calculated from regression equations for phytoplankton (31), given by,

$\log C = 0.758 \log V - 0.422$ (diatoms)
$\log C = 0.866 \log V - 0.460$ (other species)

where C is given in pg C/cell and V is the cell volume in cubic microns.
The C:N ratio for phytoplankton is used to derive nitrogen content and hence the S(DMSP):N ratio γ for each species.

3. RESULTS AND DISCUSSION

The behaviour of multi-parameter ecological models such as the system presented here can only be properly examined given a knowledge of the sensitivity of the results to changes in the rate parameters, some of which are only estimates because experimental data were not always available (22, 26). Sensitivity analysis can also yield considerable insight into the dynamics of the system by indicating which of the various processes most affect the DMS production. We have systematically conducted a empirical sensitivity analysis (29) varying each of the model parameters by plus or minus fifty percent from the 'reference' value quoted in Table 3. The metric we have used to gauge sensitivity is the absolute difference in the time-integrated DMS concentration, $X8_{av}$ calculated with upper and lower estimates of each parameter.

Figure 2 shows the values of the sensitivity metric plotted in ascending order, thus the parameters appearing at the right of the abscissa are the ones which are most determinant of time integrated DMS concentration. The time-integrated DMS prediction is most sensitive to the parameters k3, γ, K17, K11, K14, K8, K23, K9, K29 - all of which have a metric value > 8. As would be expected the algal cell DMSP concentration, reflected by γ is an important parameter. In the context of nitrogen cycling, Moloney *et al.* (22) noted that the protozoa play a pivotal role in linking the microbial loop and herbivore food chain. The most important parameters in this context are K3 and k17 which are involved in the link between the phytoplankton-protozoa and zooflagellate-protozoa fluxes. K11, K14, K8 and K9 are all associated directly with the bacterial compartment or regeneration of dissolved nitrogen which is a microbially mediated process. Clearly the bacterial compartment is central to DMS production. The V_{max} for nitrogen uptake by phytoplankton K23 is also an important parameter.

Of the abiotic processes, K29 the DMS photochemical oxidation rate is the most important parameter (metric $\cong$ 8). DMS is also clearly sensitive to k28 (metric $\cong$ 5) which describes the rate of DMS consumption by bacteria. It is worthwhile noting that DMS is relatively insensitive (metric<1) to changes in k30 the rate of ventilation to the atmosphere, which confirms the hypothesis of Kiene and Bates (18) regarding the relative importance of microbial turnover versus flux to the atmosphere.

The system of differential equations Table 2 (1 - 8) was integrated using a fourth-order Runge-Kutta scheme with time step equal to 0.01d. One of the few experimental determinations of DMS production during a monitored bloom was carried out by Nguyen *et al.* (25) in $2m^3$ tanks covered by a neutral fiberglass plate. The tanks were filled with seawater from Villefranche Bay which had been filtered through a 150 micron screen. The filter prevented mature zooplankton (copepods) from entering the tanks although juveniles could have been present. However grazing is likely to have been very much reduced compared to the natural ocean environment. The filter also prevented larger diatoms from entering the tank, thus biassing the phytoplankton population to smaller celled species. Three experiments were reported, two consisting of diatom blooms and the other an unidentified flagellate species. In the experiments the tank seawater was enriched with nitrate, phosphate and silicate and from the results presented it was evident that the phytoplankton phosphate uptake rate was slightly higher than that for nitrate or silicate. Clearly our model does not purport to reproduce the complex nutrient dynamics in these tank experiments as we have assumed nitrate to be the only limiting nutrient.

In the first diatom experiment the phytoplankton biomass reached a maximum after 83 hours and DMS peaked about fourteen hours later. Measurements were stopped when algal biomass started to decline at 97 hours, although the dissolved DMS concentration was still significant at this stage. The time scale for the experiment was thus approximately 4 days. In order to apply the model to this experiment the depth dependent parameters were re-scaled to the tank depth of 1.2 metres. There was no loss of matter from the tanks so the sedimentation parameters k7 and k22 were both set to zero. The light field will not be much attenuated over 1.2m so that the photo-oxidation rate k29 has been set to the surface value given by Brimblecombe and Shooter (9) of 2.07 d^{-1}. Ventilation of DMS depends on wind speed which will be negligible in the covered tanks so K30 has been set to zero. Initial values for the phytoplankton and DMS compartments were measured in the experiment and the other variables have been set to arbitrarily small values.

Nguyen *et al.* (25) measured total chlorophyll *a* as an indicator of phytoplankton biomass. Since there is some uncertainty in C:Chl a and N:Chl a ratios for phytoplankton it is not possible to accurately convert from nitrogen to chlorophyll *a* . Thus the normalised biomass X1(t)/X1(0) has been used for comparison purposes. Model predictions for DMS and DMSP production are shown in Figures 3a and 3b with experimental results in Figure 3c. Bearing in mind the limitations of assuming a single limiting nutrient, the model has reproduced the peak phytoplankton and DMS ratios reasonably well and the time lag between chlorophyll and DMS peaks is also clear. The time scale of the theoretical predictions is slightly compressed compared to the experiment, probably due to a longer phytoplankton growth cycle in the tanks as a result of multiple nutrient (N, P, Si) availability.

4. CONCLUSIONS

The model presented here attempts to place oceanic DMS production and its flux to the atmosphere into a proper ecological context. Ecological models can be used to gain insight into the various mechanisms that might determine DMS production in the ocean as well as predict the absolute DMS concentration for a particular site. The model presented here is meant as a heuristic aid rather than a strict predictive tool. The many processes affecting DMSP/DMS production and removal can be seen in better perspective by the model's ability to examine the totality of the system.

The observation by Kiene and Bates (18) of the central role of the microbial loop in controlling dissolved DMS concentration has been reaffirmed by our model simulations. Ventilation to the atmosphere seems to be a minor influence on the dissolved DMS concentration. This implies that the DMS flux may be significantly altered by any changes that might occur to the dynamics of the marine food web. Thus ambient factors effecting bacterial growth and biomass, such as water temperature and solar radiation, could be indirect determinants of DMS level in the ocean. It is known that bacterial metabolism is greatly suppressed at low temperatures (28), so one might expect lower bacterial consumption of DMS at high latitudes. In fact the field data records mean DMS concentrations in the Bering Sea and Gulf of Alaska of 4.7 and 5.0 nmol/l, respectively, compared with 3.1 nmol/l in tropical Pacific waters (7). Sieracki and Sieburth (30) have found that sunlight can reduce the growth rate of natural, marine bacteria with photoinibition operating even at 16% of surface light intensity - effectively over the whole of the euphotic zone. Photochemical removal of DMS will also increase under high light intensities but this process is confined to the upper metre or so of the water column (9) .Thus net DMS levels may be increased in strong light conditions. This finding may offer a plausible explanation for the apparent paradox that a linear relationship exists between solar radiation and DMS flux (6), yet correlation of DMS with phytoplankton biomass or productivity has generally been poor (3).

Ecological succession in the pelagic food web clearly influence the temporal distribution of DMS which can remain at significant levels for some days after phytoplankton biomass has declined to zero. This is because higher trophic levels such as protozoa and zooplankton will continue to release DMSP in their excreta. This has clear significance for the interpretation of field data and suggests that attempts at simple correlations of DMS concentration/flux with phytoplankton biomass or primary productivity will not be successful unless the temporal lag is included.

Table 1 . Definition of model compartments and fluxes

X1 :	Phytoplankton (mg N/m^2)
X2 :	Bacteria (mg N/m^2)
X3 :	Zooflagellate (mg N/m^2)
X4 :	Large Protozoa (mg N/m^2)
X5 :	Micro/meso Zooplankton (mg N/m^2)
X6 :	Dissolved inorganic nitrogen (mg N/m^2)
X7 :	Dissolved DMSP (mg S $(DMSP)/m^2$)
X8 :	Dissolved DMS (mg S $(DMS)/m^2$)
Fij is the flow between compartments Xi and X j .	
F12 :	bacterial decomposition of phytoplankton (mg $N/m^2/d$)
F14 :	ingestion of phytoplankton by large protozoa (mg $N/m^2/d$)
F15 :	ingestion of phytoplankton by micro/meso zooplankton (mg $N/m^2/d$)
F17 :	excretion of DMSP by phytoplankton (mg $S(DMSP)/m^2/d$)
F18 :	excretion/metabolism of DMS by phytoplankton (mg $S(DMS)/m^2/d$)
F1W :	sedimentation of phytoplankton below euphotic zone (mg $N/m^2/d$)
F23 :	ingestion of bacteria by zooflagellates (mg $N/m^2/d$)
F26 :	excretion of DIN by bacteria (mg $N/m^2/d$)
F34 :	ingestion of zooflagellates by large protozoa (mg $N/m^2/d$)
F36 :	excretion of DIN by zooflagellates (mg $N/m^2/d$)
F45 :	ingestion of protozoa by zooplankton (mg $N/m^2/d$)
F46 :	excretion of DIN by large protozoa (mg $N/m^2/d$)
F47 :	excretion of DMSP by large protozoa (mg $S(DMSP)/m^2/d$)
F56 :	excretion of DIN by m/m zooplankton (mg $N/m^2/d$)
F57 :	excretion of DMSP by zooplankton (mg $S(DMSP)/m^2/d$)
F5W :	sedimentation of zooplankton excreta (mg $N/m^2/d$)
F61 :	DIN uptake by phytoplankton (mg $N/m^2/d$)
F62 :	DIN uptake by bacteria (mg $N/m^2/d$)
F78 :	Conversion of DMSP to DMS in water column (mg S $(DMSP)/m^2/d$)
F82 :	DMS uptake/consumption by bacteria (mg S $(DMS)/m^2/d$)
F8W :	DMS photochemical oxidation, sedimentation (mg$S(DMS)/m^2/d$)
F8A :	DMS ventilation to atmosphere (mg $S(DMS)/m^2/d$)

Table 2. Model equations and fluxes

$\frac{dX1}{dt}$	=	F61 - F12 - F14 - F15 - F1W	(1)
$\frac{dX2}{dt}$	=	F12 + F62 - F23 - F26 + α F82	(2)
where α is a currency conversion factor given by α = 0.15 mg N/mg S(DMS)			
$\frac{dX3}{dt}$	=	F23 - F36 - F34	(3)
$\frac{dX4}{dt}$	=	F34 + F14 - F45 - F46	(4)
$\frac{dX5}{dt}$	=	F45 + F15 - F56 - F5W	(5)
$\frac{dX6}{dt}$	=	F26 + F36 + F46 + F56 - F61 - F62	(6)
$\frac{dX7}{dt}$	=	F17 + F47 + F57 - F78	(7)
$\frac{dX8}{dt}$	=	F18 + β F78 - F82 - F8W - F8A	(8)
with,			
F12	=	k1 * X2 [1 - exp(-k2 * X1)]	(9)
F14	=	k3 * X1 * X4	(10)
F15	=	k4 * X1 * X5	(11)
F17	=	k5 * X1* γ	(12)
where γ is the S(DMSP):N ratio for phytoplankton (species dependent) which is given in Table 4.			
F18	=	β*γ*k6 *X1	(13)
F1W	=	k7 * X1	(14)
F23	=	k8 * X3 [1 - exp(-k9 * X2)]	(15)
F26	=	k10 * X2 + k11 * [F62 + F12]	(16)
F34	=	k12 * X3 * X4	(17)
F36	=	k13 * X3 + k14 * F23	(18)
F45	=	k15 * X4 * X5	(19)
F46	=	k16 * X4 + k17*[F34 + F14]	(20)
F47	=	k18 * X4 * γ	(21)
F56	=	k19 * X5 + k20 *[F15 + F45]	(22)
F57	=	k21 * X5 * γ	(23)
F5W	=	k22 * F56	(24)
F61	=	k23 * X1 [1 - exp(-k24 * X6)]	(25)
F62	=	k25 * X2 [1 - exp(-k26 * X6)]	(26)
F78	=	k27 * X7	(27)
F82	=	k28 * X8	(28)
F8W	=	k29 * X8	(29)
F8A	=	k30 * X8	(30)

Table 3. Model parameters for a 20m ocean mixed layer

	Value	Pathway	Units	Reference
k1	4.5	P-B	d^{-1}	(22)
k2	0.0023	P-B	m^2mgN^{-1}	"
k3	0.013	P-LP	$m^2mgN^{-1}d^{-1}$	"
k4	0.006	P-Z	$m^2mgN^{-1}d^{-1}$	"
k5	0.01	P-DMSP	d^{-1}	(12)
k6	0.0085 (0.003-0.014)	P-DMS	d^{-1}	(36)
k7	0.15 (0-0.3)	P Sinking	d^{-1}	(22)
k8	17	B-F	d^{-1}	"
k9	0.0069	B-F	m^2mgN^{-1}	"
k10	0.07	B-N	d^{-1}	"
k11	0.63	B-N	-	"
k12	0.078	F-LP	$m^2mgN^{-1}d^{-1}$	"
k13	0.05	F-N	d^{-1}	"
k14	0.65	F-N	-	"
k15	0.006	LP-Z	$m^2mgN^{-1}d^{-1}$	"
k16	0.05	LP-N	d^{-1}	"
k17	0.65	LP-N	-	"
k18	0.01	LP-DMSP	d^{-1}	by comparison with k5
k19	0.05	Z-N	d^{-1}	(22)
k20	0.40	Z-N	-	"
k21	0.01	Z-DMSP	d^{-1}	by comparison with k5
k22	0.15 (0-0.3)	Z sinking	-	(22)
k23	0.9 (0.5-4.5)	N-P	d^{-1}	(21)
k24	0.0025	N-P	m^2mgN^{-1}	"
k25	0.9 (0.9-6.6)	N-B	d^{-1}	"
k26	0.0462	N-B	m^2mgN^{-1}	"
k27	0.5 (0.08-1.0)	DMSP-DMS	d^{-1}	Wakeham et al. (1987)
k28	0.95 (0.2-1.7)	DMS microbial consumption	d^{-1}	Kiene and Bates (1990)
k29	1.4	DMS photo-ox	d^{-1}	Brimblecombe and Shooter (1986)
k30	0.13	DMS - Atmos	d^{-1}	Bates et al. (1987b)

Note : k3 and k12 were incorrectly given in Moloney et al. (1986) ; k29 is scaled for the average light intensity over a 20m mixed layer

Table 4. DMSP cellular content and S:N ratio for various species from ref (17)

Class	Species	DMSP (pg/cell)	S (DMSP) (pg/cell)	Approx Volume (μ^3)	Carbon (pg/cell)	γ pgS /pgN
Dinophyceae						
	Gymnodinium sp.	24.0	5.73	1435	188	0.18
	Heterocapsa sp.	78.1	18.7	3058	362	0.31
	Scrippsiella trochidea	384	91.7	8175	847	0.64
Prymnesio-phyceae	*Chrysochromulina herdlansis*	3.62	0.86	65.3	12.9	0.40
	Emiliania huxleyi	0.75	0.18	33.6	7.27	0.14
	Phaeocystis sp.	1.0	0.24	65.7	13.0	0.11
Bacillario-phyceae	*Melosira nummuloides*	34.5	8.2	973	69.7	0.71
	Skeletonema costatum	0.50	0.12	170	18.6	0.04
	Thalassiosira sp.	5.48	1.31	22436	752	0.01

REFERENCES

1. ACKLESON, S., W.M.BALCH and P.M.HOLLIGAN (1988) White waters of the Gulf of Maine. Oceanography, November: 18-22.

2. ANDREAE, M.O. (1986) The ocean as a source of atmospheric sulfur compounds. p331-362 In, P.Buat-Menard [ed.], The role of air-sea exchange in geochemical cycling, Reidel Publishing Co.

3. ANDREAE, M.O. and W.R. BARNARD (1984) The marine chemistry of dimethylsulfide. Mar.Chem.14: 267-279.

4. AYERS, GP, J.P. IVEY and R.W. GILLETT (1991) Coherence between seasonal cycles of dimethylsulphide, methanesulphonate and sulphate in marine air. Nature, 349, 404-406.

5. AZAM, F, T FENCHEL, J.S. GRAY, L.A. MEYER-REIL and F. THINGSTAD (1983) The ecological role of water-column microbes in the sea. Mar. Ecol. Prog. Ser. 10, 257-263.

6. BATES, T.S., R.J. CHARLSON and R.H. GAMMON (1987a) Evidence for the climatic role of marine biogenic sulphur. Nature 336, 441-445.

7. BATES , T.S., J.D. CLINE , R.H. GAMMON and S.R. KELLY-HANSEN (1987b) Regional and seasonal variations in the flux of oceanic dimethylsulfide to the atmosphere. J Geophys. Res. 92, C3, 2930-2938.

8. BELVISO, S., S.-K. KIM, F.RASSOULZADEGAN, B.KRJKA, B.C.NGUYEN, N.MIHALOPOULOS, and P.BUAT-MENARD (1990) Production of dimethylsulfonium propionate (DMSP) and dimethylsulfide (DMS) by a microbial food web. Limnol. Oceanogr. 35(8): 1810-1821.

9. BRIMBLECOMBE, P. and D.SHOOTER, (1986) Photo-oxidation of dimethylsulfide in aqueous solution. Mar. Chem. 19, 343-353.

10. CHARLSON, R.J., J.E. LOVELOCK , M.O. ANDREAE and S.G. WARREN (1987) Oceanic phytoplankton, atmospheric sulphur, cloud albedo and climate. Nature, 326, 655-661.

11. COLE, J., S. FINDLAY and M.PACE (1988) Bacterial production in fresh and salt-water ecosystems: a cross-system overview. Mar.Ecol. Prog. Ser. 43: 1-10.

12. DACEY, W.H. and S.G. WAKEHAM (1986) Oceanic dimethylsulfide: Production during zooplankton grazing on phytoplankton. Science, 233, 1314-1315.

13. DUCKLOW, H. (1983) The production and fate of bacteria in the oceans. BioScience 33: 494-501.

14. FENCHEL, T. (1988) Marine plankton food chains. Ann.Rev.Ecol.Syst. 19: 19-38

15. HOLLIGAN, P.M., M. VIOLLIER, D.S. HARBOUR, P.CAMUS and M. CHAMPAGNE-PHILIPPE (1983) Satellite and ship studies of coccolithophore production along a continental shelf edge. Nature, 304: 339-342.

16. IVERSON, R., F.L. NEARHOOF and M.O. ANDREAE (1989) Production of dimethylsulfonium and dimethylsulfide by phytoplankton in estuarine and coastal waters. Limnol. Oceanogr. 34(1), 53-67.

17. KELLER, M.D., W.K. BELLOWS and R.L. GUILLARD (1989) Dimethyl Sulfide production in marine phytoplankton. In, *Biogenic sulfur in the environment,* Saltzman ES and Cooper WJ (eds), American Chemical Society, Washington, DC .

18. KIENE, R..P and T.S. BATES (1990) Biological removal of dimethylsulfide from seawater. Nature, 345, 702-705.

19. LECK, C., U. LARSSON , L.E. BAGANDER , S. JOHANSSON and S. HAJDU (1990) DMS in the Baltic Sea - Annual variability in relation to biological activity. J Geophys. Res. 95 (C3), 3353-3363.

20. LEGRAND, M.R., R.J. DELMAS and R.J. CHARLSON (1988) Climate forcing implications from Vostok ice-core sulphate data. Nature 334, 418-420.

21. MOLONEY, C.L. and J.G.FIELD (1989) General allometric equtions for rates of nutrient uptake, ingestion, and respiration in plankton organisms. Limmol. Oceanogr. 34(7): 1290-1299.

22. MOLONEY, C.L., M.O. BERGH , J.G. FIELD and R.C. NEWELL (1986) The effect of sedimentation and microbial nitrogen regeneration in a plankton community: a simulation investigation. J. Plank. Res. 8(3), 426-445.

23. MOREL, A. (1988) Optical modelling of the upper ocean in relation to its biogenous matter content (Case I waters). J. Geophys. Res. 93, 10749-10768.

24. NAGATA, T. (1986) Carbon and nitrogen content of natural planktonic bacteria. Appl. Environ. Microbiol. 52, 28-32.

25. NGUYEN, B. C., S .BELVISO, N. MIHALOPOULOS, J. GOSTAN and P. NIVAL (1988) Dimethylsulfide production during natural phytoplanktonic blooms. Mar. Chem. 24, 133-141.

26. PACE, M.L., J.E. GLASSER and L.R. POMEROY (1984) A simulation analysis of continental shelf food webs. Mar. Biol. 82, 47-63.

27. PARSONS, T.R., M. TAKAHASHI and B. HARGRAVE (1984) Biological oceanographic processes. 3rd ed. Pergamon Press, 330pp.

28. POMEROY, L.R. and W.J. WIEBE (1988) Energetics of microbial food webs. Hydrobiologia, 159: 7-18.

29. PLATT T., K.H. MANN and R.E. ULANOWICZ eds (1981) *Mathematical models in biological oceanography.* UNESCO, Paris.

30. SIERACKI, M.E. and J.MCN. SIEBURTH (1986) Sunlight-induced growth delay of planktonic marine bacteria in filtered seawater. Mar.Ecol.Prog. Ser. 33: 19-27.

31. STRATHMANN, R.R. (1967) Estimating the organic carbon content of phytoplankton from cell volume or plasma volume. Limnol. Oceanogr. 12, 411-418.

32. STRICKLAND, J.D.H. (1960) Measuring the production of marine phytoplankton. Bull. Fish. Res. Bd. Canada, 122, 1-172.

33. TAYLOR, B.F. and R.P. KIENE (1989) Microbial metabolism of dimethyl sulphide, p202-221. *In,* Saltzman E.S. and W.J. Cooper [eds.], *Biogenic sulfur in the environment.* American Chemical Society, Washington, DC.

34. THOMPSON, A.M., W.E. ESAIS and R.L. IVERSON (1990) Two approaches to determining the sea-to-air flux of dimethylsulphide: satellite ocean colour and a photochemical model with atmospheric measurements. J. Geophys. Res. 95, D12, 20551-20558.

35. TURNER, S.M., G. MALIN , P.S. LISS , D.S. HARBOUR and P.M HOLLIGAN (1988) The seasonal variation of dimethyl sulfide and dimethylsulfoniopropionate concentrations in nearshore waters. Limnol. Oceanogr. 33(3), 364-375.

36. VAIRAVAMURTHY, A., M.O. ANDREAE and R.L. IVERSON (1985) Biosynthesis of dimethylsulfide and dimethylpropiothetin by *Hymenomonas carterae* in relation to sulfur source and salinity variations. Limnol. Oceanogr. 30(1), 59-70.

37. WAKEHAM, S.G., B.L. HOWES, J.W.H. DACEY, R.P. SCHWARZENBACH and J. ZEYER (1987) Biogeochemistry of dimethylsulfide in a seasonally stratified coastal salt pond. Geochim. Cosmochim. Acta 51, 1675-1684.

38. ZEYER, J., P. EICHER, S.G WAKEHAM and R.P. SWARZENBACH (1987) Oxidation of dimethylsufide (DMS) to dimethylsulfoxide (DMSO) by phototrophic purple bacteria. Appl. Enviro. Microbiol. 53, 2026-2032.

39. ZINDER, S.H. and T.D. BROCK (1978) Production of methane and carbon dioxide from methane thiol and dimethylsulphide by anaerobic lake sediments. Nature, 273, 226-228.

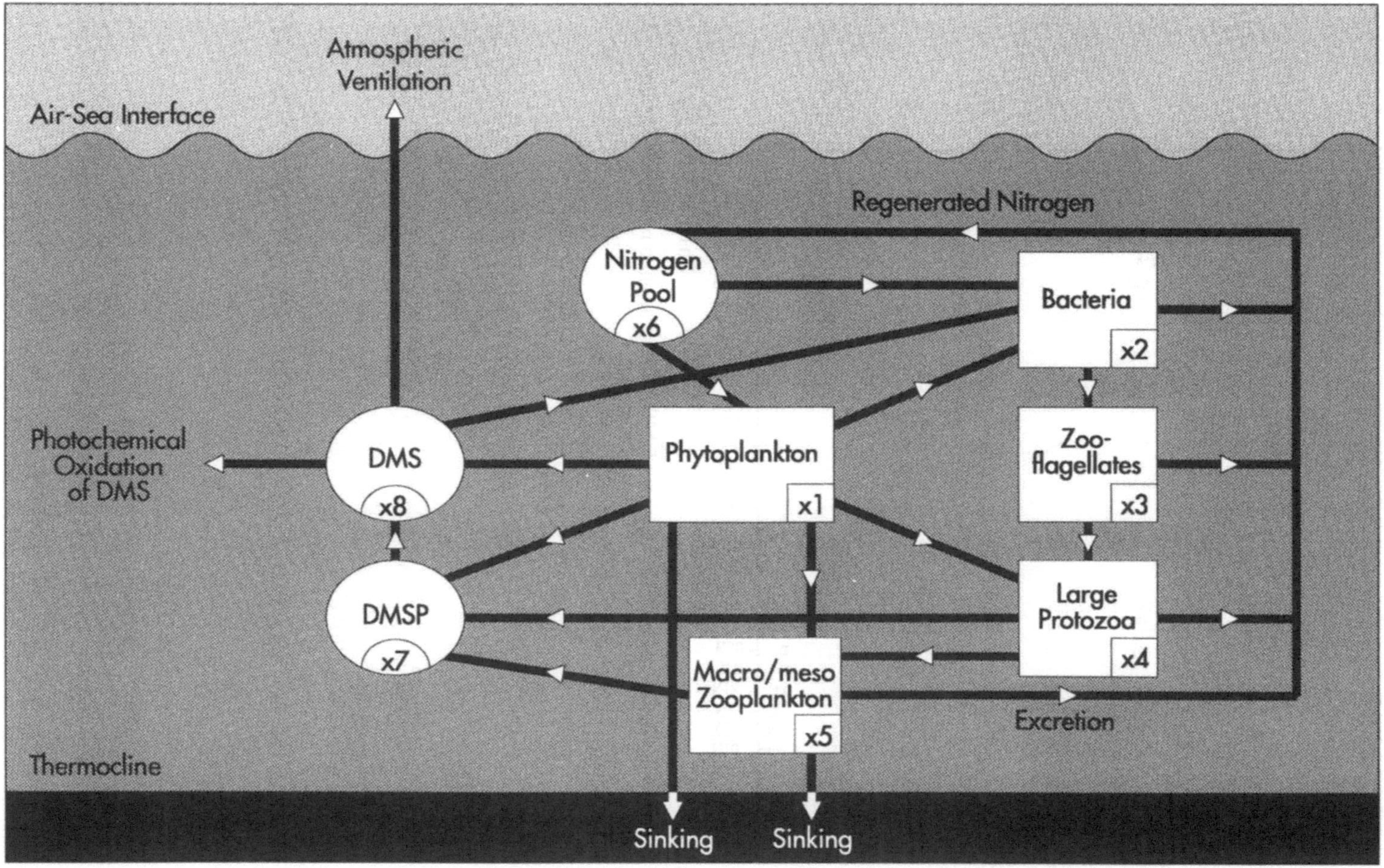

Figure 1. The model compartments and flow of N and S. Arrows indicate direction of flow.

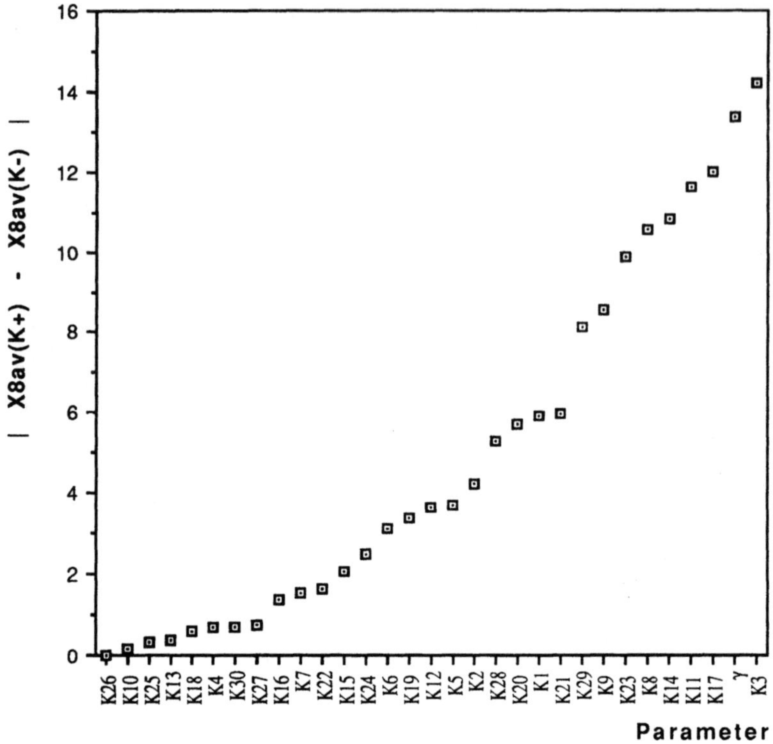

Figure 2. Sensitivity of time integrated DMS concentration to model parameters

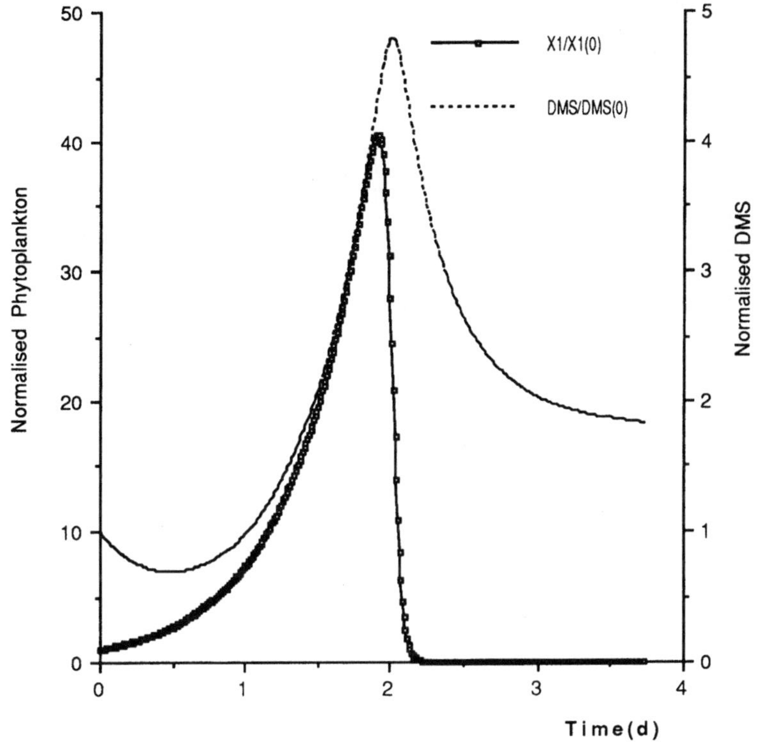

Figure 3a. Prediction of algal biomass and DMS temporal correlation

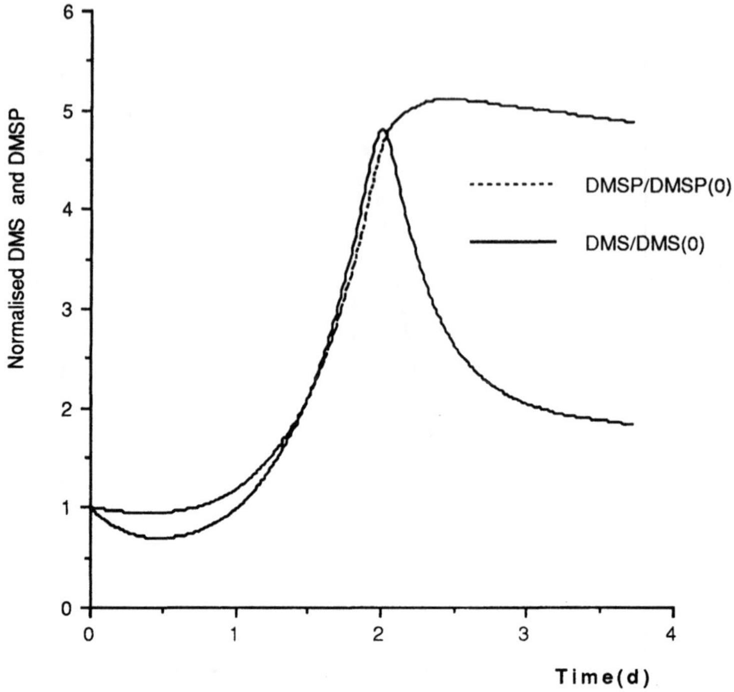

Figure 3b. Prediction of DMS and DMSP temporal correlation

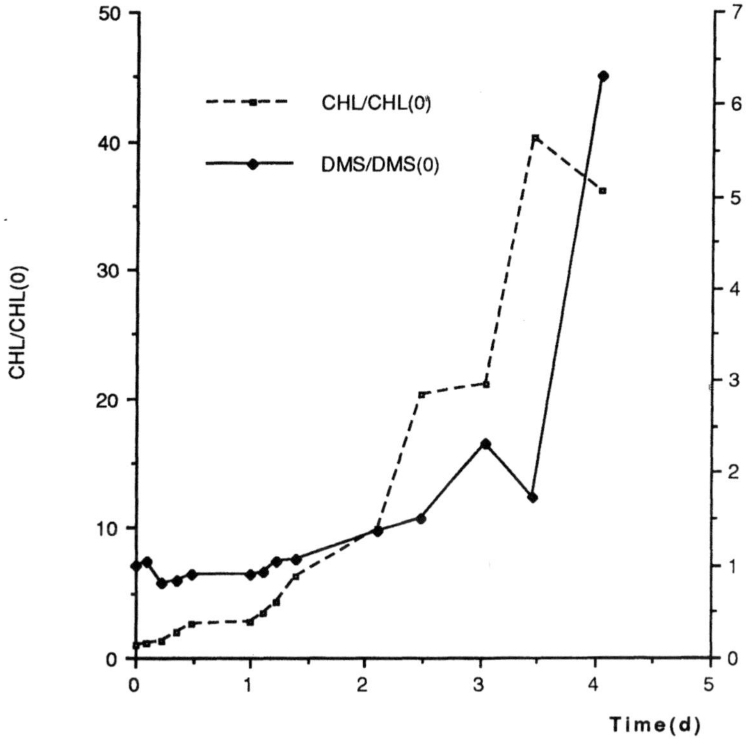

Figure 3c. Experimental observations of DMS and algal biomass

DIMETHYLSULFIDE FIELD MEASUREMENTS(*)

M.O. ANDREAE and S. RAPSOMANIKIS

Max Planck Institute for Chemistry
Biogeochemistry Department
P.O. Box 3060, 6500 Mainz, FRG

ABSTRACT

The recent hypothesis of Charlson R.J., Lovelock J.E., Andreae M.O. and Warren S.G. (Nature, Vol. 236, No. 6114, pp. 655-661, 1987) points to an important climatic role for marine biogenic sulfur. The key link is the oxidation of dimethylsulfide (DMS) to sulfur dioxide (SO_2) and eventually to water soluble sulfate aerosol particles. These particles act as cloud condensation nuclei (CCN) and their number concentration can influence cloud albedo. The connection between CCN originating from the oxidation of marine DMS, cloud albedo and climatic change may also have a feedback on the production of DMS since biological activity is responsible for its production. Measurements of the above interlinked parameters, in view of their influence on climatic change, are scarce.
We will present data from a number of cruises where measurements of some of the above parameters are made. Measurements of DMS in sea water are used in conjunction with the wind speed to estimate the sea-to-air flux of DMS. There is correlation between the emission of DMS from the sea and its atmospheric concentration. Although in a number of cases there is clear correlation between CCN and CN and atmospheric DMS concentrations, this is not always the case. Concentrations of seawater DMS correlate with the presence of certain species of phytoplankton. These concentrations also vary with geographical location and season.

(*) Only the abstract is available.

G. Restelli and G. Angeletti (eds.), Dimethylsulphide: Oceans, Atmosphere, and Climate, 83.

DIMETHYLSULFIDE AND AEROSOL MEASUREMENTS AT ROSS ISLAND, ANTARCTICA

D.J. WYLIE[1], M.J. HARVEY[2], S.J. de MORA[1], I.S. BOYD[2] and J.B. LILEY[2]

1. Chemistry Department, University of Auckland, Private Bag 92019
Auckland, New Zealand
2. The National Institute for Water and Atmospheric Research Limited
P.O. Box 3047, Wellington, New Zealand

SUMMARY

In the austral summer of 1991/92 a study was made of dimethylsulfide (DMS) and sulfur-containing aerosols in the marine boundary layer over McMurdo Sound, Antarctica. The concentration and size distribution of aerosols, and meteorological variables were also measured. The site was characterized by high concentrations of DMS (2.98 to 19.2 nmol m^{-3}, median 6.29 nmol m^{-3} STP), aerosol methanesulfonate (0.60 to 3.46 nmol m^{-3}, median 1.68 nmol m^{-3} STP) and aerosol non-sea salt sulfate (3.05 to 11.6 nmol m^{-3}, median 6.77 nmol m^{-3} STP). Windspeed had the most significant influence on the observed high concentrations of DMS. Measurements of condensation nuclei were made over a wide range of accumulation mode (0.12 to 0.5 μm diameter) and coarse mode (0.5 to 3.12 μm diameter) concentrations. The combined surface area of both modes gave a range between 1.2 and 25 $\mu m^2\ cm^{-3}$, (average 10 $\mu m^2\ cm^{-3}$). During episodes of large total aerosol surface area, the numbers of condensation nuclei recorded tended to be few and conversely, when surface area was small, the number of nuclei was largest. This observation suggests that new particle generation is suppressed when the total aerosol concentration is large as the aerosol apparently provides a surface sink for the precursor gases. New particle generation was most active following the major depletion of accumulation and coarse mode aerosol by precipitation scavenging during snow fall.

1. INTRODUCTION

The oceans surrounding Antarctica are known to be very productive and biogenic emissions of dimethylsulfide (DMS) play an important role in the biogeochemical cycling of sulfur over the continent. The Southern Ocean may contribute as much as 10 % of the global emission of DMS (1). With respect to the McMurdo Sound region, investigations suggest that extensive blooms of the micro-algae *Phaeocystis pouchetii* from the relatively sea-ice free areas of the open Southern Ocean, can move south with the Cape Bird surface current into the eastern side of McMurdo Sound (2). Barry (3) has noted that the stability of the water column in McMurdo Sound can result in an abundance of phytoplankton, however, there is a high degree of variability (4). The seasonal pattern of primary productivity is strongly correlated with the annual variation in solar insolation, with an associated rapid increase in primary production during December and January (5).

A recent study of the ecophysiology of ice-algae in Antarctica has highlighted the strong biogenic signal of DMS in both water and air, especially during periods of ice melt (6). Elevated concentrations of aerosol methanesulfonate (MSA) and aerosol non-sea salt sulfate (NSSS) have been observed in the Antarctic marine troposphere (7, 8). Aerosol concentrations follow a seasonal trend consistent with a marine biogenic source. Significant levels of MSA have also been observed in the atmosphere and snow at a remote site on the East Antarctic Plateau (9). The present study comprises an investigation of DMS and sulfur-containing aerosols in the marine boundary layer over the McMurdo Sound.

G. Restelli and G. Angeletti (eds.), Dimethylsulphide: Oceans, Atmosphere, and Climate, 85–94.

2. METHODS

This sulfur-aerosol investigation was conducted at the coastal site of McDonald Beach (77° 9'S,166° 21'E, elevation ~ 1 m) on Ross Island in McMurdo Sound, Antarctica, between January 1 and 24, 1992 (Figure 1). Meteorological information was recorded by a Campbell CR10 data logger using research grade sensors. Bulk aerosol samples of 24 h duration were collected on both Whatman-41 and Whatman-QMA filter media using a Hi-volume sampler (Sierra-Anderson) during in-sector (83° through North to 145°) winds with speeds >1.5 m s^{-1}. Both coarse and fine aerosol particles were collected using a 6-stage Anderson (Mk III) cascade impactor from which a size distribution of aerosol concentrations was obtained. Extraction and analysis by ion chromatography of the soluble aerosol ions was carried out in New Zealand.

The concentration and size distribution of aerosols were measured using a Condensation Nuclei Monitor (Environment One Model Rich 100) for particles >0.0025 μm diameter and a laser optical particle sampler (Particle Measuring Systems ASASP-100X) for particles between 0.12 and 3.12 μm in diameter.

Atmospheric and aqueous DMS was determined on site by gas chromatography with a flame photometric detector (GC/FPD) following preconcentration onto molecular sieve 5A and Tenax-GR (10). Atmospheric samples were collected at a height of 3 m. The detection limit for DMS in air (based on a 30 minute sampling time) and sea water was 0.08 nmol m^{-3} and 0.16 nmol L^{-1}, respectively.

DMS sampling was conducted during a period of high biological activity as indicated by observations of a phytoplankton bloom off shore. Sea ice coverage during the study period was variable, but generally the sea was open with light pack-ice in McMurdo Sound. As shown in Figure 1, the most frequent winds were light northerly and north-north-easterly (<5 m s^{-1}). However, the highest frequency of strongest winds (>7.5 m s^{-1}) blew from the southerly quarter, in agreement with a climatological study of the region (11). The strongest winds blew from the Ross Ice Shelf, but the most frequent wind fields, also of southerly origin, caused an eddy off the northwest coast of Ross Island which brought north-north-easterly winds to that coast. There were a number of other significant features in the weather and climate at this site. Vertical mixing in the very stable boundary layer was limited and winds tended to be terrain following (*ie* they followed the coastline). During the study period (Table I), insolation was high and it was never dark. The average daily insolation, 21.7 MJ m^{-2}, was within 10 % of the December and January climatological average at New Zealand latitudes. The temperatures

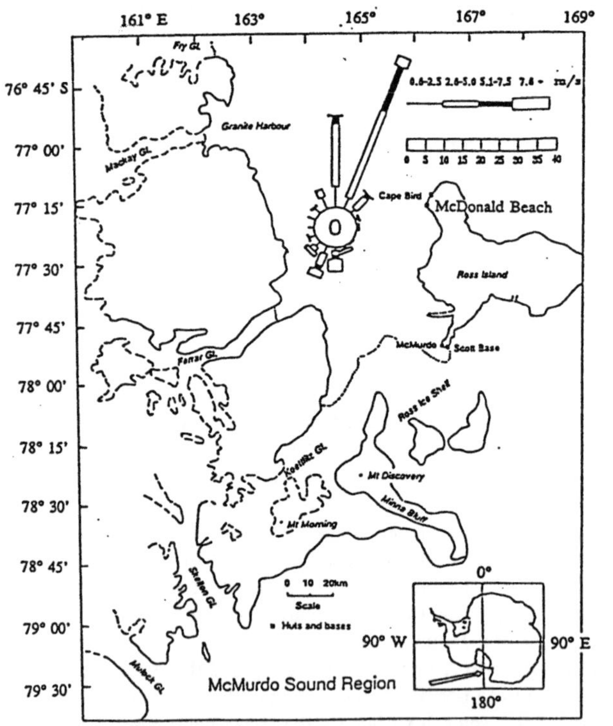

Map of Antarctica showing location of McDonald Beach

Figure 1. Map of McMurdo Sound showing McDonald Beach and wind rose.

and relative humidity were comparatively high, and much higher than experienced in continental Antarctica.

Two local sources of contamination were taken into account. The first was Cape Bird Hut, situated ~6 km north-north-east. The facility was manned during this field study and fuel was burnt for cooking and heating. However, aerosol NO_3^- concentrations (range 0.69 to 3.50 nmol m^{-3}, median 1.00 nmol m^{-3}) and gaseous NO_2 (range 4.83 to 26.1 nmol m^{-3}, median 13.5 nmol m^{-3}) at McDonald Beach were typical of remote marine regions (12). Secondly, local ponds were observed to release volatile sulfur gases, predominantly CS_2 and smaller variable quantities of DMS. CS_2 was used as an indicator of pond influences. No winds were sampled which had passed over the ponds at any stage, and therefore, no CS_2 was detected at the gas-phase sampling site.

Table I. Meteorological information summary (3 second recordings, 10 minute averages)

	Windspeed ($m\ s^{-1}$)	Temperature (°C)	Relative Humidity (%)	Solar Insolation ($W\ m^{-2}$)
No of 10 min values	2508	2073	2508	17*
Average	4.22	-0.43	69.4	21.7
Standard Deviation	2.84	1.40	11.0	7.3
Minimum	0.60	-3.97	24.3	12.4
Maximum	19.6	3.74	93.6	33.9

* Number of days

3. RESULTS AND DISCUSSION

Sulfur Gases

Atmospheric DMS concentrations (n=49) ranged from 2.98 to 19.2 nmol m^{-3} during the study period (Figure 2). No diurnal variation was observed but, of course, no periods of darkness were experienced. The highest concentrations of DMS were observed during a southerly storm (>30 knots) on January 17 and 18 which promoted a faster rate of exchange of volatile DMS across the sea to air interface. The median atmospheric DMS concentration, 6.29 nmol m^{-3} STP (Table II), was considerably higher than the average over the world's oceans of *ca.* 4.7 nmol m^{-3} (13), and higher than values of *ca.* 4.4 nmol m^{-3} reported for sub-Antarctic areas (7). The greater productivity of the waters in McMurdo Sound are thought to be responsible for the significant enhancement. Recent atmospheric DMS measurements in the Antarctic peninsula region reveal that DMS concentrations in excess of 32.2 nmol m^{-3} are not unusual for that region, especially during periods of ice melt (6).

DMS concentrations in surface sea water (n=13) ranged from 0.63 to 6.52 nmol L^{-1} (median 1.72 nmol L^{-1}). These values are comparable with the global average of 1.99 nmol L^{-1} (14) and in agreement with concentrations (0.71 to 1.35 nmol L^{-1}) reported for McMurdo Sound (10).

Several small, shallow (< 0.5 m) ponds containing cyanobacterial mats were present in the vicinity of the sampling site. Atmospheric CS_2 (not quantified) and DMS (10.2 to 10.7 nmol m^{-3}, median 10.3 nmol m^{-3}) were the only volatile sulfur gases detected in the atmosphere within the immediate vicinity of the cyanobacterial ponds. This observation is consistent with measurements made above ponds on the McMurdo Ice Shelf (15). The apparently higher relative amounts of CS_2 in air samples reflects differences in the Henry law constants; CS_2 being more volatile than DMS. CS_2 (not quantified) was detected in all pond waters. The aqueous DMS concentrations (0.30 to 0.72 nmol L^{-1},

Table II. Summary of dimethylsulfide concentrations.

	DMS nmol m^{-3} (STP) air	DMS nmol L^{-1} sea water	DMS nmol L^{-1} pond water	DMS nmol m^{-3} (STP) over ponds
Number	49	13	4	9
Median	6.29	1.72	0.45	10.3
Minimum	2.98	0.63	0.30	10.2
Maximum	19.2	6.52	0.72	10.7

median 0.45 nmol L^{-1}) did not vary significantly from pond to pond. DMS and CS_2 concentrations within the glacial melt water ponds on the McMurdo Ice Shelf ranged from 0.35 to 0.88 nmol L^{-1} and 0.07 to 0.12 nmol L^{-1}, respectively (10). Hence, both studies have now observed CS_2 to be a major volatile sulfur species present in Antarctic pond waters and in the overlying atmosphere. In Antarctic lakes studied at Davis Bay, Deprez and co-workers (16) reported high concentrations of DMS (Burton Lake, 0.81 to 22.6 nmol L^{-1}) but were unable to resolve CS_2

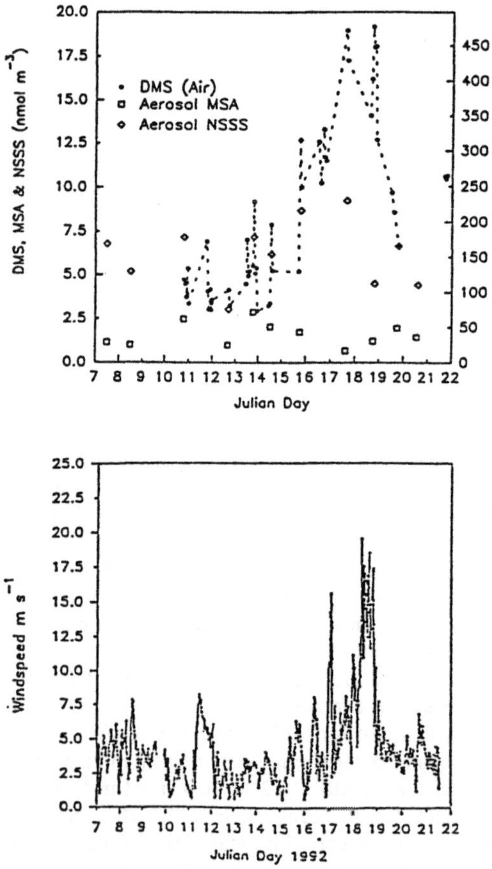

Figure 2. Time series for dimethylsulfide concentrations and windspeed .

Aerosol Physics

For most of the time the concentration of CN was less than 1000 cm^{-3}. The median concentration, 852 cm^{-3}, was almost identical to values for January recorded in long term measurement programmes at the coastal Antarctic station Georg von Neumayer (17) and during baseline conditions at Cape Grim (18).

Conditions at Antarctic coastal sites in the summer are conducive to the formation of large numbers of CN concentrations by gas-to-particle conversion. However, because of changing conditions and the great variability in weather conditions experienced during this field study, it was not possible to evaluate relationships between CN concentration and the main atmospheric variables of DMS concentration and solar flux. The variable which seemed to have most influence on the concentration of CN was the surface area of aerosols with diameters >0.1 μm. One of the functions of the CN monitor was to indicate periods of pollution from local anthropogenic sources. Contamination occurred during very light winds, wind shifts and during some of the frequent north-north easterlies (Figure 3) when wind came from the direction of Cape Bird Hut.

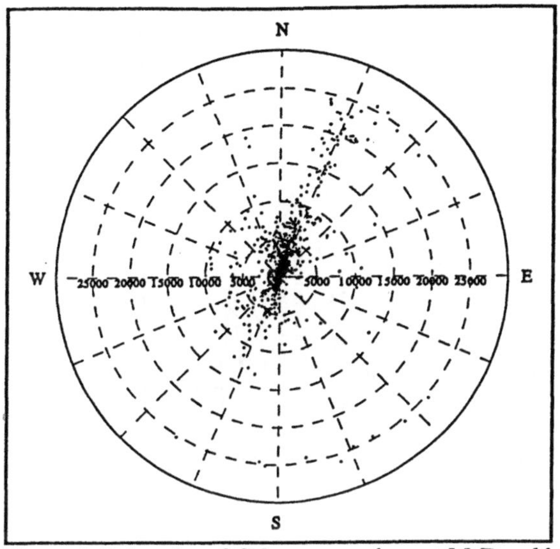

Figure 3. Polar plot of CN concentrations at McDonald Beach.

The concentration of aerosol particles >0.1 µm was large (Figure 4) in comparison with other studies in the Ross Sea region and a remote site on the East Antarctic Plateau (19, 20). These plots show average distributions for the duration of each field study and for three days of invariant meteorological conditions on the plateau. The elevated concentrations at Cape Bird were attributed in part to the high humidity which tends to shift the distribution towards larger particles. The concentration of coarse particles >1 µm in diameter was noticeably greater than in other studies and was elevated by storm events on January 17 and 18. The size distribution shows the two particle modes: the accumulation mode at 0.24 µm and the coarse mode between 1.5 and 1.8 µm, with a minimum between 0.4 and 0.5 µm. Aerosol surface area ranged between 0.7 and 25 $\mu m^2\ cm^{-3}$ and 0.4 and 29 $\mu m^2\ cm^{-3}$, for accumulation and coarse modes, respectively, with the combined range being between 1.2 and 25 $\mu m^2\ cm^{-3}$, (average 10 $\mu m^2\ cm^{-3}$).

The influence of weather on the particle size distribution is apparent in Figure 5. There was a brief increase in accumulation mode aerosol during fine weather and light winds on January 6, possibly caused by local pollution. Although snow and moderate winds on January 12 removed the majority of aerosols, concentrations recovered over a period of about half a day in light winds. Coarse mode aerosol dramatically increased during gales between 17 and 19 January. Figure 6 shows the relationship of high CN number at low aerosol surface area and vice-versa. The results suggest that the aerosol surface acts as a sink for precursor vapour molecules, thereby suppressing new particle formation. When accumulation mode aerosol is removed from the atmosphere by precipitation

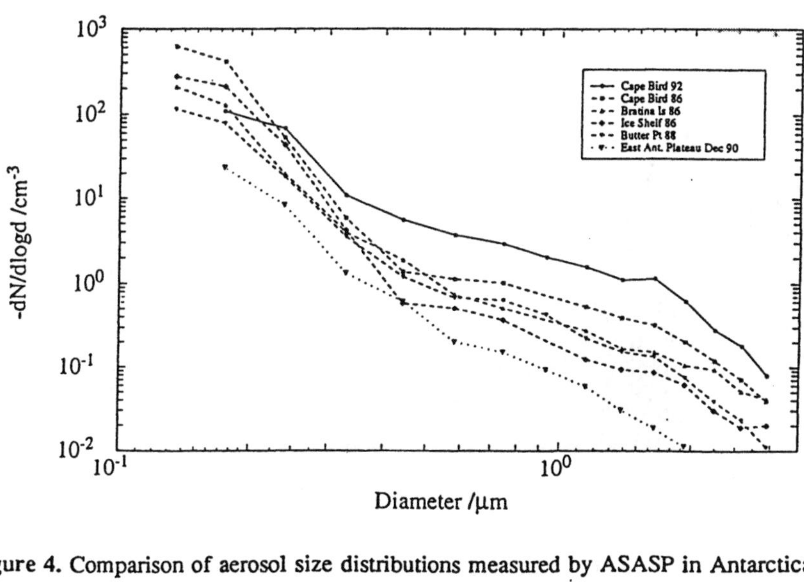

Figure 4. Comparison of aerosol size distributions measured by ASASP in Antarctica.

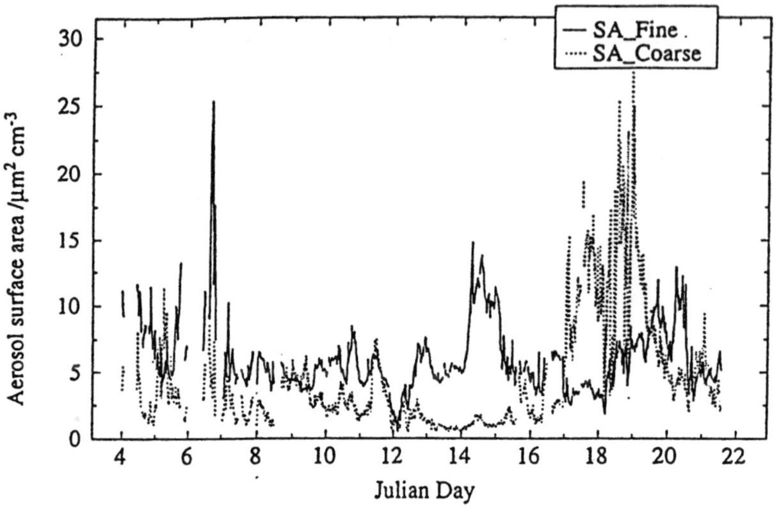

Figure 5. Time series of fine (accumulation) and coarse mode surface area.

scavenging, then new particle formation is particularly active provided there is abundant DMS or other precursor gases.

Aerosol Chemistry

Aerosol MSA and aerosol NSSS concentrations were relatively high when compared with other recent studies for Antarctica. MSA concentrations (Table III) ranged from 0.60 to 3.46 nmol m^{-3} (median 1.68 nmol m^{-3} STP). Summertime aerosol MSA concentrations measured at Mawson (21) and Cape Grim (22) were in the range 0.31 to 0.63 nmol m^{-3} STP and 0.42 to 0.52 nmol m^{-3}, respectively. NSSS concentrations at McDonald Beach (Table III) ranged from 3.05 to 11.6 nmol m^{-3} (median 6.77 nmol m^{-3}). These values were considerably higher than summertime values observed in the range

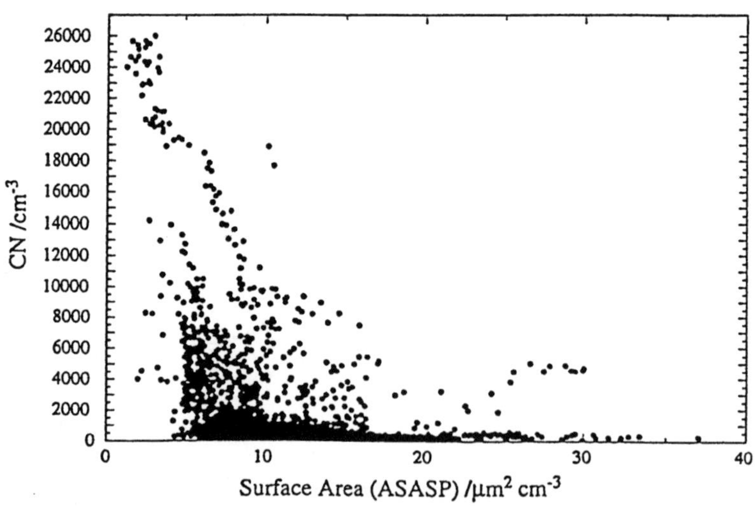

Figure 6. The relationship between CN concentration and aerosol surface area.

Table III. Summary of aerosol concentrations for McDonald Beach

Date	Na^+	Cl^-	NO_3^-	SO_4^{2-}	NSSS	MSA	MSA/ NSSS
	(nmol m^{-3},STP)						(% wt/wt)
05-06/1/92	164	37.5	1.11	9.28	7.34	2.97	40.4
06-07/1/92	154	30.8	3.50	13.2	11.6	3.46	29.8
07-08/1/92	27.8	47.3	2.87	9.21	6.77	1.08	16.2
08-09/1/92	160	58.2	1.48	8.21	5.19	0.94	18.1
09-10/1/92	104	24.6	1.18	8.23	7.17	2.43	33.9
11-13/1/92	19.5	47.9	1.00	5.53	3.05	0.93	30.4
13-14/1/92	210	8.34	0.69	7.69	7.20	2.83	39.4
14-15/1/92	9.26	25.9	1.00	7.51	6.20	1.98	31.9
15-16/1/92	150	61.6	1.55	11.9	8.70	1.68	19.3
17-18/1/92	169	110	0.76	15.0	9.24	0.60	6.54
18-19/1/92	320	269	0.92	18.4	4.51	1.16	25.6
19-20/1/92	66.4	49.9	0.74	9.24	6.66	1.90	28.5
20-21/1/92	134	63.0	0.81	7.70	4.44	1.38	31.0

2.19 to 2.60 nmol m^{-3} STP at Mawson (21) and 0.66 to 2.5 nmol m^{-3} at Cape Grim (22).

In accordance with Ayers (23), NSSS concentrations have been calculated using Cl^- as a conservative tracer of marine origin. Using Na^+ concentrations to calculate NSSS values led to negative NSSS results, most likely due to the presence of aerosol marabilite (Na_2SO_4), which forms from freezing sea water (24). Marabilite precipitation has been observed in coastal ponds on Ross Island (25).

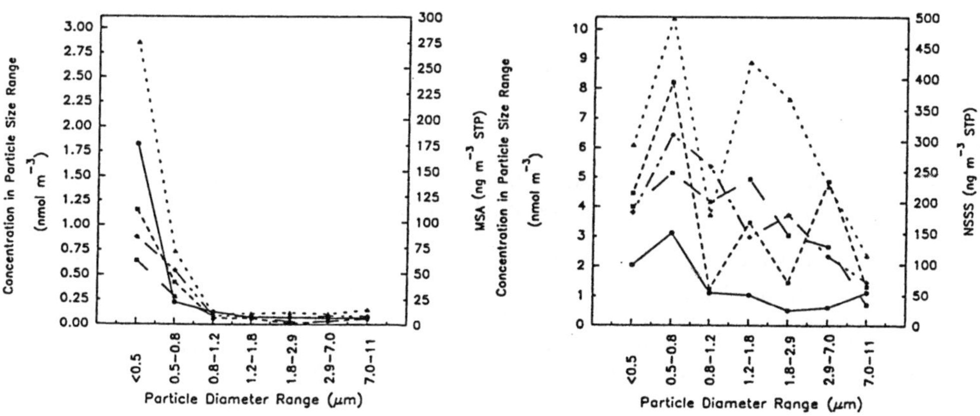

Figure 7. Size distribution of aerosol methanesulfonate and aerosol non-sea salt sulfate.

The aerosol MSA/NSSS weight ratios ranged from 6.54 to 40.4 % (median 30.4 %). In comparison, aerosol MSA/NSSS weight ratios vary in the range from 15 to 40 % (mean 22 %) at Mawson (21) and from 4.76 to 41.8 % (mean 22.2 %) at a remote summertime site on the East Antarctic Plateau (9). In contrast, aerosol MSA/NSSS weight ratios in temperate marine regions are generally <10 % (26). Elevated levels of MSA at McDonald Beach can be attributed to the high biological activity of the adjacent Ross Sea and the lack of a nocturnal DMS reaction with NO_3. Also, the temperature dependence of the DMS to MSA oxidation pathway may play a role as the efficiency of the OH· addition is enhanced in the colder Antarctic environment (27). However, recent aerosol MSA/NSSS data from Cape Grim (28) reveal that colder temperatures cannot fully explain the observed elevated ratios in Antarctica.

Size distribution data (Figure 7a) reveal that aerosol MSA is concentrated predominately on sub-micron particles (<0.5 μm diameter). In contrast, aerosol NSSS (Figure 7b) exhibits a bimodal size distribution with peaks in the 0.5 to 0.8 μm and the 1.2 to 2.9 μm diameter ranges.

4. CONCLUSIONS

In the locality of these measurements, all the necessary attributes for the rapid and abundant production of aerosols were present. These conditions included large concentrations of DMS released by abundant phytoplankton populations following sea-ice break-up during the brief Antarctic summer, 24 hour light (with daily insolation up to 33.9 MJ m^{-2}) for phytoplankton growth and atmospheric photo-oxidation of DMS and high humidity (average 69 %) promoting efficient binary nucleation between water and sulphuric acid molecules. A high aerosol MSA/NSSS weight ratio indicates that the MSA pathway of DMS oxidation is favoured in the cold Antarctic environment.

Conditions were too variable to demonstrate a clear relationship between DMS and CN concentrations or to estimate the rate of particle production. However, CN concentrations were small when the surface area of aerosol (combining the contribution of accumulation and coarse mode particle) was large. Conversely, the CN number could be very large when the surface area of aerosol was small. CN production is thought to be depressed by the absorption of precursor gases onto the surface of pre-existing aerosol.

5. ACKNOWLEDGEMENTS

We thank all staff at DSIR Antarctic Division (now MERT) and at Scott Base who made possible the successful completion of our research. We also thank VXE-6 and the RNZAF aircrews for field operations. This study was funded by the New Zealand Foundation for Research Science and Technology under the Priority Research Contracts Scheme No MET902.

6. REFERENCES

(1) GIBSON, J.A.E., GARRICK, R.C., BURTON, H.R. and McTAGGART, A.R. (1990). Dimethylsulfide and the Alga *Phaeocystis Pouchetii* in Antarctic Coastal Waters. Mar Biol. 104. 339-346.

(2) PALMISANO, A.C. and SULLIVAN, C.W. (1983). Sea Ice Microbial Communities (SIMCO) 1. Distribution, Abundance, and Primary Production of Ice Microalgae in McMurdo Sound, Antarctica in 1980. Polar Biol. 2. 171-177.

(3) BARRY, J.P. (1988). Hydrographic Patterns in McMurdo Sound, Antarctica and Their Relationship to Local Benthic Communities. Polar Biol. 8. 377-391.

(4) KNOX, G.A. (1990). Primary Production and Consumption in McMurdo Sound, Antarctica. Kerry, K.R. and Hempel, G. (eds). Antarctic Ecosystems: Ecological Change and Conservation. Heidelberg. Springer-Verlag. 115-127.

(5) EL-SAYED, S.Z., BIGGS, D.C. and HOLM-HANSEN, O. (1983). Phytoplankton Standing Crop, Primary Productivity, and near-surface Nitrogenous Nutrient Fields in the Ross Sea, Antarctica. Deep. Sea. Res. 30. (8A) 871-886.

(6) KIRST, G.O., WANZEK, M., HAASE, R., RAPSOMANIKIS, S., de MORA, S.J., SCHEBESKE, G. and ANDREAE, M.O. (1992). Echophysiology of Ice Algae (Antarctica): Dimethylsulfoniopropionate Content and Release of Dimethylsulfide During the Ice Melt. Dimethylsulfide, Oceans, Atmosphere and Climate. Belgirate, Italy.

(7) BERRESHEIM, H. (1987). Biogenic Sulfur Emissions from the Sub-Antarctic and Antarctic Oceans. J. Geophys. Res. 92. (D11) 13,245-13,262.

(8) PSZENNY, A.P.P., CASTELLE, A.J. and GALLOWAY, J.M. (1989). A Study of the Sulfur Cycle in the Antarctic Marine Boundary Layer. J. Geophys. Res. 94. (D7) 9818-9830.

(9) de MORA, S.J., WYLIE, D.J. and DICK, A.L. (1992). Methanesulfonic Acid in Aerosols and Snow at a Remote Site on the Antarctic Plateau. Dimethylsulfide, Oceans, Atmosphere and Climate. Belgirate, Italy.

(10) SHOOTER, D., de MORA, S.J., GROUT, A., WYLIE, D.J. and ZHI-YUN, H. (1992). The Chromatographic Analysis of Reduced Sulfur Gases in Antarctic Waters following the Pre-concentration onto Tenax. Intern. J. Environ. Anal. Chem. 47. 239-249.

(11) SINCLAIR, M.R. (1982). Weather Observations in the Ross Island Area, Antarctica. Technical Note No.253. New Zealand Meteorological Service. 36.

(12) GALLOWAY, J.N. (1985). Deposition of Sulfur and Nitrogen from the Remote Atmosphere. Galloway, J.N., Charlson, R.J., Andreae, M.O., and Rodhe, H. (eds). The Biogeochemical Cycling of Sulfur and Nitrogen in the Remote Marine Atmosphere. Dordrecht, Holland. D. Reidel. 143-175.

(13) ANDREAE, M.O. (1985). The Ocean as a Source of Atmospheric Sulfur Compounds. Buat-Ménard, P. (ed). The Role of Air-Sea Exchange in Geochemical Cycling. Dordrecht, Holland. D. Reidel. 331-362.

(14) BATES, T.S., CLINE, J.D., GAMMON, R.H. and KELLY-HANSEN, S.R. (1987). Regional and Seasonal Variations in the Flux of Oceanic Dimethylsulfide to the Atmosphere. J. Geophys. Res. 92. (C3) 2930-2938.

(15) de MORA, S.J., GROUT, A. and SHOOTER, D. (1990). The Analysis of Reduced Sulfur Gases in Air and Melt Waters by Gas Chromatography at Bratina Island, 78° S. N.Z. Antarctic Rec. 10. 12-21.

(16) DEPREZ, P.P., FRANZMANN, P.D. and BURTON, H.R. (1986). Determination of Reduced Sulfur Gases in Antarctic Lakes and Seawater by Gas Chromatography after Solid Adsorbent Preconcentration. J. Chrom. 362. 9-21.

(17) LEHMANN, E. and JAENICKE, R. (1991). Condensation Nuclei Measurements at the Georg von Neumayer-Station (F.R.G.), Coastal Antarctica. J. Aerosol. Sci. 22. S589-592.

(18) GRAS, J.L. (1990). Baseline Atmospheric Condensation Nuclei at Cape Grim 1977-1987. J. Atmos. Chem. 11. 89-106.

(19) LECHNER, I.S., FISHER, G.W., LARSEN, H.R., HARVEY, M.J. and KNOBBEN, R.A. (1989). Aerosol Size Distributions in the Southwest Pacific. J. Geophys. Res. 94. (D12) 14,893-14,903.

(20) HARVEY, M.J., FISHER, G.W., LECHNER, I.S., ISAAC, P., FLOWER, N.E. and DICK, A.L. (1991). Summertime Aerosol Measurements in the Ross Sea Region of Antarctica. Atmos. Environ. 25A. (3-4) 569-580.

(21) PROSPERO, J.M., SAVOIE, D.L., SALTZMAN, E.S. and LARSEN, R. (1991). Impact of Oceanic Sources of Biogenic Sulphur on Sulphate Aerosol Concentrations at Mawson, Antarctica. Nature. 350. 221-223.

(22) BERRESHEIM, H., ANDREAE, M.O., AYERS, G.P., GILLETT, R.W., MERRILL, J.T., DAVIS, V.J. and CHAMEIDES, W.L. (1990). Airborne Measurements of Dimethylsulfide, Sulfur Dioxide, and Aerosol Ions over the Southern Ocean South of Australia. J. Atmos. Chem. 10. 341-370.

(23) AYERS, G.P., IVERY, J.P. and GILLETT, R.W. (1991). Coherence Between Seasonal Cycles of Dimethylsulphide, Methanesulphonate and Sulphate in Marine Air. Nature. 349. 404-406.

(24) HERUT, B., STARINSKY, A., KATZ, A. and BEIN, A. (1990). The Role of Seawater Freezing in the Formation of Subsurface Brines. Geochim. Cosmochim. Acta. 54. 13-21.

(25) SCHMIDT, S., MOSKAL, W., de MORA, S.J., HOWARD-WILLIAMS, C. and VINCENT, W.F. (1991). Limnological Properties of Antarctic Ponds During Winter Freezing. Antarctic. Sci. 3. 379-388.

(26) SALTZMAN, E.S., SAVOIE, D.L., PROSPERO, J.M. and ZIKA, R.G. (1986). Methanesulfonic Acid and Non-Sea-Salt Sulfate in Pacific Air: Regional and Seasonal Variations. J. Atmos. Chem. 4. 227-240.

(27) HYNES, A.J., WINE, P.H. and SEMMES, D.H. (1986). Kinetics and Mechanism of OH Reactions with Organic Sulfides. J. Phys. Chem. 90. 4148-4156.

(28) LONG, A., AYERS, G., BOERS, R., FISHER, G., GRAS, J. and HARVEY, M. (1992). Australia and New Zealand Atmospheric Chemistry and Cloud Microphysics Measurements Related to Climate Change. Cloud Mircophysics and Applications to Global Change. In press.

MEASUREMENTS OF ATMOSPHERIC AND SEAWATER DMS CONCENTRATIONS IN THE ATLANTIC, THE ARCTIC AND ANTARCTIC REGION

Regina Staubes, Hans-Walter Georgii

Institute for Meteorology and Geophysics
University of Frankfurt
Feldbergstraße 47
Germany

SUMMARY

This paper reports on simultaneous measurements of Dimethyl Sulfide in surface seawater and the overlying atmosphere in several hydrographical areas of the World's oceans in a latitudinal range between 82°N and 71°S. As DMS is produced biologically in seawater the concentrations of dissolved DMS showed a reasonable seasonal and spatial variability with values ranging from 10 to 1600 ngS(DMS)/l. Maximum values were measured in highly productive upwelling and shelf areas of the temperate and high latitudes. In oligotrophic waters the concentrations varied between 5 and 60 ngS/l, largely independent from seasonal influences. The DMS mixing ratios in the marine boundary layer showed a reasonable temporal and spatial variability with values ranging from 3 to 3400 pptv depending on sea-to-air flux rates and the meteorological situation. Flux rates were calculated to range from minimum values of 5 ngS/m^{2}min from the ice-covered Weddell Sea to more than 1000 ngS/m^{2}min from highly productive areas.

Aerosol methane sulfonate (MSA) and non-sea salt sulfate (nss sulfate) concentrations ranged from 3 to 60 ppt and 44 to 270 ppt, respectively, in a latitudinal range between 50°N and 71°S. The ratio of nss sulfate to MSA measured in high southern latitudes were generally lower than those observed in temperate and tropical marine areas.

1. INTRODUCTION

Measurements in different regions of the World's oceans have shown that DMS is the predominant form of biogenic sulfur transferred from the ocean to the atmosphere. DMS is biologically produced in surface seawater by metabolic processes in certain phytoplankton species (1,2,3). This is the reason for the large range of DMS concentrations observed in surface seawater and thus the difficulty in calculating global annual DMS fluxes from the World's oceans. Andreae and Raemdonck (4) estimated a flux of 40 ± 20 TgS(DMS)/yr into the atmosphere on the basis of several shipboard measurements mainly from coastal areas. Recently Bates et al. (5) calculated a considerably lower DMS sea-to-air flux of 16 TgS/yr based on measurements in the temperate and tropical Pacific during different seasons. Emitted into the atmosphere DMS is rapidly oxidized by OH and NO_3 radicals to SO_2, methane sulfate (MSA) and - via SO_2 oxidation - to non-sea salt sulfate as major reaction products. These compounds are important factors in cloud chemistry and global climate as contributors to cloud condensation nuclei (6,7). The MSA to nss sulfate ratio was found to be highly variable resulting from the competing OH-addition and OH-abstraction pathways of the DMS oxidation, which was suggested to vary with temperature and possibly light

G. Restelli and G. Angeletti (eds.), Dimethylsulphide: Oceans, Atmosphere, and Climate, 95–102.

intensity or oxidant concentrations (8).

In September/October 1988, July/August 1989 and October/November 1990 measurements were performed in the Atlantic Ocean between Bremerhaven, Germany (52°N) and South America (Rio Grande do Sul, 32°S, 1988; Puerto Madryn, 43°S, 1989; Punta Arenas 53°S, 1990). In July and August 1990 sampling was carried out in the Arctic between Norway, Greenland and Spitsbergen. In connexion with the Atlantic transect in 1990 an additional cruise between Punta Arenas (53°S), South America and Cape Town (34°S), South Africa, crossing the Weddell Sea was conducted in November and December 1990.

2. EXPERIMENTAL

Atmospheric concentrations of DMS were measured four times a day in 6 hours intervals on the upper deck of the ship, about 21 m above the sea level. The air sampled was enriched cryogenically at the temperature of liquid argon (-185°C) in U-shaped glass sample loops. During each observation two samples were taken in parallel with volumes between 2 and 7.5 liters of air at a flow rate of 500 ml/min. Due to co-trapping of atmospheric oxidants or oxidant precursors which can cause significant DMS losses in the sample loops it was necessary to use prefilters. We used glass tubes filled with 5% Na_2CO_3 on Anakrom C22 upstream of the cryotraps. The efficiency of these scrubber tubes have been tested during the field measurements by DMS standard additions to natural air.

Simultaneously to the determination of atmospheric DMS the concentrations of seawater DMS has been examined. Seawater was obtained from the ship's continuous sea water pumping system. Samples were injected into a gas stripping column after passing through a glass fiber filter to remove algal cells. A glass frit on the bottom of the column allows the sulfur-free purge gas nitrogen to bubble through the sampled water. Thus the volatile compounds are stripped from the seawater into the nitrogen stream which subsequently passes through a cryotrap immersed in liquid argon to enrich the sulfur compounds. No detectable sulfur blanks had been shown when reanalyzing purged seawater. To avoid storage artifacts surface seawater samples were analyzed immediately after collection.

Analysis of both air and seawater samples was performed using a gaschromatographic system equipped with a flame photometric detector. The system was calibrated by using gravimetrically determined permeation tubes. The sensitivity of this system was 2 pptv in atmospheric samples and 0.5 ngS/l in seawater samples. Further details on the analytical system are given by Staubes et al. (9).

The aerosol was sampled by pumping 70 - 80 m^3 of ambient air through Microsorban 98 filters (Delbag, 47 diameter), the sample efficiency of these filters is >99 %. Pumps were controlled by a wind vane system to collect aerosol only when wind from the forward sector reach the filter to minimize the influence of sources located on the ship. The soluble compounds were extracted by flushing the filter with 4 x 5 ml of deionized water. The extracts were analyzed by ion chromatography using a Dionex HPIC AS-3 column. For further details on the aerosol analyzing system see Bürgermeister and Georgii (10).

3. RESULTS

DMS in surface seawater

The distribution of DMS in surface seawater during the five different cruises was broadly similar to that of primary productivity (Fig. 1). Thus highest DMS concentrations of several hundreds up to more than 1000 ngS/l were observed in temperate and high latitudes during seasonally elevated primary productivity particularly in shelf and upwelling zones. Maximum DMS levels of 1600 ngS/L were measured in the North Sea in July 1989, moreover high concentrations were observed along the west coast of South America during the austral spring bloom in November 1990 and in the upwelling areas along the west coast of Africa in August 1989. In oligotrophic, open ocean areas of the Atlantic the DMS concentrations varried between 5 and 60 ngS/l.

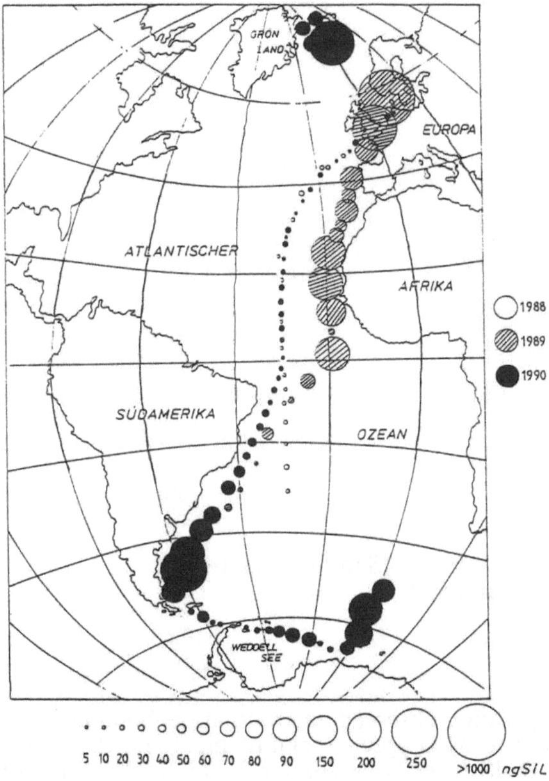

Fig.1: Distribution of DMS in surface seawater during the different cruises

Fig. 2. shows the seasonality of oceanic DMS in the Atlantic Ocean between 50°N and 50°S. For comparison the southern hemisphere values have been offset by 6 months. The shaded bars show the monthly means of marine DMS in the open ocean areas of the Atlantic. Here only a relatively weak seasonal variation of DMS in surface seawater has been found. The monthly means observed in February to April and September to November showed a high degree of constancy; they varried merely between about 35 and 45 ngS/l although they were not observed in the same area. However, in May and August the DMS values increased by 30 - 45 %. No measurements could be performed in June and July. The dark bars represent the monthly means of oceanic DMS including the biologically productive areas of the coastal and upwelling zones. In this case the DMS increase in May and August amount to 120 and 260 %, respectively. When just looking at coastal areas a summertime DMS increase by a factor of 20 was observed.

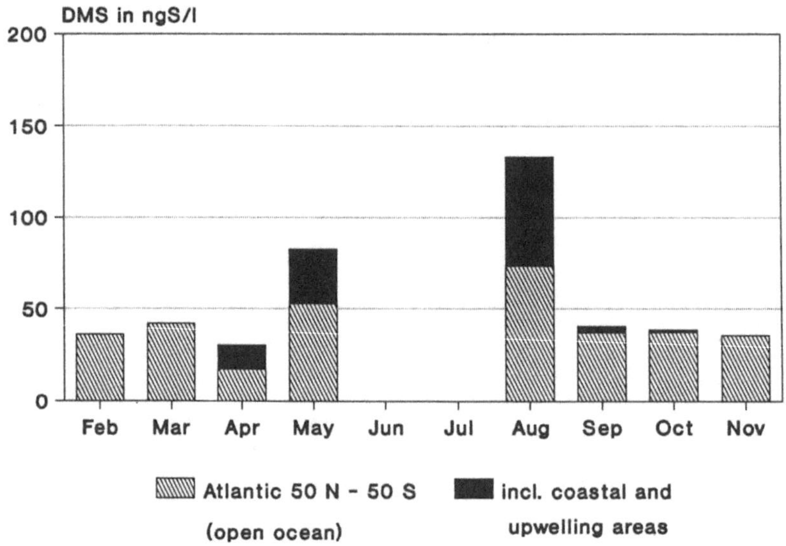

Fig.2: Monthly means of seawater DMS in the Atlantic Ocean between 50°N and 50°S

Atmospheric DMS

The DMS mixing ratios observed in the marine boundary layer between 82°N and 71°S ranged from 3 to 3400 pptv depending highly on the meteorology and the sea-to-air flux rates of DMS (Fig. 3). Maximum DMS levels of 1050 pptv on the average and peak values up to 3400 pptv were measured in the Arctic. These extremely high DMS mixing ratios were most likely caused by very high DMS emission rates in combination with reduced photochemical activity caused by fog. Moreover, elevated mixing ratios with single values of 1000 - 2000 pptv occured in 1990 over the Southern Atlantic between 40 and 70°S also at very high DMS fluxes from seawater. Minimum atmospheric DMS levels with an average of 25 pptv occured over the Weddell Sea. In this region remote maritime air masses prevailed, however, the ice coverage of the Weddell Sea had obviously minimized the gas exchange between seawater and the overlying atmosphere. In open ocean areas of the Atlantic DMS mixing ratios between 20 and 60 pptv were observed, with several maxima exceeding 100 pptv. The strong dependence of atmospheric DMS concentrations is also reflected in it's diurnal variations. Measurements performed on the research platform in the North Sea showed a more or less marked decrease of atmospheric DMS mixing ratios in the afternoon and maxima during the night in clean marine air masses. Differences between daytime and nighttime concentrations by a factor of 1.6 - 2.8 were observed. The reverse situation occured in continental air masses: maximum concentrations were observed during daylight, which were enhanced by a factor of 2.4 on the average compared to those mixing ratios measured during the dark. In these cases, the DMS oxidation by NO_3 radicals during the night was obviously much more effective than the DMS oxidation by hydroxyl radicals during daylight.

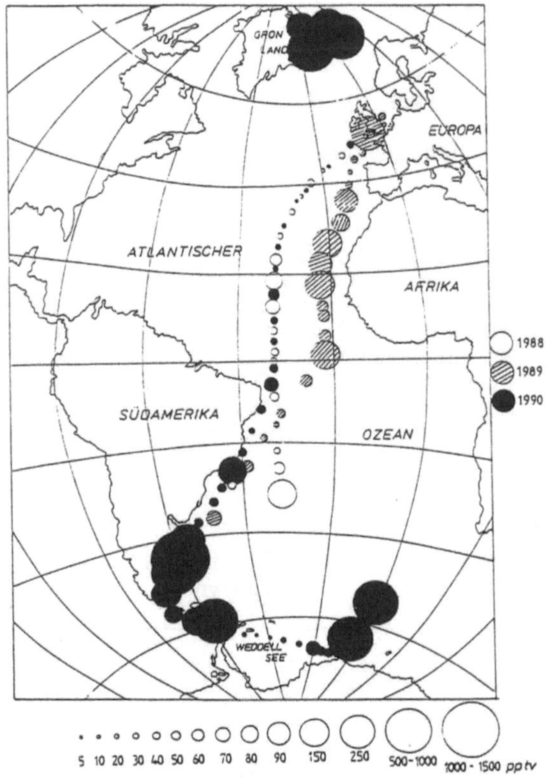

Fig.3: Distribution of atmospheric DMS in the marine boundary layer during the different cruises

Sea-to-air fluxes of DMS

The flux of a gas across an air-water interface can be calculated by using a stagnant film model described in detail by Liss and Merlivat (11), using the equation

$$F = - K (c_a/H - c_w)$$

F = flux rate
K = piston velocity
c_a = concentration in air
c_w = concentration in surface seawater
H = Henry's law constant

K is calculated in the same way as described by Berresheim (12) using an empirical formula for the radon piston velocity given by Smethie et al. (13). The radon piston velocities can be transformed to DMS piston velocities and adjusted for actual seawater temperatures by using following relationships:

$$K \sim (\mu/D)^{-2/3} \qquad \text{and} \qquad \mu{*}D/T = \text{const}$$

μ = kinematic viscosity of seawater
D = molecular diffusivity of the solute gas

The uncertainties by using the described method to calculate DMS flux rates are within a factor of 2 (13,14).

The calculated emission rates varied between less than 10 to about 500 ngS/m^2min with few values exceeding 1000 ngS/m^2min. The average flux rate from the open Atlantic amounts to 140 ngS/m^2min. Caused by considerably enhanced DMS seawater concentrations, the sea-to-air fluxes from coastal areas were much higher. For instance, the average flux rate from the North Sea and the English Channel was 700 ngS/m^2min in August 1989, the mean flux from coastal regions off Argentina amounts to 690 ngS/m^2min in austral spring 1990. Tab. 1 shows an extrapolation of these flux data to the global ocean surface. With these calculations we obtained a total DMS flux from the World's oceans to the atmosphere of 26,8 TgS/yr.

latitude	global ocean surface in 10^{12} m^2	global DMS Flux in TgS/yr
80°N - 40°N	32,30	2,90
40°N - 15°N	60,92	3,96
15°N - 15°S	101,47	6,11
15°S - 40°S	79,3	5,36
40°S - 70°S	73,04	8,47
80°N - 70°S	347,03	26,8

Tab. 1: Global sea-to-air flux rates of DMS

Measurements of nss-sulfate and methane sulfonate

During the Atlantic and the Antarctic cruise in 1990 additionally to the distribution of DMS in seawater and the atmosphere, the concentrations of the DMS reaction products aerosol nss sulfate and methane sulfonate were examined. In Tab. 2 the mean methane sulfonate, nss sulfate and atmospheric DMS concentrations in the different measuring areas are listed. The methane sulfonate concentrations over the Atlantic Ocean between 50°N and 50°S varied between 3 ppt and 13 ppt in the Southern Atlantic at very high DMS mixing ratios. Much higher methane sulfonate concentrations up to about 60 pptv were measured in higher southern latitudes. One exception is the low concentration average observed over the western part of the Weddell Sea at very low DMS mixing ratios. No significant difference between methane sulfonate levels in polluted and unpolluted air was found. In contrast, maximum nss sulfate levels of 270 ppt were measured in air masses highly influenced by continental inputs in the vicinity of Europe. Minima of nss sulfate occured over the Weddell Sea.

The ratio of nss sulfate to methane sulfonate concentrations was highly variable. In continental air masses, where a substantial portion of nss sulfate originated from continental, in particular from anthropogenic sources, the concentration ratio reached values of 60 to 300. In remote marine air masses over the temperate and tropical Atlantic the ratio averaged to 36. In Antarctic and Subantarctic air the ratio dropped to 4.9 on the average.

Region	MSA	$nssSO_4^=$ ppt	DMS
Atlantic			
50°N - 30°N	3,0	270	31
30°N - 20°S	3,2	150	43
20°S - 50°S	13,0	240	430
Drake Passage			
53°S - 62°S	19,2	80	119
Weddell Sea 60 - 71°S			
57 - 22°W	8,3	44	14
22 - 8°W	46,0	123	42
Atka B. - Cape Town			
70°S - 53°S	59,4	135	527

Tab. 2.: Mixing ratios of methane sulfonate, nss sulfate and dimethyl sulfide in the marine boundary layer between 50°N and 71°S

ACKNOWLEDGEMENTS

This work has been supported by the Deutsche Forschungsgemeinschaft and the Bundesministerium für Forschung und Technologie. We like to thank the crew of R/V Polarstern for their cooperation.

REFERENCES

(1) ANDREAE, M.O. (1985).The Emission of Sulfur to the Remote Atmosphere; in 'The biogeochemical cycling of sulfur and nitrogen in the remote atmosphere', Hrsg. Galloway, J.N. et al., D. Reidel. Publ. Comp., 5 - 25

(2) BARNARD, W.R.; ANDREAE, M.O.; WATKINS, W.E.; BINGEMER, H.; GEORGII, H.-W. (1982). The flux of dimethylsulfide from the oceans to the atmosphere; J. Geophys. Res., 87, 8787 - 8793

(3) BINGEMER, H. (1984). Dimethylsulfid in Ozean und mariner Atmosphäre - Experimentelle Untersuchung einer natürlichen Schwefelquelle für die Atmosphäre; PhD Thesis, Inst. f. Meteorology and Geophysics, Univ. of Frankfurt

(4) ANDREAE, M.O.; RAEMDONCK, H. (1983).Dimethyl sulfide in the surface ocean and the marine atmosphere: a global view; Science, 221, 744 - 747

(5) BATES, T.S.; CLINE, J.D.; GAMMON, R.H.; KELLY-HANSEN, S.R. (1987). Regional and Seasonal Variations in the Flux of Oceanic Dimethylsulfide to the Atmosphere; J. Geophys. Res., Vol. 92, No.C3, 2930 - 2938

(6) CHARLSON, R.J.; LOVELOCK, J.E.; ANDREAE, M.O.; WARREN, S.G. (1987). Oceanic phytoplankton, atmospheric sulfur, cloud albedo and climate; Nature 326, 655 - 661

(7) SCHWARTZ, S.E. (1988). Are global cloud albedo and climate controlled by marine phytoplankton?; Nature, Vol. 336, 441 - 445

(8) BERRESHEIM, H. ; ANDREAE, M.O.; AYERS, G.P.; GILLETT, R.W.; MERRILL, J.T.; DAVIS, V.J.; CHAMEIDES, W.L. (1990). Airborne Measurements of Dimethylsulfide, Sulfur Dioxide, and Aerosol Ions over the Southern Ocean South of Australia; J. Atm. Chem., Vol. 10, 341 - 370

(9) STAUBES,R. , GEORGII, H.-W., OCKELMANN, G. (1989). Flux of COS, DMS and CS_2 from various soils in Germany. Tellus 41B, 305 - 313.

(10) BÜRGERMEISTER, S., GEORGII, H.-W. (1991). Distribution of methanesulfonate, nss sulfonate and dimethylsulfide over the Atlantic and the North Sea. Atm. Env. Vol25A, 587 - 595

(11) LISS, P.S.; MERLIVAT, L. (1986). Air-Sea Gas Exchange Rates: Introduction and Synthesis; in 'The Role of Air-Sea Exchange in Geochemical Cycling', Hrsg. P. Buat-Menard, D. Reidel Publ. Comp., Dordrecht, 113 - 127

(12) BERRESHEIM, H. (1987). Biogenic Sulfur Emissions from the Subantarctic and Antarctic Oceans, J. Geophys. Res., 92, 13245 - 13262

(13) SMETHIE, W.M.; TAKAHASHI, T.; CHIPMAN, D.W. (1985). Gas Exchange and CO_2 Flux in the Tropical Atlantic Ocean Determined from ^{222}Rn and pCO_2 Measurements; J. Geophys. Res., Vol. 90, No. C4, 7005 - 7022

(14) ANDREAE, M.O. (1986). The ocean as a source of atmospheric sulfur compounds; in 'The Role of Air-Sea Exchange in Geochemical Cycling, Hrsg. P. Buat-Menard, D. Reidel Publ. Comp., Dordrecht, 331 - 362

BIOGENIC SULPHUR IN THE MARINE BOUNDARY LAYER OF THE ARCTIC- THE INTERNATIONAL ARCTIC OCEAN EXPEDITION, 1991

C. LECK, D. NILSSON, C. PERSSON
Department of Meteorology, Stockholm University, S-106 91 Stockholm, Sweden

K. BIGG
Australian Antarctic Society, Castle Hill, NSW 2154 Australia

D. C. COVERT
Department of Environmental Health, University of Washington, Seattle, WA 98195, USA

SUMMARY

A study of the natural sulphur cycle was conducted in the remote marine boundary layer of the Arctic during the International Arctic Ocean Expedition (IAOE). The expedition lasted from the end of July until the middle of October, 1991, and included open water and ice edge measurements as well as measurements far into the pack ice. Dimethyl sulphide (DMS) was quantified in the water and in the atmosphere. Particle size distribution, number concentration (CN) and total mass were determined. Chemical analysis was performed for integrated filter samples of non- seasalt (nss-SO_4^{2-}), sulphur dioxide (SO_2) and methane sulphonic acid (MSA). For interpretations of the data, the programme placed strong emphasis on meteorology. In general, the air sampled represented unperturbed marine boundary layer background conditions, which stresses the uniqueness of the expedition for the study of the natural sulphur cycle. The data showed a seasonal variation with highest values for all parameters in August during the biologically productive season and lowest values in October in the pack ice, when the polar night was approaching. This seasonality indicates a link between oceanic DMS, DMS in the air, sub µm aerosol nss-SO_4^{2-}, MSA and the CN population. The measurements are consistent with previous seasonal reports from the southern hemisphere.

1. INTRODUCTION

During recent years much concern has been expressed about the impact of human activity on the atmospheric sulphur cycle with both acidification of the environment and possible serious consequences of sulphate aerosol particles on the global climate, either directly by back scattering of solar radiation to space (2) or indirectly through their impact on cloud albedo (3). This climatic significance of sulphate aerosol particles, deriving from both anthropogenic and natural sources, has made it of considerable interest to study the atmospheric cycle. Langner et al. (7) have shown with model calculation that the entire northern hemisphere is highly conaminated by anthropogenic sulphate aerosol burden in the sources by a factor of 2 on the average with maxima up to a factor of 10 or

G. Restelli and G. Angeletti (eds.), Dimethylsulphide: Oceans, Atmosphere, and Climate, 103–115.

more in industrial regions. On the other hand, the southern hemisphere is hardly affected. It is therefore only plausible to study the natural sulphur cycle in the southern hemisphere or during the Arctic summer.

The reasons for this are that the influence from anthropogenic sources is extremely small (the transport from the polluted regions is minimal and the scavening of the pollutants on the way is very efficient) and that the natural sources, at least up to the ice edge, were as expected substantial (13). Furthermore, the climate system, including the ice edge and the degree of cloudiness, is here in a delicate balance, which potentially can be influenced by atmospheric processes involving sulphate aerosol particles.

The International Arctic Ocean Expedition (IAOE) from the end of July to the middle of October, 1991 allowed a concerted study of the production of soluble aerosol particles in the atmosphere through chemical reactions of containing gases, occurring either in the gas phase or subsequent to absorption of the gas on existing particles or cloud drops. These gases are released by marine phytoplankton in the sea surface mainly as dimethyl sulphide, DMS. The produced aerosols will become active in cloud forming processes and affect cloud reflectivity.

Comparisons between parameters (DMS in water/air, sulphur dioxide (SO_2), particulate phase methane sulphonic acid (MSA), non-seasalt sulphate (nss-SO_4^{2-}), particle number concentration, formation and size distribution) of direct relevance to these objectives will be briefly discussed.

2. METHODS

Measurements were performed aboard the Swedish icebreaker *Oden* during about 13 separate 24 hour sampling periods, and at times in between when it was possible to combine atmospheric sampling with other programmes aboard. The cruise track extended from the polar circle up to the North Pole and back again within the Atlantic sector. This enabled sampling both in open waters, along the ice edge and in the pack ice, cf. Figure 1.

A detailed meteorological programme was carried out including frequent radiosonde ascents recording the vertical distribution of relative humidity, air temperature, pressure, wind direction and wind speed (Vaisala RS80 and Digicora MW 11 Rawindsonde). From these data estimates of both the dynamic and static stability within the marine boundary layer (MBL) were possible. In addition, a complete set of meteorological surface observations, including the above mentioned, were recorded aboard the icebreaker. Five days backward, 3D, isentropic trajectories were calculated, with the position of Oden as the arrival point (11), from analysed pressure and wind fields by the European Centre for Medium-range Weather Forecasts, UK.

A size-segregating inlet for simultaneous sampling of aerosol particles (aerosol physics and chemistry) and various trace gases (DMS, Radon and Ozone) in the air was used: The air was drawn in via a three mast sampling system located at an angle of 45° to the roof of the gas and aerosol lab container. The total height above the sea surface was about 25 meters. This set up enabled sampling of air uncontaminated by the ship borne emissions provided that the wind was within the sector ± 70°of the bow. Direct contamination from the ship was excluded by a condensation nuclei control with a wind sector controller.

For sampling of aerosol particles and determination of size distribution, two out of three masts were equipped with cyclones (see below) removing particles larger than ~1 μm while the third excluded particles of diameter ~10 μm. One of the former held different filter packs for sampling of total particle mass (for further chemical analysis) and absorption of sulphur dioxide as described below, and the other had different instruments, two differential mobility particle sizers (DMPS, TSI 3071), scanning condensation particle counter (CPC, TSI 3021, 3025, 3760) and diffusion battery (DB, TSI 3040). They were used to measure continuously throughout the voyage, taking a full spectrum every 10 minutes, particle size distribution and concentration in the particle size range between 3 nm and 500 nm diameter.

Simultaneous measurements of SO_2, particulate phase MSA and nss- SO_4^{2-} were performed with a stacked sampling system (filter packs) consisting of cyclone separator (50% collection efficiency at 0.9 μm in diameter for NaCl particles at 85% RH), followed by a 47 mm Millipore Teflon aerosol particle filter (1.0 μm pore size) only/or in connection with four Whatman 41 filters coated with 0.1 M K_2CO_3 glycerol solution (12). Collection times ranged from 4 to 24 hours depending on real time concentration levels. Each sampling set consisted of two packs and a blank (average sample flow rate/cyclone = 45 slmp and blank flow rate = 0 slmp). All filters were changed in an SO_2 and particle free glove box to minimize contamination. Inside the glove box the Teflon filters were extracted with a 10 % methanol solution with a total volume of 5 ml and the Whatman filters were extracted with 5 ml of 0.1-0.01 % hydrogen peroxide solution to convert all the SO_2 to sulphate. The filter solutions were sonicated for 60 minutes and centrifuged for 10 minutes. The extracts were then analysed by Ion chromatography. Strong anions as SO_4^{2-}, NO_3^- and Cl^- were analysed with a Dionex AG4A/AS4A column. MSA^- and Na^+, NH^{4+} ions were analysed with a Dionex AG4/AS4 and a CG3/CS3 column, respectively. During the cruise levels (sample minus blank) of SO_2, particulate phase MSA and nss- SO_4^{2-} down to 0.05, 0.002 and 0.03 nmol/m^3, respectively were detected. Duplicate samples varied by less than 15 %. A more detailed description of the analytical method will be published elsewhere.

Atmospheric DMS samples were taken from the ~1 μm inlet in the three mast sampling system with a flow of 1 L/min. The samples were cryogenically preconcentrated for 20 to 60 minutes with LAr, depending on ambient air concentrations. For scavenging of oxidants a Naifion dryer was combined with a cotton wadding oxidant scrubber. The analyses were performed using gas chromatography with flame photometric detection (FPD). The sampling frequency was at least hourly. The minimum detection limit for the method was 1 ppt (v) for a 60 minutes sample. The precision and accuracy of the analysis was within ± 10 %.

Water samples for DMS analyses were collected with a water sampler (Ruttner) from a depth of between 4 and 10 meters either from the shipside, from a helicopter or with a small motorboat (for better horizontal resolution). For analyses, DMS was extracted from the water and preconcentrated, then measured using gas chromatography with flame photometric detection (FPD). The precision of the method for the environment sampled was better than ± 5%. The system had a detection limit of 0.006 nmol/L (10).

3. RESULTS AND CONCLUSIONS

During the first half of the expedition (August throughout the middle of September) we experienced weather conditions typical for the Arctic ocean summer, with advection of air masses from the Atlantic sector dominating over shorter periods of air being either advected from the continents at lower latitudes or subsiding from the free troposphere. In general this gave us weather conditions of high relative humidity (90-100%) and air temperatures in the vicinity of 0°C, as the relatively warmer maritime air was cooled when advected over the pack ice. The MBL was typically 200 to 300 m in depth. A dense and low cloud cover of stratus in combination with frequent fog episodes was present most of the time. Frontal activity was as expected low.

In the beginning of the autumn (middle of September) advection of air mass originating from the continents surrounding the Arctic ocean (north of the polar circle) became more common. Thus the air sampled was generally cooler and dryer with less cloud cover but episodes with patchy fog occurred due to low temperatures, ≤ 0°C to about - 25°C. During a few days at the end of September, we sampled warmer maritime air advected from the Atlantic Ocean giving an increase of the air temperature from -11°C to about 0°C. Radio sondes launched from the icebreaker showed a high frequency of pronounced inversions (30-200m in depth, 3-10 K/100m) caused by radiation, in the absence of low clouds, from the surface. In addition, the inversions were strengthened by periods of air subsiding from the free troposphere or/and Greenland. The ice cover was on the average 90 % during the summer period consisting of a numerous number of leads even at the North Pole. At the time when the water temperature decreased to below freezing, all open leads became covered by newly formed thin ice in strong contrast to the 2 to 4 meter thick perennial ice. The expedition also took place at the time of the year when the far north experiences 24 hour sunlight.

Figure 2 shows calculated trajectories for the major fraction (80%) of the aerosol filters sampled. They represent periods of unperturbed MBL background conditions of very "clean" periods when the air had spent a long time over the ice and open waters of the arctic region. This again stresses the uniqueness of the IAOE 1991 expedition for the study of the natural atmospheric sulphur cycle.

The existence of a seasonal cycle in sea surface DMS has been reported in the literature (8). This cycle is pronounced in temperate latitudes with peak values occurring during the summer period. Average summer time DMS concentrations in the Baltic Sea and the North Sea ranged from 3 to 8 nmol/L (9). No earlier measurements of DMS in the sea surface/ ice system have been performed in the Arctic. It was therefore of considerable interest to characterize the seasonal variation of DMS in the Arctic basin. During the expedition north of 80°N, we measured concentrations of between 0.05 and 11 nmol/L sulphur, with peak values occurring in open waters along the ice edge in the beginning of August, cf. Figure 3. The span of these measured sea surface levels is in good

agreement with reported numbers found elsewhere in the world oceans. In the period following Julian Day 247, when being close to the North Pole, the DMS levels fell below 0.1 nmol/L which is consistent with concentrations typical for the winter season in the Baltic Sea (8).
Atmospheric dimethyl sulphide concentrations ranged from 1 ppt (v) to 400 ppt (v) (0.04-16 nmol/m^3). The measured concentrations are consistent with earlier measurements over the Arctic Ocean near Barrow, Alaska (4). The highest recorded concentrations of atmospheric DMS were, as for seawater DMS, found when sampling in open waters along the ice edge. Further, low levels of DMS are found typical for air that spent at least four days over the pack ice. These low levels of DMS probably resulted from the combined effect of long distance transport scavenging in conjunction with the declined source strength of DMS as the marine biological activity tended towards zero in the middle of September.

There are basically three mechanisms for converting the atmospheric DMS derived sulphur gases (SO_2, H_2SO_4 (l) and MSA (g)) to marine particulate matter consisting of methane sulphonate (MSA^-) and non-seasalt sulphate (nss- SO_4^{2-}) :

1) nucleate spontaneously to form new ultrafine particles (*homogenous nucleation*)
2) condense onto larger existing particles followed by growth of the particle population (*heterogeneous nucleation*)
3) absorption into cloud droplets where the gases are converted to involatile compounds in the aqueous phase.

In Figure 5 we show an example of the number size distribution of the background aerosol representing the history (Δt~ hours to several days) of the aerosol, cycled through non-precipitating clouds, condensed on to preexisting particles and by local burst of new ultra fine (<15nm) particles. This tri modal (ultrafine, nuclei and accumulation mode) in situ recorded size distribution is, to our knowledge, the first time size resolved aerosol measurement below 15 nm diameter that has been reported.

Particle number concentrations (> 20nm diameter) ranged from 1 to about 1000 per cubic centimetre, cf. Figure 6. After entering the pack ice regions, aerosol concentrations were generally low compared to continental or oceanic values. Concentrations lower than any previously reported reliable measurements (6), except perhaps for some made at the South Pole, were recorded for extended periods on a few occasions. The median concentration of particles was 45 per cubic centimetre for the month of September.

On a longer time resolution, SO_2 (Δt~24h), particulate MSA and nss-SO_4^{2-} (Δt~4-12h) observations obtained along the voyage showed a similar seasonal progression with a steep fall at the end of August as the DMS concentrations in both water and air, and particle number concentration (>20nm). Concentrations of SO_2, particulate MSA and nss-SO_4^{2-} ranged from (0.7 -0.05) nmol/m^3, (1.5 -0.002) nmol m^3 and (7-0.03) nmol m^3, respectively.

One conspicuous result of this study, is that the measured SO_2 levels are substantially lower than previously reported remote marine values in the northern hemisphere (12). On the other hand, the typical strong seasonal variation for all the parameters discussed above, was in very good agreement with previous findings reported from 12 years of data from the Cape Grim baseline station on the southern hemisphere (1). This would tend to indicate that other northern hemisphere measurements are contaminated by anthropogenic emissions consistent with model results (7).

The Arctic is a region in which the atmospheric sulphur cycle is generally dominated by anthropogenic inputs from lower latitude. A great deal of the effort devoted to arctic aerosol studies has focused on the arctic haze spring phenomena (5) but much less work has been done on the summer aerosol. Only during the summer months, when the Arctic MBL to a large extent is clean from long distance transported anthropogenic sulphur, do marine biogenic sources contribute a predominant fraction of the atmospheric sulphate.

As an overall picture, all parameters showed a similar temporal progression with the highest levels found in open waters close to the ice edge (in the beginning of August) followed by an exponential decline with very low background concentrations recorded during late September to the beginning of October. This seasonal variation is consistent with biological activity.

We would like to stress that the IAOE to the summer Arctic, which took place from the end of July to the middle of October 1991, did not only provide a ship of opportunity but gave us a unique platform allowing an interdisciplinary study of the natural background atmospheric sulphur cycle, extending from the biologically active zone under the pack ice to the top of the marine boundary layer, as a part of the investigation of a possible biological control of climate.

ACKNOWLEDGEMENT

We thank the officers and crew of the icebreaker Oden (chartered from the Swedish National Maritime Administration by the Swedish Polar Research Secretariat) for their assistance and co-operation making the expedition so successful. This research was founded by the Swedish Natural Science Research Council (contract E-EG/GU 9906-30) and The Wallenberg Foundation (contract E-EG 3922-313.

REFERENCES

(1) AYERS, G.P., GILLET, R. W., IVEY, J. P., GRAS, J. L. Division of Atmospheric Research, CSIRO, Private Bag 1, Mordialloc, Australia. (1993). *Personal communication; This book.*

(2) CHARLSON, R. J., LANGNER, J., RODHE, H., LEOVY, C. B. and WARREN, S. G. (1991). Perturbation of the northern hemisphere radiation balance by back scattering from anthropogenic sulphate aerosols. *Tellus* **43 AB**, 152-163.

(3) CHARLSON, R.J., LOVELOCK, J.E., ANDREAE, M.O.and WARREN, S. G. (1987). Oceanic phytoplankton, atmospheric sulphur, cloud albedo and climate. *Nature,* **326,** 655-661.

(4) FEREK, R. J., HOBBS., P. V., RADKE ,L. F.(1993). Dimethyl sulfide in the arctic atmosphere. Submitted to *J. Atmos. Chem.*

(5) HEINTZENBERG, J. and COVERT, D. S. (1987). Chemically resolved submicrometric size distribution and external mixing of the Arctic haze aerosols.*Tellus* **39B**, 374-382.

(6) HOPPEL, W. A., FITZGERALD., J. W., FRICK, G. M., LARSSON., R. E., MACK. E. J. (1990). Aerosol size distributions and optical properties found in the marine boundary layer over the Atlantic Ocean, *J. Geophys. Res.* **95**, 3659-3686.

(7) LANGNER, J., RODHE, H., CRUTZEN, P. J.and ZIMMERMANN, P. (1992). Anthropogenic influence on the distribution of tropospheric sulphate aerosol. *Nature,* **359**. 712-715.

(8) LECK, C., LARSSON, U., BÅGANDER, L.E., JOHANSSON S. and HAJDU, S. (1990). DMS in the Baltic Sea - Annual variability in relation to biological activity. *J. Geophys. Res.* **95**, No C3, 3353-3363.

(9) LECK C. and RODHE, H. (1990). Emissions of marine biogenic sulphur to the atmosphere of northern Europe. *J. Atmos. Chem.* **12**, 63-86.

(10) LECK, C. and BÅGANDER, L.E. (1988). Determination of reduced sulfur compounds in aqueous solutions using gas chromatography flame photometric detection. *Anal. Chem..* **60**, 1680-1683.

(11) McGRATH, R. (1989). Trajectory models and their use in the Irish Meteorological Service. Memorandum No. 112/89, Irish Meteorological Service, Dublin.

(12) QUINN, P. K., BATES, T. S., JOHNSON, J. E., COVERT, D. S.and CHARLSON, R. J. (1992). Interactions between the sulfur and the reduced nitrogen cycles over the Central Pacific Ocean. *J. Geophys. Res.* **95**, 16,405-16,416.

(13) SAKSHAUG, E. (ed). (1992). Økosystem Barentshavet. *Norsk Allmennvitenskapelige Forskningsråd,* 304pp.

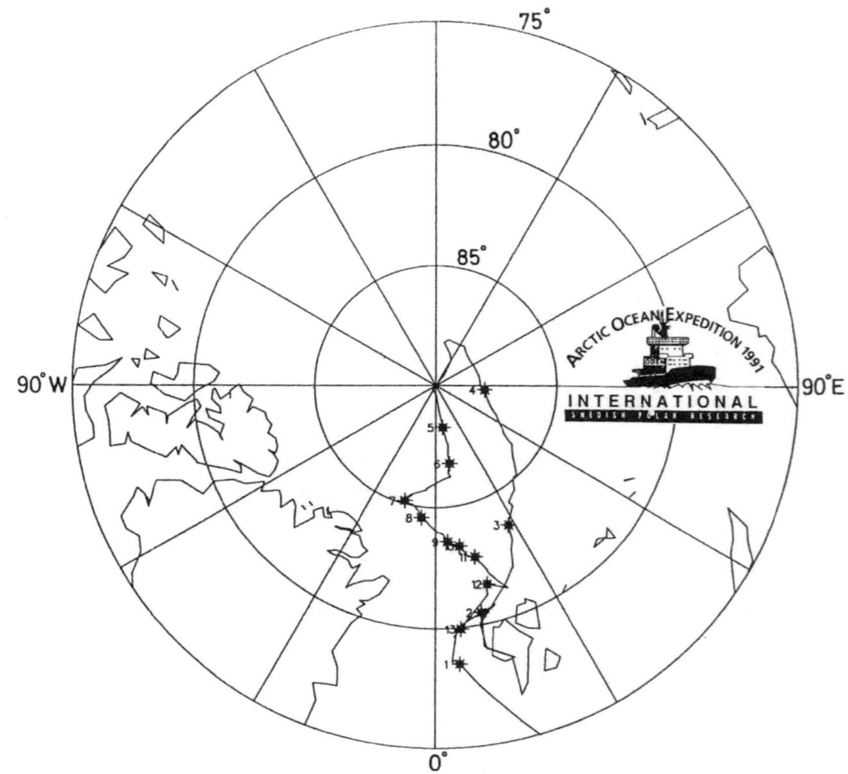

Figure 1. Map over the study area, showing the cruise track during the International Arctic Ocean Expedition, (IAOE) 1991. Stations with 24 hours continuous sampling are marked.

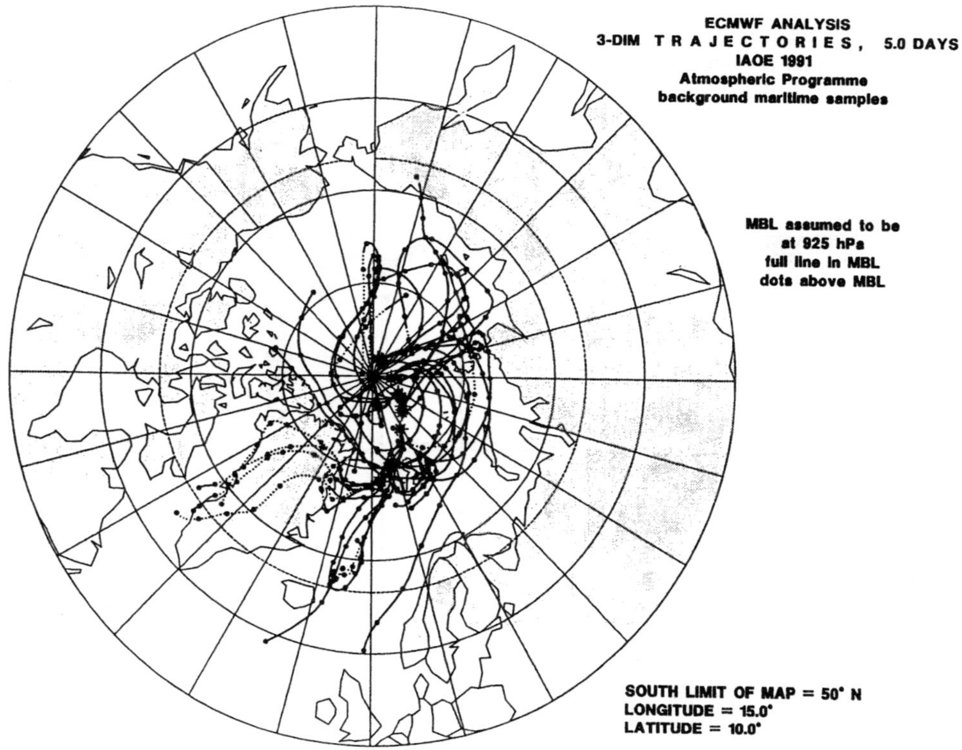

Figure 2. Calculated isentropic 5-day backward trajectories with the position of the icebreaker Oden as the arrival point.

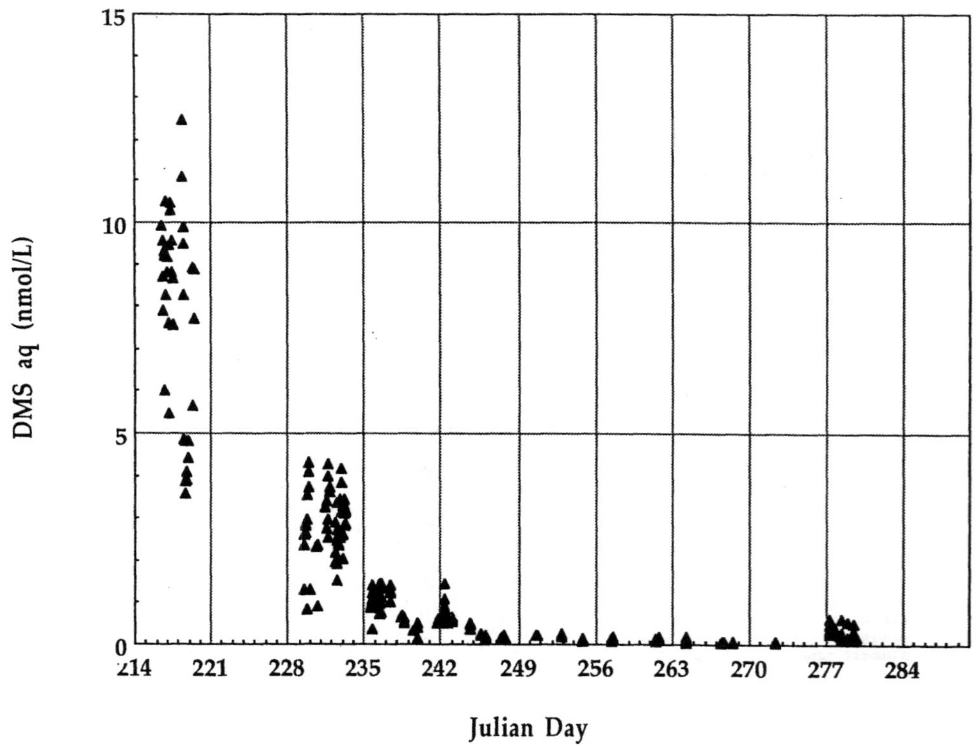

Figure 3. Surface sea water DMS concentrations in nmol/L during August to October along the International Arctic Ocean Expedition, 1991.

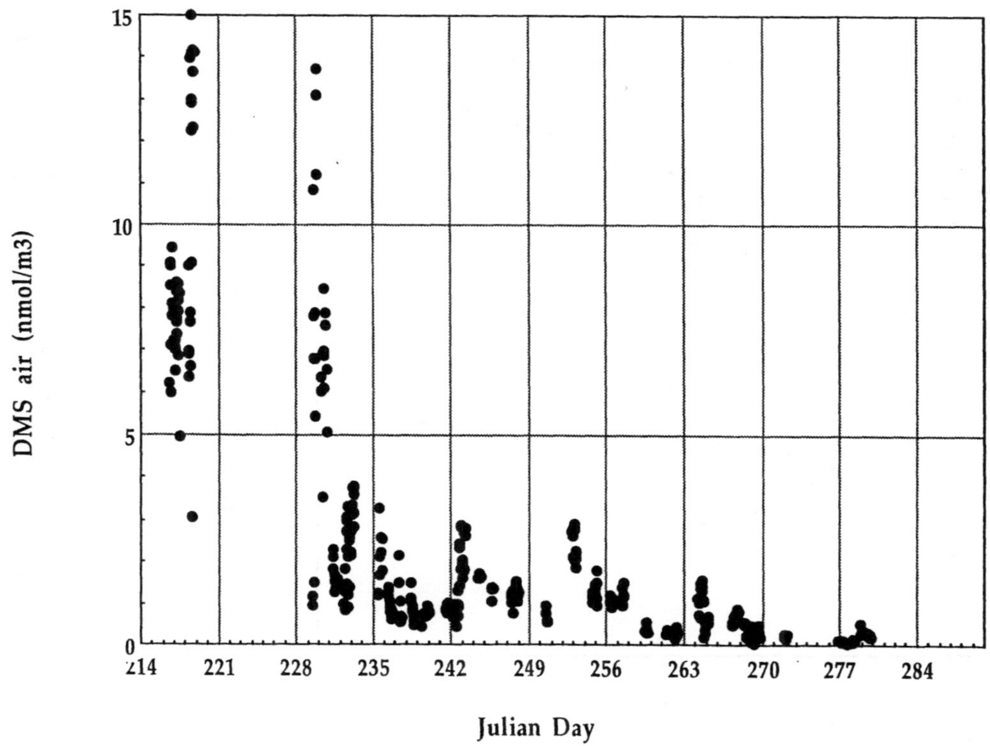

Figure 4. Atmospheric DMS concentrations in nmol/m^3 during August to October along the International Arctic Ocean Expedition, 1991.

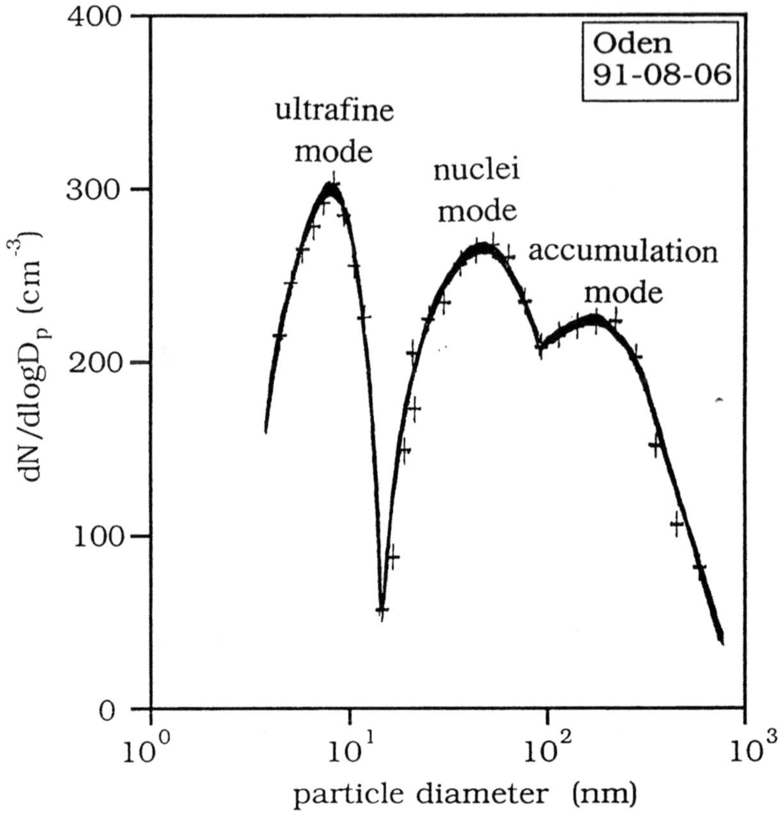

Figure 5. In situ particle number size distribution averaged over 10 minutes during the IAOE on the 8th of August 1991. The ultrafine, nuclei and accumulation mode refers to mechanisms 1, 2 and 3 respectively described in the text.

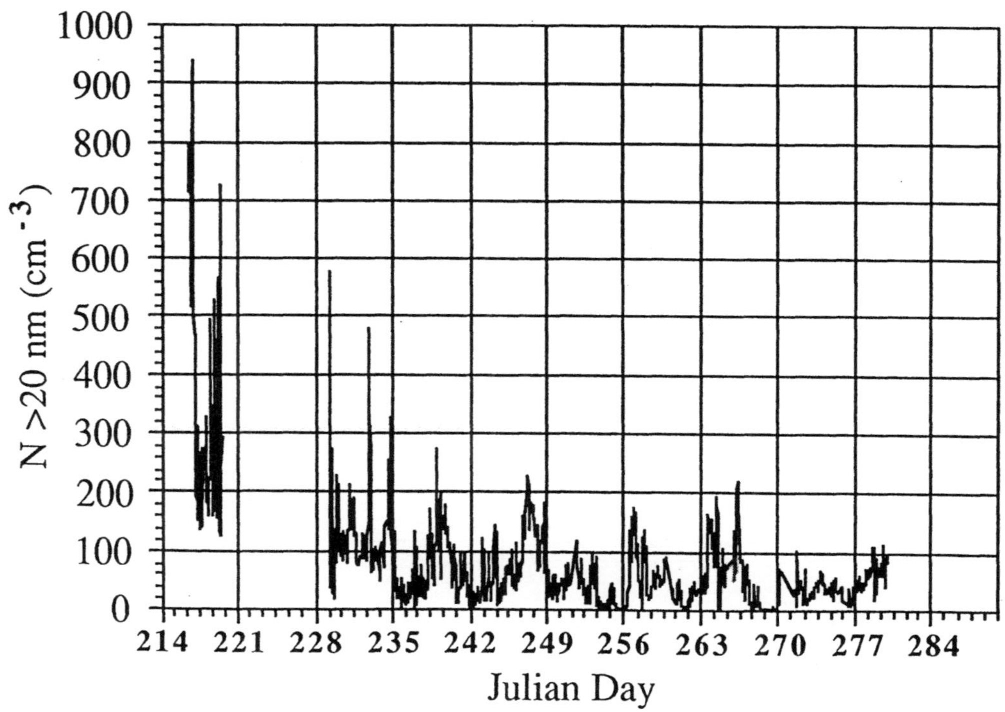

Figure 6. Particle number concentrations (>20 nm diameter) particles per cubic centimetre during August to October along the International Arctic Ocean Expedition, 1991.

MEASUREMENT OF DIMETHYL SULFIDE, SULFUR DIOXIDE, METHANE SULFONIC ACID AND NON SEA SALT SULFATE AT THE CAPE GRIM BASELINE STATION

R.W. GILLETT, G.P. AYERS, J.P. IVEY* AND J.L. GRAS

CSIRO, Division of Atmospheric Research, Private Bag No1, Aspendale, Australia, 3195.
*Australian Government Analytical Laboratory, Channel Highway, Kingston, Australia, 7050.

Summary
Results of measurements of dimethyl sulfide (DMS), methane sulfonate (MSA), nss-sulfate ($nss\text{-}SO_4$) and sulfur dioxide concentrations as well as the number concentration of cloud condensation nuclei at the Cape Grim Baseline Atmospheric Pollution Station are presented. All species, both gas and aerosol, showed clear annual cycles in concentration all with maximum concentrations in the austral summer and minima in the austral winter. The ratios of methane sulfonate : nss-sulfate also give a seasonal cycle which is very similar to that seen for the gas and aerosol species. The phase of this cycle is the opposite to that expected on the basis of the temperature dependence of the reaction rates of the competing addition / abstraction reactions. The seasonal cycle may instead may be due to the presence of another sulfur gas which also produces nss-sulfate but has no seasonal cycle, reducing the MSA : nss-sulfate during that period. Carbon disulfide is one such possible precursor of nss-sulfate.

1. Introduction

Charlson et al., (1987) proposed an hypothesis linking the emission of dimethyl sulfide (DMS) from oceanic phytoplankton with the formation of atmospheric sulfate particles which could become active as cloud condensation nuclei. Since the albedo of clouds in unpolluted regions may depend strongly on cloud condensation nucleus concentrations the proposed link constitutes an indirect effect of sulfur gas and aerosol on climate.

In addition Charlson et al., (1991) have presented modelling results which predict the direct effect of sulfate aerosol on the radiation balance of the earth. They calculate that the direct effect of sulfate forcing on the radiation balance could be about -1.1 watts m^{-2}, which is almost equivalent, but opposite in sign, to the forcing caused by the current anthropogenic perturbation to the major greenhouse gas, carbon dioxide.

The work by Charlson et al., (1987) and Charlson et al., (1991) has given increased impetus to a number of groups carrying out experimental programmes on the main sulfur gases and aerosol compounds which are central to the whole climate feedback hypothesis (Ayers et al., 1991; Berresheim et al., 1988; Ayers et al., 1986; Pszenny et al., 1989; Bandy et al., 1992; Putaud et al., 1992; Nguyen et al., 1992; Quinn et al., 1992). This paper summarises work that has been carried out as part of the experimental programme of the Baseline Atmospheric Pollution Station at Cape Grim, Australia.

2. Experimental

Figure 1 shows the location of the Cape Grim Baseline Station (40·7° S, 144.7^o E), where the various the concentrations of sulfur gases and aerosol species have been measured.

Measurements routinely carried out at the present time at Cape Grim include:

1. DMS concentrations, both daily spot values (1 hour average) and weekly average samples.
2. methane sulfonate (MSA) and non sea-salt sulfate ($nss\text{-}SO_4$) in high volume aerosol samples.
3. MSA and $nss\text{-}SO_4$ separated into size ranges above and below 1μm in radius.

G. Restelli and G. Angeletti (eds.), Dimethylsulphide: Oceans, Atmosphere, and Climate, 117–128.

Figure 1. Location of sampling sites.

4. Rainwater MSA.
5. SO_2 gas concentrations.
6. Condensation nuclei (CN) and cloud condensation nuclei (CCN), aerosol size distributions $0.0015< r<1.5$ µm.

As the experimental techniques used for the collection and analysis of the various aerosol and gas species have been published elsewhere they will only be described briefly here. The high volume MSA and nss-SO_4 measurements are obtained from ion chromatographic analysis of the aerosol collected on 10" x 8" a Pallflex teflon composite filter. The sample is collected at 700 - 800 $Lmin^{-1}$ on a weekly average basis, with the sampler activated only when the local wind direction is in the sector 190° - 280° and the CN concentration is less than 600 cm^{-3}. This is referred to as "baseline conditions".

The DMS concentrations are measured by trapping at 1 $Lmin^{-1}$ onto a gold surface which is sputter coated on glass wool. Samples have been collected since 1988 on an hourly averaged basis and more recently on a weekly averaged basis. In all cases samples are only collected in baseline conditions. Determination of the collected DMS is achieved by capillary gas chromatographic analysis with a flame photometric detector, after thermal desorption of the DMS from the gold wool trap.

Baseline concentrations of MSA and nss-SO_4 in the sub micrometre and super micrometre size ranges are measured by a dual filter pack method. Air is drawn by a diaphragm pump at 30 $Lmin^{-1}$ through an 8µm nuclepore filter, which collects particles above 1 µm radius, followed by a 1µm teflon (Fluoropore) filter where the sub micrometre aerosol is collected. After exposure, filters are removed from the holders and stored in clean plastic bags in which they are later extracted with Milli-Q water. The MSA and nss-SO_4 concentrations are then determined using chemically suppressed ion chromatography on the aqueous aerosol extract.

Sulfur dioxide concentrations are measured using a filter pack technique similar to that used for the aerosol MSA and nss-SO_4 measurements. The first stage consists of an impactor to remove particles greater than 1µm in radius, the second stage consists of a 1µm teflon (Fluoropore) filter which removes the remaining aerosol, enabling sulfur dioxide to be collected on the third stage which consists of a glass fibre filter impregnated with 0.1% Na_2CO_3 and 10% glycerol. Air is drawn through the filter pack at a flow rate of about 26 $Lmin^{-1}$ with a diaphragm pump which samples only under baseline conditions. Since the sulfur dioxide concentration is very low, especially in the winter period, a minimum of 100 m^3 of air is required to effect an adequate signal to noise ratio in the chemical analysis. As a result, most of the samples are weekly averages although some result from two or even three week collection periods. Sulfur dioxide concentrations were measured as sulfate by ion chromatography after extraction of the glass fibre filter with Milli-Q water in clean plastic bags.

CCN number concentrations were measured in situ at Cape Grim using a static thermal diffusion cloud chamber (Bartlett and Ayers, 1981).

3. Results and Discussion

Figure 2 gives the concentrations of DMS measured in hourly average samples, collected at Cape Grim between January 1989 and July 1992. This shows that at Cape Grim a distinct seasonal cycle is observed in DMS concentrations. The cycle has a clear summer maximum and a winter minimum with concentrations ranging from less than 1 nmoles m^{-3} in the winter to about 10 nmoles m^{-3} in the summer. The increase in DMS concentrations appears to begin in about November; at the end of the Austral spring. Figure 2 shows also some changes in the absolute concentrations of DMS on an interannual basis. Concentrations of DMS during the summer of 1990, with a maximum of about 6 nmoles m^{-3}, were significantly lower than those of the preceding and subsequent years, possibly reflecting changes in primary productivity in the ocean. Figure 3 shows the data overplotted on a single annual timebase to emphasise the seasonal cycle in DMS concentration.

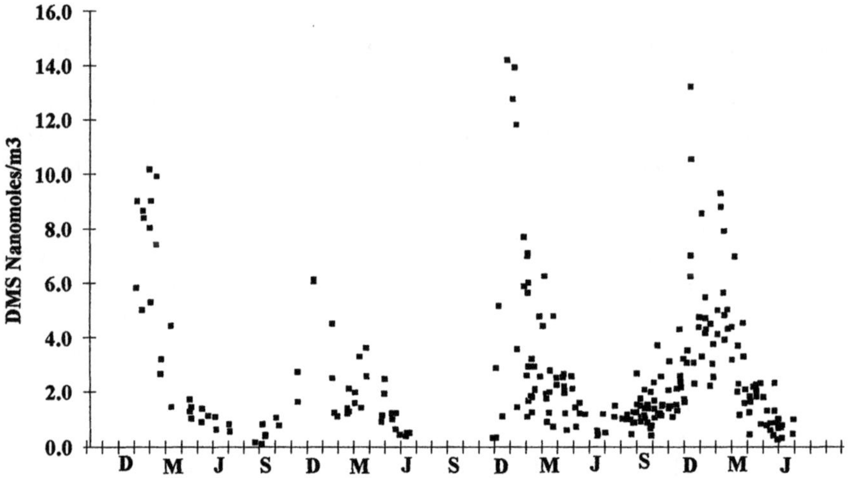

Figure 2. Hourly averaged DMS measured at the Cape Grim Baseline Station between January 1989 and July 1992.

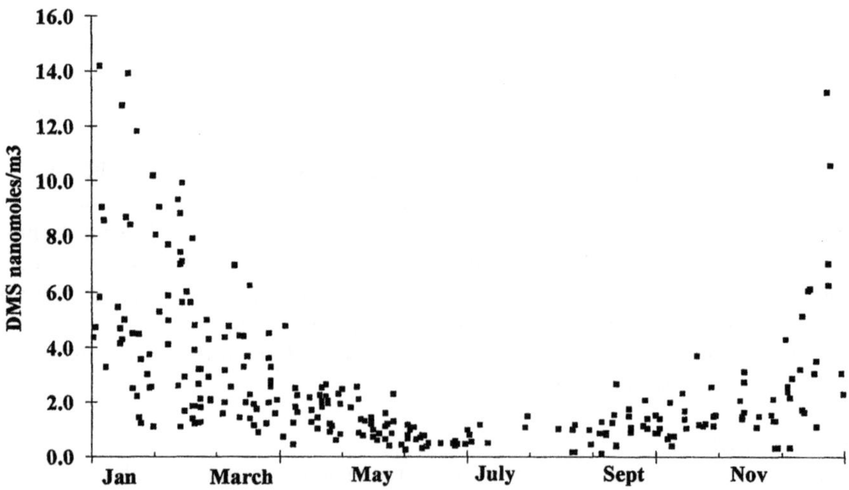

Figure 3. Hourly averaged DMS concentrations measured at the Cape Grim Baseline Station. Results for each year are overplotted to emphasise the seasonal cycle.

Since September 1991 DMS has also been measured by sampling over a one week integrated period. This allows a more detailed comparison to be made between MSA and nss-

SO_4 concentrations. Figure 4 shows the concentrations of the weekly integrated samples of DMS collected over one summer, between September 1991 and July 1992. A seasonal cycle is apparent and is similar to that observed in the hourly integrated samples. Concentrations range from less than 1 nmole m^{-3} in the austral winter to about 8 nmole m^{-3} in the austral summer. Summer concentrations in these samples are somewhat lower than the hourly integrated values but this expected given the longer averaging period.

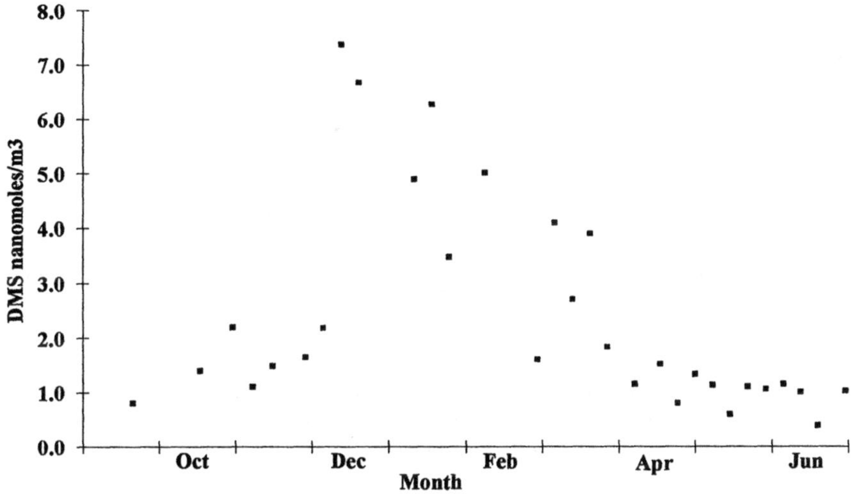

Figure 4. Weekly integrated DMS concentrations, Cape Grim Sept. 1991 - Jul. 1992.

Two major products of DMS oxidation by hydroxyl radical are MSA and nss-SO_4, both of these exist in the form of aerosol (Plane, 1989). Since a seasonal cycle in the DMS concentration has been established it would be reasonable to expect seasonal cycles in the concentrations of MSA and nss-SO_4. Figure 5 shows the concentrations of nss-SO_4 in the sub micrometre aerosol at Cape Grim. Nss-SO_4 has a similar seasonal cycle to that seen in the earlier figure for DMS concentrations with concentrations peaking in the summer to about 4 nmoles m^{-3} and decreasing in the winter period to less than 0.3 nmoles m^{-3}.

The concentrations of nss-SO_4 measured in the super micrometre size range did not demonstrate a seasonal cycle and are not included here. This is almost certainly due to the problem of separating the sea-salt sulfate from the nss-SO_4 in the samples. The sea-salt fraction was estimated by assuming all the sodium has a sea-salt origin, the sea-salt sulfate contribution was based on ion ratios provided by Millero (1974). The high sea-salt loadings at a site such as Cape Grim makes it very difficult to distinguish the nss-SO_4 component in the super micrometre size range.

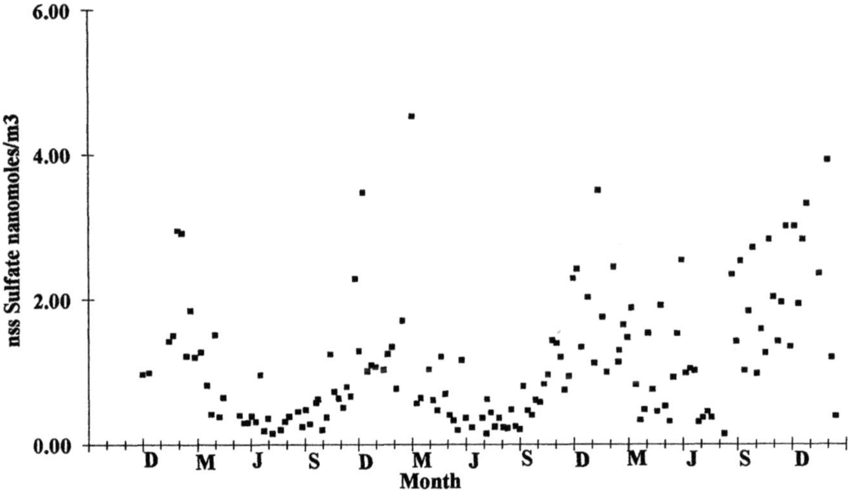

Figure 5. Concentration of nss-SO_4 measured in the sub micrometre aerosol, Cape Grim, Nov. 1988 - Feb. 1992.

Figures 8 and 9 give the concentrations of MSA measured on both the sub micrometre and super micrometre aerosol fractions. These figures show that there is a very clear seasonal cycle in the MSA concentrations in both the $<1\mu m$ and $>1\mu m$ size ranges. The seasonal cycle observed is similar both in amplitude and in period to the seasonal cycles observed for DMS and nss-SO_4.

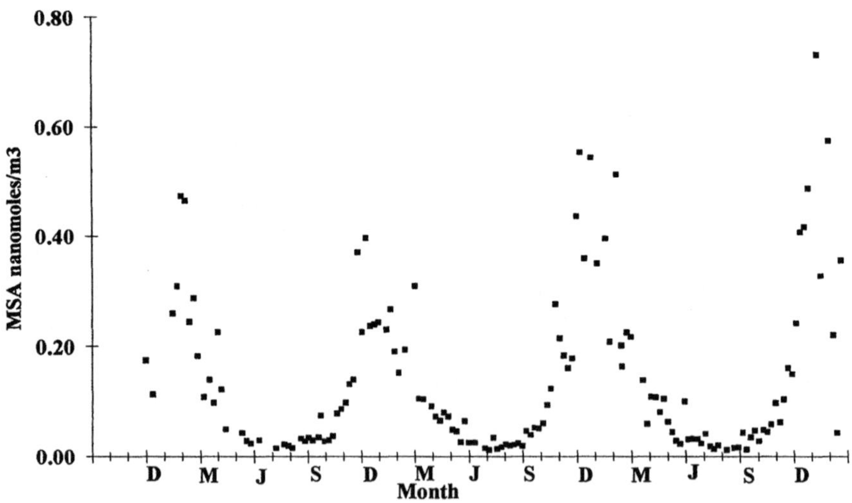

Figure 6. Concentrations of methane sulfonate measured in the sub micrometre size range at Cape Grim from November 1988 until February 1992.

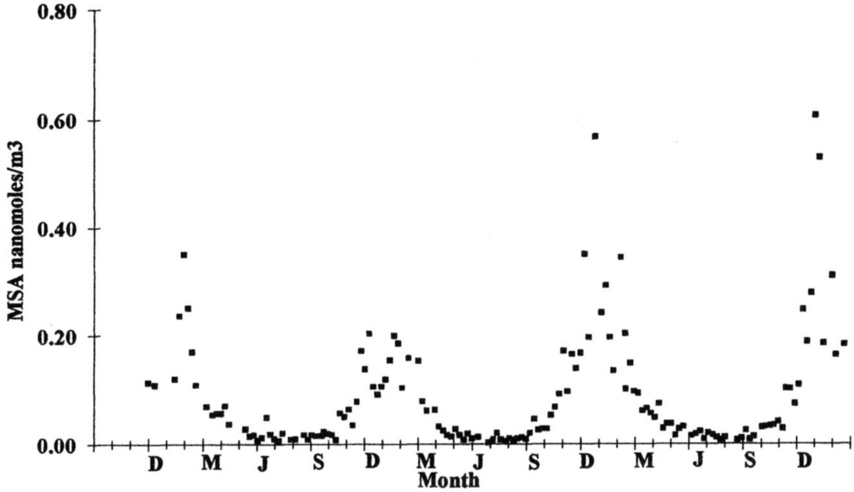

Figure 7. Concentrations of methane sulfonate in the super micrometre aerosol at Cape Grim from November 1988 until February 1992.

MSA has also been measured on Pallflex high volume filters since 1976. MSA concentrations from this record are plotted against the time of measurement, in figure 8. Even allowing for missing data, there is a distinct annual cycle with summer maxima and winter minima similar to those in the sub micrometre and super micrometre aerosol size fractions.

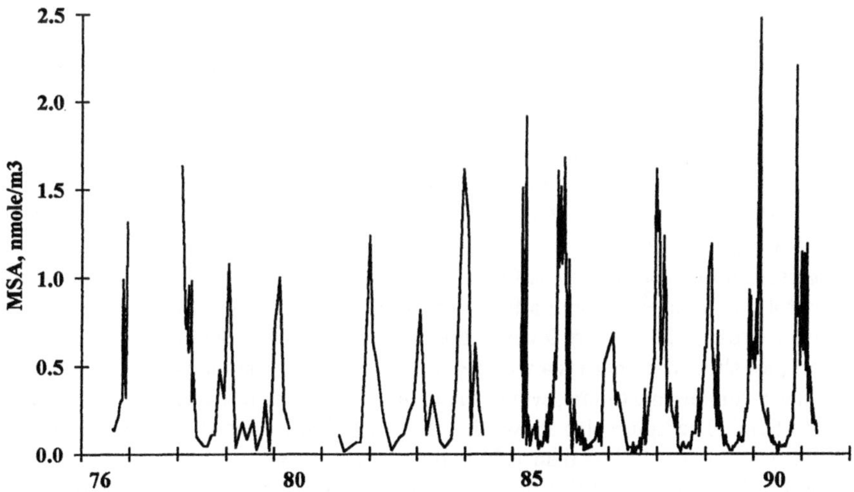

Figure 8. MSA concentrations measured on high volume samples at Cape Grim 1976 - 1991.

Sulfur dioxide concentrations have been measured at Cape Grim since October 1990. Concentrations are measured using the filter pack technique described previously. Since this technique has not been completely verified, in terms of the collection efficiency, detailed results are not presented here. Nevertheless the overall results look promising, they show an annual cycle, with a summer maximum and a winter minimum. Concentrations of sulfur dioxide range from about 0.6 nmoles m^{-3} in the summer to about 0.1 nmoles m^{-3} in winter. These sulfur dioxide concentrations measured at Cape Grim are comparable with the remote Southern Hemisphere results of Putaud et al., (1992) which also show a distinct seasonal cycle with a summer maximum and winter minimum.

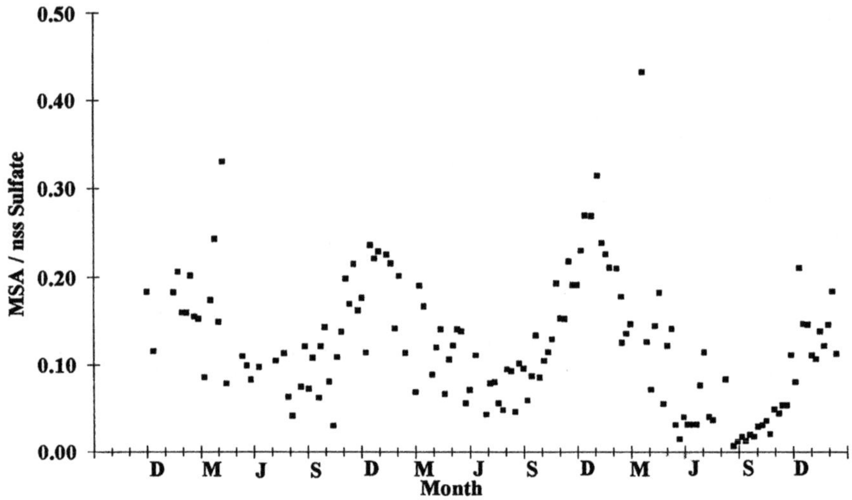

Figure 9. Ratio of methane sulfonate to nss-sulfate measured in the sub micrometre size range at Cape Grim from November 1988 until February 1992.

Figures 5, 6 and 7 show that there are seasonal cycles both in MSA and nss-SO_4 which are similar in phase to the seasonal cycle observed for DMS concentrations at Cape Grim. This provides some evidence for a link between the concentrations of DMS and the major reaction products, MSA and nss-SO_4.

A comparison of the measured nss-SO_4 concentrations with that of MSA shows that nss-SO_4, at about 4 nmoles m^{-3}, is about a factor of five times that of the MSA concentrations. This is shown more clearly in figure 9 where the ratio of MSA : nss-SO_4 is plotted against time of year. This plot show that there is a well defined seasonal cycle in the ratio of MSA : nss-SO_4 which has a summer maximum, winter minimum, not unlike the cycles observed for DMS, MSA and nss-SO_4. The figure also shows that the ratio of MSA : nss-SO_4 varies from about 25% in the austral summer to less than 10% in the austral winter.

MSA and nss-SO4, the two major products of the DMS oxidation by hydroxyl radical, are produced by one of two separate pathways. These are; hydrogen abstraction by hydroxyl radical from the DMS molecule to eventually form sulfuric acid and so nss-SO_4, and the other is the addition of the hydroxyl radical to the sulfur atom of the DMS molecule, eventually producing MSA.

A possible reason for the observed seasonal cycle in the MSA : nss-SO_4 ratio is the temperature dependence of the two competing reactions. Plane, (1987) finds that the rate constant for the addition reaction of hydroxyl to DMS, leading to MSA, increases with decreasing temperature. He finds that at 300 ^oK 25% of DMS is oxidised via the addition pathway, compared with 70% at 250 ^oK. This means that the ratio of MSA : nss-SO_4 should be about 0.25 at 300 ^oK and about 0.7 at 250 ^oK. With such a temperature dependence it might be reasonably expected that temperature dependent ratios would be observed in the measurements of MSA and nss-SO_4 at Cape Grim.

The data presented in figure 9 show that the annual cycle of MSA : nss-SO_4 appears to be completely out of phase with that expected for the temperature effects on the reaction rates of the addition abstraction reactions. It could be expected that since the addition reaction rate increases with decreasing temperature the production of MSA would be higher during the austral winter. This would result in a winter maximum, summer minimum in the ratio. The average temperatures during summer and winter at Cape Grim are only about 282 ^oK and 289 ^oK respectively and this may not be sufficient to show a significant temperature dependence on the addition, abstraction reaction rates. The other possible reason for the observed seasonal cycle may be another source of aerosol nss-SO_4, additional to DMS. If such a source of nss-SO_4 had no seasonal cycle it would produce relatively more nss-SO_4 in the austral winter and so have the effect of lowering the MSA : nss-SO4 ratio during that part of the year. Carbon disulfide (CS_2) has been observed at Cape Grim (J.P. Ivey, personal communication), albeit infrequently and this may be one possible explanation for the observed seasonal cycle.

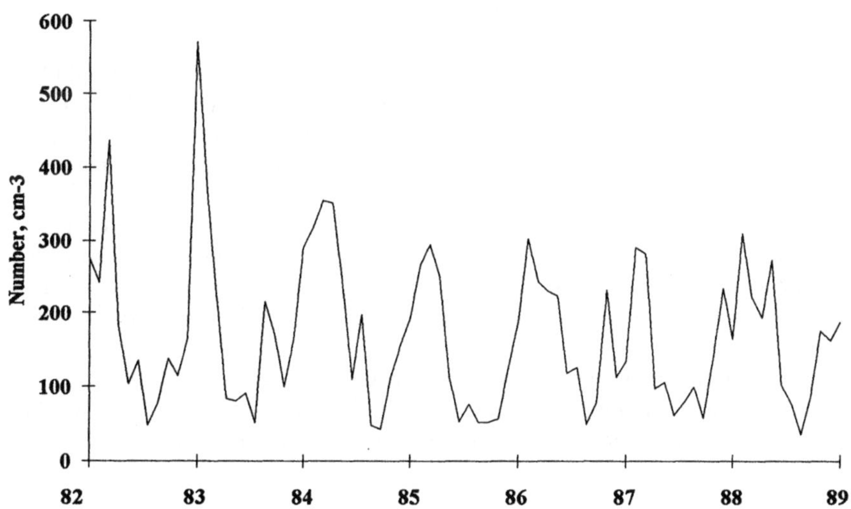

Figure 10. CCN number concentration for ccn active at 1.2% supersaturation, Cape Grim 1982 - 1989.

It is clear from our observations there are seasonal cycles the concentrations of all the major sulfur species implicated in the hypothesis of Charlson et al., (1987). The major end product of this oxidation scheme is nss-SO_4. At an unpolluted maritime site such as Cape Grim most of the nss-SO_4 is produced from the oxidisation of DMS. Further, since nss-SO_4 is active as CCN in producing clouds at a clean maritime site some seasonal cycle in CCN concentrations may be expected. Figure 10 shows a plot of concentration of CCN active at 1.2%

supersaturation against time from 1982 until 1989. The seasonal cycle observed in the CCN concentrations which with a summer maximum, winter minimum, has a similar phase as that observed for DMS, MSA and nss-SO_4.

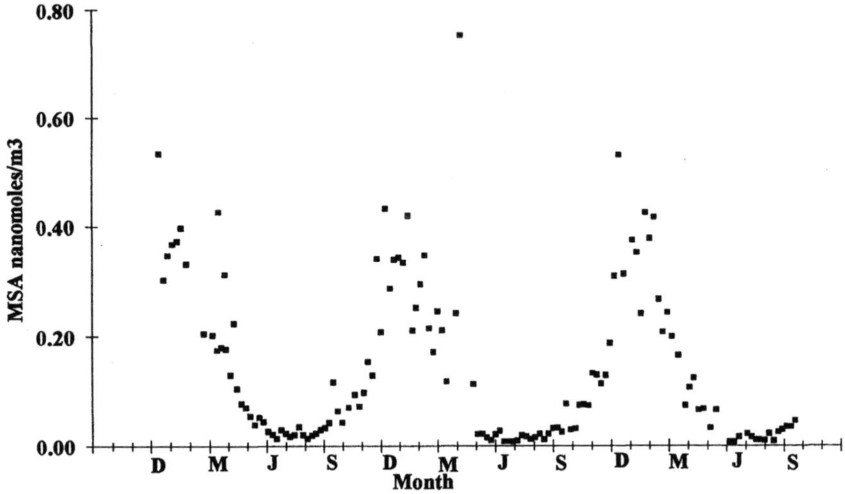

Figure 11. MSA concentrations at Macquarie Island in the sub micrometre size range, from Dec. 1988 until Sept. 1991.

In order to assess the latitudinal gradients and the effect of lower temperatures and ultra violet light, measurements of MSA and nss-SO_4 have been made at Macquarie Island (54.5° S, 158.9° E) and Mawson (67.6° S, 62.9° E) using the same filter pack described previously. Measurements began in late 1988 but since these data have not been completely validated at this stage only a few examples of the data will be presented here.

Figure 11 shows concentrations of MSA in the sub micrometre aerosol measured at Macquarie Island. This demonstrates a very distinct seasonal cycle in concentrations with a phase and amplitude similar to the species measured at Cape Grim. Concentrations range from about 0.4 nmoles m^{-3} in the austral summer decreasing to less than 0.5 nmoles m^{-3} in the austral winter.

The nss-SO_4 concentrations measured in the sub micrometre aerosol at Mawson can be seen in figure 12. They show a well defined seasonal cycle with a similar phase and amplitude as that seen in the other sulfur aerosol and gas measurements at Cape Grim.

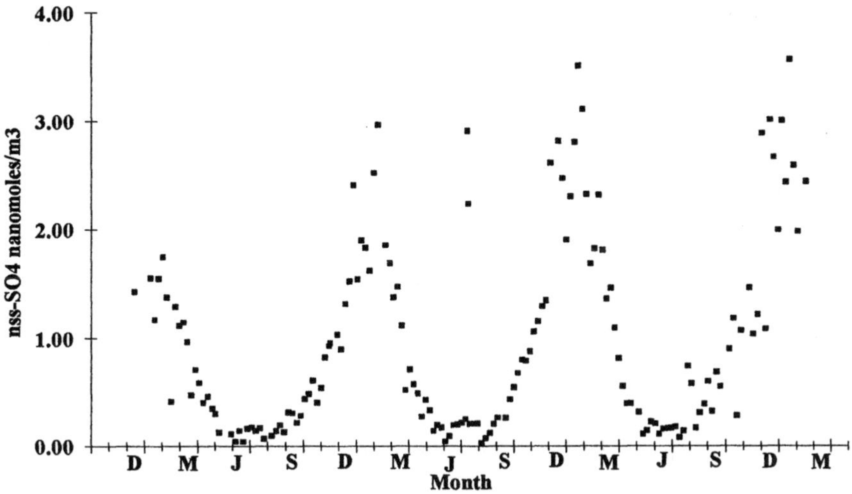

Figure 12. Non sea-salt sulfate concentrations at Mawson in the sub micrometre aerosol Dec. 1988 until Feb. 1992.

4. Conclusions

Concentrations of the sulfur gas and aerosol species DMS, MSA, nss-SO_4 and SO_2 have been measured at the Cape Grim Baseline Air Pollution Station. All these measurements show a distinct annual cycle which have similar phase and amplitude. Concentrations begin to increase in the late austral spring, peak during the austral summer and decrease to a minimum concentration in the austral winter. CCN concentrations measured over a number of years also show a similar seasonal cycle with number concentrations peaking during the austral summer. Concentrations of MSA and nss-SO_4 presented from Macquarie Island and Mawson exhibit seasonal cycles similar to those observed at Cape Grim.

The similarities in the seasonal cycles of the suite of measurements suggests that there is a linkage between DMS, MSA, nss-SO_4, SO_2 and CCN. This is expected since they form part of the oxidation scheme of DMS by hydroxyl radical ultimately leading to production of CCN and a possible increase in cloud albedo which can potentially can have an effect on climate as suggested by Charlson et al., (1987).

5. References

Ayers, G.P., Ivey, J.P. and Gillett, R.W. (1991) Coherence between seasonal cycles of dimethyl sulphide, methanesulphonate and sulfate in marine air. Nature, 349, 404-406.

Ayers G.P., Ivey, J.P. and Goodman, H.S. (1986) J. Atmos. Chem., 4, 173 - 185.

Bandy, A.R., Scott, D.L., Blomquist, B.W., Chen, S.M. and Thornton, D.C. (1992) Low yields of SO2 from dimethy sulfide oxidation in the marine boundary layer. Geophys. Res. Lett., 19, 1125 - 1127.

Bartlett, B.M. and Ayers, G.P. (1981) Static diffusion cloud chamber, J. Rech. Atmos., 15, 231 - 233. Third international cloud condensation nuclei workshop.

Berresheim, H., Andreae, M.O., Ayers, G.P. and Gillett, R.W. (1988) Distribution of biogenic sulfur compounds in the remote southern hemisphere. In: Biogenic Sulfur in the Environment (Eds. Cooper, W.J. and Saltzman E.S.) 352-366, American Chemical Society, Washington DC.

Berresheim, H., Andreae, M.O., Ayers, G.P. and Gillett, R.W., Merrill, J.T., Davis V.J. and Chameides, W.L. (1990) Airborne Measurements of dimethyl sulfide, sulfur dioxide, and aerosol ions over the Southern Ocean south of Australia. J. Atmos. Chem., 10, 341 - 370.

Charlson, R.J., Lovelock, J.E., Andreae, M.O. and Warren, S.G. (1987). Oceanic phytoplankton, atmospheric sulfur, cloud albedo and climate. Nature, 326, 655-661.

Charlson, R.J., Langer, J., Rodhe, H., Leovy, C.B. and Warren, S.G. (1991) Perturbation of the northern hemisphere radiative balance by back scattering from anthropogenic sulfate aerosols. Tellus, 43AB, 152 - 163.

Millero, F.J.A. (1974). The physical chemistry of seawater. Ann. Rev. Earth Plan. Sci., 2, 101.

Nguyen, B.C., Mihalopoulos, N., Putaud, J.P., Gaudry, A., Keene, W.C. and Galloway J.N. (1992) Covariations in oceanic dimethyl sulfide, its oxidation products and rain acidity at Amsterdam Island in the southern Indian Ocean. J. Atmos. Chem., 15, 39 - 53.

Plane, J.M.C. (1989) Gas phase atmospheric oxidation of biogenic sulfur compounds: a review. In: Biogenic Sulfur in the Environment (Eds. Cooper, W.J. and Saltzman E.S.) 404 - 423, American Chemical Society, Washington DC.

Pszenny, A.A.P., Castelle, A.J., Galloway, J.N. and Duce, R.A. (1989), A study of the sulfur cycle in the antarctic marine boundary layer. J. Geophys. Res., 94, 9818 - 9830.

Putaud, J.P., Mihalopoulos, N., Nguyen, B.C., Campin, J.M. and Belvis, S. (1992) Seasonal variations of atmospheric sulfur dioxide and dimethylsulfide concentrations at Amsterdam Island in the southern Indian Ocean. J. Atmos. Chem., 15, 117 - 131.

Quinn, P.K., Covert, D.S., Bates, T.S., Kapustin V.N., Ramsey-Bell, D.C. and McInnes, L.M. (1992) The DMS/CCN/climate system: relevant size-resolved measurements of the chemical and physical properties of atmospheric aerosol particles. J. Geophys. Res. (submitted)

CHARACTERIZATION OF PARAMETERS CONTROLLING ATMOSPHERIC CONCENTRATIONS OF BIOGENIC DIMETHYLSULPHIDE NEAR A COASTAL ALGAE FIELD

C. LUCE , S. PASHALIDIS , R. GIRARD , P. CARLIER & J. MORELLI*

Unité de Physico - Chimie et Géochimie de l' Atmosphère, associée au CNRS (URA n° 1404), Université Paris VII - 2 place Jussieu - 75251 PARIS Cedex 05 (France)

SUMMARY

Extended investigations on atmospheric DMS concentrations have been performed during three campaigns in a pristine region of the Atlantic coast of Brittany (France) , mainly in conditions of incoming oceanic winds , at a site surrounded by a large algae field uncovered at low tide .

Compared to results at a neighbouring coastal site without algae field , a very marked impact of DMS emissions has been established , with atmospheric concentrations up to 3 orders of magnitude higher than oceanic background reference values .

This powerful algal source of DMS has been used as a natural laboratory in order to distinguish and to quantifiy the relative influences of parameters controlling the fluctuations of atmospheric concentrations of this biogenic gas .

The main following features have been found :

- **the cotidal cycle has a preponderant impact , with maximal DMS atmospheric concentrations at low tide ;**
- **the temperature increase enlarges these values , mainly at low tide ;**
- **for wind speeds lower than 5-6 m/s , there is an increase of DMS atmospheric concentrations with the wind velocity .**

1. INTRODUCTION

The present research has been carried out in the framework of studies on the atmospheric sulphur cycle implemented on the Atlantic coast of Western Brittany (France) . Results related to the spatio - temporal distribution of DMS in the vicinity of a powerful natural source of this compound consisting in an algae field are presented here . They are derived from three field campaigns , a preliminary one performed in September 1983 , and the others in July 1988 and June 1989 respectively , within the European cooperative reseach project "OCEANO NOX" (CARLIER , 1988 ; BARNES et al. , 1991) .

The aim of this paper is mainly to focus attention on the specific influence of various parameters controlling DMS atmospheric concentrations in the investigated area , such as cotidal cycle , temperature and wind force .

* Present address : Institut de Biogéochimie Marine , unité associée au CNRS (URA n° 386) Ecole Normale Supérieure - 1 rue Maurice Arnoux - 92120 MONTROUGE (France)

G. Restelli and G. Angeletti (eds.), Dimethylsulphide: Oceans, Atmosphere, and Climate, 129–141.

2. PRESENTATION OF THE SAMPLING AREA

The selected source of DMS is a wide algae field uncovered at low tide and surrounding an extreme point of the Penmarc'h peninsula (see figure 1). This site, called Pointe de Penmarc'h (47° 48' N ; 4° 22' W) , is located in a pristine region submitted to the influence of prevailing oceanic winds .It was choosen on a rocky flat coast , in order to reduce the orographic effect on the air flow .

Samples were collected at the following places :

- the signal station of the French " Marine Nationale" bordering the algae field, at 5 m above the average sea level ;
- the top of the neighbouring old lighthouse of Penmarc'h , at a height of 47 m ;
- an inland station , at 4 km ENE of the Pointe de Penmarc'h in September 1983 (Kergadien) or at 2km NE of this coastal site in June 1989 (municipal stadium of Penmarc'h) .

In addition , during the campaign of September 1983 , a reference site located at the Pointe du Raz (48°02' N ; 4° 44' W), at about 36km WNW of the Pointe de Penmarc'h, has also been used , on a steep coast without algae field .

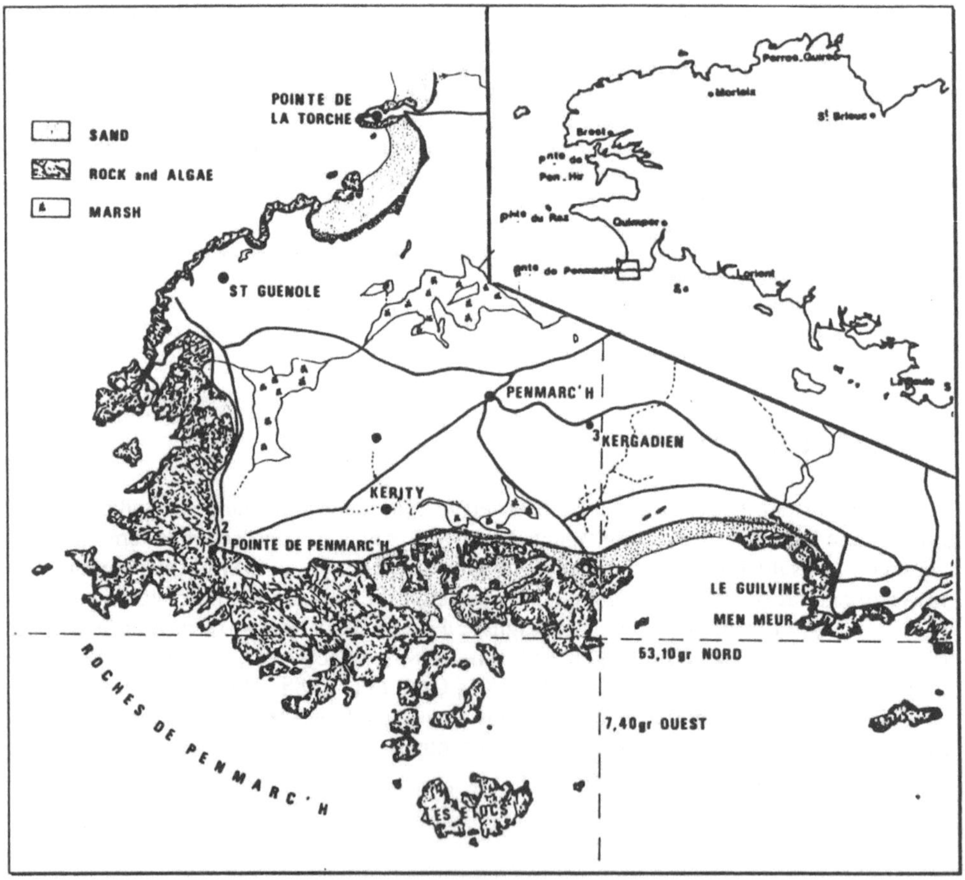

Figure 1 : Map of the sampling sites

3. EXPERIMENTALS

In 1983 , a sampling and analytical method by chemical derivatization with mercury II chloride was used . The resulting mercury complexes were submitted to analyses through regeneration by concentrated HCl of organosulphur compounds, which were then separated by gas chromatography and detected by flame photometry (LUCE et al. 1984 ; LUCE 1984) . This technique is very sensitive (detection limit ≈ 0.03 ppt), but requires extremely long times for analyses , limiting the number of collected samples . Consequently, relatively long sampling times were used (6 hours at the sites of the Penmarc'h area, 12 hours at Pointe du Raz).

In 1988 and 1989 , this technique was replaced by a preconcentration of organosulphur compounds on cooled adsorbant cartridges (Tenax T.A. at -20° C) . Then, the trapped species were thermodesorbed and again concentrated in a microtrap and injected by flash thermodesorption into a gas chromatograph fitted with a flame photometric detector (PASHALIDIS et al. 1989 a ; PASHALIDIS 1990) . This technique is less sensitive than the previous one (detection limit ≈ 0.5 ppt), but requires shorter analysis times . This has allowed to investigate the temporal fluctuations of DMS atmospheric concentrations with a sampling step of 2 hours simultaneously in 3 sampling sites .

4. IMPACT OF A POWERFUL COASTAL ALGAE FIELD ON THE OCEANIC BACKGROUND OF ATMOSPHERIC DMS

A comparison of DMS data obtained at the same time , at Pointe de Penmarc'h and at Pointe du Raz during the campaign of September 1983 shows a very marked effect of the DMS production by the investigated algae field on the concentrations of this compounds in the coastal oceanic atmosphere (table 1) . The DMS concentrations were up to 3 orders of magnitude higher at Pointe de Penmarc'h than at the Pointe du Raz , reference site without algae field . The values from the latter site correspond to prevailing oceanic wind conditions and are representative of the oceanic DMS background level in the lower troposphere over open North Eastern Atlantic (PASHALIDIS et al. 1989 b ; STAUBES et al. 1989 ; PASHALIDIS 1990)

Date	17-19 September		27-29 September	
Time intervals (GMT)	6h-18h	18h-6h	6h-18h	18h-6h
DMS average atmospheric concentrations (nmol/m3)				
Pointe du Raz	0.26	<0.02	0.69	0.12
Pointe de Penmarc'h	28.6	7.1	196.4	319.7

Table 1 : Comparison of DMS average atmospheric concentrations for simultaneous sampling at Pointe du Raz and Pointe de Penmarc'h

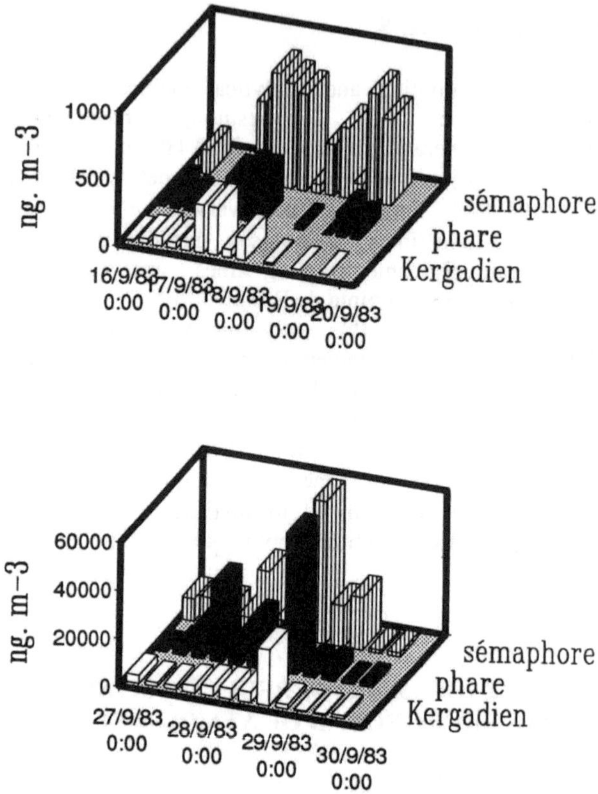

Figure 2 :Temporal fluctuation of DMS atmospheric concentration in the Penmarc'h area in September 1983

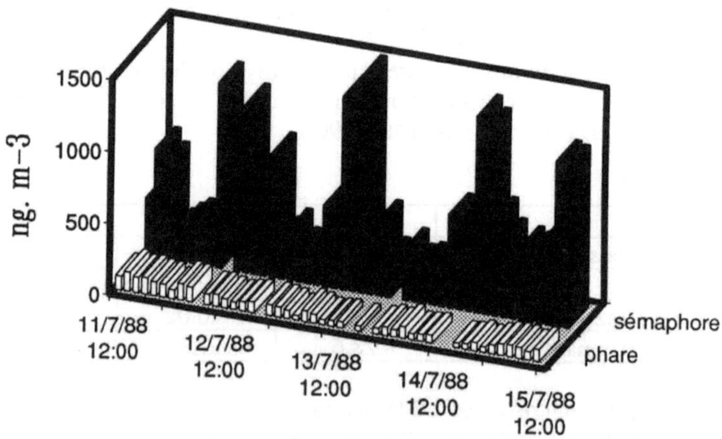

Figure 3: Temporal fluctuation of DMS atmospheric concentration in the Penmarc'h area in July 1988

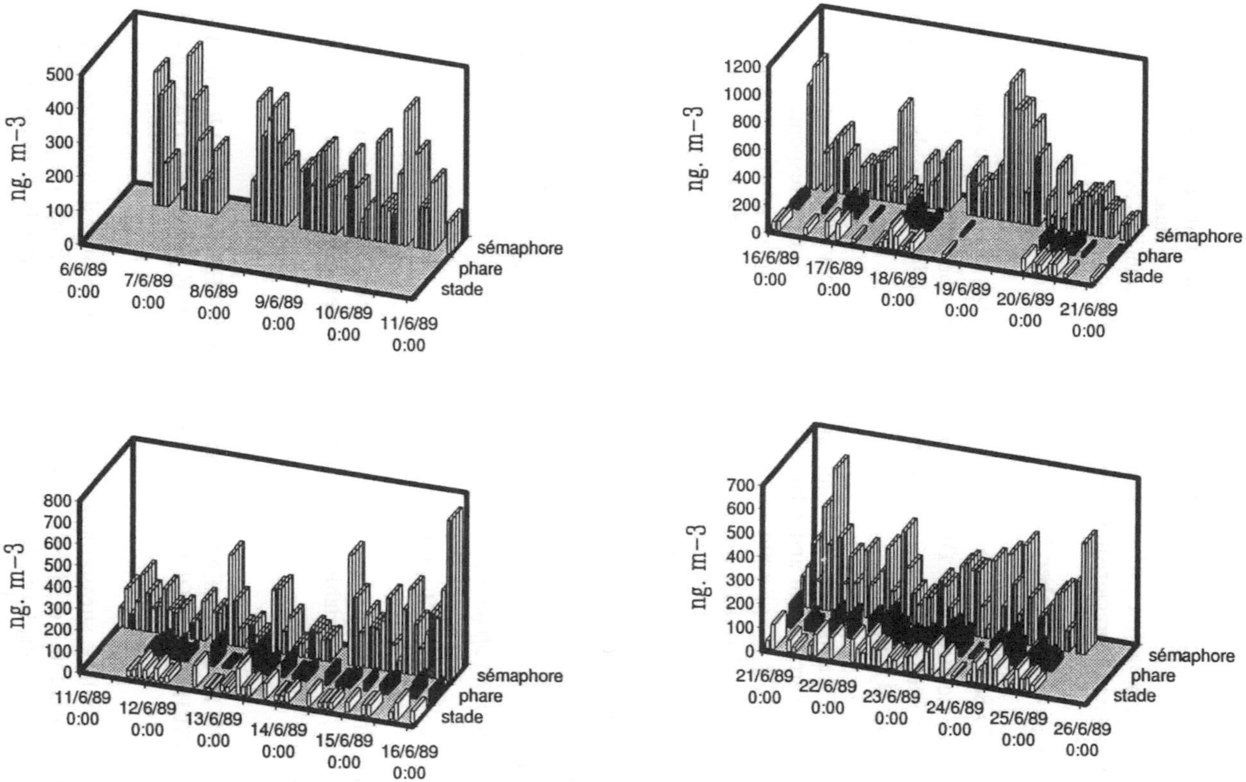

Figure 4 (a to d) : Temporal fluctuation of DMS atmospheric concentration in the Penmarc'h area in June 1989

5. FLUCTUATIONS OF ATMOSPHERIC DMS CONCENTRATIONS OVER THE PENMARC'H PENINSULA

Results of three field campaigns in the Penmarc'h area are presented on figures 2 a and b (September 1983) , 3 (July 1988) and 4 a to d (June 1989) . Some general comments are given below , in order to relate the atmospheric DMS concentrations to the main meteorological characteristics of each sampling period .

Campaign of September 1983

- 16 to 20 September : permanent influence of 3-5 m/s oceanic winds from NW to SW sector , associated with DMS concentrations in the range 1.5 - 30 nmol/m^3 and pronounced vertical gradient (lighthouse / signal station) and horizontal gradient (Kergadien / signal station).

- 27 to 30 September : anticyclonic situation with very low local winds of sector SE, maximal sunshine irradiation and relatively high temperature corresponding to very high DMS concentrations up to 1000 nmol / m^3 . These values are a consequence of the stagnation of DMS derived from intense emissions at the surface of the algae field, as shown by a very low vertical gradient of atmospheric concentrations of this compound , due to efficient vertical mixing resulting from the thermal convection .

Campaign of 11 - 15 July 1988

Under typical oceanic conditions ,the vertical gradient of DMS concentrations for this period has the same features as for the period 16 - 20 September 1983 : 1.5 -30 nmol / m^3 and a factor 20 to 50 between 5 and 47 m height .

Campaign of 6 - 25 June 1989

During this well documented period , the meteorological situation was unfortunately very complex , with variable winds , several episodes of coastal breeze and marked continental influences .

6. SPECIFIC INFLUENCE OF SOME MAIN PHYSICAL PARAMETERS ON DMS ATMOSPHERIC CONCENTRATIONS CLOSE TO THE COASTAL ALGAE FIELD.

Emphasis will be stressed now on the identification of main parameters from which depend the variations of DMS atmospheric concentrations in the vicinity of the investigated coastal algal source . In order to more accurately determine their relative impact on the short time fluctuations of these concentrations, only samples collected in conditions of incoming oceanic winds with time steps of 2 hours are considered for the discussion . Therefore, the following analysis concerns results related to the campaign of July 1988 and to a large part of June 1989 samples . Those collected during this period under the influence of local winds blowing from inland sectors (SE to N) are not taken into account , as well as samples from the preliminary campaign of September 1983, for which collection time steps of 6 hours were used.

Influence of the cotidal cycle.

For a non specialist observer, the influence of the cotidal cycle is obvious, as indicated by olfactory feeling . He can note that the algae field uncovered at low tide is a notably more active source of odorous atmospheric compounds than when it is covered at high tide . This is clearly confirmed for DMS , if we plot its average concentrations in the coastal air as a function of the cotidal phase .As an example, figure5 shows the concomitant temporal variations of the DMS atmospheric concentrations and the cotidal cycle at Pointe de Penmarc'h during the campaign of July 1988.

Influence of the temperature

DMS atmospheric concentrations are enhanced when the temperature rises . This has to be related to an intensification of the sea to air DMS transfer with increasing temperature, due to a growth of supersaturation of this compound , subsequently to a decrease of its solubility in sea water at higher temperature.

The influence of the temperature is more obvious if we discriminate between DMS values corresponding to low and high tide conditions respectively (figures 6 a and b). We can then establish that the influence of the air temperature is more marked for low tide episodes . The average slope of the DMS atmospheric concentrations as a function of the temperature increase is about 1 nmol / m^3 / °C for high tide and 2 nmol / m^3 / °C for low tide .

Influence of the wind force

For the campaign of June 1989 ,figure 7 shows an histogram giving the average atmospheric DMS concentrations near the coastal algae field for various classes of wind force. If we only consider the two lower classes (0 - 2 m/s and 2 - 4 m/s), there is a clear increase of DMS concentrations under the effect of the wind speed enhancement . This has to be related to a prevailing influence of the growth of the transfer coefficient of DMS from sea water to the atmosphere with increasing wind speed (LISS & MERLIVAT , 1986).

For intermediate ranges of wind velocities , a plateau of DMS atmospheric concentrations is observed and for higher classes (more than 8 - 10 m/s), there is a decrease of these concentrations . The contribution of two possible phenomena may explain this result : a dilution of DMS in an higher air volume for stronger winds and an anticorrelation between wind speed and temperature , higher wind speed values being generally associated with cooler temperature .

Influence of the solar irradiation

Attempts to distinguish the possible contributions of the solar irradiation to the short time fluctuations of coastal air DMS content have not been successful (see later).

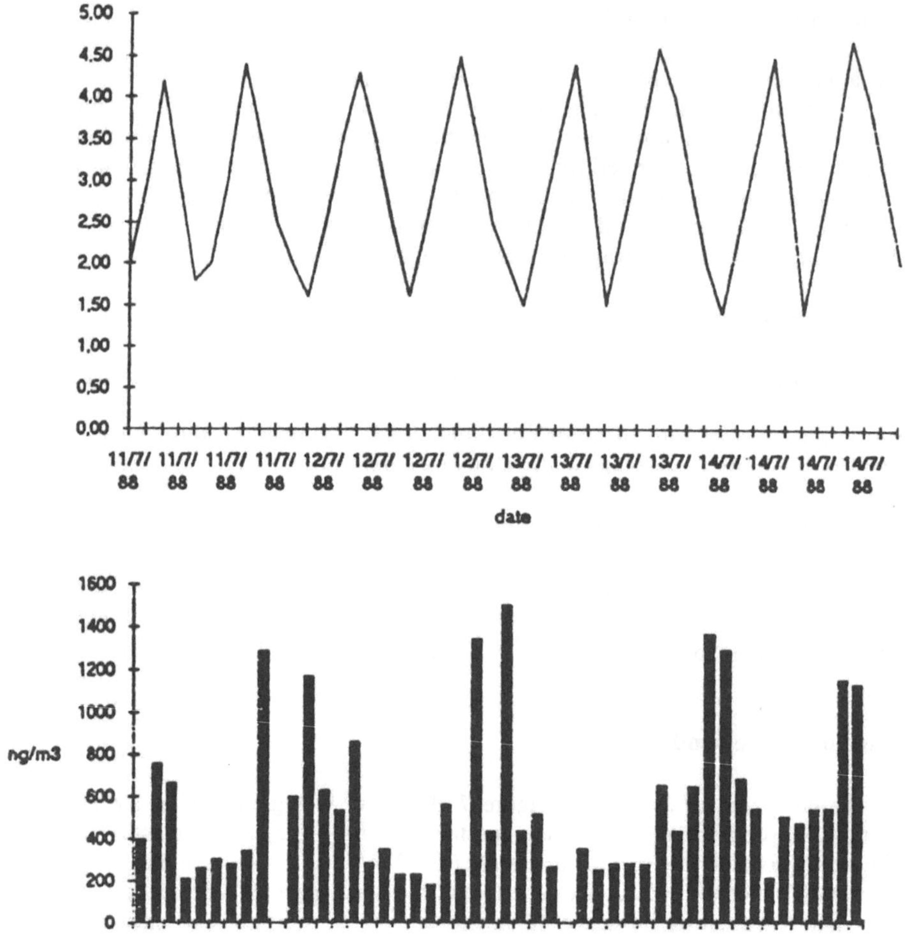

Figure 5: Co-variation of the cotidal cycle and the DMS atmospheric concentrations at Pointe de Penmarc'h during the campaign of July 1988.

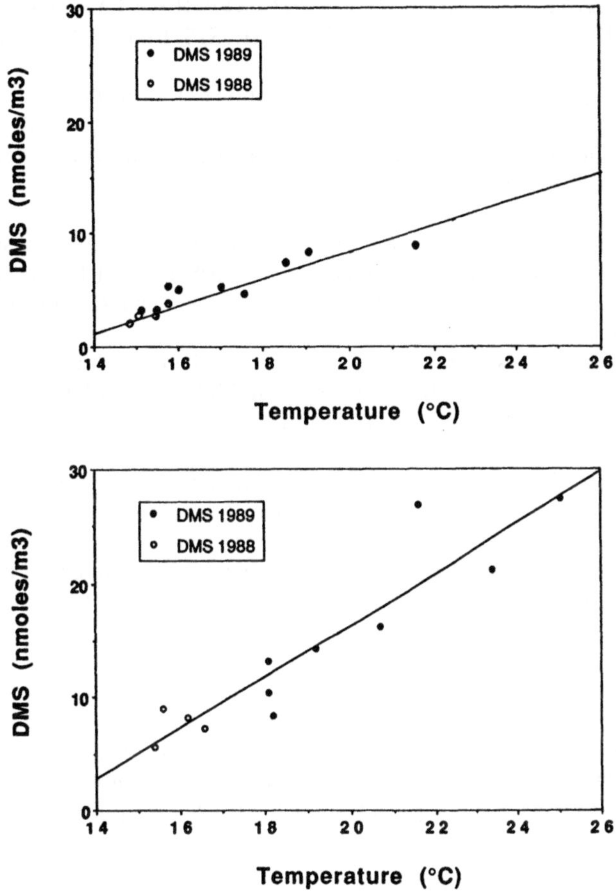

Figure 6 : Influences of the temperature on atmospheric DMS concentrations (a: at low tide ; b: at high tide)

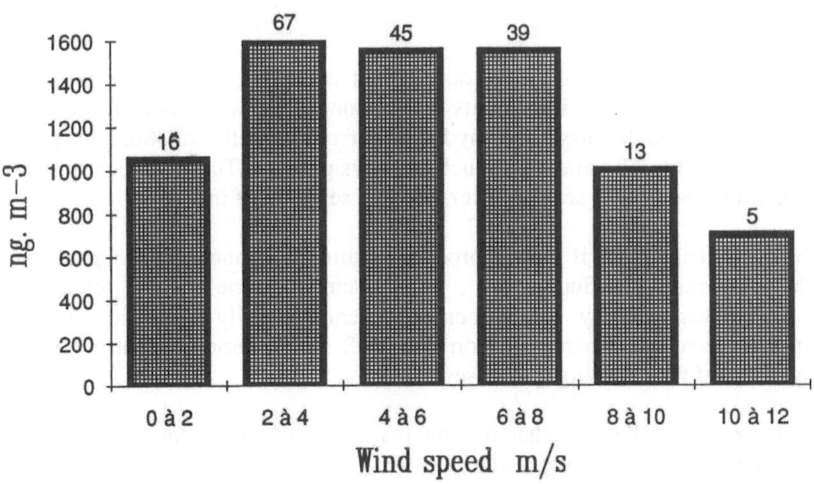

Figure 7 : Influence of the wind speed on DMS atmospheric concentrations

7. CHARACTERIZATION AND INTERPRETATION OF TYPICAL DAILY VARIATION PROFILES OF DMS ATMOSPHERIC CONCENTRATIONS

The data derived from our investigations close to the Penmarc'h coastal algae field have been carefully analyzed, with a view to reveal typical daily profiles of DMS atmospheric concentrations and to connect them to the complex combined influences of various parameters.

The characterization of typical daily profiles on the basis of the overall DMS measurements did not appear obvious . Nevertheless , it becomes only possible if DMS data are splited in two classes associated with the dephasing of the cotidal cycle relatively to the diurnal cycle :

- for the days with low tide between 9 h 00 and 15 h 00 (GMT) (figure 8 a);
- for the days with high tide between 9 h 00 and 15 h 00 (GMT)(figure 8 b) - only data from the campaign of June 1989 correspond to this case - .

In the first situation , a maximum around midday appears , due to the conjunction of low tide with the temperature maximum (for the data of 1989, this maximum is very marked) . On the other hand , minima around 6 h and 20 h are also observed (approximately close to the high tide) . The second maximum in the middle of the night is unclear and may be attributed to the compensation of low tide and nocturnal cooling effects .

In the second situation , two pronounced maxima are recorded around 4 h and 18 h (low tide) and also two minima around midnight and especially midday (high tide) . Such results reflect the fact that among the various parameters controlling the daily fluctuations of DMS levels , there is a preponderant influence of the cotidal cycle and of the resulting injection into the atmosphere of the DMS produced by the algae field .

8. CONCLUSIVE REMARKS ON THE DMS ALGAL PRODUCTION AND ITS TRANSFER INTO THE ATMOSPHERE

The transfer into the atmosphere of DMS of algal origin has to be related to the biological mechanisms responsible for its production . Following more recent concepts, DMS is derived through enzymatic hydrolysis from dimethylsulphonium propionate (DMSP), a metabolite of some algae such as polysiphonia . The biological processes of DMS production in sea water are contreversial (see section I of this issue) .

The fact remains that , if DMSP production directly connected with photosynthesis follows the solar irradiation fluctuations , the characteristic time of DMS releasing in sea water is higher than one day . It is then consistent that daily fluctuations of the solar irradiation can't have any repercussion on the DMS atmospheric concentrations , due to smoothing effect of the releasing processes .

Moreover , it is well established that the major part of the produced DMS is oxidised within sea water into dimethylsulphoxide (DMSO) and that the ratio of DMS transferred into the atmosphere is relatively small (ANDREAE , 1986 , WAKEHAM & DACEY,

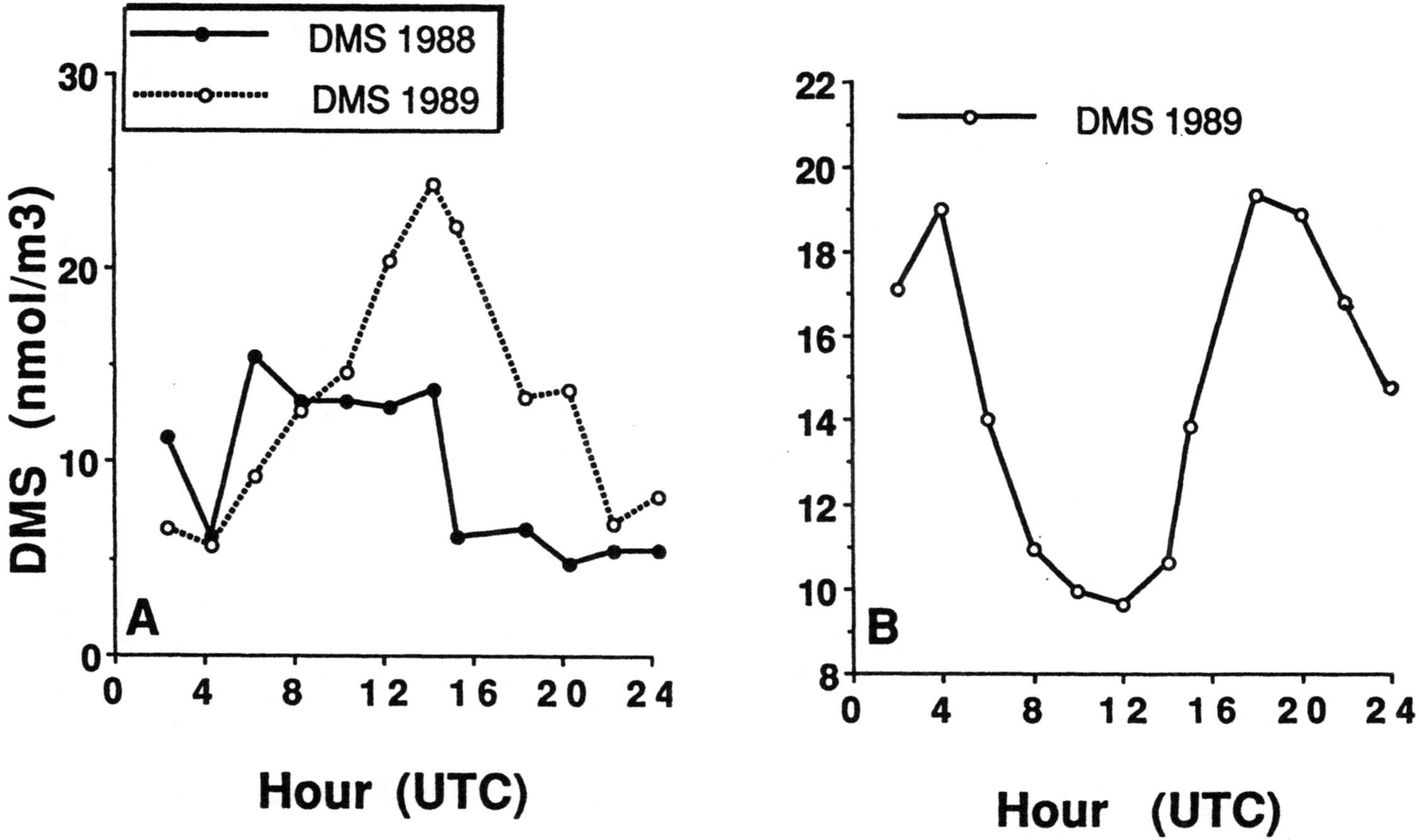

Figure 8 : Daily variation profiles of DMS atmospheric concentrations (A : for the day with low tide between 9h00 and 15h00 GMT ; B : for day with high tide between 9h00 and 15h00 GMT)

1989). There is then a quasi steady state for DMS concentrations in sea water, resulting from the compensation only between the production and the oxidation of this compound .

Consequently , the DMS flux daily fluctuations from the ocean to the atmosphere directly reflect those of its sea to air transfer coefficient (at this time scale, aqueous DMS concentrations can be assumed as stable) . Accordingly , in close vicinity to the coastal algae field , the atmospheric DMS concentrations should follow fluctuations of this transfer coefficient .

These considerations are generally invoked for the DMS production in marine superficial waters , in open sea conditions . They are also applicable in the case of coastal algae fields , even when uncovered at low tide . In this situation , the DMS released from the algal source is also injected into atmosphere by two additional processes : evaporation of residual sea water and evapotranspiration at the algae surface .

Thus, it is not inconsistent that the observed average atmospheric DMS concentrations are strongly dependent on the average irradiation during the weeks before sampling and that the fluctuations of this average irradiation can be the major cause of the seasonal fluctuations of DMS atmospheric concentrations (PUTAUD et al.; 1992). By contrast , at the time scale taken into account in our study (hour to day) , because of the integration over several days resulting from the DMS releasing processes into seawater, the daily fluctuations of photosynthesis should not significantly affect the temporal variations of the DMS flux from ocean to the atmosphere.

ACKNOWLEDGMENTS

The present study has been initiated with the support of the French "Ministère de l'Environnement " and developed within the European co-operative research project "OCEANO-NOX" supported by the Commission of the European Communities (DG XII/STEP Programme). These institutions , and also the staff of the signal-houses of the French "Marine Nationale" and "Service des Phares et Balises" are acknowledged for their help.

REFERENCES

M.O. Andreae ; 1986 ; The ocean as a source of sulfur compounds ; in : The role of air - sea exchange in geochemical cycling ; P. Buat - Menard ed. ; D. Reidel Publ. Co. , Dordrecht ; NATO ASI series **C185** p.113

I. Barnes , B. Bonsang , T. Brauers , P. Carlier , R. A. Cox , H. P. Dorn , M. E. Jenkin , G. Le Bras &U. Platt ; 1991 ; Laboratory and field studies of oxidation processes occuring in the atmospheric marine boundary layer (OCEANO NOX CEC Project) ; Air Pollution Research Report 35 ; G. Le Bras ed.

P. Carlier ; 1988 ; Synthetic Presentation of the European project OCEANO NOX in : Field Measurement and their interpretation ; Air Pollution Research Report 14 ; S. Beilke , J. Morelli & G. Angeletti eds. ; p. 55

P.S. Liss & L. Merlivat ; 1986 ; Air - sea gas exchange rate : introduction and synthesis ; in : The role of air - sea exchange in geochemical cycling ; P. Buat - Menard ed. ; D. Reidel Publ. Co. , Dordrecht ; NATO ASI series **C185** p.331

C. Luce ; 1984 ; Etude de la chimie dans la basse troposphère des composés organo-soufrés d'origine naturelle et des espèces qui en sont issues près d'une source naturelle unique intense en milieu océanique ;thèse de 3ème cycle, université Paris 7.

C. Luce , P. Carlier , R. Girard , H. Hannachi , P. Fresnet & G. Mouvier ; 1984 ; Mise au point d'une chaîne de mesure des composés dans l'atmosphère au seuil du ng/m^3 ; Analysis **12** , 350

S. Pashalidis , P. Carlier & G. Mouvier ; 1989 a ; Daily profile study of atmospheric DMS concentrations near to a intense coastal source at the "Pointe de Penmarc'h" ; Physico-Chemical Behaviour of Atmospheric Pollutants ; G. Restelli & G. Angeletti eds.; Kluwer Acad. Publ. ; Dordrecht ; p.457.

S. Pashalidis , P. Carlier & G. Mouvier ; 1989 b ; DMS distribution study over the atlantic ocean during the Polarstern crossing from Bremerhaven to Rio Grande do Sul ; Physico-Chemical Behaviour of Atmospheric Pollutants ; G. Restelli & G. Angeletti eds.; Kluwer Acad. Publ. ; Dordrecht ; p. 681.

S.Pashalidis, 1990; Etude du sulfure de diméthyle dans la basse troposhère en milieu océanique non pollué : métrologie et étude de cas concrets; thèse de doctorat en sciences, université Paris 7.

R. Staubes , H.W. Georgii & S. Bürgermeister ; Biogenic sulfur compounds in seawater and in atmosphere ; 1989 ; Physico- Chemical Behaviour of Atmospheric Pollutants ; G. Restelli & G. Angeletti eds.; Kluwer Acad. Publ. ; Dordrecht ; p.686.

J.P. Putaud , N. Mihalopoulos , B.C. Nguyen , J.M. Campin & S. Belviso ; 1992 ; Seasonal variations of atmospheric sulfur dioxide and dimethylsulfide concentrations at Amsterdam island in the southern indian ocean ; J. Atmos. Chem. , **15**, 117

S. G. Wakeham & J.W.H. Dacey ; 1989 ; Biological cycling of dimethylsulfide in marine environment ; in: Biogenic sulfur in the environment , E.S. Saltzlman & W.J. Coopman eds. ; ACS symposium series **393** , cap 10 , p.152

DIMETHYLSULPHIDE MEASUREMENTS AT BARING HEAD, NEW ZEALAND.

M.J. HARVEY[1], D.J. WYLIE[2], R.J. MARTIN[1], T.S. CLARKSON[1] & S.J. DE MORA[2].

[1] National Institute of Water and Atmospheric Research, P.O. BOX 3047, Wellington, New Zealand.

[2] University of Auckland, Dept. of Chemistry, Private Bag 92019, Auckland, New Zealand

SUMMARY

An automated gas chromatograph (GC) with cryogenic trapping and flame ionization detector (FID) has been operating since 1989, to gather a unique data-set of southern hemisphere non-methane hydrocarbons. Although not specifically designed for the purpose, this system traps dimethylsulphide (DMS) with sufficient efficiency to show that season and windspeed are the major sources of variation in atmospheric DMS concentration at the site. DMS concentrations were approximately 5 to 8 times higher in the summer than winter and tended to peak during strong southerly (onshore) winds. Such peak levels diminished slowly in summer and more rapidly in winter. Brief southerly changes with light winds also caused peaks in summer. The semi-quantitative FID data were compared with a limited number of quantitative DMS determinations from a GC with solid absorbent (molecular sieve 5A) trapping and flame photometric detector (FPD). GC/FPD data gathered so far has been limited to the winter months when DMS concentrations were small (median 0.56 nmol m^{-3}) and often below detection levels of the GC/FID system (estimated to be about 0.25 nmol m^{-3}).

1. INTRODUCTION

Baring Head atmospheric station (41° 24.6'S, 174° 52.2'E) (Figure 1) has been established as a site for the study of greenhouse gases and background chemical processes in the clean marine troposphere, typical of much of the southern hemisphere. The station was set-up in the early 1970's for baseline measurements of carbon dioxide. Facilities have been enhanced and the range of measurements increased since 1989 to include, amongst others, sulphur gases and aerosols.

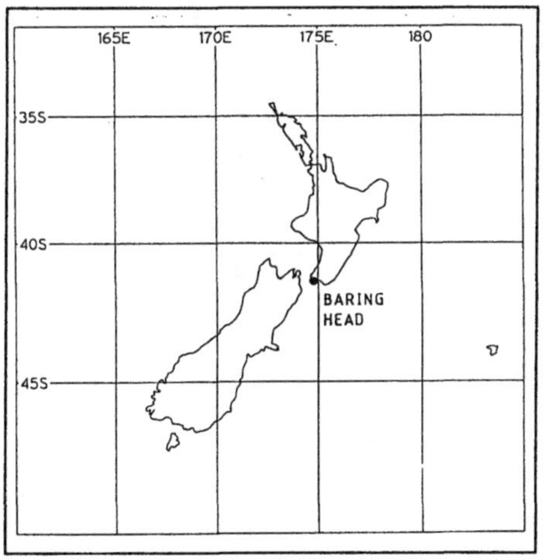

Figure 1. Location map

An automated gas chromatograph (GC) system operates continuously at Baring Head primarily to measure non-methane hydrocarbons (by flame ionisation detection [FID]) and chlorofluorocarbons (by electron capture detection [ECD]). The FID system detects dimethylsulphide (DMS) at background concentrations. This paper describes results from a recent near-continuous measurement series between December 1991 and July 1992.

A second GC (with flame photometric detector FPD), dedicated to DMS measurements was operated at Baring Head in May and June of 1992. Initial attempts to cross-calibrate the automated FID system with DMS values obtained from the FPD system are presented. The cross calibration was done during an intensive measurement period in winter when DMS values were small. A second cross-calibration exercise is planned for the next austral summer (early 1993).

G. Restelli and G. Angeletti (eds.), Dimethylsulphide: Oceans, Atmosphere, and Climate, 143–151.

2. METHODS

Sample preconcentration is used in the determination of DMS at background concentrations (sub-nmol m^{-3}). The automated (FID) system uses cryogenic trapping, the dedicated (FPD) GC system uses solid absorbent trapping.

The automated GC with FID detector operates continuously with a four hour sampling cycle. The system designed by Rudolf et al. has been extensively modified to suit local conditions (1). During each four hour cycle, an air sample of two litres is passed through an oxidant scrubber and trapped in a loop at liquid nitrogen temperature over a period of 60 minutes. Following sample collection, the loop is evacuated (at liquid nitrogen temperature) to reduce the volume of the sample and after 30 minutes it is heated to 100 °C to desorb its contents before injection. The separation column is a 30 m GS-Q Megabore (J&W Scientific, Folsom, CA) which is temperature programmed from 40 to 200 °C at 10 °C/min. For the remainder of the four hour cycle the collection system is purged with zero air and the sample loop cleaned in preparation for the next sample. This system detects dimethylsulphide at background concentrations although it contains metal components and is not considered to give high-quality quantitative results for DMS. However, this detection method is capable of showing trends in DMS concentration of a seasonal and weather dependent nature.

In May and June 1992, samples of DMS were collected in silanized pyrex tubes containing molecular sieve 5A (60/80 mesh) as a solid absorbent. Before entering the absorbent tube, water vapour was removed from the sample by passing through Nafion tubing surrounded by a drying agent and oxidants removed by passing through potassium iodide. Each sample comprised approximately 30 litres air sampled over 3 hours. The tubes were thermally preconditioned immediately before sampling and DMS was thermally desorbed at 280 °C. DMS was resolved in a 2 m teflon column packed with Carbopack B/1.5 % XE-60 and 1% H_3PO_4 with temperature programmed from 40 to 140 °C at 30 °C min^{-1} (2). Runs were completed within 10 minutes. The system was calibrated with a permeation tube source of DMS maintained at a constant temperature of 25 °C. Samples collected over the same period of time were compared in the cross calibration of the two systems.

3. RESULTS AND DISCUSSION

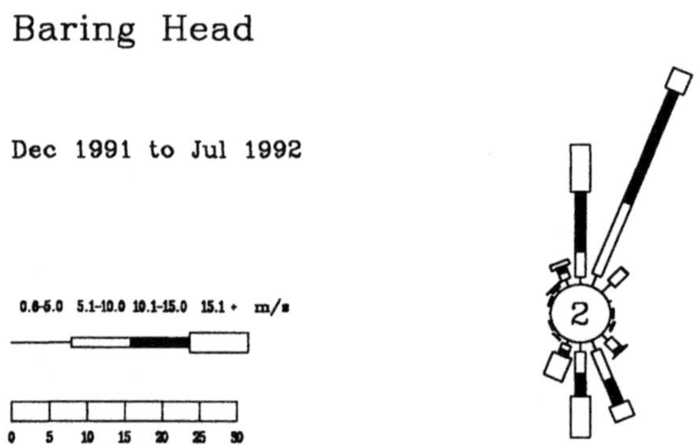

Figure 2. Wind rose frequency plot. The length of the arms corresponds to the percentage frequency determined from the frequency scale. Four classes of windspeed are determined. The percentage of 'calms' is shown in the middle of the rose.

Figure 2 shows a wind-rose frequency diagram for this period. Local orography affects the wind climatology and air is funnelled through the Cook Strait between the North and South Islands of New Zealand. Winds at the site are in either the northerly or southerly quadrant. Northerlies come off land and are often polluted (with elevated levels of hydrocarbons) from the nearby Wellington region. Southerlies arrive at the cliff top site directly off the sea and are frequently very 'clean' (with baseline concentrations of hydrocarbon gases).

A cross calibration (Figure 3a) of the FID and FPD systems was attempted on 19 and 20 June 1992. The horizontal bars show the duration of sampling. This was a period of southerly onshore winds, (Figure 3b) averaging 13 m s^{-1}. Results from the two instruments showed similar trends. However, the calculated calibration factor from this intercomparison gave DMS values that appeared to be too small. For this reason, absolute values have not been assigned to the FID data. The cross-calibration exercise and a direct calibration of the FID system need to be done in summer when background levels of the gas are larger.

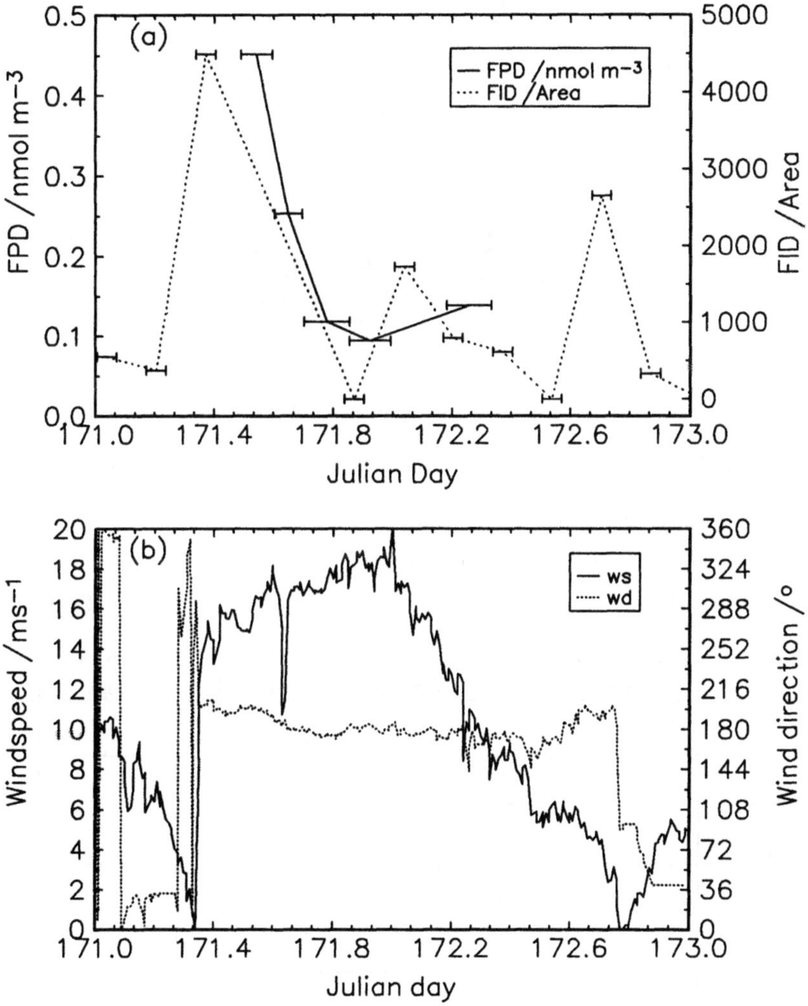

Figure 3. Comparison of DMS measurements made with two gas chromatographs (a FID system and a FPD system) during a period of onshore winds.

FID data obtained for period of almost continuous operation between December 1991 and July 1992 was put through careful quality checks. Figure 4 shows results from nine hundred and twenty-eight runs during this period for all wind directions. The range of values obtained in summer was larger than winter. Wintertime values tended to be small. Results from measurements in three seasons are compared in Table 1. The concentration of DMS in summer was about 5 to 8 times higher than in winter. Using values obtained by FPD in winter, the expected median concentration of DMS during onshore winds in summer is about 4 nmol m^{-3}.

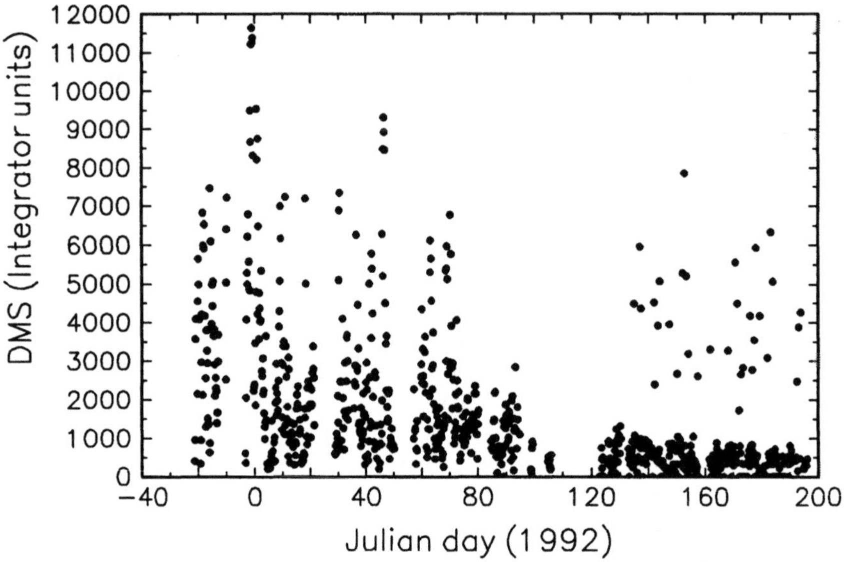

Figure 4. Time series of DMS concentrations recorded by the FID system.

Table 1. DMS values obtained at Baring Head

	FID system				FPD system	
Season	Number of readings	Median	10 percentile	90 percentile	Number of readings	Median
		Integrator Units				nmol m^{-3}
Summer	310	2050	580	5984		
(Onshore)	(90)	(3360)	(1300)	(7360)		
Autumn	330	850	150	848		
(Onshore)	(91)	(906)	(220)	(4570)		
Winter	207	360	0	1320	47	0.56
(Onshore)	(40)	(450)	(0)	(2800)		

The relationship of DMS concentration with windspeed was investigated. Figure 5 (a,b) shows DMS concentration plotted against the north-south wind vector magnitude (offshore northerly is positive and onshore southerly is negative). Because wind at the site is usually either northerly or southerly, this vector magnitude often equates to windspeed. In winter, minimum values of DMS were

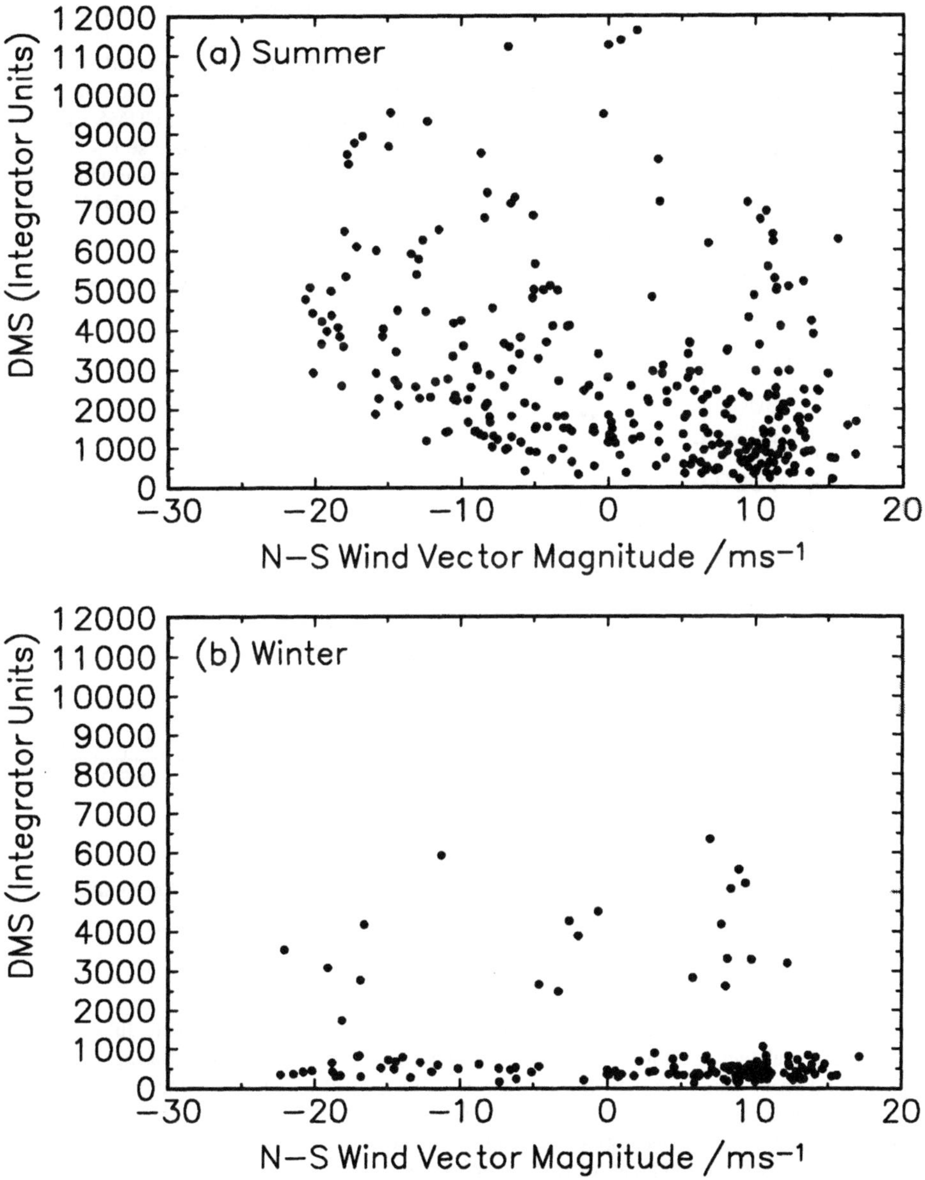

Figure 5. Relationship between DMS concentration and north-south wind vector magnitude for (a) summer (b) winter.

not windspeed dependent whereas in summer there was an envelope of minimum values which increased with the southerly wind vector indicating that the amount of DMS in the ocean was not limiting and atmospheric concentration was dependent on the transfer of gas from ocean to atmosphere.

A subset of the data from the FID system for onshore winds greater than 5 ms^{-1} is plotted in Figure 6. This shows the occurrence of 'DMS events', especially during the summer months. Two weather situations gave rise to these events. Throughout the year the occurrence of strong southerly flow and gales could cause elevated DMS concentrations. In summer DMS levels increased rapidly and diminished slowly. In contrast, there were few peaks in winter and levels diminished rapidly. For one 'DMS event', an example back-trajectory (Figure 7a) and surface wind record (Figure 7b) for a strong southerly flow shows that surface air at Baring Head had originated at the Antarctic coastline four day's previously. The second type of weather situation which gave rise to 'DMS events' was brief southerly changes in the summer. This type of event is not detected by the coarse resolution trajectory plot (Figure 7c) which shows a northerly back-track but it is shown on the surface wind record on day 18 in 1992 (Figure 7d). This second type of event would cause elevated levels of DMS to be sampled which originated from local marine sources.

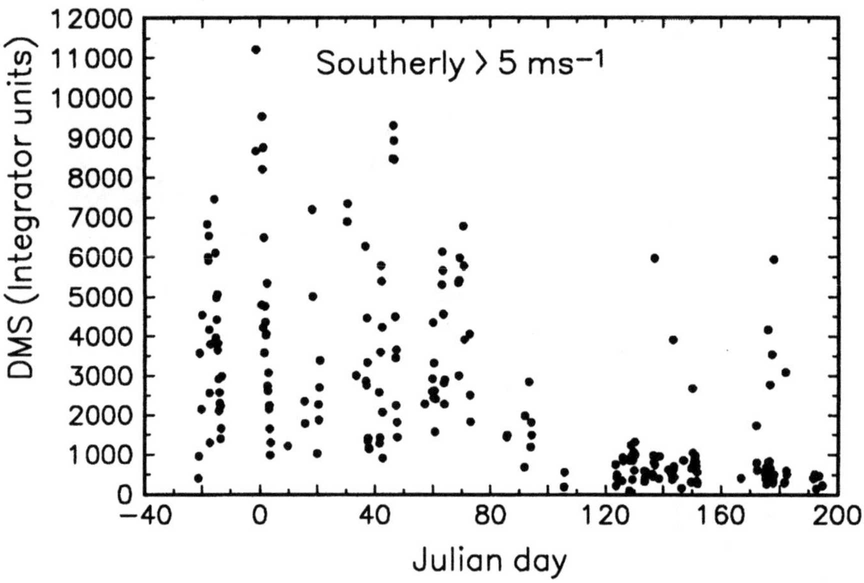

Figure 6. Time series of DMS concentrations recorded by the FID system for southerly winds above 5 m s^{-1}

4. CONCLUSIONS

From examination of data gathered between December 1991 and July 1992 the main environmental factors influencing DMS concentration in onshore winds at Baring Head were season and windspeed. Summertime values were 5 to 8 times greater than wintertime values. The minimum levels of DMS increased with windspeed. During southerly storms, events with elevated DMS occurred throughout the year but were more frequent in summer. Brief southerly changes at the site in summer also gave rise to elevated levels of DMS from local sources.

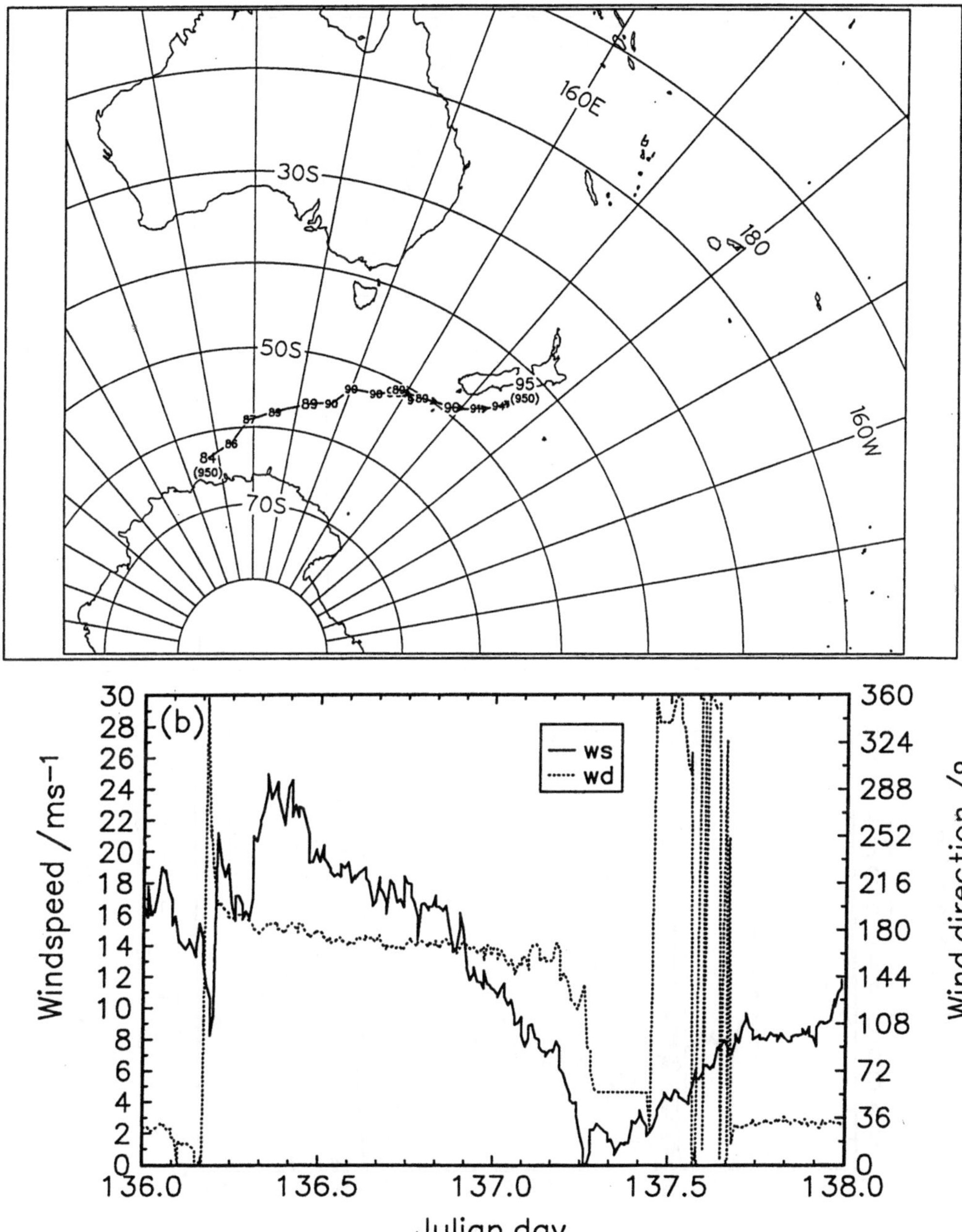

Figure 7. (a) 96 hour back-trajectory ending 12Z 15-May-1992 with six hourly intervals marked and (b) local windspeed and direction for a major southerly flow in winter with elevated DMS.

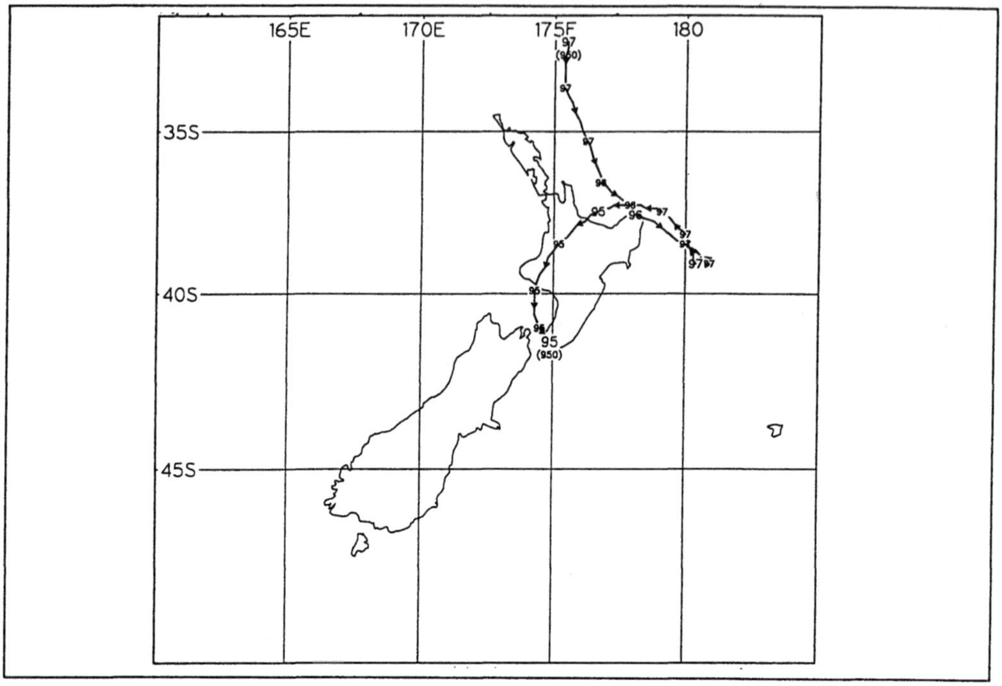

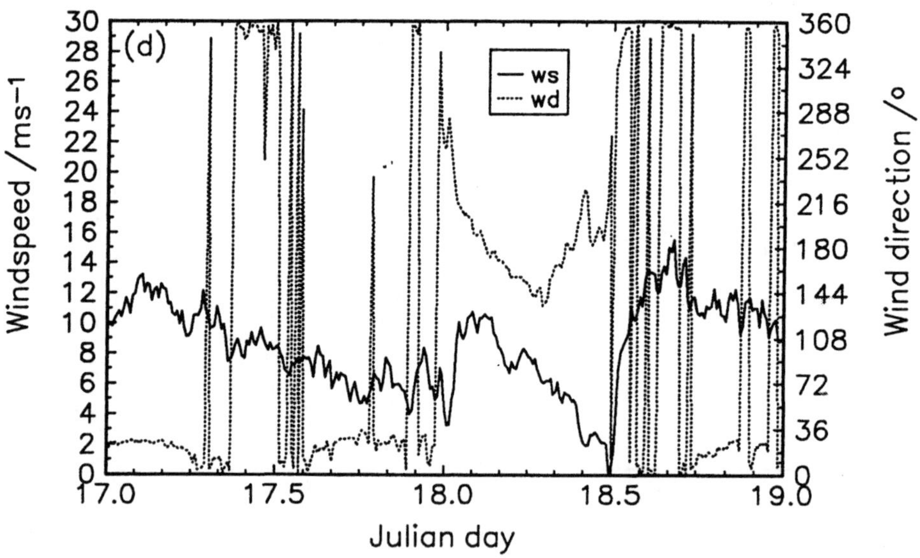

Figure 7. (c) 96 hour back-trajectory ending 12Z 17-Jan-1992 with six hourly intervals marked and (d) local windspeed and direction for a brief southerly change in summer with elevated DMS.

REFERENCES

(1) RUDOLF, J., JOHNEN, F.J., KHEDIM, A., and PILWAT, G. (1990). The use of automated on line gas chromatography for the monitoring of organic trace gases in the atmosphere at low levels. Int. J. Environ. Anal. Chem., 38, 143-145.

(2) DE MORA, S.J., GROUT, A. AND SHOOTER, D. (1990). The analysis of reduced sulphur gases in air and melt waters by gas chromatography at Bratina Island, 78°S. New Zealand Antarctic Record 10, 12-21.

THE ROLE OF METHANESULPHONIC ACID IN SNOW SAMPLES FROM TERRA NOVA BAY (ANTARCTICA).

R. UDISTI, F. CASELLA AND G. PICCARDI

Department of Public Health and Environmental Analytical Chemistry
University of Florence - Via G. Capponi 9 - 50121 Florence (Italy).

SUMMARY

Concentration profiles of Cl^-, total SO_4^{2-}, nss-SO_4^{2-} and MSA for two snowpits at Terra Nova Bay (Victoria Land - Antarctica) are reported. The spatial distribution of these compounds is correlated with the altitudo and geographical position. On the basis of a datation made by H_2O_2 determination, the seasonal trends of nss-SO_4^{2-}, MSA and MSA/nss-SO_4^{2-} ratio are discussed. A sharp seasonal trend with summer maxima and winter minima is observed and demonstrates their common origin from DMS. A lower seasonal signal and lower concentration levels are found for the higher station. The MSA/nss-SO_4^{2-} ratio at low MSA values and the correlations between MSA and nss-SO_4^{2-} may evidence the existence of a extra-DMS nss-SO_4^{2-} source, that becomes more important in the winter period, when the DMS contribute is lower.

1. INTRODUCTION

In non-polluted areas the main source of atmospheric aerosol sulphur compounds is dimethylsulphide (DMS) produced by the metabolic processes of phytoplankton blooms. In these areas DMS production plays an important role in cloud condensation nuclei (CCN) formation and has relevant climatic implications (2, 3, 5, 6, 24). DMS undergoes oxidation in the atmosphere following different reaction mechanisms, strongly dependent on temperature, generating H_2SO_4 and methanesulphonic acid (MSA) as the two main products (12).

Aerosol measurements show a strict correlation between DMS and its final oxidation products, nss-SO_4^{2-} and MSA, but their respective concentrations, and especially the ratios between these two components, vary widely with latitude and season. These variation are generally attributed to differences in the primary biological productivity of DMS in the source region and to temperatures variations (4, 5).

As DMS is a reactive species its concentration is not a suitable tracer to investigate the composition of the pre-existing aerosol; instead MSA and nss-SO_4^{2-} are stable for a long time in atmospheric precipitations. Assuming a good correlation between snow and atmospheric composition exists, the snow deposited regularly in the Antarctic ice cap provides an archive of the chemical composition of the local atmosphere over many thousands of years (9). While MSA derives solely from DMS (5, 20, 26) the nss-SO_4^{2-} can have multiple sources besides DMS oxidation (volcanic, stratospheric inputs, anthropogenic) (5, 8, 10, 15, 24).

G. Restelli and G. Angeletti (eds.), Dimethylsulphide: Oceans, Atmosphere, and Climate, 153–162.

A deeper understanding of the correlations underlying the DMS, MSA and nss-SO_4^{2-} cycles in the atmospheric aerosol at high latitudes and the nss-SO_4^{2-} and MSA concentrations in the Antarctic snow and ice will help to assess the suitability of MSA as a reliable tracer of the oceanic DMS emissions and also of the related marine biological activity, now and in the remote past.

2. EXPERIMENTAL

Instrumentation

The analysis of principal and trace anions were performed with a Dionex mod.4000i Ion Chromatograph interfaced with a MS DOS PC. The separation conditions and the method performances have been reported (29).

The H_2O_2 depth profile, used to date the snow layers, was obtained with a FIA - Spectrofluorimetric method (30).

Reagents

Fluka and Sigma reagent grade or Merck Suprapur products have been used. Standard solutions with concentrations of less than 10 mg/l were prepared daily from 1 g/l stock solutions.

Blanks and standard solutions were prepared with ultrapure water (Millipore Milli-Q system; resistivity >18 MΩ) and subsequent continuos recycling in an Elga UHQ apparatus for ion chromatography or distilled on $KMnO_4$ for H_2O_2 determination.

Sampling and Analysis

Snow-pit samples were collected in two uncontaminated areas at different altitude and geographical position during the 1990-91 Italian Antarctic Expedition to investigate the spatial distribution and temporal variability in the sulphur cycle-related components nss-SO_4^{2-} and MSA.

The sampling station coordinates are:

- Station 19 : 73° 51.9' S, 163° 41.5' E , 1700 m a.s.l.
- Station 27 : 74° 32.9' S, 162° 56.5' E , 700 m a.s.l.

The distances from the sea are, respectively, 50 and 40 Km.

A total of 226 samples were collected in snow-pits, about 2.2 m-deep, inserting pre-cleaned polyethylene vials (16 x 100 mm) in the pits' vertical walls. Strict anti-contamination procedures were adopted during sampling, storage and pre-analysis manipulations (23). Control blanks were always below the detection limits for MSA (0.6 μg/l with a 376 μl loop) and for H_2O_2 (0.01 μg/l with a 276 μl loop) and less than 1 % for the other determined anions.

The reproducibility values calculated from replicate analysis of standard solutions, at concentrations close to the minimum values measured in the real samples, were always less than 3%.

Because of the scarcity of sample volume (3-6 ml equivalent water typically) the vials were alternately analyzed for anions (inorganic and organic separately) and for H_2O_2, with a resolution of about 3.5 cm of snow layer for each type of analysis.

3. DISCUSSION

Measurements of atmospheric aerosol in southern oceans showed a good correlation between DMS, nss-SO_4^{2-} and MSA and a seasonal trend in these components' signal with summer maxima and winter minima (2, 3, 5, 24).
In Antarctica, where the day-night cycle coincides with the winter summer sequence, the seasonal pattern is particularly evident and it can be used for dating purposes (23).

Total Sulphate

Previous papers reporting data on the surface snow collected in the area around Terra Nova Bay (21, 22, 28) evidenced a strong marine spray contribution to the region' precipitation composition and an inverse relationship of this factor with the distance from the sea and the altitude. The enrichment factors (E.F.) of the total sulphates are calculated using Cl^- as the marine spray tracer as:

$$\text{E. F. } (SO_4^{2-}) = \frac{[SO_4^{2-}] / [Cl^-]_{\text{snow}}}{[SO_4^{2-}] / [Cl^-]_{\text{sea water}}}$$

Fig. 1 shows the E.F. as function of the Cl^- concentrations for both stations. The majority of samples for station 27 shows a Cl^- concentration higher than 300 μg/l. The E.F. for these samples tend toward 1 but increase, up to E.F. = 15, for samples containing small quantities of Cl^-.
Very few samples show a Cl^- concentration higher than 300 μg/l for station 19, instead, but the E.F. still tend toward 1. The majority of the samples has a low Cl^- content with an E.F.> 1.
Station 27 is more influenced by the marine contribution, i.e. there is a great number of samples with high Cl^- values, compared to station 19. However E.F. >> 1 result for both stations, implying a non-marine spray contribution to sulphates, the so-called nss-SO_4^{2-}.

Non sea salt Sulphate.

The nss-SO_4^{2-} concentrations are calculated from the chloride content, assuming a molar SO_4^{2-}/Cl^- ratio = 0.0517 in sea-water (2). The nss-SO_4^{2-} calculations sometimes result in negative values for zones with predominant marine influence, mainly in the winter period, when this component levels are quite low (1, 7, 19, 20, 21, 22, 23, 31). The interpretation of this anomaly is still debated in the literature and includes experimental errors, non-marine contributions to the Cl^- and Na^+ content, errors in the sea-water SO_4^{2-}/Cl^- or SO_4^{2-}/Na^+ ratio evaluation, fractionation effects during sea-salt aerosol formation and transport. No matter what the cause, the result is an overestimate of sea-salt sulphate, that can lead to negative nss-SO_4^{2-} values in winter and in particular atmospheric conditions. In our stations the negative values are relative to different samples belonging to the same annual layers in the two snow-pits so that instrumental errors can be ruled out. More accurate investigations of this phenomenon are under way and will be discussed later. In the present work the nss-SO_4^{2-} negative values (about 5% of the total measurements) are neglected in the mean, trends and ratios calculations.

The nss-SO_4^{2-} may have multiple sources: anthropogenic, volcanic, biogenic, stratospheric transport (5, 7, 8, 9, 14, 15, 27). In the Southern emisphere, and especially in Antarctica, the anthropogenic contribution is negligible (10) and volcanism contributes a relatively low background level during periods of reduced activity (10, 14, 15). The main source is therefore attributed to the oxidation products of DMS.
Fig. 2 and 3 show the concentration vs. depth profiles of nss-SO_4^{2-} for the two snow-pits. A seasonal pattern with summer maxima and winter minima, essentially in phase with the H_2O_2 seasonal signal reported in the same figure, is evident.
The seasonal pattern is more clearly defined for station 27, with summer maxima substantially higher than station 19. This difference in concentration can be explained by the different geographical position as to altitude and distance from the sea. This is in disagreement with the conclusions by Herron (10) as to the independence of nss-SO_4^{2-} from altitude and its inverse correlation with the mean annual accumulation rates. Station 19 has in fact a mean annual snow accumulation rate (16 g cm^{-1} Yr^{-1}) lower than station 27 (27 g cm^{-1} Yr^{-1}) (23). The nss-SO_4^{2-} concentration in the station 27 samples is on the average higher than station 19, and the same applies to the Cl^-, tot-SO_4^{2-} and MSA average concentrations. As these components are directly (marine spray) or indirectly (secondary aerosol produced by gas-to-particle conversion) linked to the oceanic source (14) the altitude must be the factor controlling these species concentration in the snow. A similar behavior was detected also by Mayewski et al. (18) in Greenland snow and by Mulvaney et al. (20) in the Antarctic Peninsula.
The components with a more defined atmospheric origin, as H_2O_2 and NO_3^-, show higher concentrations in station 19 than 27 (23).

Methansulphonic acid (MSA).

The MSA has been detected in snow and ice both in the interior and in the coastal zones of the Antarctic Continent (13, 16, 22, 25, 26, 28). Its chemical stability in the ice makes it a good tracer of the DMS oceanic emissions of remote times (17). The seasonal pattern exhibited by the MSA matches the trend already discussed for DMS and nss-SO_4^{2-}. Its spatial distribution follows the DMS, with high values found at high latitudes (2, 3, 4, 5, 24).
The correlation found between the MSA and the CCN concentration suggested that this component may play a climatic role, even if MSA seems to be mainly related to the stage of CCN grouth rather than nucleation itself (3, 5). Aerosol measurements at Mawson station, East Antarctica, show summer values about 20 times higher than the winter, with annual cycles out-of-phase with the nss-SO_4^{2-} record and in agreement with the DMS seasonal variations in coastal waters (24).
The station 27 and 19 MSA depth profiles are shown in Fig. 2 and 3. Station 27 shows a distinct seasonal pattern, practically in-phase with the nss-SO_4^{2-} and H_2O_2, with summer maxima up to 300 μg/l and winter minima in the range 1-25 μg/l. As already discussed for nss-SO_4^{2-}, a substantial decrease in the average concentration and in the summer peak values occurs in station 19. For this station a clear seasonal signal is evident for the last three years, while for higher depths the signal is blurred, and concentrations below 15 μg/l are measured even in the summer layers. These values are in agreement with the results reported for deep ice cores sampled in Antarctica both near the sea (13, 16, 26) and in the interior (17, 25, 26).

The summer values measured for the snow-pit 27 are on the other hand in agreement with the relatively high values found in surface snow samples collected in the same area, that show an inverse relationship with the altitude and the distance from the sea (28). The differences in the MSA concentrations found for the two stations can be imputed to the effects of these last two factors.

MSA/nss-SO_4^{2-} ratio

The ratio between MSA and nss-SO_4^{2-} varies in the atmosphere depending on latitude, temperature and seasonality. In uncontaminated areas the most likely DMS oxidation mechanism involves a reaction with OH radicals via addition or H+ subtraction pathways. The first path is favored by low irradiation and temperatures, and leads to an MSA enriched oxidation mixture; the second, favored by high irradiation and temperatures, produces primarily nss-SO_4^{2-} (11). As a consequence this ratio assumes the highest value (up to 1:1 in moles) for aerosol collected in the Southern Oceans and in Antarctica (4, 24).

Measurements relative to snow and ice from Antarctica show a high variability with value from 0.03 to 0.45 w/w (13, 16, 17, 20, 26).

The MSA/nss-SO_4^{2-} (w/w) ratio depth profiles are shown in Fig. 2 and 3 for station 27 and 19. A seasonal signal analogous to the one observed for nss-SO_4^{2-} can be seen, even against an increased background noise, at least for station 27 and for the first meter of station 19. In agreement with the results reported by Ayers et al. (2) for aerosol samples collected at Cape Grim, Tasmania, summer maxima and winter minima result, that cannot be explained by the inter-season differences in the temperatures. According to the reaction mechanisms proposed by Hynes et al. (11) in fact, an opposite behavior should be observed, as the winter lower temperatures favor the pathway leading to MSA in the DMS oxidation reactions. An analogous seasonal pattern is reported for aerosol measurements at low- and mid-latitudes also (5, 24).

Various explanations for this phenomenon exist: differences in the aerosol particle size distribution between summer and winter, as nss-SO_4^{2-} is transported mainly by submicrometer particles (5); winter inputs to higher latitude sites of air masses formed at mid-latitudes where the MSA/nss-SO_4^{2-} ratio is lower (24); the presence of nss-SO_4^{2-} sources not related to DMS which are obscured in the summer, when the DMS contribution is dominant (2, 5). This last explanation is the most convincing for the two sampled stations.

Fig. 4, showing the ratio nss-SO_4^{2-}/MSA (w/w) vs. MSA concentration, evidences the increasing trend of the nss-SO_4^{2-} relative content in the events characterized by a low DMS total contribution.

For station 27, where higher MSA and nss-SO_4^{2-} concentrations indicate a stronger DMS influence, the ratios level off at a value = 2 (MSA/nss-SO_4^{2-} = 0.5) for MSA values higher than 20 μg/l. For lower concentrations, winter typical values, the ratio rapidly increases to values higher than 10.

The data relative to station 19 parallel this behavior as a great number of samples have MSA concentrations lower than 10 μg/l.

The ratio variability seems too high to be accounted for only by a different partition between MSA and nss-SO_4^{2-} from DMS. A nss-SO_4^{2-} contribution extra DMS, that becomes dominant where the DMS production is low, seems more likely.

The same conclusion can be drawn plotting the nss-SO_4^{2-} concentrations vs. MSA (Fig. 5). The data relative to the same station should be well correlated assuming a common origin, even taking into account the seasonal variations in temperature. The correlation obtained for station 27, more influenced by the biogenic related components, is relatively high ($r = 0.82$, $n = 46$).

For station 19, however, the lower correlation can be interpreted as a different nss-SO_4^{2-} source. The positive intercept in the regression line at zero MSA concentration supports the hypothesis of a significant nss-SO_4^{2-} level in the absence of MSA. nss-SO_4^{2-} extra-DMS sources can be ascribed to the long-range stratospheric transport of H_2SO_4 (24) and to volcanic inputs. Legrand and Delmas (14) report a background level of nss-SO_4^{2-} of volcanic origin of about 7.5 μg/l during periods of low volcanic activity.

To conclude the MSA, directly measured and less subject to experimental errors and trend misinterpretations, seems a more reliable indicator than nss-SO_4^{2-} for the marine biological activity.

REFERENCES

(1) ARISTARAIN A.J., DELMAS R.J. and BRIAT M. (1982), Snow chemistry on James Ross Island (Antarctic Peninsula). J. Geophys. Res. 87, 11004-11012.

(2) AYERS G.P., IVEY J.P. and GILLETT R.W. (1991), Coherence between seasonal cycles of dimethyl sulphide, methanesulphonate and sulphate in marine air. Nature 349, 404-406.

(3) AYERS G.P. and GRAS J.L. (1991), Seasonal relationship between cloud condensation nuclei and aerosol methanesulphonate in marine air. Nature, 353, 834-835.

(4) BERRESHEIM H. (1987), Biogenic sulphur emissions from the Subantarctic and Antarctic Oceans. J. Geophys. Res. 92, 13245-13262.

(5) BATES T.S., CALHOUN J.A. and QUINN P.K. (1992), Variation in the Methanesulphonate to sulphate molar ratio in Submicrometer marine aerosol particles over the South Pacific Ocean. J. Geophys. Res., 97, 9859-9865.

(6) CHARLSON R.J., LOVELOCK J.E., ANDREAE M.O. and WARREN S.G. (1987), Oceanic phytoplankton, atmospheric sulphur, cloud albedo and climate. Nature 326, 655-661.

(7) DELMAS R.J. (1982), Antarctic sulphate budget. Nature, 299, 677-678.

(8) DELMAS R.J., LEGRAND M., ARISTARAIN A.J. and ZANOLINI F. (1985), Volcanic deposits in Antarctic snow and ice. J. Geophys. Res., 90, 12901-12920.

(9) DELMAS R.J. (1986), Antarctic precipitation chemistry. NATO ASI Series, Vol.G6, Chemistry of multiphase atmosferic systems. Ed. W.Jaeschke - Springer-Verlag Berlin Heidelberg 1986.

(10) HERRON M.M. (1982), Impurity sources of F^-, Cl^-, NO_3^- and SO_4^{2-} in Greenland and Antarctica precipitation. J. Geophys. Res. 87, 3052-3060.

(11) HYNES, A.J., WINE P.H. and SEMMES D.H. (1986), Kinetics and Mechanism of OH reactions with organic sulfides. J. Phys. Chem., 90, 4148-4156.

(12) HYNES, A.J. and WINE P.H. (1989), OH-initiated oxidation of biogenic sulphur compounds, in Biogenic sulphur in the environment, ACS Symp.Ser., vol. 393, ed. by Saltzman and Cooper, pp 424-436, Am. Chem. Soc., Washington, D.C.

(13) IVEY J.P., DAVIES D.M., MORGAN V. and AYERS G.P. (1986), Methanesulphonate in Antarctic ice. Tellus 38B, 375-379.
(14) LEGRAND M.R. and DELMAS R.J. (1984), The ionic balance of Antarctic snow: a 10-year detailed record. Atmos. Environ. 18, 1867-1874.
(15) LEGRAND M.R. and DELMAS R.J. (1987), A 220-year continuous record of volcanic sulphuric acid in the Antarctic ice sheet. Nature, 327, 671-676.
(16) LEGRAND M.R. and SAIGNE C. (1988), Formate, acetate and methanesulphonate measurements in Antarctic ice: some geochemical implications. Atmos. Environ., 22, 1011-1017.
(17) LEGRAND M.R., FENIET-SAIGNE C., SALTZMAN E.S., GERMAIN C., BARKOV N.I. and PETROV V.N., (1991), Ice-core record of oceanic emission of dimethylsulphide during the last climate cycle. Nature 350, 144-146.
(18) MAYEWSKI P.A., SPENCER M.J., LYONS W.B. and TWICKLER M.S., (1987), Seasonal and spatial trend in south Greenland snow chemistry. Atmos. Environ. 21, 863-869.
(19) MAUPETIT F. and DELMAS R.J., (1992), Chemical composition of falling snow at Dumont D'Urville, Antarctica. J. Atmos. Chem., 14, 31-42.
(20) MULVANEY R., PASTEUR E.C., PEEL D.A., SALTZMAN E.S. and WHUNG P.Y. (1992). The ratio of MSA to non sea-salt sulphate in Antarctic Peninsula ice cores. Tellus, in press.
(21) PICCARDI G., UDISTI R. and BARBOLANI E. (1989), Analysis of the principal components of Antarctic precipitations. Ann. Chim. (Rome) 79, 701-712.
(22) PICCARDI G., UDISTI R., BELLANDI S. and BARBOLANI E. (1990), Characterization of snow in the area surrounding Terra Nova Bay (Antarctica). Proceedings Environmental Impact in Antarctica - Rome 1990, p. 55-62.
(23) PICCARDI G., UDISTI R. and CASELLA F. (1993), Seasonal trend and chemical composition of snow at Terra Nova Bay (Antarctica). Int. J. Environ. Anal. Chem., submitted.
(24) PROSPERO J.M., SAVOIE D.L., SALTZMAN E.S. and LARSEN R. (1991), Impact of oceanic sources of biogenic sulphur on sulphate aerosol concentrations at Mawson, Antarctica. Nature 350, 221-223.
(25) SAIGNE C., KIRCHNER S. and LEGRAND M. (1987), Ion-chromatographic measurements of ammonium, fluoride, acetate, formate and methanesulphonate ions at very low levels in Antarctic ice. Anal. Chim. Acta 203, 11-21.
(26) SAIGNE C. and LEGRAND M. (1987), Measurements of methanesulphonic acid in Antarctic ice. Nature 330, 240-242.
(27) SHAW G.E. (1988), Antarctic aerosol: a review. Rewiews Geophys. 26, 89-112.
(28) UDISTI R., BARBOLANI E. and PICCARDI G. (1991), Determination of some organic and inorganic substances present at ppb level in Antarctic snow and ice by ion chromatography. Ann. Chim. (Rome) 81, 325-341.
(29) UDISTI R., BELLANDI S. and PICCARDI G. (1993), Optimization of ion chromatographic methods for the analysis of principal and trace components in rainwater and snow. J. Chromatography (in press).
(30) UDISTI R. and PICCARDI G., in preparation.
(31) WAGENBACK D., GORLACH U., MOSER K. and MUNNICH K.O. (1988), Tellus, 408, 426-436.

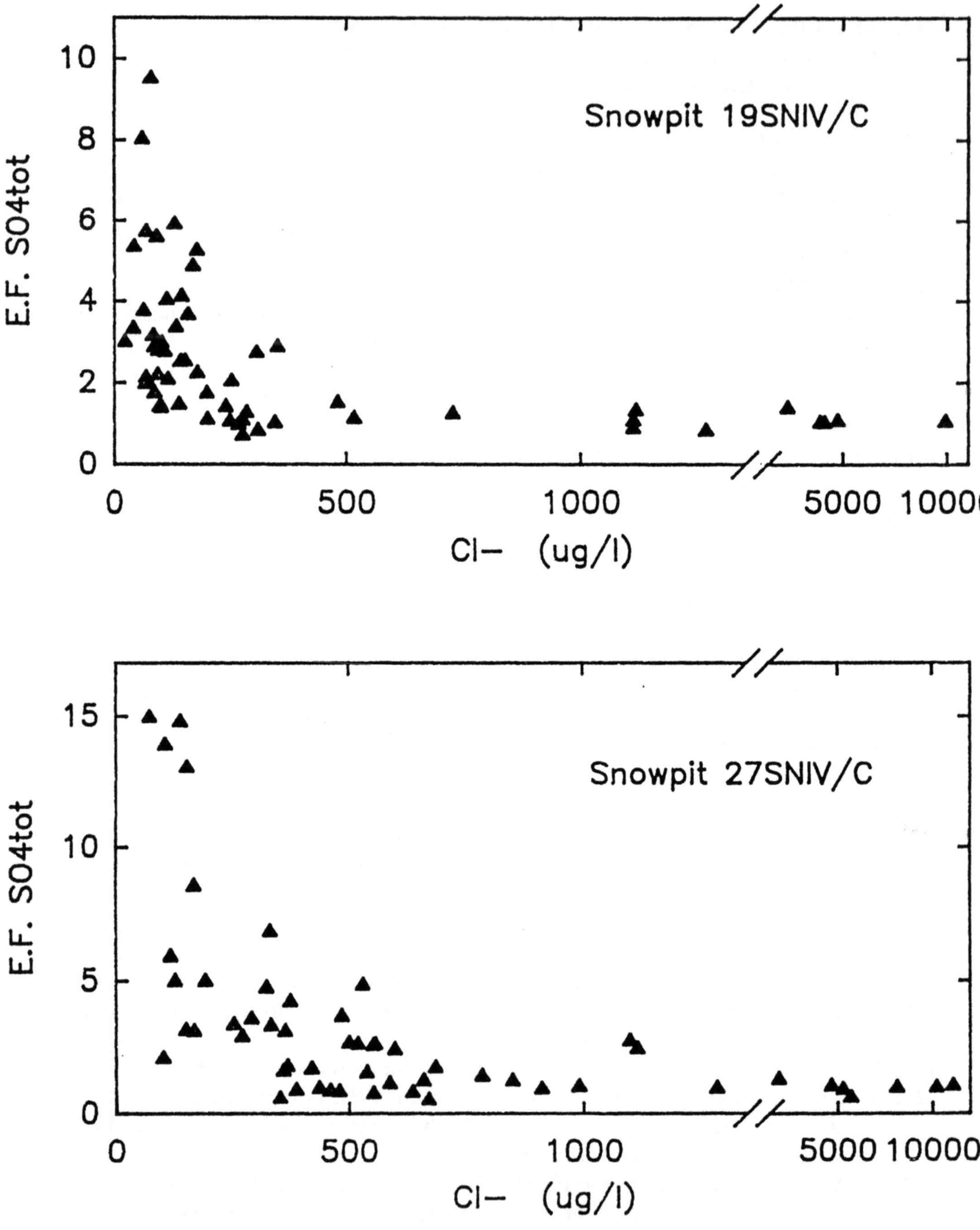

FIG. 1 - Enrichment factors of total SO_4^{2-} vs Cl^- concentration.

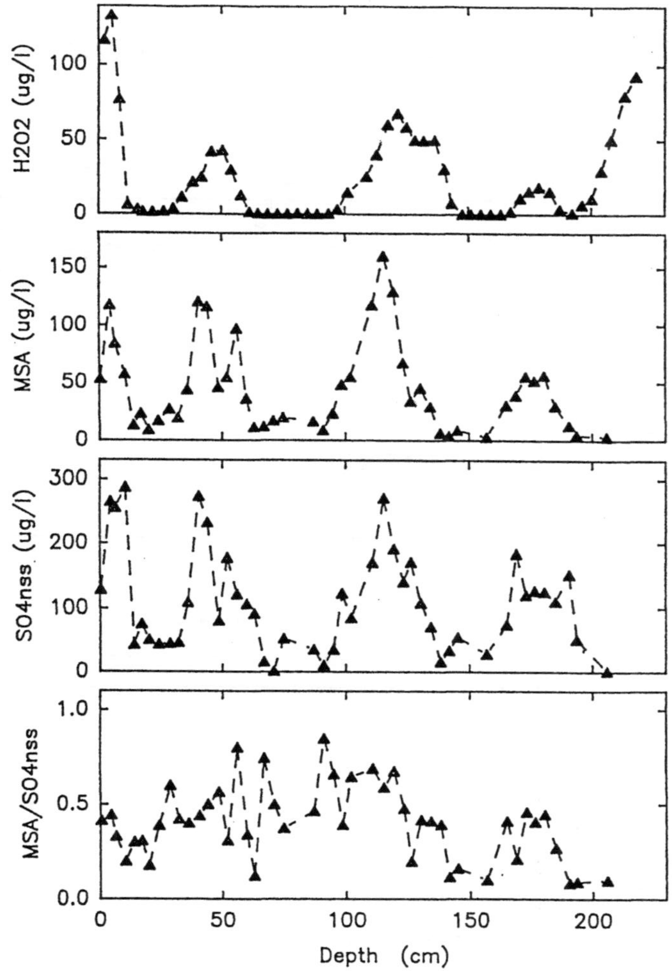

FIG. 2 - Concentration profiles for snowpit 27.

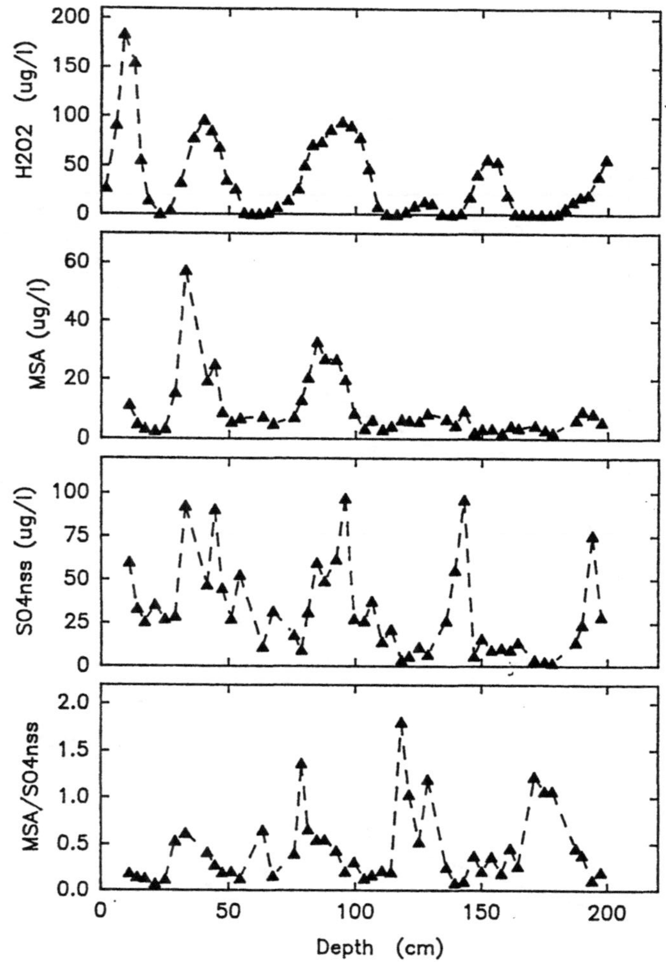

FIG. 3 - Concentration profiles for snowpit 19.

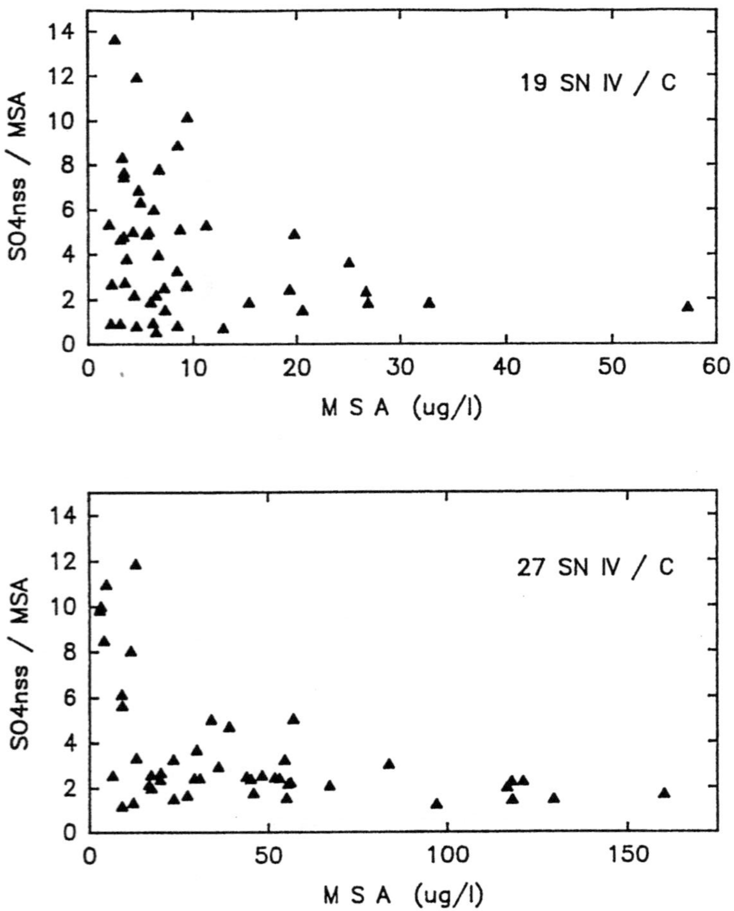

FIG. 4 - nss-SO_4^{2-} / MSA ratio vs MSA concentration.

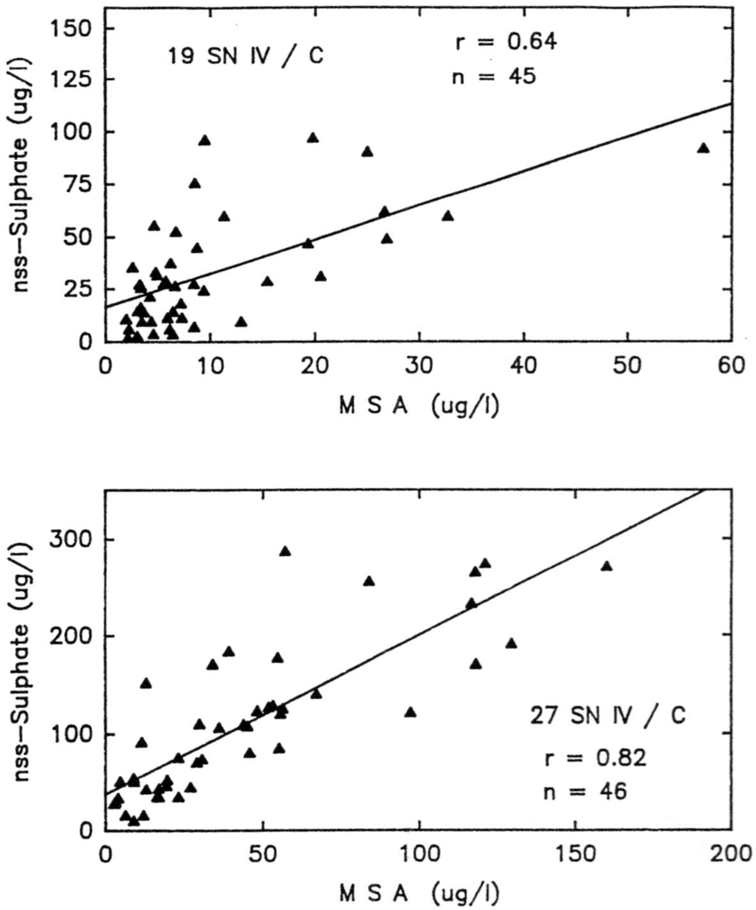

FIG. 5 - nss-SO_4^{2-} vs MSA concentration.

PRELIMINARY DATA ON DMS CONCENTRATION IN SEAWATER SAMPLES COLLECTED FROM THE LA SPEZIA GULF (LIGURIAN SEA)

R. BONIFORTI*, P. EMALDI**, R. FERRAROLI**, M. MASPERO**,
R. NAIR*, A. NOVO***

* ENEA Centro Ricerche Ambiente Marino La Spezia; ** CISE SpA Segrate (MI); *** ENEL-DSR Milano

SUMMARY

Some peculiar characteristics of the Mediterranean sea (e.g., high salinity and high solar irradiation) may make this sea a significant source of DMS, at least at a local scale. Here we present preliminary data on the concentration of DMS in water samples collected from the Gulf of La Spezia (Ligurian Sea). The concentration of DMS in the small area investigated up to now is significant even in comparison with the upwelling waters of Peru and is much higher than in various coastal/shelf waters.

1. INTRODUCTION

The climatic role of dimethylsulfide (DMS) produced by marine macroalgae and phytoplankton is not yet well understood. However, a reasonable agreement has been reached on the two following points: (i) sulfur input to the atmosphere due to DMS of marine origin [$(27\text{-}40)\times10^{12}$ g year^{-1}] may be as high as the half of the anthropogenic emissions [$(27\text{-}72)\times10^{12}$ g year^{-1}]; (ii) planktonic DMS, which is an important source of cloud-condensation nuclei in the marine troposphere, may influence the albedo because it has an effect on the scattering of solar radiation (1) and may increase the cloud cover.

Even if the Mediterranean is only a small fraction of the ocean, this sea may be a significant source of atmospheric sulfur because of its peculiar characteristics such as high salinity and high solar radiation. Since no systematic study of the presence of DMS in the Mediterranean Sea was available, an investigation was initiated with the aim of obtaining data from a coastal zone. A more systematic investigation will be performed at selected Mediterranean sites in the near future. When this task is accomplished, marine biogenic and anthropogenic sulfur fluxes to the Mediterranean atmosphere may be compared and the effect of marine DMS on the local cloud albedo can be evaluated.

2. SAMPLING AREA

Experiments were carried out during the period of April-August 1992 in the Gulf of La Spezia (Ligurian Sea). Fig. 1 shows the sampling stations. Stas. 1, 2 and 3 are within the bay, protected by a breakwater. Stas. 1, 2 and 3 were sampled regularly, while the others were occupied from 1 to 4 times. It should be noted that the inner area is highly affected by human activities.

3. METHODS

3.1 SAMPLING

Surface seawater samples were collected at a depth of 0.5-1 m in glass or polyethylene 0.25-1 L bottles, depending on subsequent measurements,

G. Restelli and G. Angeletti (eds.), Dimethylsulphide: Oceans, Atmosphere, and Climate, 163–172.

and field stored in the dark in a cooled container. Before sampling bottles were washed with seawater and filled completely without headspace. Samples were quickly transferred to the laboratory and kept at 4 °C until analysis. The DMS determination process was immediately initiated in order to avoid sample alteration. All the other analytical determinations were completed within 12-24 hours after sampling.

3.2 ANALYSIS

The DMS concentration in unfiltered samples was determined by a purge-and-trap method followed by gas chromatographic analysis similar to the technique described in (2) and (3) (Fig. 2). Comparison of DMS concentrations in the filtered and unfiltered samples showed no consistent differences if the samples were analyzed within 6-12 h, suggesting that neither disruption of algal cells nor the release of dimethylsulfoniopropionate could be presumed. If sample storage time was expected to exceed 24 hours, the samples were filtered (4, 5). 10-100 mL of seawater was purged with high-purity helium (100 mL/min) in a Dynamic Thermal Stripper (Supelco, Inc., Bellafonte, CA) for 15 min at 80 °C. DMS was trapped in a multibed (Carbotrap 300, Supelco Inc.) absorption tube kept at 85 °C. After purging the absorption tube was flushed with high-purity helium (50 mL/min) for 5 min. Under these conditions the recovery of DMS is almost quantitative (about 94%). DMS was thermally desorbed from the absorption tubes by a Thermal Desorption Unit (Supelco, Inc.) coupled via a heated transfer line to a 30 m megabore capillary column coated with a porous polymer (GS-Q, J & W Scientific, Folson, CA). The column was installed in a HRGC 5300 Carlo Erba gas cromatograph equipped with a flame photometric detector (FPD), kept at 150 °C, and a linearization module. The detection limit was estimated about 100 pg. Desorption conditions were as follows: temperature: 300 °C, heating time: 5 min, Helium flow rate: 7 mL/min. The gas chromatographic oven was kept at 70 °C during desorption, then heated to 120 °C at a rate of 2 °C/min, kept at 120 °C for 1 min, then heated to 200 °C at a rate of 20 °C/min. The chromatograms were recorded by means of a PC based integrator. Calibration curves were obtained by spiking artificial DMS-free seawater samples. High purity ethylene glycol stock solutions of DMS were gravimetrically prepared. Analyses of DMS were replicated for three samples collected at the same time from each site. So far our method has a sensitivity of 1 ng DMS/L and a precision of about 10% in the range 10-50000 ng DMS/L. Chlorophyll *a*, *b* and *c* and pheophytin were measured according to the UNESCO method (6); salinity and temperature were measured using a CDT probe (Idronaut, Milan), equipped with a polarographic sensor for oxygen determination.

4. RESULTS AND DISCUSSION

Besides the determination of DMS, other parameters which can affect the DMS concentration in SW have been determined. They are Chl *a*, *b* , and *c*, pheophytin, salinity, water temperature, oxygen, and meteorological conditions. The data are summarised in Tab. 1, which contains also the average values of the stations both inner and outer the breakwater and illustrates the global average as well. The average values are visualised in Fig. 3. The inner stations were found to have higher concentrations of DMS, chlorophyll *a* and pheophytin than the outer ones. It was found that an increase in [Chl*a*] was accompanied by an increase in [DMS]. However, as it was documented by many authors, the rates of increase are different (Fig. 4). This figure also shows that *(i)* the highest values of both [Chl*a*] and [DMS] occur at station 3 which is closest to the coast; and *(ii)* that the maximums in [Chl*a*] and [DMS] occurred in May and July, respectively.

The concentration of DMS in samples collected in May from the inner stations and samples collected in August from all stations shows a linear correlation with the concentration of chlorophyll *a*. (Fig. 5); the slope is nearly 5 times less than the slope of a similar plot for the other data. The reason for this difference is not presently understood, even if it is likely due to differences in phytoplankton and/or macroalgae species prevalent at different sampling times. Chlorophyll *a* is present in all algae, chlorophyll *b* occurs in green algae and chlorophyll *c* is found in brown algae, diatoms and dinoflagellates. The patterns of $[DMS]_{norm}$, $(Chla/Chlb)_{norm}$ and $(Chla/Chlc)_{norm}$ are given in Fig. 6 which shows that (1) the ratio $(Chla/Chlc)_{norm}$ is fairly constant, with the exception of station 5 sampled in May; (2) the ratio $(Chla/Chlb)_{norm}$ is much more variable, thus probably indicating a varying composition in green algae; and (3) $[DMS]_{norm}$ is much more correlated with $(Chla/Chlb)_{norm}$ than with $(Chla/Chlc)_{norm}$. The space-time variability of $(Chla/Chlb)_{norm}$ being much higher than that of $(Chla/Chlc)_{norm}$ may be one of the reasons for clustering of stations in two families. The speciation of algae and phytoplankton will be considered in future work. Since DMS production increases during phytoplankton senescence (7) and the senescence phase is accompanied by an increase in pheophytin, we have compared in Fig. 7 the patterns of $[DMS]_{norm}$ and $[Pheo/Chla]_{norm}$. This figure shows that in some cases (e.g., samplings 1.2, 3.2, 6.3, 7.3, 1.4) an increase or a decrease in [Pheo/Chl*a*] is accompanied by an analogous behaviour of [DMS], whereas in other cases (e.g., samplings 2.1, 3.1, 1.3, 3.3) the two curves show an opposite behaviour.

5. CONCLUSIONS

The present data set, though limited, seems to confirm that the factors influencing the marine production of DMS are difficult to define.

Table 2 shows the comparison of data from this initial survey with some of those given in the literature. It must be noted that the concentration of DMS in the small area investigated up to now is significant even in comparison with the upwelling waters of Peru and is much higher than in various coastal/shelf waters. Much more extended surveys are needed before calculating the flux of DMS to the Mediterranean atmosphere; however, this initial data set indicates that additional efforts are worthwhile since it is possible that fluxes of anthropogenic and biogenic sulfur are comparable also in this area.

6 REFERENCES

(1) Foley J.A., K.E. Taylor, S.J. Ghan (1991) Planktonic dimethylsulfide and cloud albedo: an estimate of the feedback response Climatic Change, 18,1-15

(2) Belviso, S., N. Mihalopoulos, B.C. Nguyen (1986) The supersaturation of carbonyl sulfide (OCS) in rain waters. Atmos. Environ., 21(6), 1363-1367

(3) Kim, K.H., M.O. Andreae (1987) Determination of carbon disulfide in natural waters by adsorbent preconcentration and gas chromatography with flame photometric detection. Anal Chem., 59, 2670-2673

(4) Bernard, W.R., M.O. Andreae, W.E. Watkins, H. Bingemer, H.W. Georgi (1982) The flux of dimethylsulfide from the oceans to the atmosphere. J. Geophys. Res., 87, 8787-8793

(5) Nguyen, B.C., S.Belviso, N. Mihalopoulos (1988) Dimethyl sulfide production during natural phytoplankton blooms. Mar. Chem., 24, 133-141

(6) UNESCO (1966): Determination of photosynthetic pigments in seawater. Monograph on Ocean Methodology

(7) Kaldeira, K. (1989) Evolutionary pressures on planktonic production of atmospheric sulphur. Nature, 33, 732-734

(8) Bates, T.S., J.D. Cline, R.H. Gammon, S.R. Kelly-Hansen (1987) Regional and seasonal variations in the flux of oceanic dimethylsulfide to the atmosphere. J. Geophys. Res., 92(C3), 2930-2938

(9) Iverson, R.L., F.L. Nearhoof, M.O. Andreae (1989) Production of dimethylsulfonium propionate and dimethylsulfide by phytoplankton in estuarine and coastal waters. Limnol. Oceanogr., 34(1), 53-67

(10) Leck, C., U. Larsson, L.E. Bagander, S. Johansson, S. Hajdu (1990) Dimethyl sulfide in the Baltic Sea: annual variability in relation to biological activity. J. Geophys. Res., 95(C3), 3353-3363

(11 Turner, S.M., G. Malin, P.S. Liss, D.S. Harbour, P.M. Holligan (1988) The seasonal variation of dimethyl sulfide and dimethylsulfoniopropionate concentrations in nearshore waters. Limnol. Oceanogr., 33(3), 364-375

sampling date	sta.		DMS (ng/L)	DMS nmol/L	chl a (ug/L)	chl b (ug/L)	chl c (ug/L)	pheo (ug/L)	temp. (°C)	salin.	oxyg. (sat %)	global radiation* (MJ/m2/d)	wind* (m/s)
04.28.92	1.1		1300	21.0	1.240	0.055	0.586	1.080	15.6	36.34		21,9 - 18,8	2,5 - 2,9
	2.1		650	10.5	0.490	-0.036	-0.027	0.570	15.65	36.4	87.9		
	4.1		3150	50.8	5.680	0.191	3.210	4.850	15.62	36.19	91.4		
	4.1		800	12.9	0.720	0.049	0.335	0.670					
		avg. 1	**1700**	**27.4**	**2.470**	**0.070**	**1.256**	**2.167**					
		avg. 2	**800**	**12.9**	**0.720**	**0.049**	**0.335**	**0.670**					
		avg. 3	**1475**	**23.8**	**2.033**	**0.065**	**1.026**	**1.793**					
05.26.92	1.2		208	3.4	1.883	0.14	0.747	1.169	18.46	37.23	97.2	23,9 - 25,7	3,4 - 3,7
	2.2		558	9.0	4.310	0.248	1.588	4.433	18.47	37.12	83.9		
	3.2		2903	46.8	25.676	7.552	6.893	31.506	18.75	37.24	92.4		
	4.2		603	9.7	0.539	0.055	0.137	0.514					
	5.2		822	13.3	0.351	-0.01	0.006	0.504					
		avg. 1	**1223**	**19.7**	**10.623**	**2.647**	**3.076**	**12.369**					
		avg. 2	**713**	**11.5**	**0.445**	**0.023**	**0.072**	**0.509**					
		avg. 3	**1019**	**16.4**	**6.552**	**1.597**	**1.874**	**7.625**					
07.21.92	1.3		647	10.4	0.313	0.033	0.172	0.530	23.14	36.21	90.8	26,7 - 24,9	2,5 - 2,4
	2.3		1168	18.8	1.805	0.183	0.665	2.040	23.24	36.15	98.1		
	3.3		3832	61.8	6.979	3.097	14.013	3.740					
	4.3		1022	16.5	1.012	0.107	0.454	1.002					
	5.3		1094	17.6	1.003	0.178	0.721	1.147					
	6.3		396	6.4	0.415	0.109	0.379	0.339					
	7.3		678	10.9	0.526	0.754	1.592	0.740					
	8.3		194	3.1	0.200	0.037	0.252	0.236					
		avg. 1	**1882**	**30.4**	**3.032**	**1.104**	**4.950**	**2.103**					
		avg. 2	**677**	**10.9**	**0.631**	**0.237**	**0.680**	**0.693**					
		avg. 3	**1129**	**18.2**	**1.532**	**0.562**	**2.281**	**1.222**					
08.25.92	1.4		<20	<0,3	1.079	0.167	0.614	1.083				21,6 - 21,7	2,4 - 3,1
	2.4		194	3.1	2.014	0.287	1.285	2.522	24.75	36.28			
	3.4		781	12.6	5.752	1.005	2.843	4.112					
	5.4		94	1.5	0.845	0.087	1.381	0.691					
		avg. 1	**332**	**5.3**	**2.948**	**0.486**	**1.581**	**2.572**					
		avg. 2	**94**	**1.5**	**0.845**	**0.087**	**1.381**	**0.691**					
		avg. 3	**106**	**1.7**	**2.423**	**0.387**	**1.531**	**2.102**					

avg. 1=average of stations inside the La Spezia bay in a semiclosed area separated from open sea by a breakwater (stas. 1,2,3)
avg. 2=average of stations outside the La Spezia bay in an area not protected by a breakwater (stas. 4, 5, 6, 7, and 8)
avg. 3=average of all stations
*values of the sampling day and the day before

Table 1. Data on DMS concentration and other parameters measured in the La Spezia Gulf

	min.	max.	mean
FLORIDA (8) (coastal waters)			0.44
PERU (8) (coastal waters)	0.94	43.75	6.88
DELAWARE BAY (9) (shelf waters)	2	13.6	
CHESEPEAKE BAY (9) (shelf waters)	2.0	10	
OCHLOCKONEE BAY (9) (shelf waters)	12	14	
BALTIC SEA (10) (sal. 0.6-0.8 ppt)	0.06	6.25	
MAINLAND BRITAIN (11) (coastal waters; summer)			6.87
MAINLAND BRITAIN (11) (coastal waters; winter)			0.12
MAINLAND BRITAIN (11) (coastal waters; overall)	0.03	34.37	
LA SPEZIA (this work) (coastal waters)	<0.3	61.8	16.21

TABLE 2. Sulfur concentration in SW from the La Spezia Gulf and other coastal and shelf areas (nmol/L)

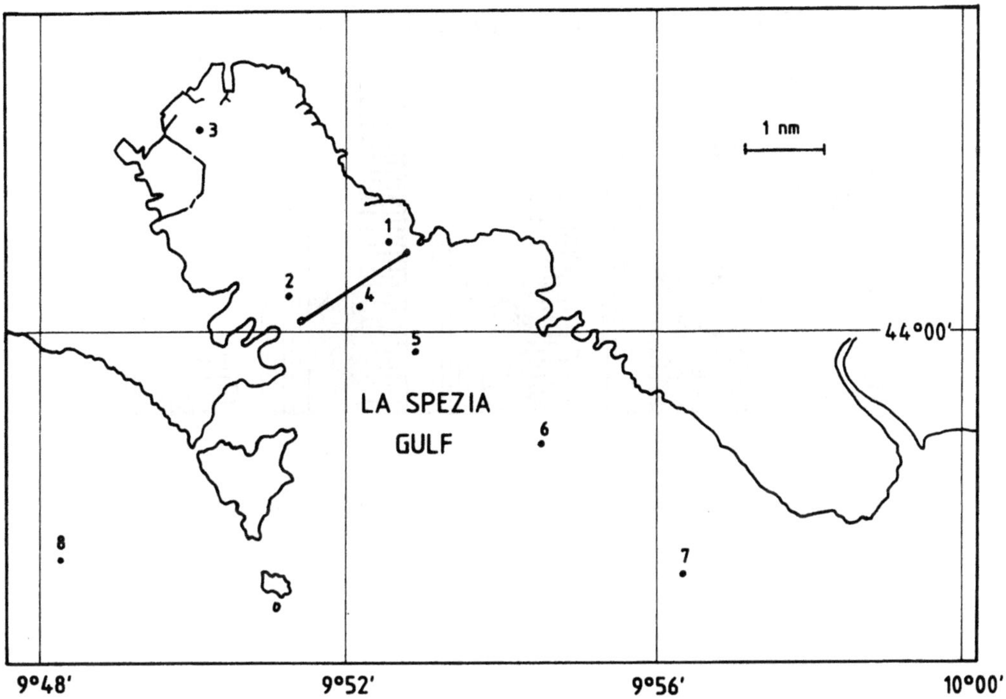

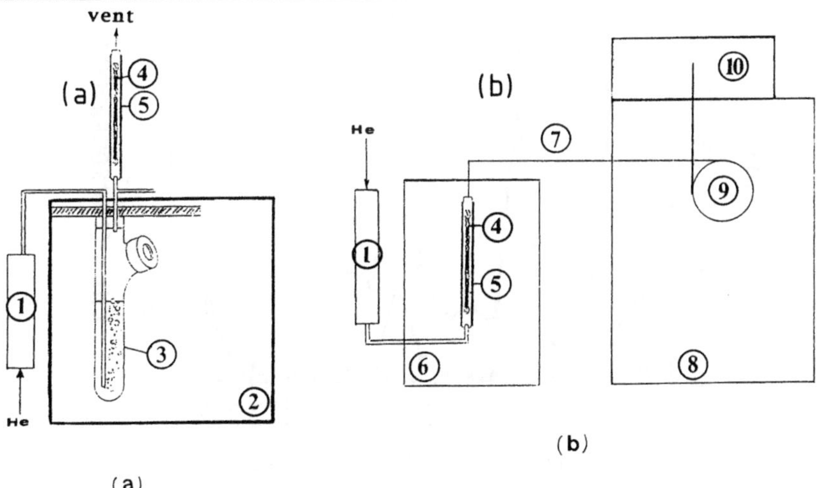

Fig. 2. (a) Field apparatus (b) Analytical apparatus

1 - gas scrubber
2 - dynamic thermal stripper
3 - sample vial
4 - thermal desorption tube
5 - heated jacket
6 - thermal desorber
7 - transfer line
8 - gas chromatograph
9 - GC column
10 - detector FPD

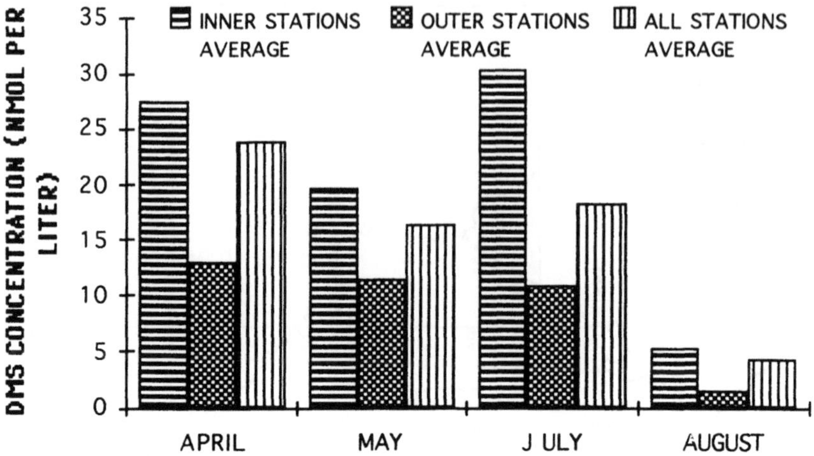

Fig. 3. DMS concentration vs. sampling month

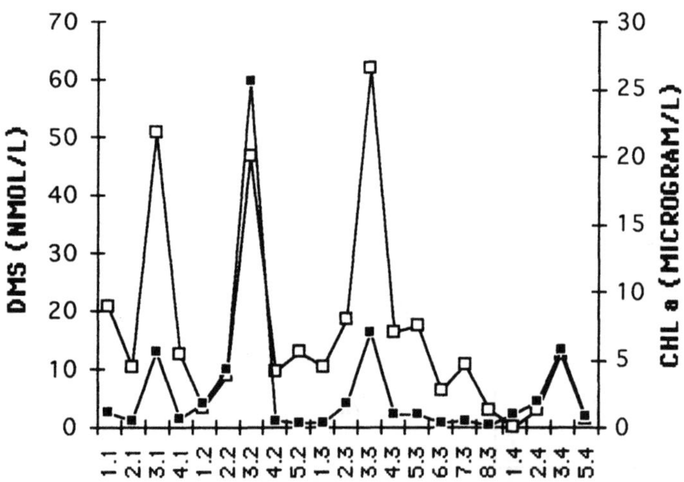

FIG. 4 Plot of [DMS] vs. [Chl *a*]

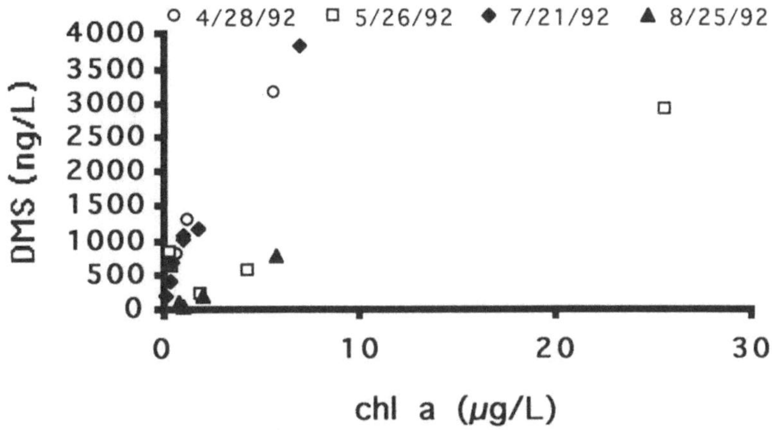

Fig. 5 Correlation between [DMS] and [Chl *a*]

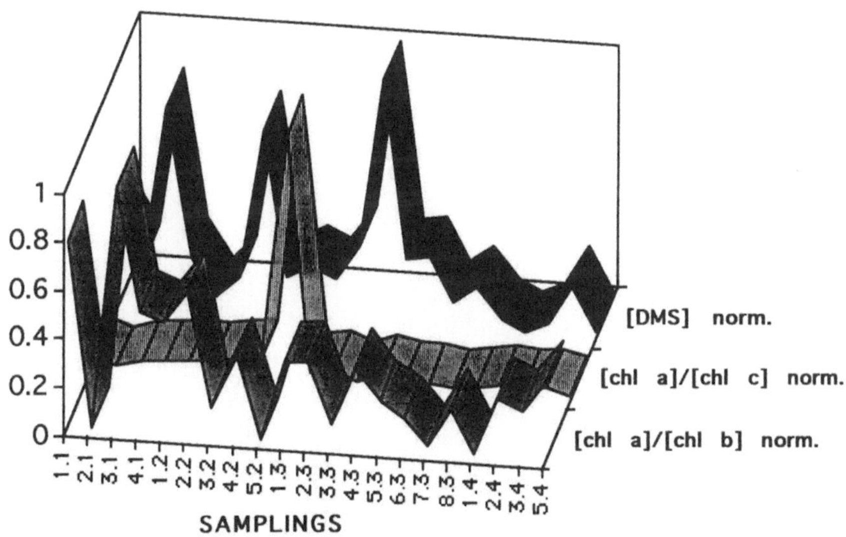

Fig. 6 Patterns of normalized [DMS], Chl *a*/Chl *c* and Chl *a*/Chl *b*

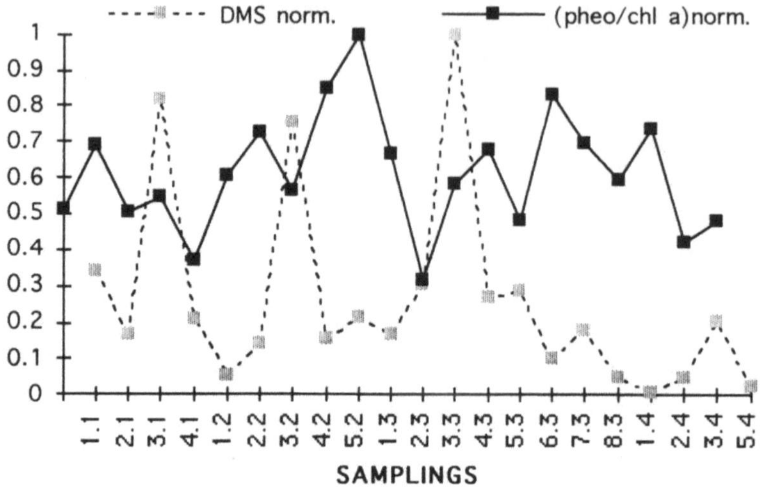

Fig. 7. Pattern of [DMS]norm. and [Pheo/Chl *a*] norm.

DIMETHYLSULPHIDE AND OTHER VOLATILE ORGANIC SULPHUR COMPOUNDS IN SOME NEGLECTED ECOSYSTEMS: A STUDY IN EVAPORITIC ENVIRONMENTS AND IN SULPHATE-RICH KARSTIC LAKES.

R. SIMó, R. DE WIT, J. O. GRIMALT and J. ALBAIGÉS.

Dept. of Environmental Chemistry. C.I.D.-C.S.I.C.
Jordi Girona 18-26. 08034 Barcelona, Catalonia, Spain.

SUMMARY

Dimethylsulphide (DMS) and other volatile sulphur compounds were studied in shallow coastal hypersaline water bodies and in a sulphate-rich karstic lake. DMS was the predominant volatile sulphur compound in the hypersaline environment. Its concentration was lowest in ponds with well established microbial mats of the cyanobacterium *Microcoleus chthonoplastes*. Apart from this drop, no clear trend with salinity was observed and the concentrations in the evaporating ponds were similar to values found in the oceanic environment. On the other hand, carbonylsulphide and methanethiol were predominant in a karstic lake. The concentrations of these compounds in the deep anoxic water layers were one to two orders of magnitude higher than DMS in the coastal solar salterns. Nevertheless, most of these volatile sulphur compounds do not escape to the atmosphere when the lake ecosystem is stratified.

1. INTRODUCTION.

A comprehensive understanding of the global sulphur cycle, and of the relations between oceanic dimethylsulphide (DMS) emissions and climate will not be achieved if the sulphur emissions from anthropogenic sources (1) or from various continental and coastal ecosystems are neglected. Reliable estimates of the anthropogenic sulphur emissions are now available (1,2). Conversely, it has been stated that "in contrast to the oceanic emissions, our knowledge about biogenic sources of volatile sulphur compounds from the continents is extremely limited" (3). Bates *et al.* (4) have elaborated a global sulphur emission inventory using the data base available so far. Their inventory contains some estimates for continental and coastal aquatic ecosystems, but, unfortunately, several aquatic ecosystems that are common on a global scale, have not been studied in this respect.

Wetlands have been generally recognized as potential sources of biogenic volatile sulphur compounds. Hence, the review of Andreae (5) considered coastal wetlands, but most studies cited therein have been performed on intertidal wetlands in temperate climates. Thus, the hypersaline evaporitic environments, which are common ecosystems in semi-arid climates, have not been studied. Examples of such ecosystems are represented by the solar salterns that are common along the Mediterranean coast (Figure 1). Typically, these solar salterns encompass an array of shallow evaporating ponds of increasing salinities ranging from sea-water up to halite precipitation. In each pond, the salinity of the water column only exhibits minor fluctuations. Hence, in solar salterns, one can study the relation between water column salinity, and the emission of the volatile sulphur compounds.

So far, the continental wetlands and lakes have received little attention. Moreover, the wetlands (6) and lakes (7) studied on the continents comprised fresh water environments with low sulphate concentrations. Especially, an important omission in the

G. Restelli and G. Angeletti (eds.), Dimethylsulphide: Oceans, Atmosphere, and Climate, 173–181.

Fig. 1: Map of the Iberian Peninsula indicating the distribution of ▼, Mediterranean coastal solar salterns; ⁝⁝⁝, regions where karstic lakes are common; ▒, regions were inland (hyper-)saline lakes are common. The sampling sites are indicated: 1, solar saltern ponds in "les Salines de la Trinitat" (Ebre Delta, S. Catalonia, Spain); 2, karstic lake Cisó (Banyoles, N.E. Catalonia, Spain).

global sulphur emission inventories published by Bates *et al.* (4) involve the continental (hyper-)saline lakes. Continental saline lakes comprise 104 x 10^3 km^3 water volume representing a quantity that is only slightly inferior to the volume of the fresh water lakes and rivers which equals 126 x 10^3 km^3 (8). The saline lakes are predominant in the semi-arid climates of Northern Africa, The Middle East, Central Asia, Northern America and Australia. Furthermore, they commonly occur in the Andes mountains and in the East-African Rift-Valley. However, within the European Community, the inland saline lakes are exclusively confined to the Iberian Peninsula (9, 10); their distribution in this region is depicted in Figure 1.

As mentioned above, another type of neglected aquatic ecosystem comprise the fresh water lakes with high sulphate concentrations. In the eastern part of the Iberian Peninsula, karstic lakes are common (11), (Figure 1). Most of these contain high sulphate concentrations caused by the dissolution of the gypsum present in the parent material. These lakes are stratified during a large part of the year, consisting of an oxic epilimnion and an anoxic hypolimnion. In the hypolimnion and in the sediments, bacterial sulphate reduction predominates over methanogenesis. Hence, the stratification of the bacterial populations and processes are similar as those found in stratified marine environments (12).

In this study, we explore the occurrence of volatile sulphur compounds in the shallow hypersaline coastal ecosystems. Secondly, we have studied a stratified karstic lake as a model system for a stratified community with high sulphate reduction rates in the bottom anoxic habitats.

2. MATERIALS AND METHODS.

The field sites studied are shown in Figure 1 and were, respectively: 1- the salt-evaporating ponds in the Salines de la Trinitat (Ebre Delta, South Catalonia, Spain) and 2- the karstic lake Cisó (Banyoles, N.E. Catalonia, Spain). The microbial communities present in these systems, were studied by optic phase-contrast microscopy.

Water samples for volatile organic sulphur compounds analysis were collected in glass bottles, stored at 4°C in darkness and measured close to the field site within 1-12 hours after sampling. The samples were purged with nitrogen for 15 to 20 min at room temperature and the volatile compounds were collected by cryogenic trapping (liquid argon). These compounds were analysed by gas-chromatography (GC) on a field apparatus using the procedures described by Haunold *et al.* (13).

In the solar saltern ponds, the salinities were estimated by density measurements using a conversion table between salinities and densities expressed in degrees Baumé. In the karstic lake, total dissolved sulphide was measured colorimetrically (14); oxygen and temperature were measured with a combined WTW oxygen/temperature probe.

3. RESULTS AND DISCUSSION.

Solar saltern.

The volatile sulphur compounds in the water column of six salt-evaporating ponds with different salinities have been studied. The samples were taken at midday. The concentration of the compounds as a function of the salinity in each pond is shown in Figure 2. DMS was the most abundant compound, generally, its concentration was within the range of 1 to 4 nmol/l, as reported for many oceanic environments (5). However, a much higher DMS concentration (33 nmol/l) was found in the pond with 140 g/l salinity, whereas very low DMS concentrations (< 0.4 nmol/l) did occur in ponds with 110 to 130 g/l salinity.

Both carbonylsulphide (COS) and carbondisulphide (CS_2) were also present over the full salinity range, but their concentration was one order of magnitude lower than DMS. Similarly as for DMS, the lowest concentrations of COS and CS_2 were found in ponds with 110-130 g/l salinity. Methanethiol (MeSH) and hydrogen sulphide (H_2S) were only present in the water column of the first concentrating pond (55 g/l salinity).

These differences can be compared with the biota present in the different ponds. The first concentrating pond is dominated by large areas of muddy black sediment very rich in organic material. Microbial mats of the cyanobacteria *Phormidium valderianum*, *Microcoleus chthonoplastes* and *Lyngbya* spp. are found mainly along the borders, whilst patches of filamentous algae, e.g., *Cladophora* spp., are also present.

The ponds of 110-130 g/l dissolved salt, which had the lowest concentrations of the volatile sulphur compounds, are characterised by well established mats of the cyanobacterium *M. chthonoplastes*, that cover the pond sediments over their entire surface. Disintegrating mats of *M. chthonoplastes* were found in the pond at 140 g/l salinity and a large surface of its sediment was devoid of these mats. Thus, it appears that the remarkably low concentrations of the volitile sulphur compounds in the 110-130 g/l salinity ponds, are related directly to the presence of well established *M. chthonoplastes* -mats, rather than to a direct effect of the salinity. In this respect, a high capacity for metabolising DMS has been reported for other marine environments with well established *M. chthonoplastes* -mats (15).

The evaporating pond at 210 g/l dissolved salts had mats of the cyanobacterium *Aphanothece halophytica* incrustated with gypsum crystals that formed undulating

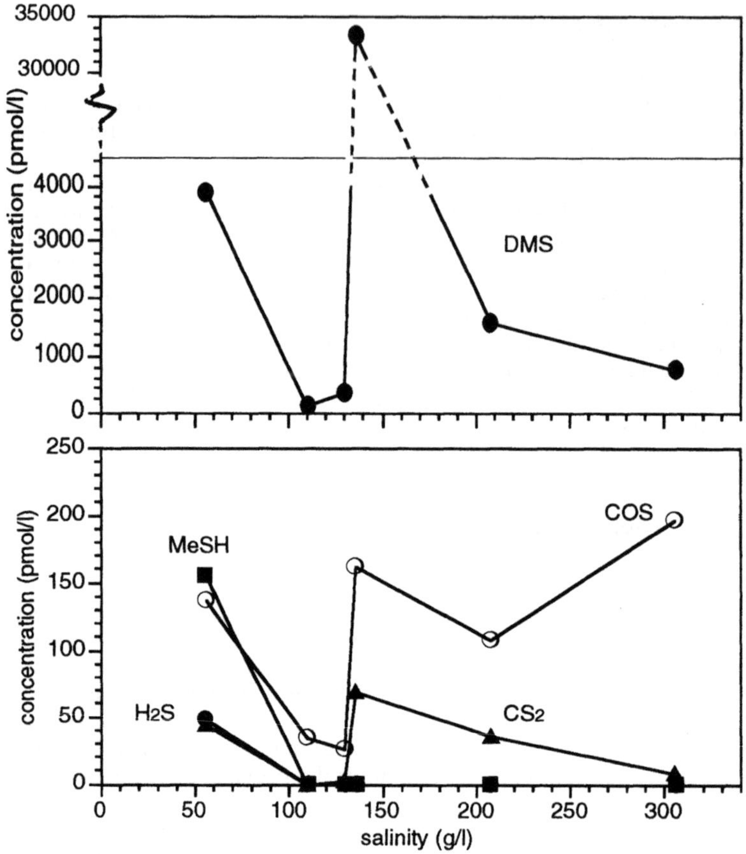

Fig. 2: Relation between the concentrations of the volatile sulphur compounds and the salinity in the water column of the evaporating ponds in the coastal solar saltern measured 27 May 1992.

structures covering the whole surface of the sediment. In comparison to the ponds with the *Microcoleus*-mats, DMS, COS and CS_2 concentration were high. Above 300 g/l the extreme halophilic archeabacteria dominated.

In short, DMS was the most abundant volatile sulphur compound in the hypersaline environment studied, followed by COS and CS_2. This composition is probably related to the fact that most DMS originates from the hydrolysis of the osmolyte DMSP which is present in both planktonic and benthic (16) phototrophic microorganisms. COS may originate from photochemical reactions of organic sulphur compounds including DMS and DMSP (17). Neither DMS, nor COS concentrations did show a clear trend with the salinity. It seems that in these shallow aquatic environments, the benthic communities influence the volatile sulphur compounds present in the water column.

Another source of the volatile sulphur compounds represent the anaerobic microbial degradation processes that take place in the sediment. These processes may give rise to volatile sulphur compounds including H_2S, MeSH, DMS and dimethyldisulphide (18-20). Hence, the variety of volatile sulphur compounds found in the water column of the first concentrating pond suggest an effect of these anaerobic degradation processes.

Sulphate-rich karstic lake.

This lake was studied as a model system for a stratified community with high sulphate-reduction rates in the bottom parts. From the microbiological point of view, such a lake bears resemblance to benthic microbial mat communities, although the scale of the stratification is very different. Hence, in a microbial mat, the different populations comprise millimetre fine laminae whilst in the lake, the planktonic populations form decimetre thick plates (12). Lake Cisó was sampled on 17-18 September 1992. On this date, a thin plate of purple sulphur bacteria, mainly *Chromatium minus* and *Amoebobacter purpureus*, were found at 1 m depth. Below this layer, high densities of the consortium *Chlorochromatium* were found. The alga *Cryptomonas phaseolus*, which is a characteristic species in this lake, was only found in extremely low densities. The surface of the lake was covered with a film comprising organic material and neustonic bacteria.

The depth profiles of temperature, oxygen, total sulphide and volatile organic sulphur compounds are shown in Figure 3. The lake was thermally stratified; the thermocline was from about 80 cm to 2 m depth. Very low oxygen concentrations were found in the epilimnion, because the densities of the oxygen-producing algae were extremely low and the superficial neustic layer strongly reduced oxygen penetration. In the first metre, the coexistence of oxygen and hydrogen sulphide was observed. Due to the very high hydrogen sulphide concentrations, it was not possible to obtain reliable simultaneous measurements of this compound and the other volatile sulphur compounds. Therefore, no efforts were made to quantify H_2S. Moreover, in the deeper layers, the elevated hydrogen sulphide concentrations gave rise to chromatographic interferences which resulted in increased detection limits for the other volatile sulphur compounds.

In contrast with the hypersaline environment, not DMS, but rather COS and MeSH were the most abundant volatile sulphur compounds in the lake. The highest concentrations of COS and MeSH were found below 1 m water depth. Strikingly, the deep anoxic water layers in the lake did contain one to two orders of magnitude higher concentrations of volatile organic sulphur compounds than found in the water column of the solar salterns. However, the top oxic layers of the lake were characterised by much lower concentrations.

In the karstic lake, probably, DMSP is not an important DMS precursor and most volatile organic sulphur compounds are generated in the anoxic layers by anaerobic microbial degradation processes. Dead planktonic microorganisms, and dead plant material including fallen trees, are captured in the anoxic parts of the lake and provide substrates for these microbial processes. Thus, volatile sulphur compound may be generated from sulphur containing cell constituents (18-20). Furthermore, the recently discovered sulphide-dependent bacterial demethoxylation of lignin residues (21, 22) is perhaps important in the lake. This process results in mixtures of DMS and MeSH (21, 22). In lake Cisó, MeSH is about one order of magnitude higher than DMS which does not correspond to other observations in fresh water lakes where DMS predominated as the end product (21, 22). Probably, this is related to the high sulphide concentrations in the lake, since sulphide and MeSH are competing substrates for the methyltransferases involved in the demethoxylation process.

In the photic zone, it was found that DMS concentrations are higher at night than during daytime. This trend may reflect the activity of phototrophic purple bacteria which are able to photooxidize this compound to dimethylsulphoxide (23).

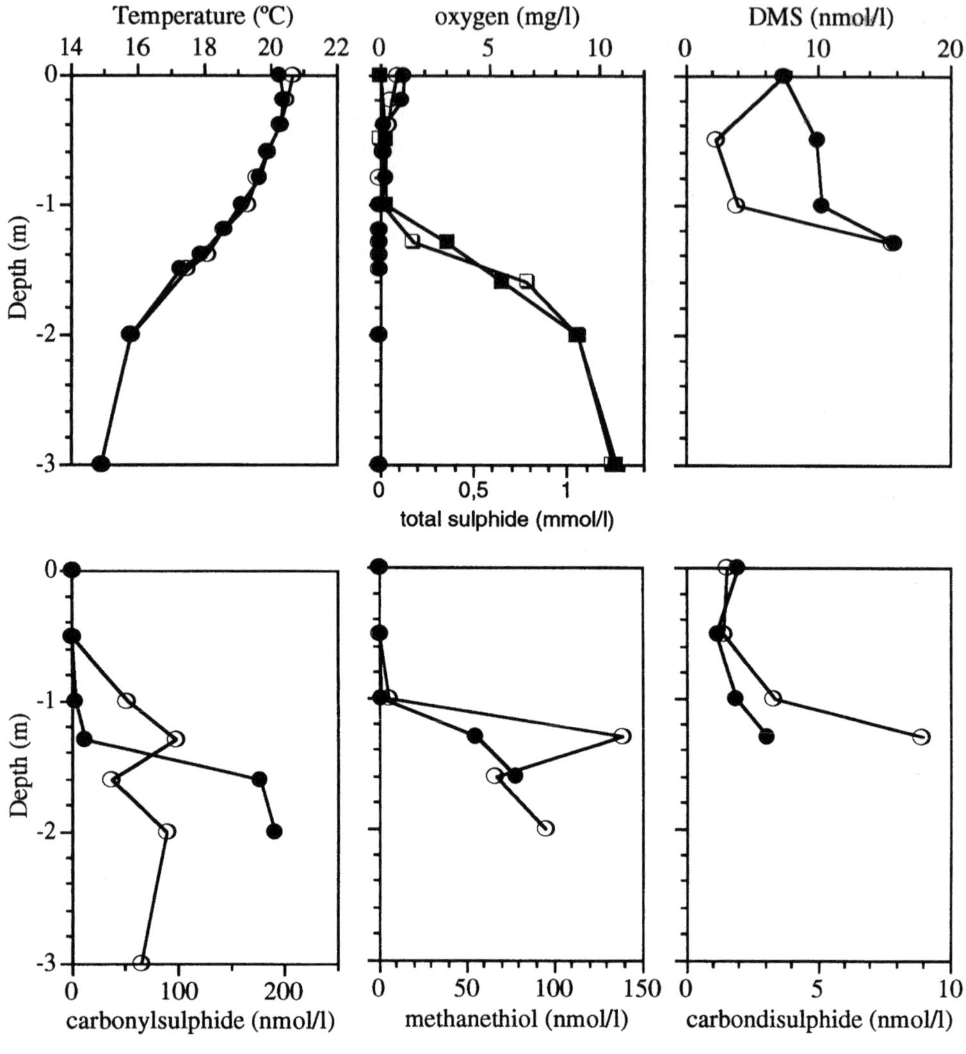

Fig. 3: Depth profiles of temperature, oxygen, total sulphide and different volatile organic sulphur compounds in the karstic lake Cisó measured on 17-18 September 1992. Open symbols indicate day time profiles (17 September 16.00 h) and closed symbols indicate night time profiles (18 September 5.00 h).

The hollow profiles of COS and MeSH indicate that both compounds are consumed in the epilimnion. A drastic decrease of the slope of the MeSH profile does occur from 1.3 m upto 50 cm depth which is coincident with the plate of the purple sulphur bacteria.

A preliminary experiment was performed on 18 September 1992, to measure the potential COS and MeSH disappearance rates at 1 m depth. Lake water samples from the

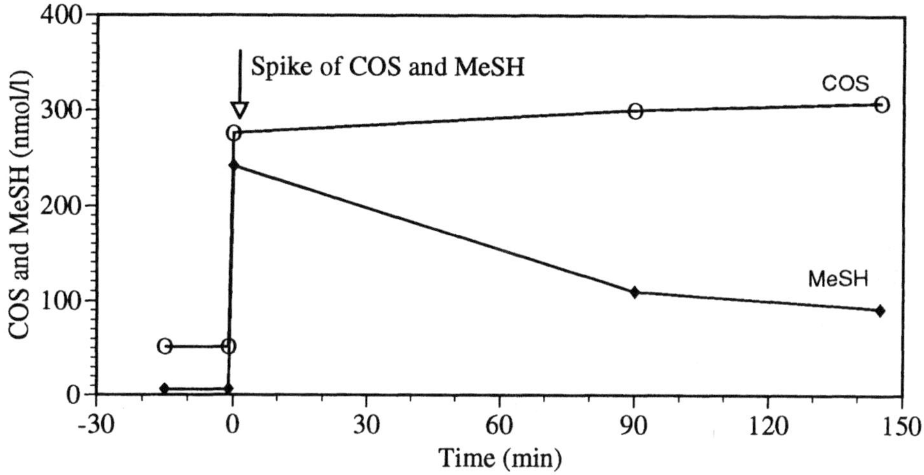

Fig. 4: Time course of an *in situ* incubation experiment to determine the potential disappearence of MeSH and COS at 1 m depth, measured on 18 September 1992 from 16.00 to 19.00 h. On time zero, COS and MeSH were added to the incubation vial to the final concentrations indicated.

layer of interest were transposed into incubation vials, spiked with COS and MeSH and incubated *in situ* at the same depth in the lake. At different times, small aliquots were withdrawn for the analyses of volatile sulphur compounds. The time series of this experiment is shown in Figure 4. It was found that MeSH disappeared, although the rate declined to low values when the MeSH concentrations had decreased to approximately 100 nmol/l. In contrast, during the time period studied, COS did not disappear and even a slight increase was observed. MeSH disappearance is probably due to microbial activity. It has been found that MeSH is degraded by chemotrophic sulphur bacteria (24). The absence of COS disappearance in the presence of MeSH may perhaps be related to diauxy, i.e. its consumption does not take place unless MeSH levels have dropped to low values. Another possibilty is that COS is formed from MeSH. Obviously, more experiments are required to study these possibilities.

4. CONCLUSION

We conclude that more studies in different coastal and continental aquatic environments are necessary to get a better understanding of their contribution to biogenic emissions of volatile sulphur compounds. In this respect, within the European Community, the Iberian Peninsula offers excellent opportunities to study (hyper-) saline and karstic aquatic ecosystems that are common in more remote and less accessible areas.

More detailed studies are required to understand the microbial degradation of volatile organic sulphur compounds. The studies in stratified lakes may be very useful to bridge the gap between physiological studies with isolated microbial cultures in the laboratory and environmental studies. In this respect, sulphate-rich karstic lakes are appropriate for developing methods to study the *in situ* metabolism.

ACKNOWLEDGEMENTS.

Financial support from DGICYT (MEC), Project No. PB90-0094, is acknowledged. R. Simó and R. de Wit have been supported by grants from the Spanish Ministry of Education and the European Environmental Research Organisation (E.E.R.O.), respectively. We are grateful to INFOSA for access to the solar saltern ponds in the Ebre Delta. Thanks are due to Dr. Carles Abella (University of Girona, Council of Banyoles) for accommodation and field lab facilities close to lake Cisó.

REFERENCES.

(1) SCHWARTZ, S.E. (1988). Are global cloud albedo and climate controlled by phytoplankton? Nature (London) **336**, 441-445

(2) LANGNER, J., ROHDE, H., CRUTZEN, P.J. and ZIMMERMAN, P. (1992). Anthropogenic influence on the distribution of tropospheric sulphate aerosol. Nature **359**, 712-715

(3) ANDREAE, M.O. and JAESCHKE, W.A. (1992). Exchange of sulphur between biosphere and atmosphere over temperate and tropical regions. *In:* R.W. Howarth, J.W.B. Stewart and M.V. Ivanov (Eds.). Sulphur cycling on the continents. SCOPE 48. pp. 27-61. Wiley, Chichester.

(4) BATES, T.S., LAMB, K.B., GUENTHER, A., DIGNON, J. and STOIBER, R.E. (1992). Sulfur emissions to the atmosphere from natural sources. J. Atmospheric Chem. **14**, 315-337

(5) ANDREAE, M.O., (1990). Ocean-atmosphere interactions in the global biogeochemical sulfur cycle. Mar. Chem. **30**, 1-29

(6) COOPER, D.J., COOPER, W.J., DE MELLO, W.Z., SALTZMAN, E.S. and ZIKA, R.G. (1989). Variability in biogenic sulfur emissions from Florida Wetlands. *In:* E.S. Saltzman and W.J. Cooper (Eds.) Biogenic sulphur in the Environment. pp 31-43. Am. Chem. Soc. Symp. Ser. 393, Am. Chem. Soc., Washington.

(7) RICHARDS, S.R., KELLY, C.A. and RUDD, J.W.M., (1991). Organic sulfur in lakes of the Canadian Shield and its loss to the atmosphere. Limnol. Oceanogr. **36**, 468-482

(8) WILLIAMS, W.D. (1986). Limnology the study of inland waters: a comment on perceptions of studies on salt lakes, past and present. In: De Decker, P. and Williams, W.D. Limnology in Australia. pp. 471-486. Dr. W. Junk Publishers, Dordrecht.

(9) COMIN, F. and ALONSO, M. (1988). Spanish saline lakes: their chemistry and biota. Hydrobiologia **158**, 237-245

(10) GUERRERO, M.C. and DE WIT, R. (1992). Microbial mats in the inland saline lakes of Spain. Limnetica **8**, 197-204

(11) MIRACLE, M.R., VICENTE, E. and PEDROS-ALIO, C. (1992). Biological studies of Spanish meromictic and karstic lakes. Limnetica **8**, 59-77

(12) JØRGENSEN, B.B. (1982). Ecology of the bacteria of the sulphur cycle with special reference to the anoxic/oxic interface. Phil. Trans. R. Soc. London **B298**, 543-561

(13) HAUNOLD, W., GEORGII, H.W. and OCKELMANN, G. (1992). Gas chromatographic analysis of atmospheric sulfur dioxide and reduced sulfur compounds. LC-GC Int. **5(10)**, 28-35

(14) TRÜPER, H.G. and SCHLEGEL, H.G. (1964). Sulphur metabolism in Thiorhodaceae. I. Quantitative measurements on growing cells of *Cromatium okenii*. Antonie van Leeuwenhoek J. Microbiol. Serol. **30**, 225-238

(15) VISSCHER, P.T., QUIST, P. and VAN GEMERDEN, H. (1990). Methylated sulfur compounds in microbial mats: *in situ* concentrations and metabolism by a colorless sulfur bacterium. Appl. Environ. Microbiol. **57**, 1758-1763

(16) VISSCHER, P.T., and VAN GEMERDEN, H. (1991). Production and consumption of dimethylsulfoniopropionate in marine microbial mats. Appl. Environ. Microbiol **57**,

(17) FEREK, R.J. and ANDREAE, M.O., 1984: Photochemical production of carbonyl sulphide in marine surface waters. Nature (London) **307**, 148-150

(18) DROTAR, A., BURTON, G.A., TAVERNIER, J.E. and FALL, R., (1987). Widespread occurrence of bacterial thiol methyltransferases and the biogenic emission of methylated sulfur gases. Appl. Environ. Microbiol. **53**, 1626-1631

(19) KIENE, R.P. and TAYLOR, B.F., (1988). Biotransformations of organosulfur compounds in sediments via 3-mercaptopropionate. Nature (London) **332**, 148-150

(20) TOMITA, B., INOUE, H., CHAYA, K., NAKAMURA, A., HAMAMURA, N., UENO, K., WATANABE, K. and OSE, Y. (1987). Identification of dimethyl disulfide-forming bacteria isolated from activated sludge. Appl. Environ. Microbiol. **53**: 1541-1547

(21) BAK, F. , FINSTER, K and ROTFUß, F., (1992). Formation of dimethylsulfide and methanethiol from methoxylated aromatic compounds and inorganic sulfide by newly isolated anaerobic bacteria. Arch. Microbiol. **157**, 529-534

(22) FINSTER, K., KING, G.M., and BAK, F., (1990). Formation of methylmercaptan and dimethylsulfide from methoxylated aromatic compounds in anoxic marine and fresh water sediments. FEMS Microbiol. Ecol. **74**, 295-302

(23) ZEYER, J., EICHER, P., WAKEHAM, S.G. and SCHWARZENBACH, R.P. (1987). Oxidation of dimethyl sulfide to dimethyl sulfoxide by phototrophic purple bacteria. Appl. Environ. Microbiol. **53**, 2026-2032

(24) KELLY, D.P. and SMITH, N.A., 1990: Organic sulfur compounds in the environment. Biogeochemistry, microbiology, and ecological aspects. *In:* K.C. MARSHALL. Advances in Microbial Ecology **11**, 345-385

STABLE SULFUR ISOTOPE RATIOS: SOURCE INDICATORS(*)

J. CALHOUN(*), R. DELMAS(*), H. NIELSEN(**), G. GRAVENHORST(**) and J. JOUZEL(***)

(*) LGGE, BP 96, 38402 Saint Martin d'Heres, France
(**) Institute für Geochemie/Bioklimatologie, respectively, Göttingen, Germany
(***) LMCE/CEA/DMS CEN, Saclay, France

ABSTRACT

Understanding the relative importance of the sulfur sources to the remote atmosphere is crucial for determining the environmental and possible climatic consequences of anthropogenic sulfur emissions. The methane sulfonate (MSA) to non-seasalt (nss) sulfate molar ratio (R) has been thought to be useful for estimating the relative amounts of nss sulfate from continental and marine sources, based on a near constant ratio of 6.5% of the first reported ratios from the lower-latitude marine sites. However, during more recent studies, large variations in R have been observed suggesting that other factors such as temperature, light intensity, oxidant concentration, and the size range of the sampled particles may affect the observed molar ratio (Bates et al., in press). Therefore, the use of R as a marine indicator is limited.
Stable isotope ratios of sulfur (δ^{34} S) are another potential source indicator based on the large variability of the isotopic composition of sulfur in the atmosphere and its source dependency (Nielsen, 1974). During a recent study in the remote South Pacific, the average δ^{34} S value of the nss sulfate in submicrometer marine aerosol particles (+ 15.6 +/- 3.1°/$_{oo}$, relative to CDTS) was consistently lighter than sulfate derived from seawater but isotopically heavier than sulfate derived from most continental sources (Calhoun et al., 1991). These results support the use of the δ^{34} S in nss sulfate as a marker for biogenic nss sulfate derived from DMS.
In the framework of the Eurocore (CCE) project, a group of scientists is currently preparing to make measurements of the δ^{34} S of sulphate from polar ice, with the hopes of obtaining additional information about marine, volcanic, and anthropogenic sources. Such studies have never been performed because of the low concentrations of sulfur in polar ice and the analytical limits imposed by the chemical pretreatment and/or mass spectral analysis. Selection and sub-sampling of the ice cores, preconcentration and chemical conversion of sulfate samples to Ag_2S will be performed at L.G.G.E. (J. Calhoun and R. Delmas). This compound will be further processed to SF_6 for mass spectral analysis according to the method developed by H. Nielsen and G. Gravenhorst. The group has access to a new powerful MS (J. Jouzel) where it is hoped that samples in the sub micromole range can be run routinely with a precision of +/- 0.2°/$_{oo}$.

(*) Only the abstract is available.

G. Restelli and G. Angeletti (eds.), Dimethylsulphide: Oceans, Atmosphere, and Climate, 183.

THE ATMOSPHERIC OXIDATION OF DIMETHYL SULFIDE: ELEMENTARY STEPS IN A COMPLEX MECHANISM

A. A. Turnipseed and A. R. Ravishankara*
National Oceanic and Atmospheric Administration, Aeronomy Laboratory, R/E/AL2, 325 Broadway, Boulder, CO 80303 and The Cooperative Institute for Research in Environmental Sciences, University of Colorado, Boulder, CO 80309
* Also associated with the Department of Chemistry and Biochemistry, University of Colorado, Boulder CO 80309.

SUMMARY

The current understanding of the atmospheric oxidation of dimethyl sulfide (DMS) is based on atmospheric measurements, smog chamber studies, and laboratory studies of elementary reactions. Synthesis of the oxidation mechanism from elementary reaction studies is reviewed. The reactions that may initiate the atmospheric oxidation of the DMS and the products of such reactions is discussed. The role of CH_3S and CH_3SO_x, the key intermediates in the oxidation mechanisms, is described in an attempt to connect their oxidation with the formation of SO_2 and/or other products. Special emphasis is placed on understanding the yield of H_2SO_4 in these oxidation mechanisms and how the yield may vary with temperature, pressure, and composition.

1. INTRODUCTION

Organo-sulfur compounds from natural sources account for ~60% of the total sulfur flux into the atmosphere in the southern hemisphere and ~15% in the northern hemisphere,[1] mostly in the form of Dimethyl Sulfide (DMS) from oceanic phytoplankton.[2] It is postulated that if DMS is oxidized to gas-phase H_2SO_4, new condensation nuclei (CN) can be formed which leads to the formation of aerosols and cloud condensation nuclei (CCN). These, in turn, produce clouds. Both aerosols and clouds can affect the heat balance of the atmosphere and possibly influence climate.[3] Smog chamber experiments indicated that DMS oxidation produces mostly SO_2,[4,5] which is further oxidized in the atmosphere to sulfate in the time scale of weeks; the homogeneous oxidation of SO_2 is expected to produce $H_2SO_{4(g)}$. Formation of $H_2SO_{4(g)}$ is crucial for the production of new particles. Also, if SO_2 is produced, it would be widely dispersed before it is oxidized to H_2SO_4, since the SO_2 lifetime is much longer than that of DMS. However, recent field[6] and modeling studies[7] suggest that SO_3, rather than SO_2, may be formed in DMS oxidation. If true, the formation of the particles would be restricted to the regions where DMS is oxidized and the fraction of DMS that leads to particles would also be changed. Alternatively, if DMS is oxidized to species such as methanesulfonic acid (CH_3SO_3H, MSA), dimethyl sulfoxide, or dimethyl sulfone, these species can be lost from the atmosphere with no new aerosol particle production. Furthermore, if SO_2 or SO_3 is oxidized heterogeneously to $SO_4^=$, new CNs are not produced because it takes place on a pre-existing particle. Therefore, it is important to understand the kinetics and mechanisms of oxidation of DMS, as well as that of SO_2 and SO_3, to determine the final oxidation products, atmospheric lifetimes, and particle production potentials of these compounds.

In this paper we attempt to construct an overall mechanism for the oxidation of organo-sulfur compounds using the evaluated data base[8,9] on the elementary reactions. Using this information, a few observations regarding the temperature dependence of the MSA formation and the likely atmospheric oxidation paths are made.

2. INITIATION REACTIONS

Table I lists various reactions that may remove DMS from the atmosphere along with the lifetimes, τ, for each of the loss reactions. It is clear that reactions with OH (a daytime species) and NO_3 (a nighttime species) are the major loss processes for DMS, with the possible exception of BrO reactions in the Arctic. Figure 1 shows a schematic of the major processes which initiate the oxidation of DMS and the expected products. The reactions of DMS with O_3, NO_2 and HO_2 are too slow to be important.[10-12] The table shows that the atmospheric lifetime of DMS is a day or so, and the relative importance of OH and NO_3 depends on the concentrations of NO_x,

G. Restelli and G. Angeletti (eds.), Dimethylsulphide: Oceans, Atmosphere, and Climate, 185–195.

Table I. Reactions that could contribute to the removal of DMS from the atmosphere, along with their relative contributions to the overall lifetime of DMS

X	[X] (a)	k(298K) (b)	τ (c)	COMMENTS	REF.
OH	2.5×10^6 (d) 6×10^5 (e)	$\cong 5 \times 10^{-12}$	20 80	Very variable.At high lats the variation is very large	8
NO_3	7×10^6	$\cong 1 \times 10^{-12}$	30	Variable Very latitude dep.	8
HO_2	$\cong 1 \times 10^8$	$< 2.5 \times 10^{-15}$	>1000	UNIMPORTANT	12
IO	$<10^7$	$\cong 1 \times 10^{-14}$	>2500	UNIMPORTANT	32-34
Cl	$<10^3$	$\cong 2 \times 10^{-10}$	>1300	UNIMPORTANT	36
Br	$<10^3$	$\cong 2 \times 10^{-10}$	>1300	UNIMPORTANT	36
ClO	$<10^7$	$\cong 1 \times 10^{-14}$	>2500	UNIMPORTANT	34
BrO	$<10^6$	$\cong 2.6 \times 10^{-13}$	>2500	POSSIBLY IMPORTANT IN THE ARCTIC	34,35

(a) Diurnally averaged, Units are cm^{-3}, assuming that NO_X is 5 pptv, $[O_3]$ = 20 ppbv & [CO] = 90 ppbv; (b) Units: cm^3 $molecule^{-1}$ s^{-1}; (c) lifetime in hours; (d) for summer conditions; (e) Globally averaged.

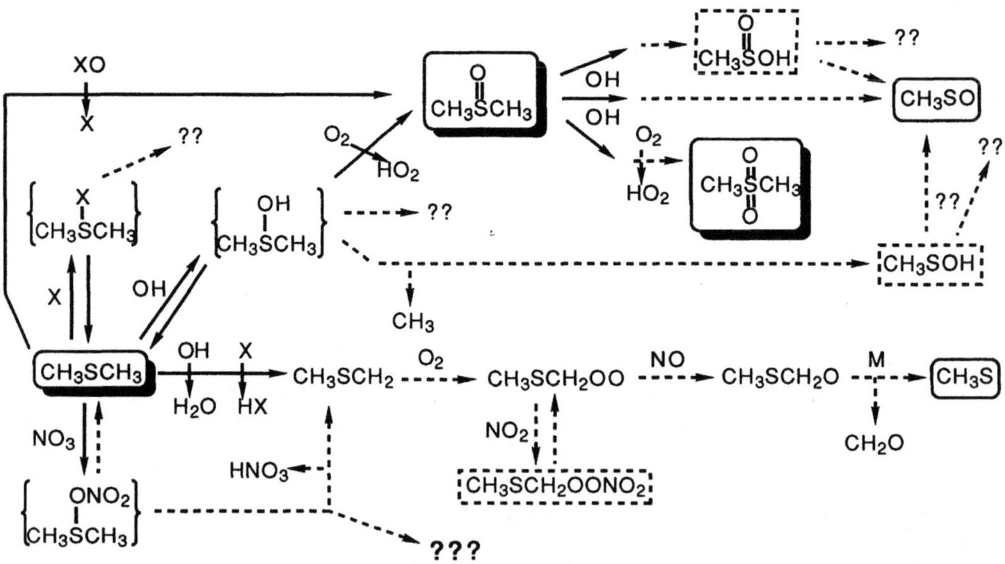

Figure 1. Schematic of the initiation of the atmospheric oxidation of DMS to what is thought to be the key intermediates, CH_3S and CH_3SO (shown in solid boxes). Shaded boxes represent stable species which have been observed in either the atmosphere or in end product studies; dashed boxes indicate possible stable species. Solid arrows are reaction steps which have been observed directly, whereas dashed arrows represent postulated steps.

O_3, H_2O, temperature, and solar flux, which will determine the concentrations of OH and NO_3. Regardless of the oxidant (OH, NO_3, X, or OX, where X is a halogen) which initiates the DMS oxidation, the mechanism of the initiation reactions have been suggested to occur through hydrogen abstraction and an electrophilic addition to the sulfur atom. In the latter pathway, the adduct can decompose to different products or react further with the atmospheric constituents. The mechanisms of these initiation reactions are briefly described below.

Reaction with OH: There is a general agreement that the reaction of OH with DMS proceeds by two different pathways: (1) hydrogen abstraction, which is dominant in the absence of oxygen, and (2) addition of OH to the sulfur atom to form a weak adduct which can then react with O_2,[13]

$$OH + CH_3SCH_3 \rightarrow H_2O + CH_3SCH_2 \quad (1a)$$
$$OH + CH_3SCH_3 \rightleftharpoons CH_3S(OH)CH_3 \quad (1b, -1b)$$
$$CH_3S(OH)CH_3 + O_2 \rightarrow \text{Products} \quad (2)$$

Recently, Wine and co-workers[14] have observed the reversible addition of OH to CD_3SCD_3 (with a DMS-OH bond energy of ~14 kcal mol^{-1}). Gu and Turacek[15] concluded that the DMS-OH adduct was not observable; however, their experiment was ill-suited to make such a determination. Further work on identifying the short-lived DMS-OH adduct is worthwhile. At room temperature and one atmosphere of air, Hynes et al.[13] report a branching ratio of 0.7 for the abstraction channel. Under lower atmospheric temperatures of the mid-troposphere, the addition and abstraction channels are likely to be of equal importance. Martin et al.[16] have some evidence for the formation the abstraction product, CH_3SCH_2, and the adduct.

The bifurcation of the rate coefficient of reaction 1 into addition and abstraction channels must be viewed with caution. The abstraction pathway may involve the initial addition of OH to DMS, followed by intramolecular H-atom abstraction. The OH-DMS adduct could also decompose to products such as CH_3 + CH_3SOH (1c) and CH_3S + CH_3OH (1d). Such channels would appear to be an "abstraction" channel. Dominé et al.[17] observed the production of CH_3SOH in the reaction of OH with $CH_3SC_2H_5$, which suggests that channel 1c is possible. Nielson et al.[18] report very little production of CH_3 from reaction 1c. In their system the observed production of CH_3 from reaction 1c may be due to the reaction of H + DMS, which is known to give CH_3 radicals.[19] End-product studies have not observed evidence for CH_3OH formation,[20] and recent preliminary results in our laboratory suggest that CH_3S is not produced from reaction 1d (<10%).[21] Therefore, channel 1d may be only of minor importance. The observed kinetic isotope effects[13] and the observation of HDO in OH reaction with deuterated DMS[22] are consistent with an abstraction channel. However, the reaction of OH with CH_3SH has been shown to yield CH_3S (along with H_2O) with unity yield,[41] but the reaction is thought to occur via addition and not abstraction. It would be very surprising if two independent channels are active in a reaction. Previous bifurcation between addition and abstraction in hydrocarbon radical reactions have turned out to be over-simplifications and erroneous.[23]

The eventual products of the addition channel in the case of DMS (reaction 1b) is unclear. Recent work from Wine's group suggests that reaction 2 produces HO_2 (most likely with DMSO) approximately 50% of the time.[14] There has been no observation of DMSO in any end-product studies which may be due to the rapid oxidation of DMSO by OH (rate coefficient (0.6-1.0) x 10^{-10} cm^3 $molecule^{-1}$ s^{-1}).[14,24] The eventual product of DMSO + OH reaction in air is thought to be $(CH_3)_2SO_2$ (dimethyl sulfone), at least part of the time. This product has been observed in an end-product study by Barnes et al.[4,24] It should be noted that oxidation of DMS via the OH-DMS adduct (or DMSO oxidation) will not produce CH_3S. It may produce intermediates such as CH_3SOH or CH_3SO.

In summary, even though the reaction of OH with DMS appears to proceed via both addition and abstraction pathways, the products of this reaction in the presence and the absence of O_2 have not been fully identified. This crucial gap should be an area for active research in the near future.

Reaction with NO_3: The rate coefficients for the reaction,

$$NO_3 + CH_3SCH_3 \rightarrow \text{Products} \quad (3)$$

is very rapid and has a negative or zero activation energy, which suggests the formation of a complex.[8] It is likely that NO_3 adds to the sulfur atom. However, Daykin and Wine and Jensen et al. have observed a large kinetic isotope effect (ca. 3.5-3.8) upon deuteration of DMS,[25,26] and Jensen et al.[26] have measured large yields of HNO_3 (with CH_3SCH_3) and DNO_3 (with CD_3SCD_3) in end product studies. Jensen et al. have also observed formation of an intermediate peroxy nitrate compound in their experiments, which they tentatively assign to $CH_3SCH_2OONO_2$; this suggests an abstraction pathway (followed by O_2 and NO_2 addition).[26] Both groups argue that an intramolecular hydrogen abstraction within the NO_3-DMS complex is responsible for the observation. Dlugokencky and Howard[27] and Tyndall et al.[28] did not detect NO_2 in this reaction; an upper limit of about 5% for the channel which forms NO_2 and dimethyl sulfoxide (DMSO) was suggested, and DMSO has not been observed in end-product studies.[26] Also, there is no evidence for a reaction between a NO_3-DMS adduct with O_2.[25] However, the formation of an adduct followed by its decomposition via C-S scission cannot be ruled out. In H_2S, where C-S bond cleavage is not possible, the NO_3 reaction is 10^4 times slower and suggests that the C-S bond plays some role in the reactivity of DMS with NO_3.

Even though the rate coefficients for the reaction of NO_3 with DMS, which may be the dominant loss process for DMS, is well defined, the products of the reaction are not clear. The eventual products of this reaction may depend on atmospheric conditions. Efforts to determine the products of this reaction in the presence and the absence of O_2 is essential.

Reactions with halogen atoms and halogen oxides: The original suggestion that the observed short lifetime of DMS in coastal areas was due to its reaction with iodine monoxide (IO)[29] sparked the interest in halogen and halogen oxide reactions in DMS oxidation. The availability of halogen species, especially Br, BrO, Cl, and ClO, in the troposphere is based on the laboratory finding that XNO_2 products are produced in the heterogeneous processing of N_2O_5 on salt.[30] Except for the recent tentative measurements[31] of BrO in the Arctic troposphere, none of the inorganic halogen species have been observed in the troposphere, and hence, the role played by these species is rather uncertain.

Recent measurements of the rate coefficient for the reaction of IO with DMS (Table I) is 3 orders of magnitude smaller than those reported earlier.[32-34] There are good reasons to suspect the earlier measurements and the recent determinations are to be preferred. The reactions of DMS with other halogen oxides (Table I) are unlikely to be important. Product studies of Barnes et al.[34] indicate that DMSO is produced with unit yield in the BrO + DMS reaction. The reaction of Cl and Br with DMS has been discussed by Wine and coworkers.[36]

In summary the reactions of OH and NO_3 are the most likely pathways for DMS oxidation. Even in the clean upper tropospheric air, NO_3 concentrations are sufficient[37] to play a major role in the oxidation of DMS. Diurnal variations of DMS measured by Andreae et al.[38] show a maximum at night and minimum during the afternoon hours, which is to be expected if OH is the dominant oxidizer, and the flux is independent of sun light. However, when oceanic air which had been transported from continental regions was sampled, a smaller diurnal variation was noted; this suggests a nighttime sink for DMS, presumably due to its reaction with NO_3.[38] The reaction of BrO with DMS could be important in polar regions if the measured [BrO], ≅10-20 pptv,[31] are correct.

3. CHEMISTRY OF CH_3S

As noted above, even though the rate coefficients for the initiation reactions are well defined, the products of these reactions are unclear. Therefore, the assessment of the eventual products of DMS oxidation relies heavily on end-products studies carried out in chambers or photochemical reactors, where the intermediates involved are not directly detected,[4,5,9,20,26,39] and the reactions responsible for the observed end products are not always clear. Furthermore, many end products studies have employed high NO_x levels, usually to initiate DMS degradation via OH and NO_3 reactions. The chemistry that occurs in these high NO_x systems can be very different from those in low NO_x systems where reactants such as O_3 can be more important. Therefore, the details of the elementary steps in the atmospheric DMS oxidation mechanism are very unclear. Even the eventual products, their yields, and their variations with the ambient concentrations, temperature, and pressure are uncertain. The mechanisms are based, to a certain extent, on speculations and analogies. Yet, there is reason to believe that CH_3S is an important intermediate and, hence, its chemistry is discussed next.

The reasons for expecting CH_3S to be a key intermediate in DMS oxidation are as follows: The mechanism for production of CH_3S from the OH + DMS reaction is thought to be[9,13,40]

$$OH + CH_3SCH_3 \rightarrow H_2O + CH_3SCH_2 \quad (1a)$$
$$CH_3SCH_2 + O_2 + M \rightarrow CH_3SCH_2OO + M \quad (4)$$
$$CH_3SCH_2OO + NO \rightarrow CH_3SCH_2O + NO_2 \quad (5)$$
$$CH_3SCH_2O + M \rightarrow CH_3S + CH_2O \quad (6)$$

The mechanisms for NO_3 initiated oxidation of DMS is thought to be analogous, if H-abstraction occurs. The observation of CH_2O and CH_3SNO in the OH initiated systems,[20,39] and CH_3SNO_2 and $CH_3SCH_2OONO_2$ in the NO_3 initiated studies[26] supports the contention that CH_3S is a key intermediate in DMS oxidation. Since the C-S bond is intact in CH_3S, the mechanism of CH_3S oxidation may hold the key to the identities and yield of products such as SO_2, SO_3, and MSA.

Reactions of CH_3S: The proposed reaction pathways for the loss of CH_3S are summarized in Figure 2 and the rate coefficients for the reactions listed in Table II. The figure also shows the reactions of the products of CH_3S reactions, both observed and hypothesized, which may lead to the final products observed in the field and in chamber studies.

The major loss processes for CH_3S are its reactions with O_2, O_3, and NO_2. Based on the

Table II. Rate Coefficients for the Atmospherically Important Reactions of CH_3S, CH_3SO, and CH_3SOO

Reaction	k_{298K} $cm^3 molecule^{-1} s^{-1}$	A $cm^3 molecule^{-1} s^{-1}$	E/R K	Reference
$CH_3S + O_2$[a]	$< 2 \times 10^{-17}$	-	-	53
	$< 2 \times 10^{-16}$	-	-	54
	$< 3 \times 10^{-18}$	-	-	44
	$< 6 \times 10^{-18}$	-	-	b
$CH_3S + O_3$	$< 8 \times 10^{-14}$	-	-	54
	5.1×10^{-12}	-	-	41
	5.6×10^{-12}	-	-	17
	5.2×10^{-12}	2.0×10^{-12}	-290 ± 40	43
$CH_3S + NO_2$	1.1×10^{-10}	8.3×10^{-11}	-80 ± 100	53
	6.1×10^{-11}	-	-	44
	5.1×10^{-11}	-	-	45
	6.1×10^{-11}	2.1×10^{-11}	-320 ± 40	43
$CH_3SOO + O_2$	$< 4 \times 10^{-17}$	-	-	42[c]
	$< 6 \times 10^{-18}$	-	-	b
$CH_3SOO + O_3$	$< 8 \times 10^{-13}$	-	-	43[d]
$CH_3SOO + NO_2$	2.2×10^{-11}	-	≈ 0	43[e]
$CH_3SOO + NO$	1.1×10^{-11}	-	≈ 0	43[f]
$CH_3SO + O_3$	6×10^{-13}	-	-	17
$CH_3SO + NO_2$	1.0×10^{-11}	-	-	44
	1.2×10^{-11}	-	-	45

(a) Rate Coefficients are for the reaction of $CH_3S + O_2$ going to products other than CH_3SOO. These upper limits are not corrected for the $CH_3S + O_2 \rightleftharpoons CH_3SOO$ equilibrium (c) Calculated from the data on $CH_3S + O_2$ reported from Tyndall and Ravishankara[44] and the $CH_3S + O_2 \rightleftharpoons CH_3SOO$ equilibrium data of Turnipseed et al.[42] (b) At T= 258 K. (d) At T = 227 K. (e) At T = 227-242 K.
(f) At T = 227-255 K.

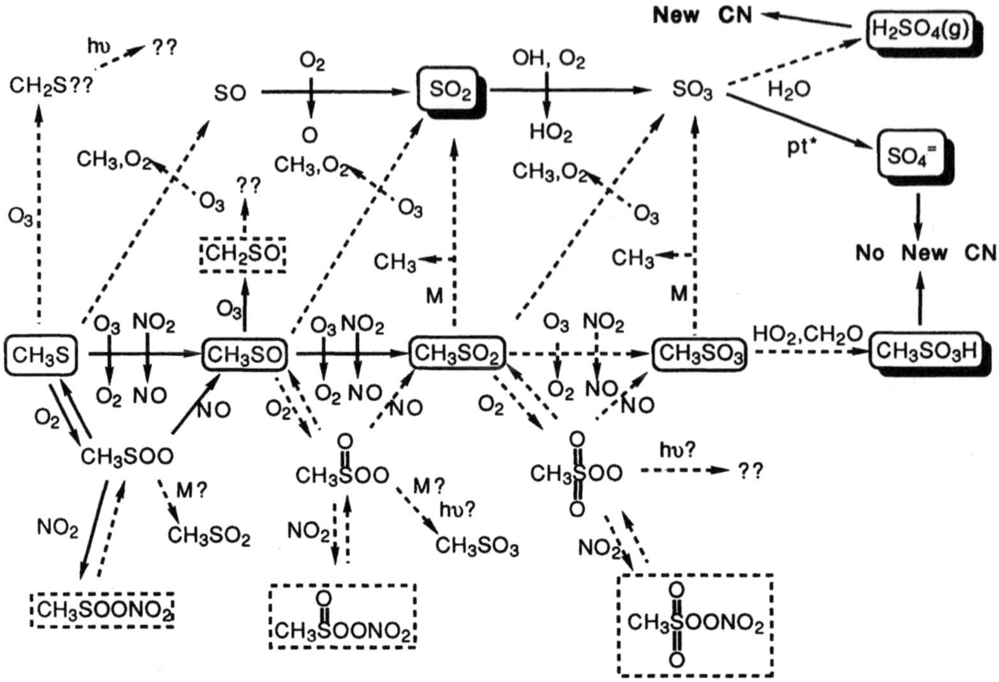

Figure 2. Schematic of the oxidation of CH_3S to final products. Species in shaded boxes are stable products which are observed in the atmosphere. Solid boxes are key radical intermediates and dashed boxes are possible stable products. Solid arrows are steps which have been observed directly, whereas dashed arrows are postulated steps.
*pt = particle

rate coefficient data (Table II), it was concluded that CH_3S reaction with O_3 should be the major pathway for its removal.[8,41] However, recent work in our laboratory[42] has directly observed the reversible addition of CH_3S to O_2,

$$CH_3S + O_2 \rightleftharpoons CH_3SOO \qquad (7a,-7a)$$

which has been previously postulated.[9a] The measured bond strength, 11 kcal mol^{-1},[42] shows that the equilibrium will be established rapidly and the fraction of the adduct tied up as CH_3SOO will be strongly temperature-dependent. Under typical atmospheric conditions, anywhere from 20-80% of CH_3S will be tied up as the CH_3SOO adduct. The rate constant for the reactions of CH_3SOO with O_2, O_3, NO_2, and NO have been indirectly determined (Table II), and are described elsewhere.[43] It is clear that the oxidation of CH_3S via reactions of the CH_3SOO species will also play a role in the overall oxidation scheme. Previous measurements were unable to observe the rapid equilibration of reactions 7a, -7a, and the upper limits reported the rate coefficients for the reaction of CH_3S with O_2 were actually for the equilibrium mixtures of CH_3S and CH_3SOO.[42,44] Using the data reported by Tyndall and Ravishankara and the equilibrium data of Turnipseed et al., we calculate $k_{7b} < 6 \times 10^{-18}$ and $k_8 < 6 \times 10^{-18}$ cm^3 $molecule^{-1}$ s^{-1} for the reactions

$$CH_3S + O_2 \rightarrow \text{Products other than } CH_3SOO \qquad (7b)$$
$$CH_3SOO + O_2 \rightarrow \text{Products other than } CH_3S \qquad (8)$$

at 298 K. Although these rate coefficients are small, these reactions cannot be ruled out as a possible loss processes for CH_3S (or CH_3SOO), due to the overwhelming O_2 concentration in the atmosphere. As it now stands, the fate of CH_3S in the atmosphere is not clear. Either reactions of CH_3S or the adduct with O_2 or NO_2 could be the important pathway. In either case the reactions are fast in the atmosphere, but the products could be very different depending on the reaction that is important. Also, the branching between the four possibilities will be very temperature sensitive.

Even though it is clear that the reaction of CH_3S with NO_2 yields CH_3SO and NO,[44,45] the products of other reactions of CH_3S and CH_3SOO with the atmospheric reactants are not clear. Dominé et al.[17] reported a 15% yield for CH_3SO in the reaction of CH_3S with O_3 in their low-pressure (0.7-5.0 Torr) flow tube experiments. Since the channel to produce CH_3SO {and $O_2(^3\Sigma)$} is exothermic by 80 kcal mol^{-1}, CH_3SO may initially be formed hot and subsequently decompose to other products (such as CH_3 + SO). Therefore, the CH_3SO yields at such low pressure may not be indicative of the true yield under atmospheric conditions. Collisional and radiative stabilization of CH_3SO may be important. Dominé et al. measured the yield of CH_2S to be < 4 % and they were unable to quantify the observed yield of CH_3.[17] Observation of CH_3 suggests that C-S bond cleavage occurs to some extent in the CH_3S reaction with O_3. Since $CH_3S + O_3$ may be the dominant atmospheric loss process for CH_3S, further information on the product yields is imperative. Products of the reaction between CH_3S and O_2, if it occurs at all, are unknown. The products of the reactions of CH_3SOO with O_2, O_3, NO_2, and NO have not been identified. The reaction with NO_2 may proceed via an addition pathway or bimolecularly to form NO_3. The reaction of CH_3SOO with NO is assumed to produce NO_2 based largely on the analogy with hydrocarbon peroxy species, and also because of the large rate coefficient for this reaction which is temperature independent. The possibility of the isomerization of CH_3SOO to CH_3SO_2 is worth investigating. If this reactions occurs on the time scale of seconds, the isomerization may dominate all other bimolecular reactions. Further investigations on the products of these reactions are imperative to fully understand the products of DMS oxidation. Yield of different products in the reactions of CH_3S and CH_3SOO could be a major branching point in the oxidation scheme of DMS.

4. CHEMISTRY OF CH_3SO_x

The observations of CH_3SO as a product in the reactions of CH_3S with both O_3 and NO_2[17,45] suggest that CH_3SO and higher oxides (CH_3SO_x) may be important in DMS oxidation. It may also be formed directly from the reactions of OH-DMS adduct. The rate coefficients for the reactions of CH_3SO with O_3 and CH_3SO with NO_2 are currently the only directly studied reactions of this species,[17,45] (Table II). Based on this limited data set, the reaction of $CH_3SO + O_3$ should control the loss of CH_3SO in the clean marine troposphere. However, the products of both of these reactions are not well known. Dominé et al. observed CH_2SO (10%) and CH_3S (13%) as products of this reaction.[17] As with the $CH_3S + O_3$ reaction, $CH_3SO + O_3$ is highly exothermic (for the channel producing CH_3SO_2 and O_2 ($^3\Sigma$), ΔH = - 75 kcal mol^{-1}) and the observed products may be due to the dissociation of the energy rich product. This species may undergo C-S bond scission to produce SO_2. The reaction of $CH_3SO + NO_2$ produces NO,[44] and SO_2 was observed by mass spectrometry by Mellouki et al. from the $CH_3S + NO_2$ system.[46] It was assumed to be from unimolecular decomposition of CH_3SO_2.

Detection of peroxy sulfinyl nitrate ($CH_3S(O)OONO_2$) in end-product studies of organo-sulfide oxidation (by both OH and NO_3) has been reported by several groups.[26,47,48] This observation suggests that CH_3SO can add to O_2, as in the case of CH_3S

$$CH_3SO + O_2 \rightleftharpoons CH_3S(O)OO \qquad (9).$$

This new bond is expected be stronger than that for $CH_3S\text{-}O_2$.[9a] Equilibration should be rapid (a μs to ms time scale) and, depending upon the thermal stability of $CH_3S(O)OO$, this reaction may dominate CH_3SO chemistry in the atmosphere and bears further investigation.

Yin et al.[9a] have proposed that unimolecular decomposition of CH_3SO_2 to SO_2 via

$$CH_3SO_2 + M \rightarrow CH_3 + SO_2 + M \qquad (10)$$

competes with O_2 addition

$$CH_3SO_2 + O_2 \rightleftharpoons CH_3S(O)_2OO \quad (11)$$

Further oxidation of $CH_3S(O)_2OO$ produces CH_3SO_3H as a final product. The C-S bond energy in CH_3SO_2 is ~ 20 kcal mol^{-1}; therefore it is expected to have a relatively short atmospheric lifetime for the unimolecular dissociation (a few seconds). However, depending on the thermal stability of the $CH_3S(O)_2OO$, O_2 addition may be much faster (μs time scale as is $CH_3S + O_2$) and force the production of this species. If this happens, SO_3 and MSA will be preferrentially formed over SO_2.

Recent studies[6] suggest that the major crossroads for formation of CH_3SO_3H versus H_2SO_4 is the reactions of CH_3SO_3. It is likely that the production of CH_3SO_3 will involve NO rather than the sequence indicated by Bandy and coworkers.[6] Since the sulfur valence is full in CH_3SO_3, further addition of O_2 is unlikely. The C-S bond energy is ~22 kcal mol^{-1} and, thus, is expected to dissociate readily under atmospheric conditions to generate CH_3 and SO_3. The bond scission would most likely be in competition with abstraction of a hydrogen from a species such as HO_2 to generate CH_3SO_3H.[7] The beauty of such a mechanism is that it involves a unimolecular reaction with a very large activation energy and could explain the variation in the observed MSA to NSS (non-sea salt sulfate) with temperature changes.[49,50] For example, if the activation energy is 25 kcal mol^{-1}, the fraction that undergoes thermal decomposition could change by a factor of 2 when the temperature changes from 274 to 270K. Therefore, the observed changes in the MSA to NSS ratio can easily be accounted for by small changes in temperature. Even the fraction of CH_3S that is tied up as CH_3SOO would show sufficient change for the observed temperature changes to account for the observed MSA to NSS ratio.

6. CONCLUSIONS

The atmospheric lifetimes of DMS is the best determined parameter in its oxidation process. The oxidation is initiated mostly by the reactions of OH and NO_3 radicals. The products of these reactions are unclear. The subsequent reaction which lead to the end-products are also poorly defined. Based largely on end-products studies, it is assumed that CH_3S is a key intermediate in the DMS oxidation and the chemistry of this intermediate determines the final products.

Field[49] and ice-core data[50] indicate that the MSA to (MSA + $SO_4^{=}$) ratio increases as the temperature is lowered, suggesting a change in the oxidation mechanism. This change could be due to different initiation reactions, branching in the reaction pathways with temperature, or occurrence of different atmospheric reactions due to changes in composition. Since NO_x levels during the pre-industrial era were bound to be low, NO_x cannot be responsible for this change in the MSA concentration in the ice-core data. However, NO_x chemistry may influence MSA to NSS ratios in coastal regions where polluted air can be transported over the oceans.[51]

It has been suggested that addition of OH to DMS (channel 1b) becomes dominant at lower temperature relative to the abstraction channel, and that this leads to larger yields of MSA.[49] However, the fractional change in the branching with temperature is quite small and is unlikely to be the reason for the changes in MSA yield. The elevated MSA concentrations may also be due to competition of the $CH_3SO_x + O_2 \rightleftharpoons CH_3S(O)_xOO$ equilibria or unimolular dissociations, which are likely to be highly temperature-dependent, relative to the reactions of CH_3SO_x with O_3, RH, and NO_x species, as discussed earlier.

The major current concern regarding DMS is the fraction of the emitted DMS that turns in to particles and to H_2SO_4. In the case of the particle formation, the key question is what fraction of DMS leads to SO_2 and SO_3, and what fraction of these species lead to H_2SO_4 in the gas phase. Recent modeling and field studies suggest that SO_2 levels are lower than would be expected if DMS were primarily oxidized to form this species.[6,7] It is also been suggested that these low levels of SO_2 would be insufficient in generating new CN due to the inefficiency of the OH + SO_2 reaction.[7] Bandy et al. have suggested that DMS oxidation does not lead to SO_2, but to SO_3.[6] We have shown here that there are many pathways for the direct production of SO_3 in DMS oxidation. Lin and Chameides have also shown that direct SO_3 formation in DMS oxidation could account for the measured efficient production of new particles.[7] One major effect of forming SO_2 as opposed to SO_3 is that in the former case the lifetime of SO_2(many days) is sufficient to be widely disperse SO_2. If SO_3 is formed directly, it would be converted to either H_2SO_4 or sulfate in a matter of seconds, and hence will be constrained to the location where DMS oxidation was initiated. Currently, it is not clear as to how much gas phase H_2SO_4 can be formed. The gas phase oxidation of SO_2 also proceeds via SO_3 formation and the fate of SO_3 is the key unknown. If it reacts homogeneously in the gas phase, it could produce H_2SO_4. If it is lost via uptake on a particle, it is equivalent to SO_2 oxidation In the condensed phase. The upper limit for the rate coefficient for the reaction of SO_3 with H_2O is the determining factor in this assessment. The upper limit of $< 6 \times 10^{-15}$ cm^3 $molecule^{-1}$ s^{-1} reported by Wang et al.[52] for this reaction is not sufficient to clearly distinguish between gas phase H_2SO_4 formation and condensation of SO_3 on pre-existing particles. In either case, it is very likely that most of DMS ends up as sulfate and hence contribute to the pH of precipitation in areas where DMS is emitted. If SO_2 is produced, as mentioned earlier, the contribution will not be localized.

Many questions still remain concerning the oxidation of organo-sulfur compounds in the atmosphere. The key uncertainties are in the formation and reactions of O_2 adducts, products of the CH_3S and CH_3SO reactions with O_3, and the rate coefficient for the reaction of SO_3 with H_2O in the gas phase. The chemistry of organo-sulfur compounds has provided many unexpected discoveries in the past, and future studies should reveal even more complex and interesting phenomena. Further detection of the crucial intermediates should better define the oxidation mechanism as well as contribute to the overall knowledge of physical chemistry and reaction kinetics.

ACKNOWLEDGEMENTS

This work was carried out as part of NOAA's Climate and Global Change Program

REFERENCES

(1) Bates, T.S., Lamb, B.K., Guenther, A., Dignon, J. and Stoiber, R.E. (1992) Sulfur Emissions to the Atmosphere from Natural Sources, *J. Atmos. Chem.* **14**, 315-337.

(2) Andreae, M.O. (1985) The Emission of Sulfur to the Remote Atmosphere, in *The Biogeochemical Cycling of Sulfur and Nitrogen in the Remote Atmosphere*, J.N. Galloway et al., Eds., D.Reidel Publishing Co., 5-25.

(3) Charlson, R.J., Lovelock, J.E., Andreae, M.O. and Warren, S.G. (1987) Oceanic Phytoplankton, Atmospheric Sulphur, Cloud Albedo and Climate, *Nature* **326**, 655-661.

(4) Barnes, I., Bastian, V. and Becker, K.H. (1988) Kinetics and Mechanisms of the Reaction of OH Radicals with Dimethyl sulfide, *Int. J. Chem. Kinet.* **20**, 415-431.

(5) Grosjean, D. (1984) Photooxidation of Methyl Sulfide, Ethyl Sulfide, and Methanethiol, *Environ. Sci. Technol.* **18**, 460-468.

(6) Bandy, A.R., Scott, D.L., Blomquist, B.W., Chen, S.M. and Thornton, D.C. (1992) Low Yields of SO_2 from Dimethyl Sulfide Oxidation in the Marine Boundary Layer, *Geophys. Res. Lettrs.* **11**, 1125-1127.

(7) Lin, X. and Chameides, W.L. (1992) CCN Formation from DMS Oxidation with SO_3 Instead of SO_2 Acting as an Intermediate, submitted to *Geophys. Res. Lettrs.*

(8) Tyndall, G.S. and Ravishankara, A.R. (1991) Atmospheric Oxidation of Reduced Sulfur Species, *Int. J. Chem. Kinet.*, **23**, 483-527.

(9) a. Yin, F., Grojean, D., and Seinfeld, J.H. (1990) Photooxidation of Dimethyl Sulfide and Dimethyl Disulfide, I : Mechanism Development, *J. Atmos. Chem.* **11**, 309-364. (b) Yin, F., Grosjean, D., Flagan, R.C. and Seinfeld, J.H. (1990) Photooxidation of Dimethyl Sulfide and Dimethyl Disulfide, II : Mechanism Evaluation, *J. Atmos. Chem.* **11**, 365-399.

(10) Martines, R.I. and Herron, J.T. (1978) Stopped-flow Study of the Gas-phase Reaction of Ozone with Organic Sulfides: Dimethyl Sulfide, *Int. J. Chem. Kinet.* **10**, 433-452.

(11) Balla, R.J. and Heicklen, J. (1984) Oxidation of Sulfur Compounds. II. Thermal Reactions of NO_2 with Aliphatic Sulfur Compounds, *J. Phys. Chem.* **88**, 6314-6317.

(12) Mellouki, A. and Ravishankara, A.R. (1991) Upper Limits for the Rate Constants for the Reactions of HO_2 with H_2S, CH_3SH, and CH_3SCH_3, presented at the *American Geophysical Union Fall Meeting*, San Francisco.

(13) Hynes, A.J., Wine, P.H. and Semmes, D.H. (1986) Kinetics and Mechanism of OH Reactions with Organic Sulfides, *J. Phys. Chem.* **90**, 4148-4156.

(14) Hynes, A.J., Pounds, A.J., McKay, T., Bradshaw, J.D. and Wine, P.D. (1992) Detailed Mechanistic Studies of the OH-initiated Oxidation of Biogenic Sulfur Compounds under Atmospheric Conditions, at *the 12th International Symposium on Gas Kinetics*, University of Reading, England, Abstract G6.

(15) Gu, M. and Turecek, F. (1992) The Elusive Dimethylhydroxysulfuranyl Radical. An Intermediate or a Transition State?, *J. Am. Chem. Soc.* **114**, 7146-7151.

(16) Martin, D., Jourdain, J.L. and LeBras, G. (1985) Kinetic Study for the reactions of OH Radicals with Dimethylsulfide, Diethylsulfide, Tetrahydrothiophene, and Thiophene, *Int. J. Chem. Kinet.*, **17**, 1247-1261.

(17) Dominé, F., Ravishankara, A.R. and Howard, C.J. (1992) Kinetics and Mechanisms of the Reactions of CH_3S, CH_3SS, CH_3SO and CH_3SSO with O_3 at 300 K and Low Pressures, *J. Phys. Chem.* **96**, 2171.

(18) Nielson, O.J., Sidebottom, H.W., Nelson, L. Treacy, J.J. and O'Farrel, D.J. (1989) An Absolute and Relative Rate Study of the Reaction of OH Radicals with Dimethyl Sulfide, *Int. J. Chem. Kinet.* **21**, 1101-1112.

(19) Yokota, T. and Strausz, O.P. (1979) Reaction of Hydrogen Atoms with Dimethyl Sulfide, *J. Phys. Chem.* **83**, 3196-3202.

(20) Hatakeyama, S. and Akimoto, H. (1983) Reactions of OH Radicals with Methanethiol, Dimethyl Sulfide, and Dimethyl Disulfide in Air, *J. Phys. Chem.* **87**, 2387-2395.

(21) Turnipseed, A.A., Barone, S.B. and Ravishankara A.R., (1992) unpublished results.

(22) Stickel, R. e., Zhao, Z., and Wine, P.H., (1992) private communications.

(23) Wagner, A.F., Slagle, I.R., Sarzynski, D. and Gutman, D. (1990) Experimental and Theoretical Studies of the $C_2H_5 + O_2$ Reaction Kinetics., *J. Phys. Chem.* **94**, 1853-1868.

(24) Barnes, I., Bastian, V., Becker, K.H. and Martin, D. (1989) Fourier Transform IR Studies of the Reactions of Dimethyl Sulfioxide with OH, NO_3, and Cl Radicals, in *Biogenic Sulfur in the Environment, ACS Symposium, Series 393*, 476-488.

(25) Daykin, E.P. and Wine, P.H. (1990) A Study of the Reactions of NO_3 Radicals with Organic Sulfides: Reactivity Trend at 298 K, *Int. J. Chem. Kinet.* **22**, 1083-1094

(26) Jensen, N.R., Hjorth, J., Lohse, C., Skov, H. and Restelli, G. (1992) Products and Mechanisms of the Gas Phase Reactions of NO_3 with CH_3SCH_3, CD_3SCD_3, CH_3SH, and CH_3SSCH_3, *J. Atmos. Chem.* **14**, 95-108.

(27) Dlugokencky, E.J. and Howard, C.J. (1988) Laboratory Studies of NO_3 Radical Reactions with some Atmospheric Sulfur Compounds, *J. Phys. Chem.* **92**, 1188-1193.

(28) Tyndall, G.S., Burrows, J.P., Schneider, W. and Moortgat, G.K. (1986) Rate Coefficient for the Reaction between NO_3 Radicals and Dimethyl Sulphide, *Chem. Phys. Lett.* **130**, 463-466.

(29) Carlier, P. (1985) L'Ozone serait-il l'Oxidant Principal du Sulfure de Dimethyle en Milieu Océanique?, in *Atmopsheric Ozone*, C.S. Zerefos and A. Ghazi (eds.), D. Reidel, 815-819.

(30) Zetzsch, C. and Behnke, W. (1992) Heterogeneous Photochemical Sources of Atomic Cl in the Troposphere, *Ber. Bunsenges. Phys. Chem.* **96**, 488-493.

(31) Hausman, M., Rudolph, T., and Platt, U. (1992) presented at *TheTropospheric Chemsitry of Ozone in the Polar Regions,* Wolfville, Nova Scotia, Canada, August 23-28, 1992.

(32) Daykin, E.P. and Wine, P.H. (1990) Rate of Reaction of IO Radicals with Dimethyl Sulfide, *J. Geophys. Res.* **95**, 18547-18553.

(33) Maguin, F., Mellouki, A., Laverdet, G., Poulet, G. and Le Bras, G. (1991) Kinetics of the Reactions of the IO Radical with Dimethyl Sulfide, Methanethiol, Ethylene, and Propylene, *Int. J. Chem. Kinet.* **23**, 237-245.

(34) (a) Barnes, I., Bastian, V., Becker, K.H. and Overaty, R.D. (1991) Kinetic Studies of the Reactions of IO, BrO, and ClO with Dimethyl Sulfide, *Int. J. Chem. Kinet.* **23**, 579-591. (b) Barnes, I, Bonsang, B., Bravears, T., Carlier, P., Cox, R.A., Dorn, H.P., Jenkin, M.E., Le Bras, G. and Platt, U. (eds.) Laboratory and Field Studies of Oxidation Processes Occurring in the Atmospheric Marine Boundary Layer, in *Air Pollution Research Report 35, Commission of the European Communities*, ISBN 2-87263-064-3.

(35) Le Bras, G., Maguin, F., Mellouki, A. Lancar, I.T., and Balestra, C. (1991) in Laboratory and Field Studies of Oxidation Processes Occurring in the Atmospheric Marine Boundary Layer, Barnes, I, Bonsang, B., Bravears, T., Carlier, P., Cox, R.A., Dorn, H.P., Jenkin, M.E., Le Bras, G. and Platt, U. (eds.) , *Air Pollution Research Report 35, Commission of the European Communities*, ISBN 2-87263-064-3, 106-116.

(36) (a)Stickel, R. E., Nicovich, J. M., Wang, S., Zhao, Z., and Wine, P. H., Kinetic and Mechanistic study of the reaction of atmic chlorine with dimethyl sulfide, *J. Phys. Chem.*, in press 1992.; (b) Wine, P.H., private communications.

(37) Noxon, J.F.J. (1983) NO_3 and NO_2 in the Mid-Pacific Troposphere, *J. Geophys. Res.* **88**, 11017-11021.

(38) Andreae, M.O., Ferek, R.J., Bermond, F., Byrd,K.P., Engstrom, R.T., Hardin, S., Houmere, P.D., LeMarrec, F., Raimdonck, and Chatfield, R.B. (1985) Dimethyl Sulfide in the Marine Atmosphere, *J. Geophys. Res.* **90**, 12891-12900.

(39) Niki, H., Maker, P.D., Savage, C.M., and Breitenbach, L.P. (1983) An FTIR Study of the Mechanism for the Reaction HO + CH_3SCH_3, *Int. J. Chem. Kinet.* **15**, 647-654.

(40) Atkinson, R. (1986) Gas Phase Reactions of the Hydroxyl Radical, *Chem. Rev.* **86**, 69-201.

(41) Tyndall, G.S. and Ravishankara, A.R. (1989) Kinetics of the Reaction of CH_3S with O_3 at 298K, *J. Phys. Chem.* **93**, 4707-4710.

(42) Turnipseed, A.A., Barone, S.B. and Ravishankara (1992) Observation of CH_3S Addition to O_2 in the Gas Phase, *J. Phys. Chem.* 96, 7502-7505.

(43) Turnipseed, A.A., Barone, S.B. and Ravishankara, A.R. (1992) The Reactions of CH_3S and CH_3SOO with O_3, NO_2 and NO, manuscript in preparation.

(44) Tyndall, G.S. and Ravishankara, A.R. (1989) Kinetics and Mechanisms of the Reactions of CH_3S with O_2 and NO_2 at 298 K, J. *Phys. Chem.* **93**, 2426-2435.

(45) Dominé, F. Murrells, T.P. and Howard, C.J. (1990) Kinetics of the Reactions of NO_2 with CH_3S, CH_3SO, CH_3SS, and CH_3SSO at 297 K and 1 Torr, *J. Phys. Chem.* **94**, 5839-5847.

(46) Mellouki, A., Jourdain, J.L. and Le Bras, G. (1988) Discharge Flow Study of the CH_3S + NO_2 Reaction Mechanism using Cl + CH_3SH as the CH_3S Source,*Chem. Phys. Lett.* **148**, 231-236.

(47) Barnes, I., Bastian, V., Becker, K.H. and Niki, H. (1987) FTIR Spectroscopic Studies of the CH_3S + NO_2 Reaction under Atmospheric Conditions, *Chem. Phys. Lett.* **140**, 451-457.

(48) Hatekeyama S. (1989) Mechanisms for the Reaction of CH_3S with NO_2, in *Biogenic Sulfur in the Environment, ACS Symposium, Series 393,* 459-463.

(49) Berrsheim, H., Andreae, M.O., Ayers, G.P., Gillett, R.W., Merrill, J.T., Davis, V.J. and Chameides, W.L. (1990) Airborne Measurements of Dimethylsulfide, Sulfur Dioxide, and Aerosol Ions over the Southern Ocean South of Australia, *J. Atmos. Chem.* **10**, 341-370.

(50) LeGrand, M., Feniet-Saigne, C., Saltzman, E.S., Germain, C., Barkov, N.I. and Petrov, V.N. (1991) Ice-core Record of Oceanic Emissions of Dimethylsulphide During the Last Climate Cycle, *Nature* **350**, 144-146.

(51) Mihalopoulos, N., Nguyen, B.C., Boissard, C., Campin, J.M., Putaud, J.P., Belviso, S., Barnes, I. and Becker, K.H. (1992) Field Study of Dimethyl Sulfide Oxidation in the Boundary Layer: Variations of Dimethyl Sulfide, Methanesulfonic Acid, Sulfur Dioxide, Non-sea-salt Sulfate and Aitken Nuclei at a Coastal Site, *J. Atmos. Chem.* **14**, 459-477..

(52) Wang, X., Jin, Y.G., Suto, M. and Lee, L.C. (1988) Rate Constant of the Gas Phase Reaction of SO_3 and H_2O, *J. Chem. Phys.* **89**, 4853-4860.

(53) Balla, R.J., Nelson, H.H. and McDonald, J.R. (1986) Kinetics of the Reactions of CH_3S with NO, NO_2, and O_2, *Chem. Phys.* **109**, 101-107.

(54) Black, G. and Jusinski, L.E. (1986) Laser-induced Fluorescence Studies of the CH_3S Radical, *J. Chem. Soc., Faraday Trans. 2,* **83**, 2143-2151.

FTIR PRODUCT STUDY OF THE PHOTOLYSIS OF CH_3SSCH_3: REACTIONS OF THE CH_3S RADICAL

I. BARNES, K.H. BECKER and N. MIHALOPOULOS*

Bergische Universität Gesamthochschule Wuppertal
Physikalische Chemie FB 9, Gaußstr. 20,
5600 Wuppertal 1, Germany

*** Present address: SOAP Laboratory, Engine Depot, Elefsis Airforce Base, Greece**

SUMMARY

Presented here are results from studies on the 254 nm photolysis of DMDS at 760 Torr total pressure ($N_2 + O_2$) and 298 ± 2 K which has been used as a source of methyl thiyl radicals (CH_3S). At low levels of DMDS the major observed products were SO_2, HCHO and methane sulphonic acid (CH_3SO_3H, MSA) while at higher concentrations minor contributions from methane sulphonic acid anhydride ($(CH_3SO_2)_2O$) and methyl methanethiolsulphonate ($CH_3SO_2SCH_3$) were observed. The behaviour of the products as a function of the O_2 partial pressure suggests that the further reactions of CH_3SOH (methyl sulphenic acid) formed in the reaction of OH radicals with DMDS are probably responsible for the majority of the MSA formation in the reaction system. The results imply that the reaction of both O_2 and O_3 with CH_3S will result in early cleavage of the C-S bond and consequently SO_2 formation.

The results of the study have implications for the atmospheric photooxidation mechanism of DMS and these are briefly considered.

1. INTRODUCTION

Dimethyl sulphide (CH_3SCH_3, DMS) is probably the most important organic sulphur specics released to the atmosphere. Half of the global flux of sulphur into the the atmosphere is thought to be biogenic in origin (1,2) and approximately 50 % of this flux is comprised of DMS volatilised from the ocean (3,4). Although only indirect evidence is available, the methyl thiyl radical (CH_3S) is firmly believed to be one of the key intermediates of the OH initiated oxidation of DMS (5, and references therein). A complete understanding of its atmospheric chemistry is thus essential in order to correctly evaluate the role of DMS not only in the tropospheric sulphur cycle but also the contribution it makes to the atmospheric O_3, NO_x and HO_x cycles.

The 254 nm photolysis of dimethyl disulphide (DMDS, CH_3SSCH_3) represents a

G. Restelli and G. Angeletti (eds.), Dimethylsulphide: Oceans, Atmosphere, and Climate, 197–210.

convenient and clean method for the generation of CH_3S radicals. Although there have been previous studies of the photolysis of DMDS most of these have been carried out at high DMDS concentrations or the product analyses were not particularly comprehensive or conclusive (5,6). The products of both the 254 nm photolysis and the OH initiated oxidation of low levels of DMDS are currently being studied in this laboratory as a function of the O_2 partial pressure at a total pressure of 760 Torr ($N_2 + O_2$) and room temperature in large reaction chamber using in situ long path FTIR spectroscopy. The aim of the experiments is to better define and quantify the reaction pathways of CH_3S which result ultimately in the formation of SO_2 and CH_3SO_3H (methane sulphonic acid, MSA). The preliminary results of these investigations are presented their possible implications for the photooxidation mechanism of DMS is discussed.

2. EXPERIMENTAL

FTIR product studies of the photolysis of 1-2 ppm levels of DMDS were carried out as a function of O_2 partial pressure at a total pressure of 760 Torr ($O_2 + N_2$) in a 1080 l quartz glass reaction chamber by 298 ± 2 K. A schematic diagram of the apparatus is shown in Figure 1. Further details of the experimental set-up can be found in Barnes et al. (7). Photolysis was carried out using up to 16 germicidal lamps (Philips, TUV 40 W, λ_{max}= 254 nm). The cell contains a White mirror coupled to a Bruker IFS 88 FTIR spectrometer which was operated at a pathlength of 498 m. IR spectra were recorded in the 600-4000 cm^{-1} spectral range with 1 cm^{-1} resolution by co-adding 30 interferograms over 24 s. In typical experiments 25 to 35 such spectra were recorded over a period of 9 to 14 min. Concentrations of reactants and products were determined by computer-aided substraction of calibrated reference spectra. The concentration of gas phase MSA was determined using the absorptions given in Mihalopoulos et al. (8).

Some product analyses using higher levels of DMDS (10 to 80 ppm) were carried out in another 420 l reaction chamber also equipped with an FTIR facility. Details of this experimental set-up can be found in Barnes et al. (9).

3. RESULTS

The products of the 254 nm photolysis of 1-2 ppm levels of DMDS as a function of the O_2 partial pressure at a total pressure of 1000 mbar ($N_2 + O_2$) as measured at 298 K after 10 min irradiation in a 1080 l reaction chamber by FTIR are listed in Table 1. The major sulphur containing compounds are SO_2 and CH_3SO_3H and the major hydrocarbons products are CO, HCHO, CH_3OH and CH_3OOH (methyl hydroperoxide). The SO_2 yield (% S) falls from 90 % to 65 % on reducing the O_2 partial pressure from 200 to 13 mbar, however, for MSA the same variation increased the yield from 7 to 23%. Increasing the initial concentration of DMDS from 2 to 65 ppm caused the SO_2 yield to decrease from 90 to approximately 45 %

(Figure 2). This was accompanied by a corresponding increase in the yield of MSA mainly in the form of aerosol and also the formation of $(CH_3SO_2)_2O$ (methane sulphonic acid anhydride) and $CH_3SO_2SCH_3$ (methyl methanethiolsulphonate) (Figure 3). The yields of MSA and the two organic sulphur compounds at the higher DMDS concentrations still have to be quantified.

During the course of the investigations, tests with the 1080 l quartz-glass reactor showed that low concentrations of O_3 are produced when only pure synthetic air is irradiated in the reactor with the mercury lamps. In the presence of 150 Torr O_2 the O_3 concentration reached steady state levels of approximately 20 ppb after 15 min. At a partial pressure of 10 Torr O_2 the O_3 reached steady state levels of 3 ppb. Further tests with added hydrocarbons showed that OH radicals are also produced. Although the concentrations of O_3 and OH radicals produced are small they still can contribute to the decay of DMDS and effect the product distributions and thus the interpretation of the results. The discovery of the low levels of O_3 and OH radicals led to additional experiments being performed in the presence of elevated concentrations of both these species to assess their role in the mechanism of the DMDS photooxidation.

O_3 was never observed above the 1 ppb detection limit in the experiments with DMDS. However, additional experiments were carried out with an added initial excess of O_3 (4 ppm) the presence of partial pressures of 150 and 10 Torr O_2 in order to evaluate any posssible influences. The addition of excess O_3 had no effect upon the product distribution in the presence of 150 Torr O_2 while with a partial pressures of 10 Torr O_2 the addition of excess O_3 was found to increase the yield of SO_2 and decrease the MSA yield compared to the system containing no initial O_3 (Table 2).

Several different routes for the formation of OH radicals are possible in the reactor : a) generation at the reactor surface, b) generation via reactions of O atoms from the photolysis of O_3 produced in the reactor, c) OH generation during the photolysis of DMDS by reaction of CH_3S or CH_3SO with O_3 as discussed by Dominé et al. (10) and d) reaction of CH_3S with HO_2 radicals as suggested by Hynes and Wine (11) in their study of the reaction of OH radicals with CH_3SH. From the decay of the two alkenes an upper limit of about 0.5×10^6 radicals cm^{-3} was estimated for the OH concentration with just hydrocarbons and synthetic air in the chamber. With DMDS in the system an OH concentration of 1.6×10^6 radicals cm^{-3} was estimated. This value is higher than that observed with just hydrocarbons present in the system and probably indicates the generation of additional OH by another channel such as the reaction of CH_3S radicals with O_3 or HO_2. Since the concentration of O_3 was always under the 1ppb detection limit when DMDS was present in the reaction system the reaction of CH_3S with HO_2 is considered the likely additional OH radical source. This is supported by computer similutations of the reaction system and also unpublished results from this laboratory on the 254 nm photolysis of CH_3SH in which OH concentrations up to 10^7 radicals cm^{-3} were observed and are attributed to the reaction sequence:

$$CH_3SH + h\nu\,(254\,nm) \longrightarrow CH_3S + H$$
$$H + O_2 + M \longrightarrow HO_2 + M$$

$$CH_3S + HO_2 \longrightarrow CH_3SO + OH$$

If an OH concentration of 1.6 x 10^6 radicals cm^{-3} is being formed during the photolysis of DMDS alone in 150 of O_2 it will account for approximately 12% of the measured DMDS decay. Similar experiments performed in the presence of 10 Torr O_2 showed that the OH levels are $\leq 10^6$ radicals cm^{-3} and could account for ≤50% of the measured DMDS decay. As discussed above the yield of MSA is found to increase from 7% to 23% on decreasing the partial pressure of O_2 from 150 to 10 Torr. Therefore, parrallel to the O_2 decrease the possible percentage contribution of OH to the DMDS decay is increasing.

4. DISCUSSION

The following discussion is not intended to be an in-depth analysis of the results, only the major points will be outlined here. A full analysis and discussion of the data can be found in Barnes et al. (12). Figure 4 shows a simplified mechanism for the atmospheric oxidation of DMDS in the absence of NO_x. This scheme is based largely on that suggested by Yin et al. (5) and is intended to facilitate comprehension of the following discussion.

The yields of CH_3OH and CH_3OOH: This is the first quantitative observation of both CH_3OOH and CH_3OH in the photolysis of DMDS and provides strong circumstantial evidence for the formation of methyl radicals CH_3 in the system and also consequenly implies a cleavage of the C-S bond during the course of the oxidation of CH_3S. Both CH_3OH and CH_3OOH are expected to be formed in the sequence of reactions involving the addition of O_2 to CH_3 to form methyl peroxy radicals (CH_3OO) followed either by the self reaction of these radicals or their reaction with HO_2 (13):

$$CH_3 + O_2 \longrightarrow CH_3O_2$$
$$CH_3OO + CH_3OO \longrightarrow 2\ CH_3O + O_2 \quad (22\%)$$
$$\longrightarrow CH_3OH + HCHO + O_2 \quad (69\%)$$
$$CH_3O + O_2 \longrightarrow HCHO + HO_2$$
$$CH_3OO + HO_2 \longrightarrow CH_3OOH$$

Failure to detect CH_3OH in other studies (14) has been taken as an indication that generation of CH_3 is negligible, however, Yin et al. (5) have pointed out that this is probably due to the competeing reaction:

$$CH_3OO + CH_3S \longrightarrow CH_3O + CH_3SO$$

The observation of this work that the yields of CH_3OH and CH_3OOH increase substantially on adding O_3 to the reaction system (Table 2) which scavenges the CH_3S radicals (10) lends credibility to the hypothesis of Yin et al. (5).

Yields of SO_2 and MSA: The 90% yield of SO_2 in the photolyis of low concentrations of DMDS in synthetic air (150 Torr O_2 and 610 Torr N_2) is in agreement with other studies (5,14) . It is interesting to note that the addition of O_3 to the synthetic air system does not significantly effect either the SO_2 or the MSA yield. Preliminary experiments on the photolysis of methyl methanethiolsulfonate ($CH_3SO_2SCH_3$) as a source for CH_3SO_2 and CH_3S radicals show that MSA and SO_2 are formed with approximately 60 and 40 % yields, respectively. The experiments suggest that the major fate of CH_3SO_2, at least at room temperature, is not decomposition to CH_3 and SO_2 but rather further oxidation to MSA. Therefore, once CH_3SO_2 is produced in an organic sulphur reaction system high yields of MSA formation can be expected. This further implies that the oxidation of CH_3S by either O_2 or O_3, which the results show is not very conducive to MSA production, does not apparently result in the oxidation of CH_3S to CH_3SO_2 but must result in early cleavage of the C-S bond. This observation is also supported by the recent product study of the $CH_3S + O_3$ reaction by Dominé et al. (10).

Increasing the initial concentration of DMDS results in large decrease in the SO_2 yield and an increase in MSA aerosol and also the formation of $(CH_3SO_2)_2O$ and $CH_3SO_2SCH_3$ (Figure 2 and 3). The use of high starting concentrations of DMDS give large steady state concentrations of CH_3S radicals. It would appear that the large concentrations of CH_3S radicals promote the oxidation of the CH_3SOO adduct to CH_3SO_3 and hence MSA formation possibly by reactions of the type shown below and also indicated in Figure 4:

$$CH_3S + O_2 \longrightarrow CH_3SOO$$
$$CH_3SOO + CH_3S \longrightarrow CH_3SO + CH_3SO$$
$$CH_3SO + O_2 \longrightarrow CH_3S(O)OO$$
$$CH_3S(O)OO + CH_3S \longrightarrow CH_3SO_2 + CH_3SO$$
$$CH_3SO_2 + O_2 \longrightarrow CH_3SO_2OO$$
$$CH_3SO_2OO + CH_3S \longrightarrow CH_3SO_3 + CH_3SO$$
$$CH_3SO_3 + HO_2 \longrightarrow CH_3SO_3H + O_2$$

Adduct formation between CH_3S and O_2 has recently been observed by Turnipseed et al. (15) at low temperatures. $(CH_3SO_2)_2O$ and $CH_3SO_2SCH_3$ are presumably formed in the reactions such as,

$$CH_3SO_2 + CH_3SO_2O \longrightarrow (CH_3SO_2)_2O$$
$$CH_3SO_2 + CH_3S \longrightarrow CH_3SO_2SCH_3$$

and provide evidence for the involvment of CH_2SO_2 and CH_3SO_3 radicals in the reaction mechanism. Balla and Heicklen (16) in a study of the photolysis of Torr concentrations of DMDS have detected products with mass spectral peaks at m/e values of 126 and 174 which they could not identify. These peaks correspond to the masses of $CH_3SO_2SCH_3$ and $(CH_3SO_2)_2O$, respectively, observed in this study. The peak observed at m/e 126 by Balla and Heicklen (16) could also belong to $(CH_3SO)_2$ and was actually assigned by these authors to this compound.

At least one relatively major sulphur containing product still remains to be identified in the high concentration DMDS photolysis experiments. Possible candidates apart from $(CH_3SO)_2$ include CH_3SOSCH_3, $CH_3SOSO_2CH_3$ and $(CH_3SO_2)_2$. However, positive identification is not presently possible since IR spectra of such compounds are not available.

This represents the first study in which MSA has been detected directly in the gas phase during the oxidation of an organic sulphur compound. At 1-2 ppm initial levels of DMDS the MSA yield is ~7% in air and increases with decreasing partial pressure of O_2 in the reaction system. This observed increase with the decrease in O_2 partial pressure is contary to the oxidation scheme outlined in Figure 4. The observation that OH radicals are being formed in the sytem and that their probable contribution to the decay of DMDS increases with decreasing partial pressure of O_2 i.e. from ~12 % with 150 Torr O_2 to ~50 % with 10 Tor O_2 suggests that MSA formation is strongly correlated to the percentage OH radical contribution to the DMDS decay. If this is the case then MSA formation could be the result of the further reactions of methyl sulphenic acid (CH_3SOH) formed in the reaction of OH radicals with DMDS.

$$CH_3SSCH_3 + OH \longrightarrow CH_3SOH + CH_3S$$

CH_3SOH is known to be a product of this reaction although the absolute yield is still uncertain (17). To test this hypothesis experiments were performed in which H_2O_2 was added as an additional OH source to 2 ppm of DMDS in synthetic air. The yield of MSA formation was found to be 30% compared to the 7 % measured in the absence of the extra OH radical source. All the information supports that the further reactions of methyl sulfenic acid (CH_3SOH) formed in the reaction of OH with DMDS are the main source of MSA in the photolysis of 1-2 ppm levels of DMDS. One possible mechanism for MSA formation is:

$$OH + CH_3SSCH_3 \longrightarrow CH_3SOH + CH_3S$$
$$CH_3SOH + CH_3SO_3 \longrightarrow CH_3SO + CH_3SO_3H$$

The OH + DMDS reaction could also contribute to the enhanced levels of MSA aerosol formation observed in the experiments using high concentrations of DMDS but simulations indicate that the high steady state CH_3S radical concentrations in these systems promote the oxidation of CH_3SOO to CH_3SO_2 and thus MSA in this case.

Implications for the atmospheric DMS oxidation mechanism: The results presented above have a number of important implications for the oxidation of DMS which are outlined below:

- CH_3SOH has been postulated as one of the intermediates in the OH initiated oxidation of DMS (5).

$$OH + CH_3SCH_3 \longrightarrow CH_3SCH_2 + H_2O$$
$$\longrightarrow CH_3S(OH)CH_3 \longrightarrow CH_3SOH + CH_3$$
$$\longrightarrow CH_3OH + CH_3S$$

If CH_3SOH is being formed in the reaction of OH with DMS the results from the present work suggest that high yields of MSA should be observed. However, the available product data from chamber studies (5, and references therein) show that high yields of MSA are only obtained in the presence of NO_x. In NO_x free systems the detected yields of MSA are ≤4% (18 -20). This would suggest that CH_3SOH is probably not, therefore, a major intermediate in the OH initiated oxidation of DMS otherwise high yields of MSA would have been detected.

- The high yields of MSA generally observed in smog chamber studies of the OH initiated oxidation of DMS in the presence of NO_x are probably due to effective oxidation of CH_3S by NO_2 to CH_3SO_2 or CH_3SO_3. Preliminary experiments in this laboratory suggest the further oxidation of CH_3SO_2 results in high yields of MSA even in the absence of NO_x.

- The addition of O_3 to the reaction system has very little influence on the product distribution. Since the addition of O_3 effectively scavages the CH_3S radicals one might expect high yields of MSA if CH_3S is oxidised through to CH_3SO_3 as in the reaction scheme in Figure 4. However, the yield of MSA remains low which suggests that the C-S bond is being cleaved early in the oxidation chain. This observation is in agreement with the recent results of Dominé et al. (10).

5. CONCLUSIONS

The oxidation of CH_3S, produced in the photolysis of 1-2 ppm levels of DMDS, with either O_2 or O_3 do not result in high yields of MSA. The level of MSA formed in the system appears to be regulated by the steady state concentration of OH radicals in the reaction system. At higher concentrations of DMDS the reactions of CH_3S radicals appear to control the formation of MSA. Further, the oxidation of CH_3S with O_2 and O_3 probably mainly results in cleavage of the C-S bond early in the oxidation chain which precludes the formation of MSA.

The results from the studies on the photolysis products of DMDS imply that CH_3SOH is not a major intermediate in the OH initiated oxidation of DMS, at least at room temperature. Further, once CH_3S has been oxidised to CH_3SO_2 formation of MSA rather than decompostion to CH_3 and SO_2 appears to be the dominant pathway at room temperature. The presence of NO_x appears to promote the oxidation of CH_3S to CH_3SO_2 and thus MSA formation. Therefore, in the atmosphere if CH_3S is the main intermediate in the OH initiated oxidation of DMS high yields of MSA are to be expected in areas with high NO_x. Recent field measurements performed within the OCEANO-NOX project (19,21) lend support to this hypothesis, however, further measurements are required before any positive conclusions can be drawn.

REFERENCES

(1) CULLIS, C.F. and HIRSCHLER, M.M. (1980) Atmospheric sulfur: natural and man-made sources. Atmos. Environ. 14, 1263-1278.

(2) SCHWARTZ, S.E. (1988) Are global cloud albedo and climate controlled by marine phytoplankton. Nature 336, 441-445.

(3) ANDREAE. M.O. (1986). The ocean as a source of atmospheric sulfur. In: P. Buat-Menard ed), The Role of Air-Sea Exchange in Geochemical Cycling, D. Reidel, Dordrecht, pp 331-362.

(4) BATES, T.S., CLINE, J.D., GAMMON, R,H. and KELLY-HANSEN, S.R. (1987). Regional and seasonal variation of the flux of oceanic dimethylsulfide to the atmosphere. J. Geophys. Res. 92, 2930-2938.

(5) YIN, F., GROSJEAN, D. and SEINFELD, J.H. (1990). Photooxidation of dimethyl sulfide and dimethyl disulfide. I: Mechanism Development. J. Atmos. Chem. 11, 309-364.

(6) HEICKLEN, J. (1985). The gas-phase oxidation of organic sulfur compounds. Reviews of Chemical Intermediates 6, 175-196.

(7) BARNES, I., BECKER, K.H., MIHALOPOULOS, N. (1992). An FTIR Study of the Photooxidation of Dimethyl Disulfide. J. Atmos. Chem., to be submitted.

(8) MIHALOPOULOS, N., BARNES, I. and BECKER, K.H. (1991). Infrared absorption spectra and integrated band intensities for gaseous methanesulphonic acid (MSA). Atmos. Environ. 26A, 807-812.

(9) BARNES, I., BASTIAN, V., BECKER, K.H., FINK, E.H., KLEIN, Th., KRIESCHE, V., NELSEN, W., REIMER, A. and ZABEL, F. (1984). Untersuchung der Reaktionssysteme $NO_x/ClO_x/HO_x$ unter troposphärischen und stratosphärischen Bedingungen. GSF München, BPT-Bericht 11/84, ISSN 0176/077.

(10) DOMINÉ, F., RAVISCHANKARA, A.R. and HOWARD, C.J. (1992). Kinetics and mechanisms of the reactions of CH_3S, CH_3SO, CH_3SS with O_3 at 300 K and low pressure. J. Phys. Chem. 96, 2171-2178.

(11) HYNES, A. J. and WINE, P.H. (1987). Kinetics of the OH + CH_3SH reaction under atmospheric conditions. J. Phys. Chem.. 91, 3672-3676.

(12) BARNES, I., BECKER, K.H. and MILALOPOULOS, N. (1992). An FTIR product study of the photooxidation of dimethyl disulfide. Atmos. Environ. (to be submitted).

(13) ATKINSON, R., BAULCH, D.L., COX, R.A., HAMPSON, R.F., Jr., KERR, J.A. and TROE, J. (1989). Evaluated kinetic and photochemical data for atmospheric chemistry: Supplement II. J. Phys. Chem. Ref. Data 18, 881-1097.

(14) HATAKEYAMA, S. and AKIMOTO, H. (1983). Reactions of OH radicals with methanethiol, dimethyl sulfide and dimethyl disulfide in air. J. Phys. Chem. 87, 2387-2395.

(15) TURNIPSEED, A.A., BARONE, S.B. and RAVISHANKARA, A.R. (1992). Observation of CH_3S addition to O_2 in the gas phase. J. Phys. Chem. 96, 7502-7505.

(16) BALLA, R.J. and HEICKLEN, J. (1985). Oxidation of sulfur compounds III: The photolysis of $(CH_3S)_2$ in the presence of O_2. J. Photochem. 29, 297-310.

(17) DOMINÉ, F. and RAVISHANKARA, A.R. (1992). The yield of CH_3S from the reaction of OH with CH_3SSCH_3. Int. J. Chem. Kinetics 24, 943-951.

(18) BARNES, I., BASTIAN, V. and BECKER, K.H. (1988). Kinetics and mechanisms of the reaction of OH radicals with dimethyl sulfide. Int. J. Chem. Kinetics 20, 415-431.

(19) BARNES, I., BONSANG, B., BRAUERS, T., CARLIER, P., COX, R.A.,DORN, H.P., JENKIN, M.E., LE BRAS, G. and PLATT, P. (1991). Laboratory and field studies of the oxidation processes occurring in the atmospheric marine boundary layer (OCEANO-NOX CEC Project) Air Pollution Research Report 35, Commission of the European Communties, Brussels 1991, ISBN 2-87263-064-3.

(20) MARTINET, A. (1989). Etudes cinetiques, sous Conditions atmospheriques simulees, de Reactions elementaires intervenant dans la Photooxydation de Precurseurs d´acidite en Milieu marin. PhD. Thesis, l'Université Paris VII.

(21) MIHALOPOULOS, N., NGUYEN, B.C., BOISSARD, C., CAMPI, J.M., PUTAND, J.P., BELVISO, S., BARNES, I. and BECKER, K.H. (1992). Field study of dimethyl sulfide oxidation in the boundary layer: Variations of dimethylsulfide, methane sulfonic acid, sulfur dioxide, non-sea-salt sulfate and aitken nuclei at a coastal site. J. Atmos. Chem. 14, 459-477.

Table 1. Products observed in DMDS/O_2/N_2 photolysis systems after 10 min irradiation as a function of the O_2 partial pressure at 760 Torr total pressure and 298 K. The yields are in % S or % C.

O_2 (Torr)	N_2 (Torr)	DMDS Δppm	SO_2 ppm	SO_2 %	MSA[a)] ppm	MSA[a)] %	HCHO[a)] ppm	HCHO[a)] %	CH_3OOH ppm	CH_3OOH %	CH_3OH ppm	CH_3OH %
150	610	1.52	2.69	**88**	0.21	**7**	2.26	**90**	0.30	**10**	0.07	**2**
100	660	1.21	2.19	**90**	0.21	**9**	1.99	**82**	0.22	**9**	0.05	**2**
50	710	1.08	1.75	**81**	0.35	**16**	1.70	**79**	0.14	**6**	0.05	**2**
25	735	0.76	1.15	**76**	0.27	**18**	1.17	**77**	0.06	**4**	0.05	**3**
10	750	0.20	0.26	**65**	0.09	**23**	0.28	**70**	0.02	**5**	0.01	**3**

a) Corrected for wall loss

Table 2. Percentage yields (% S or C) of products from DMDS/O_2/N_2 photolysis mixtures with and without the addition of 4 ppm O_3 for O_2 partial pressures of 150 and 10 Torr O_2 and a total pressure of 760 Torr .

O_2 (Torr)	O_3 ppm	DMDS Δppm	SO_2 ppm	SO_2 %	MSA ppm	MSA %	HCHO ppm	HCHO %	CH_3OOH ppm	CH_3OOH %	CH_3OH ppm	CH_3OH %
150	0	1.52	2.69	**88**	0.21	**7**	2.26	**90**	0.30	**10**	0.07	**2**
150	4	0.90	1.65	**92**	0.10	**6**	1.06	**59**	0.40	**22**	0.20	**11**
10	0	0.20	0.26	**65**	0.09	**23**	0.28	**70**	0.02	**5**	0.01	**3**
10	4	1.70	2.80	**82**	0.50	**15**	2.20	**65**	0.60	**18**	0.20	**6**

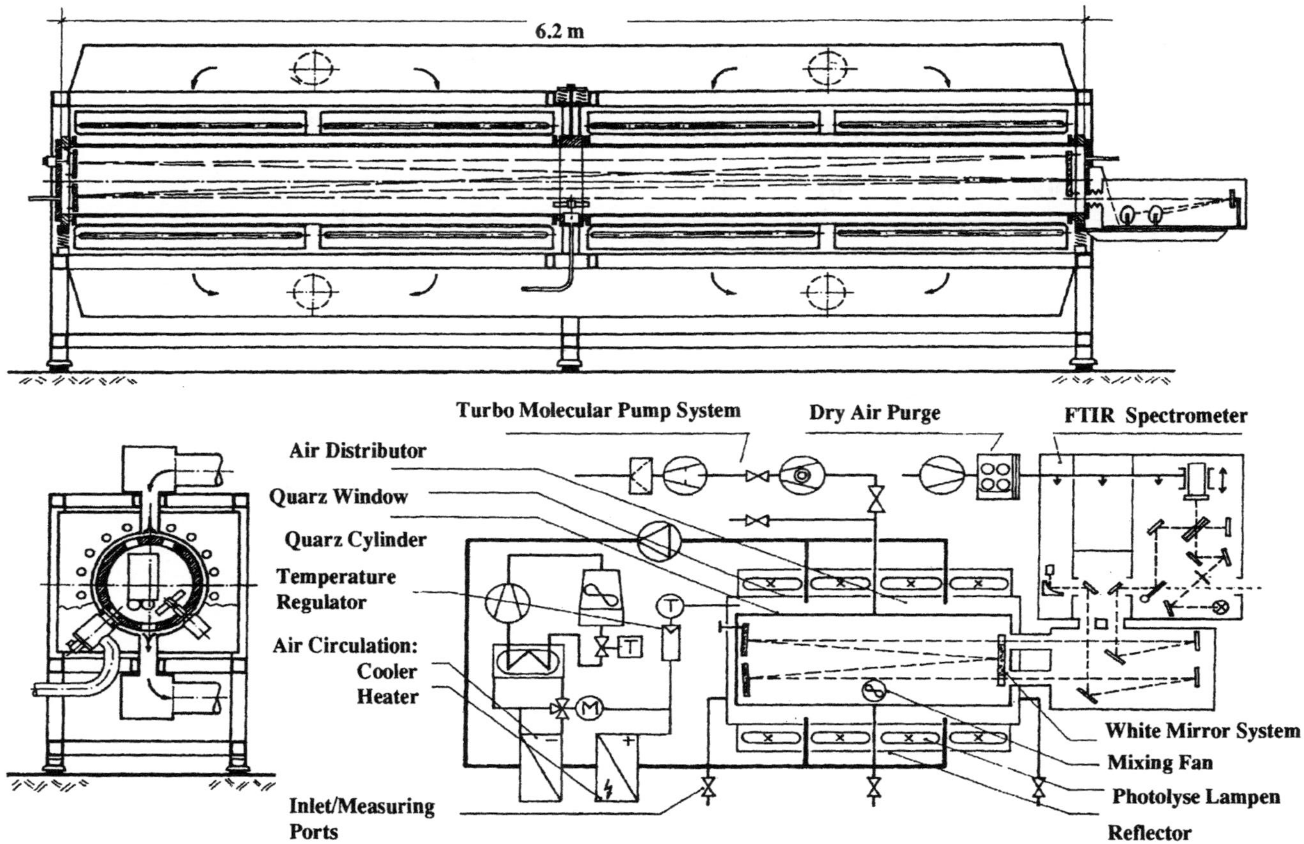

Figure 1. Schematic diagram of the 1080 l quartz glass reaction chamber with FTIR facility.

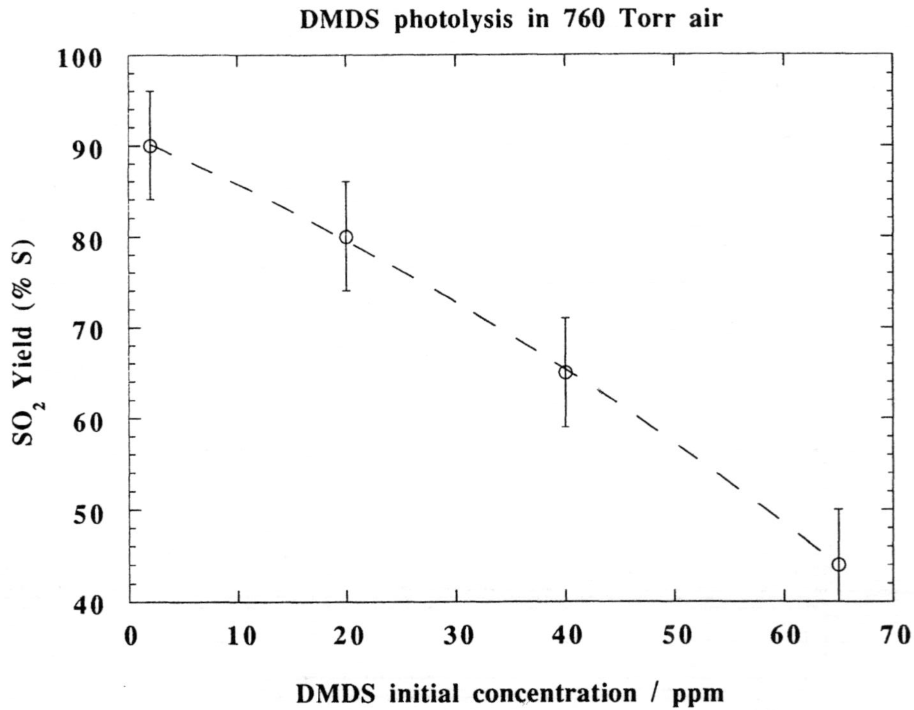

Figure 2. SO_2 yields (% S) as a function of the initial dimethyl disulphide concentration.

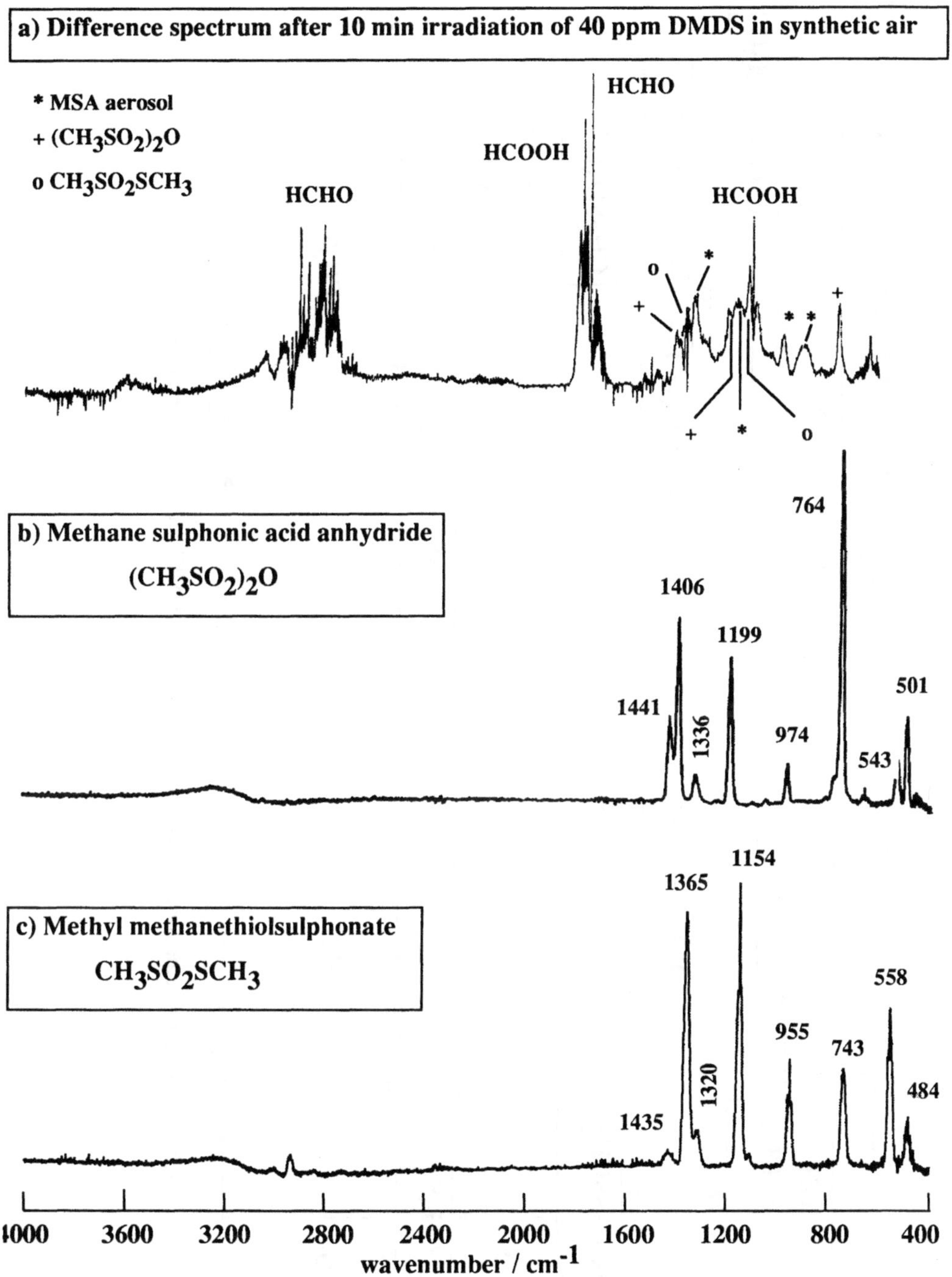

Figure 3. Trace a) is a product spectrum of a DMDS (40 ppm)/ air (760 Torr) recorded after 10 min irradiation. SO_2 has been subtracted. Traces b) and c) are spectra of methane sulphonic acid and methyl methanethiolsulphonate, respectively.

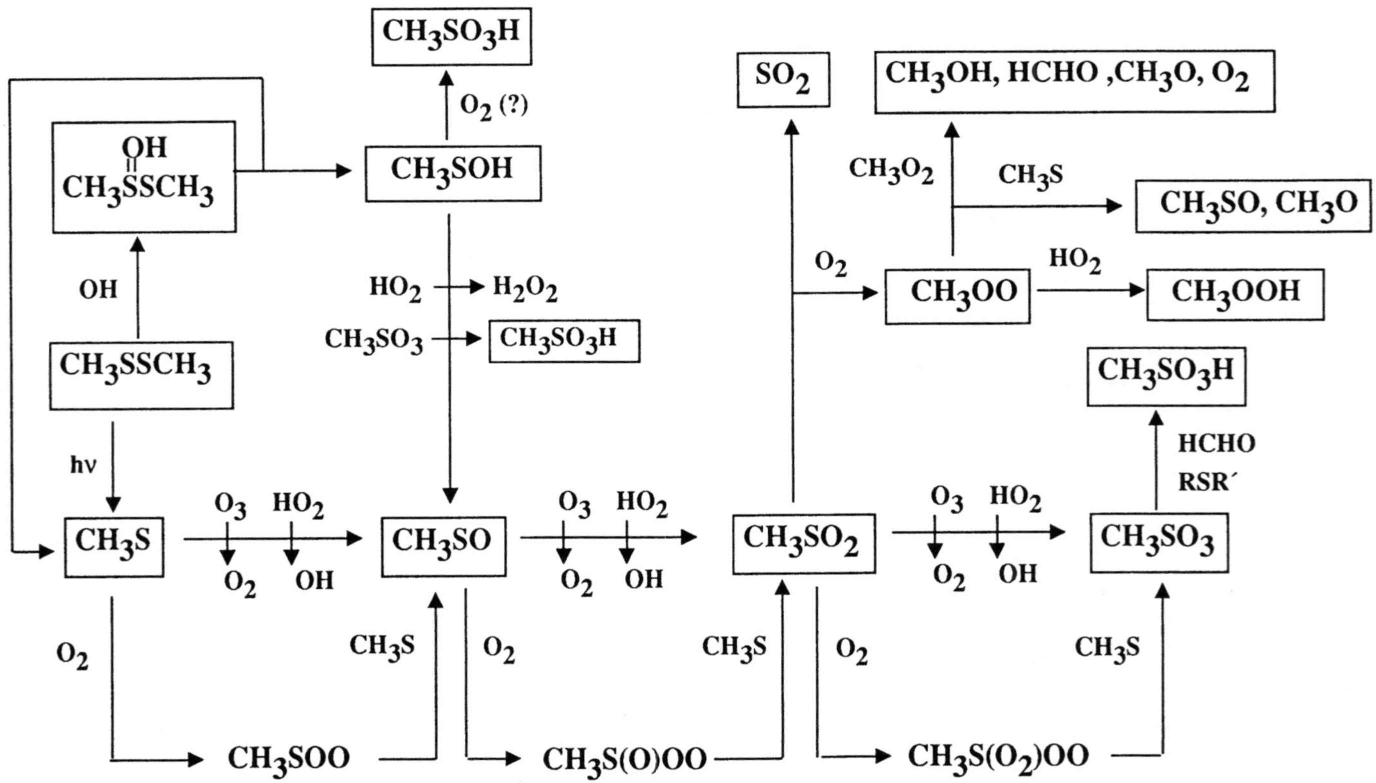

Figure 4. Simplified reaction scheme for the atmospheric oxidation of DMDS in the absence of NO_x

MECHANISTIC STUDIES OF THE OH-INITIATED OXIDATION OF DIMETHYLSULFIDE

A. J. Hynes[a,d], R. E. Stickel[a], A. J. Pounds[b], Z. Zhao[c], T. McKay[a], J. D. Bradshaw[c] and P. H. Wine[a,b,c]

Georgia Institute of Technology
Atlanta, Georgia 30332, U.S.A.

a. Physical & Atmospheric Chemistry Branch, EOPSL-GTRI
b. School of Chemistry & Biochemistry
c. School of Earth & Atmospheric Sciences
d. Current address: RSMAS, University of Miami
4600 Rickenbacker Causeway, Miami, FL 33149

SUMMARY

Laser flash photolysis has been used to study the mechanism of the OH initiated oxidation of dimethylsulfide (DMS) and its deuterated analog. Pulsed laser induced fluorescence was employed to monitor OH decays and tunable diode laser absorption spectroscopy was used to monitor yields of HOD. The formation of an OH-DMS-d_6 adduct was directly observed and its binding energy determined. The rate coefficient for the reaction of the adduct with oxygen was determined. Evidence for production of HO_2 in the reaction of the adduct with oxygen was obtained. Direct observation of the HOD product of the reaction of OD with DMS in high yield supports H atom abstraction as the dominant reaction pathway in the absence of oxygen. The observations confirm our previous hypothesis that the OH + DMS reaction under atmospheric conditions proceeds via a multi-channel mechanism involving (at least) H-atom abstraction, reversible adduct formation and adduct reaction with oxygen.

1. INTRODUCTION

Estimates of global sulfur emissions indicate that natural and anthropogenic sources make roughly equal contributions to the atmospheric sulfur burden. Recent detailed estimates (1) indicate that oceanic emissions account for approximately 43% of the emissions budget in the southern hemisphere and about 15% of the global total. The oceanic emissions consist almost exclusively of dimethylsulfide (DMS, CH_3SCH_3) which is emitted by phytoplankton. DMS has long been thought to be the precursor of the non-seasalt sulfate/methanesulfonic acid aerosol which is ubiquitous in the marine boundary layer. Sulfur containing aerosols are thought to act as cloud condensation nuclei (CCN), and Charlson, et al. (2) have argued that DMS, via its role in promoting cloud formation and altering the earth's albedo, could play a major role in global climate regulation. Chameides and Stelson (3) have suggested that if SO_2 is an important intermediate in DMS oxidation, then the yield of gas phase H_2SO_4 (the probable precursor to new aerosol formation) from DMS oxidation will not be large enough for homogeneous nucleation of H_2SO_4(g) to be an important process in the marine boundary layer. Recently, Bandy, et al. (4) have suggested, based on a lack of correlation between levels of DMS and SO_2 measured in the marine boundary layer, that DMS is not converted efficiently to SO_2; these authors suggest that H_2SO_4(g) may be generated in large yield from DMS oxidation via a mechanism which does not involve SO_2 as an intermediate (4). Clearly a detailed understanding of the rates, mechanisms and ultimate products of the atmospheric degradation of DMS are required in order to assess its contribution to aerosol formation.

G. Restelli and G. Angeletti (eds.), Dimethylsulphide: Oceans, Atmosphere, and Climate, 211–221.

Previous Work

To clarify the motivation for the present study we briefly review relevant previous work. Tyndall and Ravishankara (5) have reviewed the checkered history of laboratory measurements which address this problem and we only discuss more recent measurements here. It is generally accepted that, except in polluted environments which facilitate the build up of appreciable levels of NO_3 at night, attack by the OH radical constitutes the initial step in the atmospheric oxidation of DMS.

$$CH_3SCH_3 + OH \rightarrow \text{products} \quad (1)$$
$$CD_3SCD_3 + OH \rightarrow \text{products} \quad (2)$$

In previous work (6) we have shown that the effective rate coefficient for reaction 1 and its deuterated analog, reaction 2, is dependent on oxygen concentration. In the absence of oxygen, the reaction is independent of pressure, shows a positive activation energy and a kinetic isotope effect. The reaction shows enhanced reactivity for sulfides with secondary hydrogen atoms and, as Atkinson points out (7), the observed deuterium isotope effect is similar in magnitude to that observed for H-atom abstraction from secondary C-H bonds in alkanes. The C-H bonds in DMS have a similar bond dissociation energy to the alkane secondary C-H, suggesting that the mechanism of the oxygen independent channel is H abstraction. However, in spite of these indications, the rate coefficient increases as the partial pressure of oxygen is increased and the rate of increase is the same for both DMS and its deuterated analog. The effective rate coefficient of reaction 2 in one atmosphere of air shows a dramatic negative temperature dependence increasing from 3.4 to 11.6×10^{-12} $cm^3molecule^{-1}s^{-1}$ as the temperature decreases from 300K to 252K. Our observations were consistent with the reaction proceeding via two reactive channels. Our proposed mechanism (written for DMS-d_6) involves a direct abstraction reaction, together with reversible adduct formation followed by adduct reaction with molecular oxygen.

$$CD_3SCD_3 + OH \rightarrow CD_3SCD_2 + HOD \quad (2a)$$
$$CD_3SCD_3 + OH + M \rightleftarrows CD_3S(OH)CD_3 + M \quad (2b,-2b)$$
$$CD_3S(OH)CD_3 + O_2 \rightarrow \text{Products} \quad (3)$$

We were unable to directly observe the equilibration process. However, by assuming that the limiting rate of k_2 in oxygen at 261K corresponded to the forward addition rate, k_{2b}, we were able to extract elementary rate coefficients and estimate an adduct binding energy of 13±3 kcal/mole. A recent competitive rate study of reaction 1 by Barnes, et al. (8) also shows an "oxygen dependence" in the rate coefficient but the absolute values are somewhat higher than those obtained in our direct study; their result in 760 Torr of air is in reasonable agreement with another competitive rate study by Wallington, et al. (9). In spite of our inability to directly observe adduct formation, the two channel mechanism has been widely accepted as the operative mechanism. In a recent study, however, Gu and Turecek (10) claim to have both experimental and theoretical evidence which indicates that the adduct does not exist. Also Ravishankara and coworkers (5,11) have cited evidence that the O_2-independent channel may correspond to an irreversible decomposition of the energized adduct, perhaps breaking a C-S bond to produce methyl radicals (written below for reaction 2):

$$CD_3SCD_3 + OH \rightarrow CD_3SOH + CD_3 \quad (2c)$$

Objectives of the current study

The objectives of the current work were to use the Laser Flash Photolysis-Pulsed Laser Induced Fluorescence (LFP-PLIF) technique to attempt to: a) directly observe the formation of the OH-DMS adduct and determine values for the elementary rates, to perform these measurements as a function of temperature and, from the temperature dependence of the equilibrium constant, determine the adduct binding energy; b) determine the dependence of the observed rate on oxygen concentration in order to both assess the significance of reaction channels regenerating OH, and to determine the rate coefficient for the reaction of the adduct with oxygen; and c) attempt to identify HO_2 as a product of reaction 3 by back titration with NO to regenerate OH. In a separate series of experiments Laser Flash Photolysis-Tunable Diode Laser Absorption (LFP-TDLA) was used to monitor the HOD formed in the reaction

$$CH_3SCH_3 + OD \rightarrow CH_3SCH_2 + HOD \tag{4}$$

to confirm that the OH + DMS reaction does indeed proceed via a H atom abstraction route.

2. EXPERIMENTAL TECHNIQUES

LFP-PLIF Experiments

The LFP-PLIF apparatus has been described in detail elsewhere (6). A brief review of its operation is given below. The pyrex reaction vessel had an internal diameter of 4 cm and a length of 50 cm. Two side arms, 4 cm i.d. and 14 cm long, which terminated in Brewster angle windows, were attached to the center of the cell. The photolysis laser passed through these side arms across the direction of gas flow while the probe laser passed along the length of the cell. Fluorescence was detected through a third side arm, 4 cm i.d. and 5 cm long, perpendicular to the photolysis and probe beams. The central 25 cm length of the cell was jacketed to permit the flow of heating or cooling fluid from a thermostated bath. A copper-constantan thermocouple with a stainless steel jacket was inserted into the reaction zone through a vacuum seal, allowing measurement of the gas temperature under the precise pressure and flow conditions of the experiment.

OH was produced by the pulsed laser photolysis of H_2O_2, using the 266nm 4th harmonic output from a Nd-Yag laser. Pulsed laser induced fluorescence using a Nd-Yag-laser pumped, frequency doubled dye laser was used for OH detection. Excitation was via the $Q_1 1$ line of the $A^2\Sigma^+$-$X^2\Pi$(1-0) transition at ≈282nm. Fluorescence in the 0-0 and 1-0 bands was detected by an EMI 9813QB photomultiplier after passing through collection optics and filters to discriminate against Rayleigh scattering and Raman scattering from N_2 and/or O_2. The photomultiplier output was appropriately terminated and fed to a 100-MHz waveform analyzer to obtain the peak voltage averaged for (typically) 100 laser shots. Kinetic information was obtained by varying the delay between the photolysis and probe lasers using a digital delay generator. Signal was collected for 15-20 delays, varying from 0-20 ms, in order to map out an OH decay profile over three 1/e times. In order to avoid the accumulation of photolysis or reaction products all experiments were carried out under "slow flow" conditions with the linear flow rate through the reactor varied between 1 and 5 cm s^{-1}. Concentrations of DMS-d_6 were measured by *in situ* uv absorption at 228.8nm.

LFP-TDLA Experiments

The LFP-TDLA apparatus has also been described in detail elsewhere (12). Briefly, the reaction cell was a pyrex cylinder 25 mm in diameter and 1.2 m in length with angled calcium fluoride windows epoxied to the ends. Axial photolysis of DNO_3 by a KrF excimer laser (248 nm) produced OD radicals. Second harmonic detection of HOD infrared absorption at 1383.61 wavenumbers was used to follow the number density of HOD before and after photolysis. The measurement was performed twice for each set of experimental conditions: once with DMS and once with cyclohexane or n-hexane as the reaction partner (the yield of HOD from OD + alkane reactions is unity). The yield of HOD from the OD + DMS reaction was taken as the ratio of the observed absorbances after normalization for variations in photolysis energy and DNO_3 concentration and after correction for background OD losses. Total pressure was varied from 10 to 30 Torr and temperature was varied from 296K to 348K.

3. RESULTS AND DISCUSSION

LFP-PLIF studies in the absence of oxygen

All of the LFP-PLIF studies reported in this paper used DMS-d_6 as the reactant. Because the deuterated compound has a slower reaction rate in the absence of O_2, the effects of the adduct equilibration process are enhanced (5) and our attempt to identify HO_2 as a product of the adduct reaction with oxygen is possible. Figure 1 shows the variation of pseudo-first-order decay rate of OH, k′, as a function of DMS-d_6 concentration at 267K in 100 Torr of N_2. Note that for a single elementary reaction we would expect k′ to be a linear function of DMS concentration with the slope giving us the rate coefficient. The curvature in the k′ versus [DMS] plot is a clear indication of a complex mechanism. If adduct removal is much faster than adduct formation, then the steady state approximation can be applied to the adduct, i.e., it can be assumed that d[adduct]/dt = 0. For a mechanism consisting of reactions 2a, 2b, -2b, 2c, and 3, the steady state approximation (SSA) leads to the following relationship:

$$k_{obs} = (k_{2a} + k_{2c} + Xk_2) / (1 + X) \quad \text{(I)}$$

where

$$X = k_3[O_2] / k_{-2b} \quad \text{(II)}$$

$$k_2 = k_{2a} + k_{2b} + k_{2c} \quad \text{(III)}$$

Equation I indicates that in the absence of oxygen and at relatively low DMS-d_6, then $k_{obs} = k_{2a} + k_{2c}$. As we increase the concentration of DMS we would expect deviations from the SSA to occur, and if we can shift the equilibrium 2b, -2b sufficiently to the right and our time resolution is adequate, we would expect to see a decay into equilibrium. At 267K and at the highest concentration of DMS-d_6 employed we measured rates in excess of 10^6 s^{-1} and observed no deviation from exponential behavior. However at 258K we obtained the data shown in Figure 2. In this case we can clearly resolve the initial rapid equilibration step.

As shown elsewhere (6), under pseudo-first-order conditions the rate equations for the suggested reaction mechanism can be solved analytically; a double exponential OH decay is predicted. From the best fit of the data in Figure 2 to the analytical solution (6), values for $k_{2a} + k_{2c}$, k_{2b}, and k_{-2b} can be obtained. Although we cannot directly observe equilibration at 267K, we can use the double exponential solution to model the "roll-off" at high DMS-d_6. The "slow" component of the double exponential decay gives the pseudo-first-order decay rate, k′, which we measure experimentally. Figure 1 shows the calculated variation in k′ as a function of DMS-d_6 concentration as we vary k_{-2b} with $k_{2a} + k_{2c}$ and k_{2b} fixed. This demonstrates that the observed roll-off behavior is totally consistent with our assumed mechanism and allows us to extract a value of k_{-2b}. By using our measured values of k_{2b} and k_{-2b} we can determine the equilibrium constant as a function of temperature.

$$K_p\,(RT) = K_c = k_{2b} / k_{-2b} \quad \text{(IV)}$$

From the van't Hoff relationship, the slope of a plot of ln K_p vs T^{-1} gives the exothermicity of reaction 2b, i.e., the adduct binding energy.

$$\ln K_p = (\Delta S/R) - (\Delta H/RT) \quad \text{(V)}$$

A preliminary analysis of experiments completed thus far allows us to construct the van't Hoff plot shown in Figure 3. We obtain an adduct binding energy of approximately 14 kcal/mole which is good agreement with our previous estimate (6). A recent *ab initio* calculation (13) gives an adduct binding energy of 10 kcal/mole in reasonable agreement with this value. It is not possible to give a detailed critique of the work of Gu and Turecek (10) here except to note that,

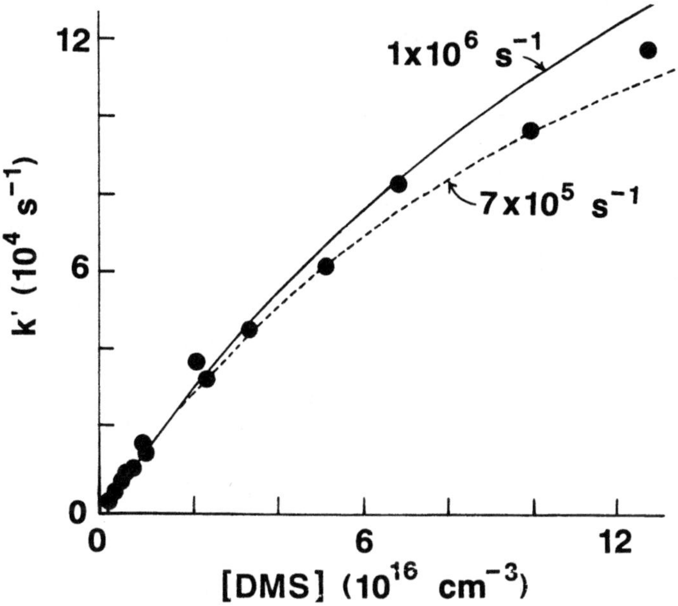

Figure 1. Variation of pseudo-first order decay rate of OH as a function of DMS-d_6 concentration at 267K in 100 Torr N_2. Solid and dotted lines show the calculated variation for different values of k_{-2b}. Values assumed for $k_{2a} + k_{2c}$ and k_{2b} in units of 10^{-12} $cm^3molecule^{-1}s^{-1}$ are 1.6 and 4.0, respectively.

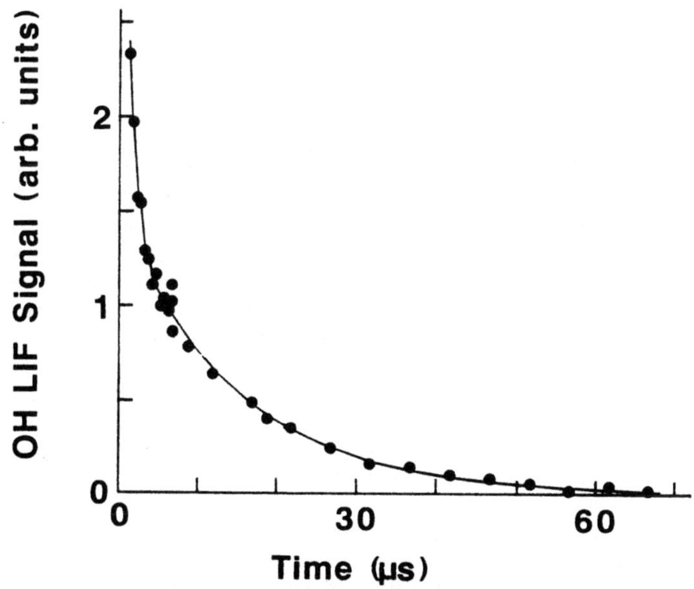

Figure 2. OH double exponential temporal profile observed at 258K in 100 Torr of N_2. Solid line shows the best fit to the data. [DMS] = 1.01×10^{17} cm^{-3}. Best fit values $k_{2a} + k_{2c}$ and k_{2b} in units of 10^{-12} $cm^3molecule^{-1}s^{-1}$ are 1.69 and 4.74, respectively; best fit value for k_{-2b} is 3.65×10^5 s^{-1}.

in our view, neither their experimental nor their theoretical approaches were adequate to address this problem and do not justify their reported conclusion that a thermalized DMS-OH adduct does not exist.

Rate Enhancement in the Presence of Oxygen

Figure 4 shows the enhancement in the observed rate of reaction 2 at 258K as a function of $[O_2]$ at a constant total pressure of 110 Torr. We can use the rate constants $k_{2a} + k_{2c}$, k_{2b}, and k_{-2b} in equation I to model the oxygen rate enhancement and extract a value for k_3. Figure 4 shows the calculated variation of k_{obs} with oxygen concentration for two values of k_3 and gives a "best fit" of $k_3 \simeq 8 \times 10^{-13}$ cm^{3}molecule^{-1}s^{-1}. This is somewhat slower than our estimate of k_3 at 261K and 700 Torr total pressure (6). This measurement should be more accurate because we have direct measurements of the elementary rate constants; however it could be indicative of a pressure dependence in the reaction of the adduct with oxygen.

Mechanism of the Reaction of the Adduct with Oxygen

A number exothermic channels exist for the reaction of the adduct with oxygen.

$$CD_3S(OH)CD_3 + O_2 \rightarrow (CD_3)_2SO + HO_2 \quad (3a)$$
$$\rightarrow CD_3SOH + CD_3O_2 \quad (3b)$$
$$\rightarrow (CD_3)_2SO_2 + OH \quad (3c)$$

In our previous work (6) we identified reactions 3a and 3b as possibilities. Barnes at al. (8) identified dimethylsulfone (DMSO2; $(CH_3)_2SO_2$) as a reaction product and suggested the possibility of 3c. In fact, observation of DMSO2 could be consistent with either 3a or 3c because the DMSO produced in 3a would react rapidly with OH, and DMSO2 is a likely product of that reaction in the presence of oxygen (14,15).

To understand its atmospheric significance it is important to determine the mechanism of reaction 3 and the ultimate stable products. A detailed mechanistic understanding of reaction 3 is also potentially relevant to the discrepancies between our direct study, which involves determination of the rate of loss of OH in a large excess of DMS, and the competitive studies which monitor the rate of loss of DMS relative to a standard in the presence of a photolytically generated steady state concentration of OH. Because our studies monitor OH loss we cannot measure a contribution to the oxygen rate enhancement by a process such as channel 3c which regenerates OH. If channel 3c proceeds at a significant rate then our "effective rate coefficients" in 760 Torr of air (6) would underestimate the true loss rate of DMS which is appropriate for atmospheric models. However competitive rate studies, if free of complications from secondary chemistry, will measure the total rate. Unfortunately, competitive studies, particularly of sulfur reactions, are often subject to complications which lead to erroneous results. Hence it is unclear whether the discrepancies between the direct and competitive studies can be readily assigned to the fact that a reactive channel which regenerates OH proceeds at a significant rate. In fact, in our direct studies which monitor OH we can exclude the possibility of OH regeneration under the following conditions: a) if the equilibration process 2b, -2b can be directly observed and the elementary rate coefficients k_{2b},k_{-2b}, can be determined; b) if we can determine a limiting value of the oxygen rate enhancement. From equation I it can be seen that when $[O_2] = 0$, $k_{obs} = k_{2a} + k_{2c}$ and as $[O_2] \rightarrow \infty$, $k_{obs} \rightarrow k_2$, i.e., the limiting value of the oxygen rate enhancement is k_{2b}. In the absence of oxygen the thermalized adduct is in equilibrium with DMS and OH, and OH is lost via reactions 2a and/or 2c. As we add oxygen some of the adduct reacts with oxygen and this constitutes an additional OH loss process; hence we observe an increase in the rate. The rate increases as a function of oxygen concentration until we reach a point at which every adduct molecule which forms reacts with oxygen. No adduct molecules decompose back to OH and DMS. This represents the maximum observable rate enhancement, k_{2b}, and further addition of oxygen produces no further increase in the rate. The above considerations offer two independent routes to the determination of k_{2b}, and if they are in good agreement then this indicates that no OH regeneration channels proceed at a significant rate. If the limiting value

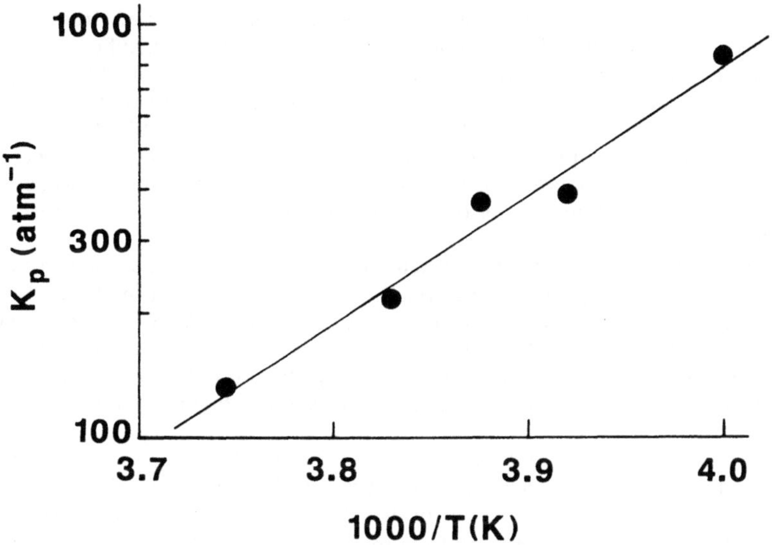

Figure 3. Van't Hoff plot for the reaction OH + DMS-d_6 → OHDMS-d_6. Solid line is obtained from a linear least squares analysis and gives ΔH = -14.2 ± 2.6 kcal mole^{-1} (error is 2σ).

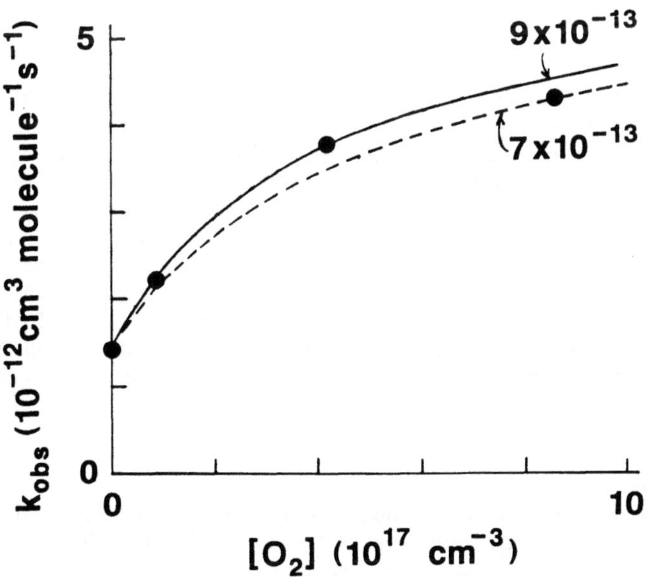

Figure 4. Variation of k_{obs} as a function of oxygen partial pressure at 258K and 110 Torr total pressure. Solid and dotted lines show the calculated variation for two values of k_3 (in units of cm^{3}molecule^{-1}s^{-1}). Assumed values for k_{2a} + k_{2c} and k_{2b} in units of 10^{-12} cm^{3}molecule^{-1}s^{-1} are 1.45 and 4.50, respectively; assumed value for k_{-2b} is 3.50 x 10^5 s^{-1}.

of the rate enhancement is less than the value of k_{2b} determined from equilibration measurements, then this is an indication that OH regeneration may be significant. If the limiting value of the rate enhancement is greater than the value of k_{2b} determined from equilibration measurements, then this is an indication that one or both of the measurements is in error or that we do not fully understand the reaction mechanism. Consequently these independent measurements represent an opportunity to confirm the internal consistency of our measurements and obtain new information concerning mechanistic detail. Also, as shown above, by modelling the oxygen rate enhancement we can determine the rate coefficient for the reaction of the adduct with oxygen.

From results at 267K and 258K we can attempt to estimate the significance of reaction channels such as 3c which regenerate OH. At 267K, $k_2 = 5.7 \times 10^{-12}$ cm^{3}molecule^{-1}s^{-1}, k_{obs}(100 Torr O_2) = $(4.9 \pm 0.7) \times 10^{-12}$ cm^{3}molecule^{-1}s^{-1}. At 258K, $k_2 = 5.9 \times 10^{-12}$ cm^{3}molecule^{-1}s^{-1}, k_{obs}(100 Torr Air) = $(4.3 \pm 0.3) \times 10^{-12}$ cm^{3}molecule^{-1}s^{-1}. The quoted errors in the measurements of k_{obs} are 2σ and reflect precision only. We do not have good estimates of the statistical precision of our k_{2b} values at this time because our fitting procedure does not provide this parameter.

The values of k_{obs} are somewhat lower than the sum $k_2 \equiv k_{2a} + k_{2b} + k_{2c}$; however, it is not clear that we have reached the limiting value of the oxygen enhancement at either temperature, and within their combined error bars the values overlap. Hence OH regeneration cannot be excluded and there are indications that OH regeneration could be of some significance accounting for perhaps 20-30% of the adduct plus oxygen rate. High precision measurements of k_{2b} are necessary to reduce this uncertainty.

HO_2 Production in the Adduct Reaction with Oxygen

If HO_2 is a product of the reaction of the adduct with O_2 via a reaction such as 3a, then addition of NO to the reaction mixture should back titrate the HO_2 to OH via the fast reaction

$$HO_2 + NO \rightarrow OH + NO_2 \tag{5}$$

Figure 5 shows OH profiles with and without added NO at 267K, 100 Torr of O_2 and [DMS-d_6] = 1×10^{16} cm^{-3}. It is clear that addition of NO regenerates OH, an observation which is consistent with HO_2 production from the reaction of the adduct with oxygen. As we noted above, the fact that we are studying reaction 2 rather than reaction 1 allows us to draw this conclusion. For example if the CH_3SCH_2 radical is generated by the abstraction component of reaction 1, it could conceivably react further in the presence of oxygen to regenerate OH. If this happens in reaction 2 then OD not OH is regenerated. It is important to note that laser excitation of OH and OD are isotope specific and allow either species to be monitored in the presence of the other. Figure 5 shows OH profiles calculated using a GEAR integration and assuming that reaction 3a accounts for 100% and 50% of the overall reaction. The experimental results are consistent with $k_{3a}/k_3 = 0.5$. Clearly, direct identification of DMSO and HO_2 is required to confirm this result.

HOD Yield Study

Because the TDLA detection scheme is somewhat less sensitive than the PLIF technique, higher radical concentrations were employed in the TDLA experiments than in the PLIF experiments; hence, radical-radical secondary reactions could play a role in the TDLA studies of HOD production from reaction 4. In Figure 6, measured HOD yields at 298K are plotted as a function of $[OH]_o/k'$ (in the back-to-back experiments used to obtain a single yield measurement, the concentrations of DMS and hexane were adjusted to give approximately equal values of k' in the two experiments); the yield in the limit where $[OH]_o/k' \rightarrow 0$ should be free of complications from HOD production via radical-radical secondary reactions. As shown in Figure 6, the 298K data support a HOD yield of 0.81 ± 0.11 (error is 2σ). At 348K (where the adduct lifetime toward unimolecular decomposition is very short), a somewhat higher HOD yield

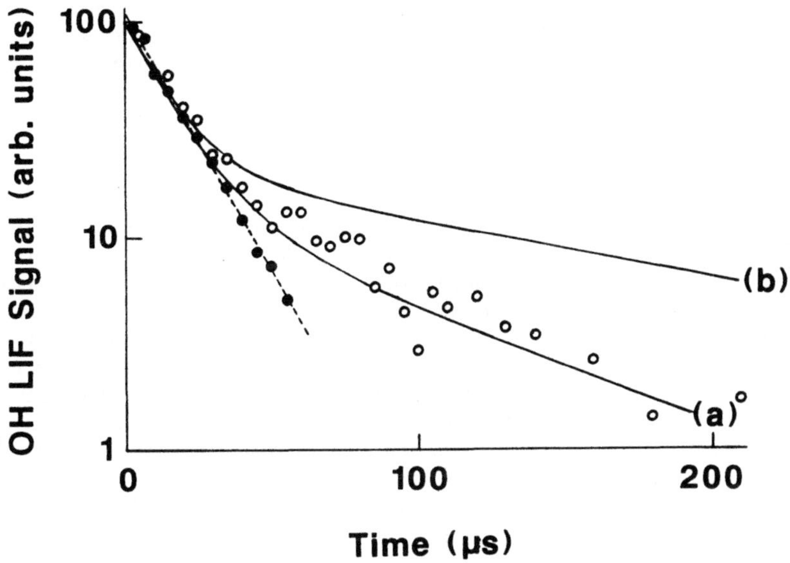

Figure 5. Modification of OH temporal profile by the addition of NO. Experiment performed in 100 Torr of O_2 at 267K. Closed circles are data obtained with [NO] = 0; dashed line is best fit to a single exponential decay. Open circles are data obtained with [NO] = 2.2×10^{15} cm^{-3}. Solid lines are based on simulations (see text) which assume an HO_2 yield from reaction 3 of (a) 0.5 and (b) 1.0.

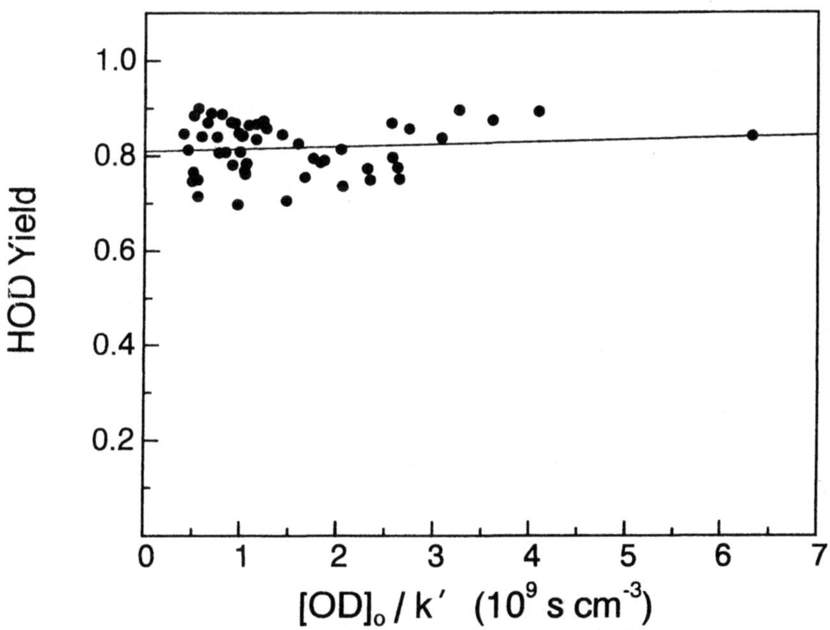

Figure 6. Plot of apparent HOD yield from reaction 4 as a function of the parameter $[OH]_o/k'$. The intercept, 0.81 ± 0.11 (error is 2σ) is taken to be the actual HOD yield.

of 0.91 ± 0.07 is measured. Planned experiments to directly measure the methyl radical yield are needed to quantify the possible contribution of a minor C-S cleavage channel which, based on the HOD yield studies discussed above, cannot account for more than 30% of overall OD + DMS reactivity.

4. CONCLUSIONS

While the data presented here are somewhat preliminary in nature, the results clearly confirm the basic hypothesis of our previous work, a multi-channel mechanism involving both hydrogen abstraction and reversible adduct formation. Direct observation of adduct formation together with a determination of the adduct binding energy demonstrate that the mechanism of "oxygen rate enhancement" is via the reaction of oxygen with a DMS-OH adduct which is in equilibrium with OH and DMS. The data are consistent with the reaction of the adduct with oxygen producing HO_2 with a yield of 0.5, and the possibility of a channel which regenerates OH cannot be excluded. The observation of HOD production in high yield indicates that the dominant channel for reaction 4 in the absence of oxygen is H-atom abstraction, a result consistent with our previous observations of the enhanced reactivity of secondary C-H bonds and the magnitude of the kinetic isotope effect; a minor channel producing CH_3 + CH_3SOD cannot be excluded at this time.

REFERENCES

(1) BATES, T. S., LAMB, B. K., GUENTHER, A., DIGNON, J. and STOIBER, R. E. (1992). Sulfur emissions to the atmosphere from natural sources, Journal of Atmospheric Chemistry, Vol. 14, 315-337.

(2) CHARLSON, R. J., LOVELOCK, J. E., ANDREAE, M. O. and WARREN, S. G. (1987). Oceanic phytoplankton, atmospheric sulphur, cloud albedo and climate, Nature, Vol. 326, 655-661.

(3) CHAMEIDES, W. L. and STELSON, A. W. (1992). Aqueous-phase chemical processes in deliquescent seasalt aerosols, Berichte Bunsengesellschalt fur Physikalische Chemie, Vol. 96, 461-470. Journal of Geophysical Research, in press.

(4) BANDY, A. R., SCOTT, D. L., BLOMQUIST, B. W., CHEN, S. M. and THORNTON, D. C. (1992). Low yields of SO_2 from dimethyl sulfide oxidation in the marine boundary layer, Geophysical Research Letters, Vol. 11, 1125-1127.

(5) TYNDALL, G. S. and RAVISHANKARA, A. R. (1991). Atmospheric oxidation of reduced sulfur species. International Journal of Chemical Kinetics, Vol. 23, 483-527.

(6) HYNES, A. J., WINE, P. H. and SEMMES, D. H. (1986). Kinetics and mechanism of OH reactions with organic sulfides, Journal of Physical Chemistry, Vol. 90, 4148-4156.

(7) ATKINSON, R. (1989). Kinetics and mechanisms of the gas phase reactions of the hydroxyl radical with organic compounds., Journal of Physical and Chemical Reference Data, Monograph No. 1.

(8) BARNES, I., BASTIAN, V. and BECKER, K. H. (1988). Kinetics and mechanisms of the reaction of OH radicals with dimethyl sulfide, International Journal of Chemical Kinetics, Vol. 20, 415-431.

(9) WALLINGTON, T. J., ATKINSON, R., TUAZON, E.C. and ASCHMAN, S.M. (1986). The reaction of OH radicals with dimethyl sulfide. International Journal of Chemical Kinetics, Vol. 18, 837-846.

(10) GU, M. and TURECEK, F. (1992). The elusive dimethylhydroxysulfuranyl radical. An intermediate or a transition state?, Journal of the American Chemical Society, Vol. 114, 7146-7151.

(11) DOMINE, F., RAVISHANKARA, A. R. and HOWARD, C. J. (1992). Kinetics and mechanism of the reactions of CH_3S, CH_3SS, CH_3SO and CH_3SSO with O_3 at 300K and low pressure, Journal of Physical Chemistry, Vol. 96, 2171-2178.

(12) STICKEL, R. E., NICOVICH, J. M., WANG, S., ZHAO, Z. and WINE, P. H. (1992). Kinetic and mechanistic study of the reaction of atomic chlorine with dimethylsulfide, Journal of Physical Chemistry, Vol. 96, in press.

(13) McKEE, M. L. (1992). Private communication.

(14) BARNES, I., BASTIAN, V., BECKER, K. H. and MARTIN, D. (1989). Fourier transform IR studies of the reactions of dimethyl sulfoxide with OH, NO_3 and Cl radicals, in Biogenic Sulfur in the Environment, edited by E. S. Saltzman and W. J. Cooper, ACS Symposium Series No. 393, 476-488.

(15) HYNES, A. J. and WINE, P. H. (1993). To be published.

OVERVIEW AND ATMOSPHERIC SIGNIFICANCE OF THE RESULTS FROM LABORATORY KINETIC STUDIES PERFORMED WTHIN THE CEC PROJECT "OCEANO-NOX"

I. BARNES

Bergische Universität Gesamthochschule Wuppertal
Physikalische Chemie FB 9, Gaußstr. 20,
5600 Wuppertal 1, Germany

SUMMARY

The OCEANO-NOX was a combined field and laboratory study (1987-89) funded by the CEC and directed principally but not only at understanding the chemistry of DMS in air parcels advecting over Western Europe. Of the 6 institutions involved in the project the major results from the 3 involved in the laboratory studies are presented and discussed within the context of the allied field measurements.

1. INTRODUCTION

The Commission of the European Communities (CEC) funded project OCEANO-NOX entitled "Laboratory and Field Studies of the Oxidation Processes Occurring in the Atmospheric Marine Boundary Layer" was carried out jointly by 6 research groups during the period 1987 - 1989. The project was aimed specifically at attaining a better comprehension of the chemical processes involving reactions of HO_x, NO_x and XO_x (X = Cl, Br, and I) occurring in Altantic air masses advecting over large regions of Western Europe with particular reference to the potential oxidation of acid precursors such as DMS. To this end the methodology of the OCEANO-NOX project involved interactive laboratory, field and modelling studies. Chemical modelling has been used to integrate and interpret the laboratory kinetic studies and field measurement data of species concentrations of relevance in coastal oceanic air masses.

The laboratories involved in the kinetic measurements were the Harwell Laboratory (UK), CNRS Orléans (F) and the Bergische University Wuppertal (FRG). The laboratories carrying out the field measurements were CNRS Gif-sur-Yvette (F), Université de Paris VII (F) and KFA Jülich (FRG). The two field campaigns were carried out in the summer of 1988 and 1989 at the Pointe de Penmarc'h, a coastal area of Brittany in France which has large natural emissions of DMS.

It is well establised that the oxidation of DMS leads to the formation of gaseous pro-

G. Restelli and G. Angeletti (eds.), Dimethylsulphide: Oceans, Atmosphere, and Climate, 223–237.

ducts such as methanesulphonic acid (CH_3SO_3H, MSA) and sulphur dioxide (SO_2) and ultimately acidic sulphate aerosol. Dimethyl sulphoxide (CH_3SOCH_3, DMSO), dimethyl sulphone ($CH_3SO_2CH_3$, $DMSO_2$) and dimethyl disulphide (CH_3SSCH_3, DMDS) are also possible products. Although much progress has been made in recent years in unravelling the complex oxidation mechanisms of DMS, major areas of uncertainty still remain particularly in the relative yields of the products for different atmospheric conditions such as different NO_x levels and also temperatures. It is particularly difficult to juxtapose the results obtained in laboratory studies with those of the real atmosphere and vice versa since laboratory experiments are often complicated by secondary chemistry and field experiments by meteorology. A schematic diagram of reactions currently thought to be of possible importance in the OH initiated oxidation of DMS is given in Figure 1. For many of the reactions shown neither kinetic data nor branching ratios were known and many were just speculation at the time of the project. For an up to date survey of the state of knowledge on the oxidation of DMS the reader is referred to the review article by A.R. Ravishankara in these Proceedings.

Here the major results from the laboratory and model studies of chemical reactions relevant to the photooxidation of DMS will be summarised and their significance for the related field measurements discussed. The laboratory studies focussed to a large extend on the role of halogen oxides, in particular IO, and also OH radicals on the oxidation of DMS, whereas the modelling studies dealt more with the possible involvment of iodide in the atmospheric cycles of O_3, NO_x and HO_x.

The results from the entire project including the field measurements have recently been published by the CEC (1). The work reported here reflects the state of knowledge at the time of the project and although much progress has been made in the chemistry since then attempts to update the data have only been made in instances where it was considered to be either particularly pertinent for our understanding of the atmospheric chemistry of DMS or the present article.

2. RESULTS AND DISCUSSION

The rate coefficients determined for the reactions of halogen oxides with DMS in the OCEANO-NOX project are listed in Table 1 were they are compared with other literature data.

IO + DMS reaction:

A re-investigation of the rate coefficient for the reaction of IO radicals with DMS from the laboratories in Wuppertal (2) and Orléans (3) both using the DF-MS technique have confirmed the findings of another direct determination by Daykin and Wine (4) using the LFP-LPA that the reaction is slow. The studies have shown that the reaction of IO with DMS results in quantitative conversion of DMS in DMSO. The reaction has a rate coefficient of $k \sim 1 \times 10^{-14}$ cm^3 $molecule^{-1}$ s^{-1} which when taken in conjunction with typical calculated ambient levels of IO of 1×10^6 molecule cm^{-3} leads to an atmospheric residence time for DMS due to reaction with IO of $\tau = 1/k[IO] \sim 1157$ days. This can be compared to the residence time of $\tau = 46$ h calculated for the daytime oxidation of DMS via OH radicals using an aver-

Table 1. Experimental and literature values for the reactions of IO, BrO and ClO radicals with DMS at 298 K

Reaction	k [cm^3 $molecule^{-1}$ s^{-1}]	Method	Reference
IO + DMS	(3.0 ± 1.5) x 10^{-11}	FT - IR	Barnes *et al.* (6)
	(1.5 ± 0.5) x 10^{-11}	DF - MS	Martin *et al.* (7)
	≤ 3.5 x 10^{-14}	LFP-LPA	Daykin and Wine (4)
	(1.5 ± 0.2) x 10^{-14}	DF - MS	OCEANO-NOX (1) / Maguin *et al* . (3)
	(8.8 ± 2.1) x 10^{-15}	DF - MS	OCEANO-NOX (1) / Barnes *et al.* (2)
	(1.2 ± 0.3) x 10^{-14}	FT-IR	Barnes *et al.* (8, 9)
BrO + DMS	(2.65±0.65) x 10^{-13}	DF - MS	Barnes *et al.* (11)
	(2.40±0.6) x 10^{-13}	DF - EPR	OCEANO-NOX (1, Le Bras)
	(2.7 ± 0.5) x 10^{-13}	DF - MS	OCEANO-NOX (1) / Barnes *et al.* (2)
ClO + DMS	(3.9 ± 0.5) x 10^{-14}	DF - MS	Barnes *et al.* (11)
	(9.5 ± 2.0) x 10^{-15}	DF - MS	OCEANO-NOX (1) / Barnes *et al.* (2)

DF - MS : discharge flow - mass spectrometry

FT - IR : Fourier transform infrared spectrometry

LFP - LPA : Laser flash photolysis - long pathlength absorption aged OH radical concentration of $1x10^6$ molecules cm^{-3} and a rate coefficient $k = 6x10^{-12}$ cm^3 $molecule^1$ s^{-1} (5). The results show that because the rate coefficient for IO + DMS is slow and the atmospheric concentration of IO is probably less than 10^7 molecule cm^{-3} (1) the reaction will not play a significant role in marine atmospheric chemistry or at least in the coastal continental regions. The high values of the rate coefficient from previous determinations by the two laboratories, Barnes *et al.* (6) and Martin *et al.*(7), are now thought to be in error because of heterogeneous effects and secondary chemistry. The reason for the high value in the steady state photolysis study of Barnes *et al.* (6) is not entirely clear and warrants further investigation to eliminate any posssible relevance for the atmospheric oxidation of DMS. Similar experiments have subsequently been performed in the Wuppertal laboratory using the photolysis of methylene iodine in the presence of O_3 as an NO_x free source for IO radicals (6):

$$CH_2I_2 + h\nu \longrightarrow CH_2 + I$$
$$I + O_3 \longrightarrow IO + O_2$$

$$IO + IO + M \longrightarrow \text{iodine oxides}$$
$$CH_2I + O_2 + M \dashrightarrow (CH_2IO_2) \dashrightarrow HCHO + I$$

$$\text{net reaction: } CH_2I_2 + 2O_3 \longrightarrow \text{products}$$

Two molecules of O_3 were consumed for each molecules of CH_2I_2 photolyzed. The reactions of CH_2I leading to the formation of HCHO and I atoms are not presently clear. Computer simulations of the decay of DMS using this IO source lead to a value of $k = (1.2\pm0.3) \times 10^{-14}$ cm^3 $molecule^{-1}$ s^{-1} as rate coefficient for the reaction of IO with DMS in good agreement with the recent values in Table 1. Examples of the data are shown in Figure 2. Recently work on the temperature dependence of the reaction of IO with DMS which was commenced during the OCEANO-NOX project has been completed (8,9). Using the DF-MS technique the following Arrhenius expression has been obtained for the reaction of IO with DMS:

$$k(IO+DMS) = (1.77^{+0.28}_{-0.22}) \times 10^{-13} \exp[-(775\pm532)/T] \text{ cm}^3 \text{ molecule}^{-1} \text{ s}^{-1}$$

The individual rate coefficients are listed in Table 2 and the data is plotted in Figure 3a.The rate coefficient was found to be virtually independent of temperature at least under the experimental conditions employed, i.e. temperatures between 265-338 K and total pressures of 0.5-6.8 mbar He.

The laboratory studies have also shown that the reactions of IO radicals with other reduced sulphur compounds such as CH_3SH, DMDS, DMSO, CS_2 etc. are also slow and thus of no consequence for marine atmospheric chemistry (1,3, 10).

Table 2. Rate coefficents for the reaction of IO and BrO radicals with DMS as a function of the temperature and total pressures of 2-6.7 mbar He (Overath (8)).

Temperature [K]	k_{IO+DMS} [10^{-14} cm^3 $molecule^{-1}$ s^{-1}]	$k_{BrO+DMS}$ [10^{-13} cm^3 $molecule^{-1}$ s^{-1}]
262		3.50 ± 0.42
264		3.38 ± 0.62
269	0.91 ± 0.08	
273	1.28 ± 0.18	3.02 ± 0.28
283	1.09 ± 0.05	3.19 ± 0.16
298	1.18 ± 0.16	2.76 ± 0.22
323	1.43 ± 0.08	2.72 ± 0.12
338	2.05 ± 0.10	2.46 ± 0.08
350		2.35 ± 0.20
353		2.54 ± 0.26

ClO/BrO + DMS reaction

The reactions of ClO (Wuppertal) and BrO (Orléans, Wuppertal) with DMS have also been investigated and rate coefficients of $(9.5 \pm 2.0) \times 10^{-15}$ and 2.55×10^{-13} cm^3 $molecule^{-1}$ s^{-1} (middle value Orléans/Wuppertal) have been measured (2,11). Similar to IO + DMS an Arrhenius expression is now available for the reaction of BrO with DMS (8,9):

$$k(BrO+DMS) = (9.17^{+2.54}_{-2.24}) \times 10^{-14} \exp[(343 \pm 78)/T] \; cm^3 \; molecule^{-1} \; s^{-1}$$

The individual rate coefficients are listed in Table 2 and are plotted in Figure 3b. As for the reaction of IO with DMS the temperature dependence for the reaction with BrO is very slight. Efforts to measure the temperature dependence of the ClO + DMS reaction have as yet been unsuccessful due to complications with the chemical system (8).

Both the DF-MS and FTIR studies of the reactions of IO, ClO and BrO showed that DMSO was the major product The FTIR studies give yields of 90±20% for DMSO in the IO and BrO reactions, however, in the case of ClO quantification is difficult.

The reactivity of XO towards DMS takes the order BrO > ClO ≈ IO. However, from the exothermicities of the reactions the reactivity order IO > BrO > ClO would be expected. The reason for this trend is not presently clear. As for IO the reaction of ClO with DMS will be of negligible importance as an atmospheric sink for DMS because of its low atmospheric concentration and the slow rate coefficient (9). It has, however, been calculated (12,13) that the BrO concentation could be as high as 10 ppt in the lower Arctic atmosphere which when taken in conjunction with the rate coefficient for its reaction with DMS would give a residence time τ of ~4.2 h for DMS. Very recently the group of Platt, University Heidelberg (14) have made direct measurements of BrO in the Arctic (Alert, Canada during April 1992) using long path differential absorption spectroscopy (LP-DOAS) as part of the Polar Sunrise experiment !992. Bromine oxide levels ranging from the detection limit of approximately 3 ppt up to a maximum of 16.9 ppt were observed. Although more measurements are needed to confirm the observations it seems a real possibility that reaction of with BrO radicals could be a significant sink for DMS in the Arctic atmosphere. Further, under the conditions which exist in the springtime Arctic marine boundary layer it is possible that reactions of bromine and chloride atoms with DMS may also play an important role in coupling the halogen and sulphur cycles in this region (12,15,16).

Reaction of OH radicals with DMS

The kinetics and products of the reaction of OH radicals with DMS was investigated in an NO_x-free system using the steady state photolysis of H_2O_2 as the OH source (1,17). The study represents the only product investigation to date on the reaction carried out under NO_x-free conditions. The results demonstrated, in agreement with other literature studies by Hynes et al. (18) and Wallington et al. (19), that the overall rate coefficient for the reaction of OH with DMS increases on increasing the O_2 partial pressure and indicates the active involvement of O_2 in the reaction mechanism. Although the exact nature of the interaction between OH and DMS is still not clear the observed products indicate that two channels are operative which can be ascribed to an overall addition and abstraction mechanism.

(addition): $OH + CH_3SCH_3 \longrightarrow CH_3S(OH)CH_3 \dashrightarrow$ products (possibly DMSO)
(abstraction): $OH + CH_3SCH_3 \longrightarrow CH_3SCH_2 \dashrightarrow$ products (SO_2 and MSA)

In the absence of NO_x, SO_2 and $DMSO_2$ have been found as products with yields of ~70 and ~20%, respectively (1,17). $DMSO_2$ is thought to be formed from the further reaction of DMSO formed in the reaction of O_2 with the OH-DMS adduct. Work on the rate coefficient for the reaction of OH with DMSO has shown it be nearly gas kinetic (1,20) and this has been confirmed recently by Wine (21). MSA has also been found as a minor product with a yield of <4% by Martinet (22). Work on the 254 nm photolysis of $DMDS/O_3$/air reaction mixtures has shown that the yield of MSA is <6% in this system where the initiating reactions will be,

$$CH_3SSCH_3 + h\nu \longrightarrow CH_3S + CH_3S$$
$$CH_3S + O_3 \longrightarrow CH_3SO + O_2$$

In the presence of NO_x the products of the OH + DMS reaction change dramatically, the yield of SO_2 falls sharply and that of MSA rises. Depending on the experimental conditions and size of reactor used the SO_2 yield can be of the order of 20-30% and that of MSA 50-70% (17, 23-15). The reaction(s) controlling the branching ratio for SO_2 and MSA formation are still obscure, particularly when NO_x is present. However, it is now clear from the laboratory studies that CH_3S is a key radical in the mechanism and that the presence of NO_2 greatly influences the SO_2/MSA branching ratio. Therefore, a knowledge of the CH_3S reactions is essential before the DMS oxidation mechaism can be completely understood.

CH_3S chemistry:

The CH_3S radical is known to react very slowly with O_2 and an upper limit of $k \leq 3 \times 10^{18}$ cm^3 $molecule^{-1}$ s^{-1} has been reported for the rate coefficient at room temperature by Tyndall and Ravishankara (26). Very recently Turnipsipseed et al. (27) have succeeded in observing adduct formation between CH_3S and O_2 and measuring the equilibrium constant for the reaction in the range 216 to 258 K. They report an upper limit of $k < 4.0 \times 10^{-17}$ cm^3 $molecule^{-1}$ s^{-1} for the reaction at 258 K. The mechanism of the reaction of CH_3S with NO_2 was studied at Orléans and Wuppertal and it was shown that the reaction results in the formation of CH_3SO and NO (1,28,29). The rate coefficient for this reaction was not determined, however, from the experimental data a rate coefficient of $(3\pm2) \times 10^{-11}$ cm^3 $molecule^{-1}$ s^{-1} was derived for the reaction $CH_3SO + NO_2 \longrightarrow CH_3SO_2 + NO$ (28). Whether the reaction will be of importance for the atmosphere depends on the other possible reactions of CH_3SO (see Figure 1). Irrespective of its atmospheric importance, it will be of use in interpretating laboratory experimental data.

The rate coefficients for the reactions of NO_2 and O_3 with CH_3S are known from other studies reported in the literature and have values of approximately 6×10^{-11} and 6×10^{-12} cm^3 $molecule^{-1}$ s^{-1}, respectively (24). Despite the low rate coefficient for the reaction of O_2 with CH_3S it could still be of possible significance for the atmosphere due to the large atmospheric concentration of O_2. Table 3 shows the loss rates for CH_3S radicals in the atmosphere due to reaction with O_2, O_3 and O_3 calculated using typical concentrations of O_3 and NO_2 observed during the OCEANO-NOx campaign. If the rate coefficient for the reaction of O_2 with CH_3S is much lower than reported, as laboratory studies seem to suggest, then it is evident from Table 3 that under pure oceanic conditions where the concentration of nitrogen oxides are very low the reaction of CH_3S with O_3 will dominate. Only under relatively polluted conditions (high NO_x) will the reaction of CH_3S with NO_2 be of significance. The results when taken in conjunction with the product studies suggest that in the atmosphere high yields of SO_2 and low yields of MSA should be observed under NO_x-poor conditions and that with increasing NO_x the yield of SO_2 should decrease and that of MSA increase.

Table 3. Loss rates for CH_3S due to reaction with O_2, O_3 and NO_2

Species (X)	Rate coefficient (k)[a] (cm^3 $molecule^{-1}$ s^{-1})	[X] (ppm)	Loss Rate (k[X]) (s^{-1})
O_2	< 3×10^{-18}	0.2×10^6	< 14.7
O_3	6×10^{-12}	20×10^{-3} (b) $20 - 90 \times 10^{-3}$ (c) 45×10^{-3} (d)	2.9 2.9 - 13.2 6.6
NO_2	6×10^{-11}	$0.01 - 0.1 \times 10^{-3}$ (b) $0.1 - 8.5 \times 10^{-3}$ (c) 2×10^{-3} (d)	0.0015 - 0.015 0.015 - 12.5 3.5

a) Tyndall and Ravishankara [26] and Yin *et al.* [24]; b) Typical pure oceanic values; c) Range observed during the campaign; d) Mean value observed during the campaign.

Short summary of field measurement results

During the field campaigns DMS and its oxidation products, SO_2, MSA and non-sea-salt sulfate (nss-SO_4^{2-}) were monitored. The meteorological situtaion during the 1989 campaign which was the most informative show that for most of the time the air masses were of continental origin and containe high levels of NO_x The NO_2 concentrations as measured by the KFA (Jülich, Germany) and the laboratoire d´Aérologie (Toulouse, France) ranged from 0 to 8 ppb with a mean value of 2.0 ppb. The O_3 concentration during the same period ranged from 20 up to 100 ppb with a mean value of 45 ppb [1,30]. The observed concentration range of DMS (n=157) was 46 - 1593 pptv (1 nmoles m^{-3} = 24.5 pptv) and for its oxidation products SO_2 (n=50), nss-SO_4^{2-} (n=44) and MSA (n=44) the ranges were 14.7 - 2308, 1029 - 8575 and 14.7 - 284 pptv respectively, where "n" is the number of samples. Details can be found in Barnes et al. (1) and also in Mihalopoulos *et al.* (31,32).

When contential conditions predominated, the measured SO_2 concentrations were lower than those expected from the observed DMS concentration under such conditions. Also the MSA concentrations are at least 3 to 100 times higher than those observed in other locations worldwide. Although it is known that MSA shows a pronounced seasonal cycle the MSA concentrations observed during the campaign are still at least 3 to 10 times higher than the values observed during summertime at other sites. Further the MSA/DMS molar ratio was 0.33 which is 2 to 4 times higher than the ratios observed in remote areas or under pure oceanic conditions (33-35). MSA showed a well defined diurnal variation which appears to be anticorrelated with that of DMS. The results are in line with the postulation that gas phase oxidation of DMS is probably the exclusive source of MSA and also that under the conditions of the campaign (high NO_x) enhanced production of MSA in comparison to other locations is occurring. From a consideration of the observed decrease in the DMS with the parallel increase in MSA it was estimated that MSA accounted for at least 20-25% of the oxidized

DMS under the conditions of the campaign.

Another observation consistent with increased production of MSA under the conditions of the campaign is the observed MSA/nss-SO_4^{2-} ratio of 6 ± 3%. This ratio is similar to the average value of 6.9% observed in remote areas at sea level by Berresheim (34) and Saltzman *et al.* (36) which is not what would be expected since during the 1989 campaign the air was continental in origin and contained high levels of nss-SO_4^{2-} which would be expected to decrease the ratio considerably. Indeed, the nss-SO_4^{2-}/SO_2 ratio during the campaign was 4.5 which is much higher the value of 1.0 observed by Bonsang *et al.* (37) in the same region and by Bates *et al.* (35) in the Pacific Ocean under marine conditions and serves to underline the strong continental influence on the nss-SO_4^{2-} concentration. Thus the observed MSA/nss-SO_4^{2-} ratio can only be rationally explained by enhanced levels of MSA production during the conditions of the campaign.

Comparison of laboratory and field measurements

It has been unequivocially established from the laboratory and field studies that the reaction of IO with DMS is not important for marine chemistry. The laboratory studies show that the reaction is slow and the field studies that the the atmospheric conentration of IO is less than 10^7 molecule cm^{-3}. The question of a possible atmospheric significance for the reaction of BrO with DMS still remains open . The laboratory studies show that the reaction could be of importance if the atmopheric concentration of BrO is greater than ~10^7 molecule cm^{-3} which recent model calculations predict and recent field measurements confirm for the Arctic.

Both the laboratory data and field observations indicate that the NO_x concentration is a critical parameter in determining the SO_2/MSA branching ratio in the oxidation of DMS. From Figure 2 it can be seen that the reason for this importance is the competition for the CH_3SO and CH_3SO_2 radicals between O_2, O_3 and NO_2.

The laboratory experiments show that in the absence of NO_x the SO_2/DMS yield is ~70% and the MSA/DMS yield is ≤6%. According to the mechanism in Figure 1 this implies that 6% of the oxidation proceeds through the oxygen transient reactions of sulphur containing radicals with O_3 or HO_2. It is still not clear from either the laboratory or field studies which of the competing reaction pathways in Figure 1 are responsible for the change in the SO_2/MSA ratio. In the field experiments where continental conditions predominated the observed SO_2 concentrations were consistently very much lower than those expected for the observed DMS concentrations. Similarly, higher concentrations of MSA were observed than expected from the measured DMS concentrations. The present of enhanced levels of MSA was also endorsed by the observation that the measured mean MSA/nss-SO_4^{2-} ratio of 6 ± 3% was similar to the mean value of 6.9% measured in pure oceanic regions.

In conclusion, although much progress has been made in the OH initiated daytime oxidation of DMS, particularly with regard to the role of NO_x, it has become evident during the course of the joint field and laboratory studies that the rate coefficients for the key competition steps are not well enough known. The laboratory studies indicate that the reactions of CH_3S and CH_3SO_x radicals need to be extensively investigated since the further reactions of these radicals with NO_x appear to control the MSA/SO_2 branching ratio. The field measurements show that it may be possible to obtain these ratios under real atmospheric conditions. DMS shows a sharp vertical gradient in the troposphere and, therefore, simultaneous field measurements of the concentrations of DMS, SO_2 and possibly MSA under different atmospheric conditions (high and low NO_x) as a function of altitude would allow a direct determination of the SO_2/DMS and MSA/DMS branching ratios. Such information taken in conjunction with laboratory measurements would be invaluable in the furtherance of our understanding of the atmospheric sulphur cycle and its possible consequences for the "climatology" of the troposphere.

REFERENCES

(1) BARNES, I., BONSANG, B., BRAUERS, T., CARLIER, P., COX, R.A.,DORN, H.P., JENKIN, M.E., LE BRAS, G. and PLATT, P. (1991). Laboratory and field studies of the oxidation processes occurring in the atmospheric marine boundary layer (OCEANO-NOX CEC Project) Air Pollution Research Report 35, Commission of the European Communties, Brussels 1991, ISBN 2-87263-064-3.

(2) BARNES, I., BASTIAN, V., BECKER, K.H. and OVERATH, R.D. (1991). Kinetic study of the reactions of IO, BrO and ClO with dimethylsulfide. Int. J. Chem. Kinetics, 23, 579-591.

(3) MAGUIN, F., MELLOUKI, A., LAVERDET, G., POULET, G. and LE BRAS, G., (1991). Kinetics of the reactions of the IO radical with dimethyl sulfide, methanethiol, ethylene and propylene. *I*nt. J. Chem. Kinetics, 23, 237-245.

(4) DAYKIN, E.P. and WINE, P.H. (1990). Rate of reaction of IO radicals with dimethyl suldide. J. Geophys. Res. 95, D11, 18.547-18.553.

(5) ATKINSON, R., BAULCH, D.L., COX, R.A., HAMPSON, Jr., R.F., KERR R.A. and TROE, J. (1989). Evaluated kinetic and photochemical data for atmospheric chemistry: Supplement III. J. Phys. Chem. Ref. Data 18, 881-1097.

(6) BARNES, I., BECKER, K.H., CARLIER, P. and MOUVIER, G. (1987). FTIR study of the DMS/NO_2/I_2/N_2 photolsis system: The reaction of IO radicals with DMS. Int. J. Chem. Kinetics 19, 489-501.

(7) MARTIN, D., JOURDAIN, J.L., LAVERDET, G. and LE BRAS, G. (1987). Kinetic study of the reaction of IO with CH_3SCH_3. Int. J. Chem. Kinetics 19, 503-512.

(8) OVERATH , R. (1991). Kinetische Untersuchungen der Reaktionen von Organo schwefelverbindungen mit Halogenoxid-Radikalen im Strömungssystem. PhD Thesis, Physikalische Chemie / FB 9, Bergische Universität - Gesamthochschule Wuppertal, FRG.

(9) BARNES, I., BECKER, K:H. and OVERATH, R.D. (1992). Oxidation of organic sulphur compounds. In Proceedings of the NATO Advance Research Workshop " Tropospheric Chemistry of Ozone in the Polar Regions", NATO ASI subseries "Global Environmental Change", Springer-Verlag, in press.

(10) BARNES, I., BASTIAN, V. and BECKERr, K.H. (1990). Reactions of IO radicals with sulphur containing compounds, in Physico-Chemical Behaviour of Atmospheric Pollutants, eds. G. Restelli and G. Angeletti, Kluwer Academic Publishers, Dordrecht, pp 166-171.

(11) BARNES, I., BECKER, K.H., MARTIN, D., CARLIER, P., MOUVIER, G., JOURDAIN, J.L., LAVERDET, G. and LE BRAS, G. (1989). Impact of halogen oxides on dimethyl sulfide oxidation in the marine atmosphere. In: Biogenic Sulfur in the Environment, E.S Saltzman and W.J. Cooper, Eds., ACS Symposiun Series No. 393, 464-475.

(12) BARRIE, L.A., BOTTENHEIM, J.W., SCHELL, R.C., CRUTZEN, P.J. and RASMUSSEN, R.A. (1988). Ozone destruction and photochemical reactions at polar sunrise in the lower Artic atmosphere. Nature 334, 138-141.
(13) McCONNEL, J.C., HENDERSON, G.S., BARRIE, L., BOTTENHEIM, J., NIKI, H., LANGFORD, C.H. and TEMPLETION, E.M.J. (1992). Photochemical bromine production implicated in Arctic boundary-layer ozone depletion. Nature 355, 150-152.
(14) HAUSMANN, M., RUDOLPH, T., and PLATT, U. (1992). Spectroscopic measurement of bromine oxide, ozone and nitrous oxide in Alert. In Proceedings of the NATO Advance Research Workshop " Tropospheric Chemistry of Ozone in the Polar Regions", NATO ASI subseries "Global Environmental Change", Springer-Verlag, in press.
(15) STURGES, W.T. and BARRIE, L.A. (1988). Chloride, Bromine and Iodine in Arctic Aerosols. Atmos. Environ. 22, 1179-1194.
(16) STURGES, W.T. (1992). Halocarbons in the Arctic and ANtarctic troposphere. In Proceedings of the NATO Advance Research Workshop " Tropospheric Chemistry of Ozone in the Polar Regions", NATO ASI subseries "Global Environmental Change", Springer-Verlag, in press.
(17) BARNES, I., BASTIAN, V. and BECKER, K.H. (1988). Kinetics and mechanisms of the reaction of OH radicals with dimethyl sulfide. Int. J. Chem. Kinetics 20, 415-431.
(18) HYNES, A.J., WINE, P.H. and SEMMES, D.H. (1986). Kinetics and mechanism of OH reactions with organic sulfides. J. Phys. Chem. 90, 4148-4156.
(19) WALLINGTON,T.J., ATKINSON, R., TUAZON, E.C. and Aschmann, S.M. (1986).The reaction of OH radicals with dimethyl sulfide. Int. J. Chem. Kinetics 18, 837-846.
(20) BARNES, I., BASTIAN, V., BECKER, K.H. and MARTIN, D. (1989). FT-IR Studies of the reactions of OH, NO_3 and Cl radicals. In: Biogenic Sulfur in the Environment, E.S. Saltzman and W.J. Cooper, Eds., ACS Symposiun Series No. 393, 477-488.
(21) WINE, P.H. (1990). Kinetics and mechanism of the reaction of hydroxyl radicals with dimethylsulfoxide under atmospheric conditions. XIXth Informal Conference on Photochemistry, Ann Arbor, MI, June.
(22) MARTINET, A. (1989). Etudes cinetiques, sous Conditions atmospheriques simulees, de Reactions elementaires intervenant dans la Photooxydation de Precurseurs d'acidite en Milieu marin. PhD. Thesis, l'Université Paris VII.
(23) YIN, F. GROSJEAN, D. and SEINFELD, J.H. (1986). Analysis of Atmospheric Photooxidation Mechanisms for organosulfur compounds. J. Geophys. Res. 91, D13, 14,417-14,438.
(24) YIN, F., GROSJEAN, D. and SEINFELD, J.H. (1990). Photooxidation of dimethyl sulfide and dimethyl disulfide. I: Mechanism Development. J. Atmos. Chem. 11, 309-364.
(25) YIN, F., GROSJEAN, D., FLAGAN, R.C. and SEINFELD, J.H. (1990). Photooxidation of dimethylsulfide and dimethyl disulfide. II: Mechanism Evaluation. J. Atmos. Chem. 11, 365-399.
(26) TYNDALL, G.S. and RAVISHANKARA, A.R. (1989). Kinetics and mechanism of the reactions of CH_3S with O_2 and NO_2 at 298 K. J. Phys. Chem. 93, 2426-2435.
(27) TURNIPSEED, A.A., BARONE, S.B. and RAVISHANKARA, A.R. (1992). Observation of CH_3S addition to O_2 in the gas phase. J. Phys. Chem. 96, 7502-7505.

(28) MELLOUKI, A., JOURDAIN, J.L. and LE BRAS, G. (1988). Discharge flow study of the $CH_3S + NO_2$ reaction mechanism using $Cl + CH_3SH$ as the CH_3S source. Chem. Phys. Letters 148, 231-236.

(29) BARNES, I., BASTIAN, V., BECKER, K.H. and NIKI, H. (1987). FTIR spectroscopic studies of the $CH_3S + NO_2$ reaction under atmospheric conditions. Chem. Phys. Letters 140, 451-457.

(30) CARLIER. P , FRESNET, P., PASHALIDIS, S., PETITET, F., TSETSI, M., LES-COAT, V., DUPONT, B., RITZ, D., NGUYEN, B.C., BONSANG, B., MIHALO POULOS, N., KANAKIDOU, M., PUTAND, J.P., BERTRAND, I., BOISSARD,C., CAMPI, J.M., PLATT, U., DORN, H.P., BRAUERS, T., KOCH, T., NEUROTH, R., MARENCO, A., NEDELEC,P. and PERROS, P.(1990). Overview of the atmospheric field campaign at the "Pointe de Penmarc'h". Paper presented at the "7th International Symposium of the Commission on Atmospheric Chemistry and Global Pollution", 5-11 Sept, Chamrousse, France.

(31) MIHALOPOULOS, N., NGUYEN, B.C., BOISSARD, C., CAMPI, J.M., PUTAND, J.P., BELVISO, S., BARNES, I. and BECKER, K.H. (1992). Field study of dimethyl sulfide oxidation in the boundary layer: Variations of dimethylsulfide, methane sulfonic acid, sulfur dioxide, non-sea-salt sulfate and aitken nuclei at a coastal site. J. Atmos. Chem. 14, 459-477.

(32) MIHALOPOULOS, N., NGUYEN, B.C., PUTAND, J.P., PASCHALIDIS, S. and CARLIER, P. (1990). Distribution of DMS concentration near to an intense coastal source at the "pointe de Penmarc'h" during the OCEANO-NOX field campaign. Paper presented at the "7th International Symposium of the Commission on Atmospheric Chemistry and Global Pollution", 5-11 Sept, Chamrousse, France.

(33) ANDREAE. M.O. (1986). The ocean as a source of atmospheric sulfur. In: P. Buat-Menard ed), The Role of Air-Sea Exchange in Geochemical Cycling, D. Reidel, Dordrecht, pp 331-362.

(34) BERRESHEIM, H. (1987). Biogenic sulfur emissions from the subantarctic and antarctic oceans. J. Geophys. Res. 92, 13,245-13,262.

(35) BATES , T.S., JOHNSON, J.E., QUINN, P.K., GOLDAN, P.D., KUSTER, W.C., COVERT, D.C., and HAN, C.J. (1990). The biogeochemical sulfur cycle in the marine boundary layer over the Northeast Pacific Ocean. J. Atmos. Chem. 10, 59-81.

(36) SALTZMAN, E.S., SAVOIE, D.L., PROSPERO, J.M. and ZIKA, R.G. (1986). Methanesulfonic acid and non.sea-salt sulfate in Pacific air: regional and seasonal variations. J. Atmos. Chem. 4, 227-240.

(37) BONSANG, B., NGUYEN, B.C. and PAUGAM, J.Y. (1975). Sur la formation de gaz sufurés et de noyaux d'Aitken dans l'atmosphère le long du littoral. C.R. Acad. Sc. Paris 283, 1285-1288.

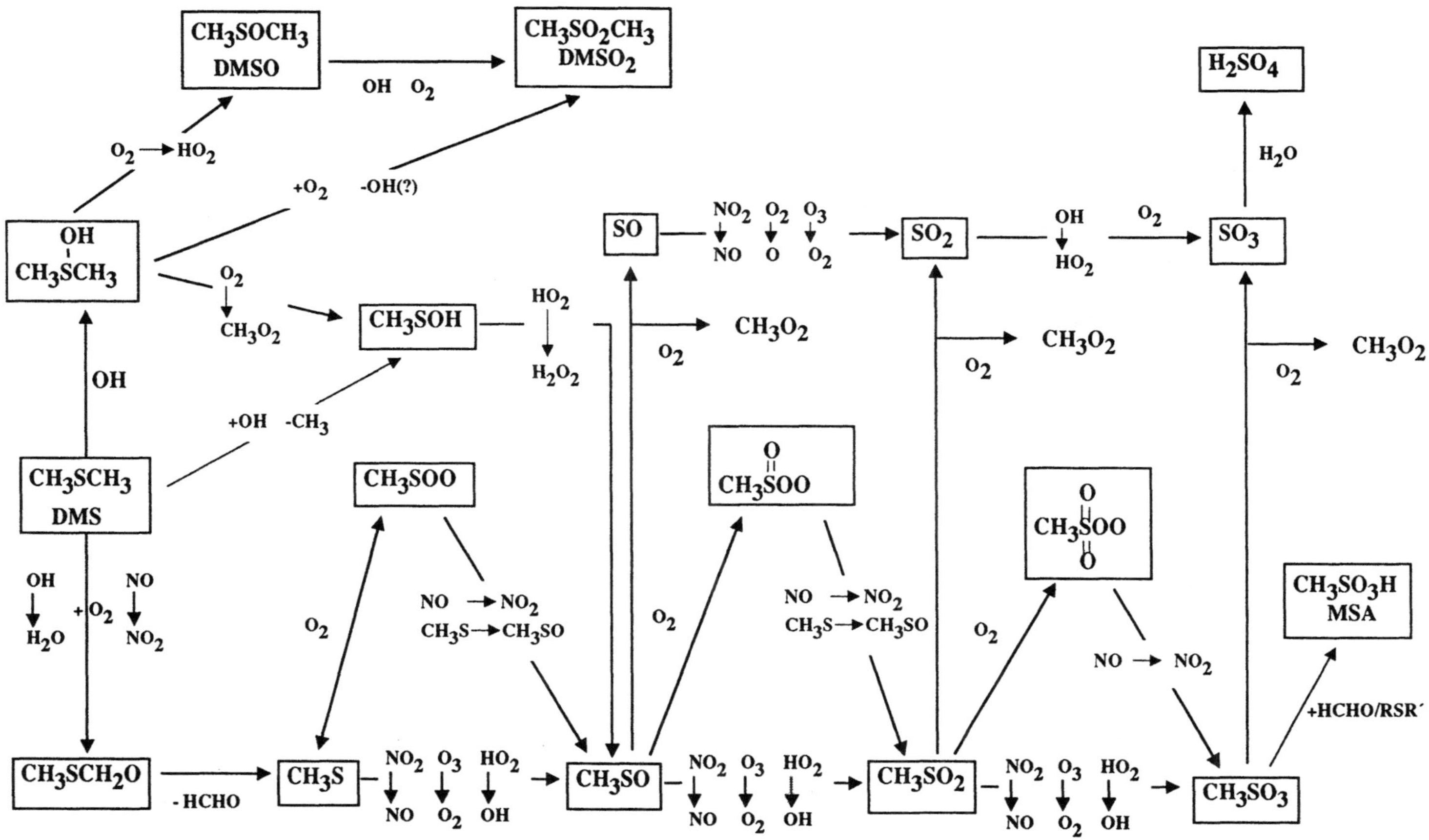

Figure 1. Possible reactions occurring in the OH initiated oxidation of DMS. The scheme is based largely on the mechanism proposed by Yin *et al*. (24).

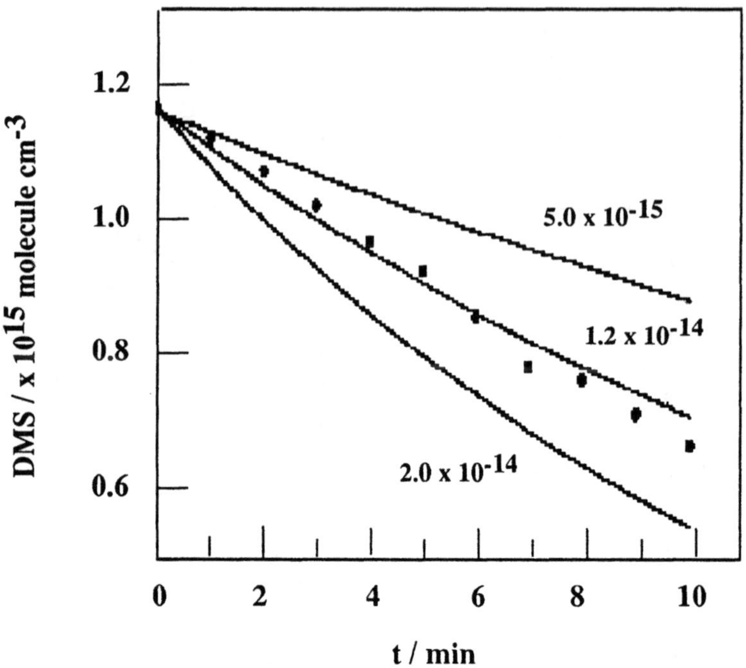

Figure 2. Decay of DMS in a CH_2I_2/O_3/DMS/air photolysis system. The filled circles are the experimental points and the solid lines are computer simulations of the reaction system for different values of the rate coefficient for the reaction IO + DMS –> DMSO + I

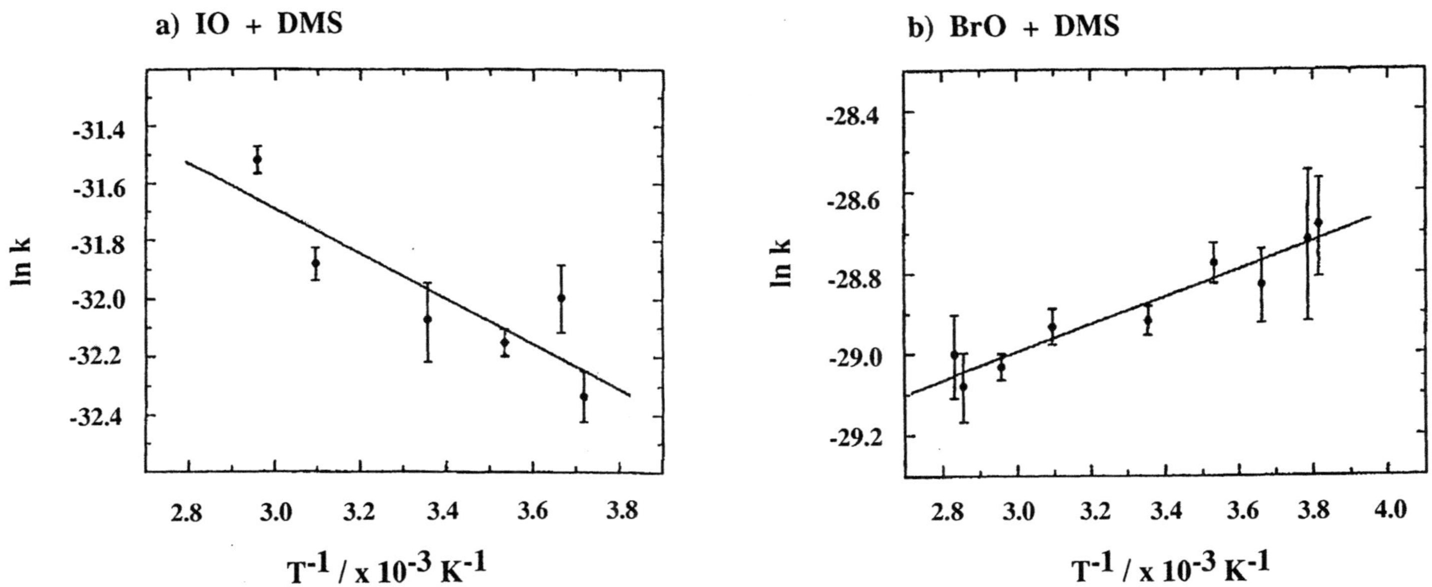

Figure 3. Arrhenius plots of the data obtained for the reactions a) IO + DMS and b) BrO + DMS

FIELD STUDIES OF ATMOSPHERIC DMS CHEMISTRY USING SELECTED ION CHEMICAL IONIZATION MASS SPECTROMETRY

H. BERRESHEIM, F.L. EISELE, and D.J. TANNER

Georgia Institute of Technology, Atlanta, GA 30332, USA.

SUMMARY

In this study we present results obtained from the first real time measurements of dimethylsulfoxide (DMSO) in ambient air using selected ion chemical ionization mass spectrometry, SI/CI/MS. Due to its extreme sensitivity this technique allows for atmospheric measurements of DMSO and other trace species such as dimethylsulfide (DMS), SO_2, H_2SO_4, and gas phase methanesulfonic acid, MSA(g), to be made without sample preconcentration. The estimated detection limit for DMSO at 60 s signal integration is 0.5 pptv. The present field measurements were made during the 1991 PSI-3 study at a Pacific coastal site near Cape Flattery, Washington. During this study both SI/CI/MS and the gold wool adsorption/GC/FPD technique were used to measure DMS, DMSO, H_2SO_4, and MSA(g) concentrations simultaneously in marine air. The detection limit for the two latter species was approximately 0.001 pptv. This paper presents some of the most important results of these measurements.

1. INTRODUCTION

One of the major uncertainties in our present understanding of the DMS-cloud-climate relationship is due to the complexity of reaction mechanisms involved in atmospheric DMS oxidation (1). Specifically, the relative yields of the major oxidation products H_2SO_4, MSA(g) (methanesulfonic acid), SO_2, DMSO (dimethylsulfoxide), and $DMSO_2$ (dimethylsulfone) are highly uncertain. The concentrations of these species in the marine atmosphere are either estimated or known to be extremely low (lower ppt to sub-ppt range). Only recently, with the development of selected ion chemical ionization mass spectrometry, SI/CI/MS, has it become possible to measure most of these products in real time (i.e., without preconcentration) in the ambient atmosphere (2-4). As part of the April 1991 Pacific Sulfur/Stratus Investigation (PSI-3) experiment SI/CI/MS was used for the first time to make comprehensive real time measurements of DMSO, H_2SO_4, and MSA(g) in the marine atmosphere. DMS was measured simultaneously using the gold wool adsorption/ GC/FPD (GWA/GC/FPD) technique (5). The measurements were conducted at Cheeka Peak Research Station (480 m asl), about 2 km inland from the Pacific coast near Cape Flattery, Washington.

2. EXPERIMENTAL

Details of the SI/CI/MS apparatus and its applications for real-time measurements of DMS, SO_2, DMSO, H_2SO_4, and MSA(g) have already been published elsewhere (2-4). Thus, only a brief description is given here. The apparatus consists of three major sections. The first section consists of a laminar flow reactor tube in which the neutral sample molecules are ionized at atmospheric pressure by charge transfer reactions with specifically selected reactant ions. These reactant ions are produced in an ultra-clean carrier gas (usually N_2) by either alpha or beta bombardment from a radioactive ^{241}Am or ^{63}Ni source, respectively. The production of reactant ions is spatially separated from the production of sample product ions in the reactor in order to minimize the interference of radical and metastable species in the chemical ionization of the sample molecules. The sample product ions are then guided by electrical fields into the second part of the apparatus, a collision-dissociation chamber operated at 0.1 torr, in which neutral cluster molecules are removed from the core ions. The ions are then guided into a quadrupole mass spectrometer (Extrel) in which they are finally mass-separated and detected. In the present study DMSO was detected in the positive mode as $DMSO(H^+)$ at mass 79 after chemical ionization with NH_4^+ ions. H_2SO_4 and

G. Restelli and G. Angeletti (eds.), Dimethylsulphide: Oceans, Atmosphere, and Climate, 239–242.

MSA(g) were detected in the negative mode as HSO_4^- and MSA^-, respectively. NO_3^- was used as the corresponding reactant ion species. The respective detection limits are 0.5 pptv for DMSO, and 0.001 pptv for the latter two species at 60 s signal integration. Currently, we are also developing a similar technique for real-time measurement of $DMSO_2$ in ambient air.

3. RESULTS AND DISCUSSION

Relatively clean marine air was advected to the study site on April 24, 25, 27, and 29. The corresponding results obtained during this period are shown in Figure 1. The data obtained for the DMS oxidation products have been smoothed by 10 min (DMSO) and 20 min averaging, respectively. DMSO measurements ended on April 25. H_2SO_4 and MSA(g) measurements were temporarily interrupted on April 26 and early April 28 during precipitation events. The data in Figure 1 clearly show pronounced diel variations for all four sulfur compounds. The diel changes in DMS levels were dominated by changes in the mixed layer height and the advection of DMS from the coastal marine boundary layer. By contrast, the diel variations of the DMS oxidation products were dominated by their in situ photochemical production from DMS and their scavenging by aerosol particles and droplets. DMSO levels were significantly higher than those of H_2SO_4 and MSA(g) which is consistent with the relatively higher DMSO vapor pressure (6). However, in view of the rapid further oxidation of DMSO by OH (7) the observed DMSO levels may also suggest that a significant fraction of DMS was converted to $DMSO_2$ through the OH addition channel.

Statistical regression analyses of the data showed significant correlations between individual sulfur species (95% or higher confidence level) which are described by the following equations:

$$[DMS] = 11.98\ [DMSO] + 2.81\quad (r = 0.72,\ N = 14)$$
$$[DMSO] = 10.91\ [H_2SO_4] + 0.30\quad (r = 0.62,\ N = 45)$$
$$[MSA(g)] = 0.071\ [H_2SO_4] + 0.008\quad (r = 0.77,\ N = 171).$$

The slope of 0.07 resulting from the correlation of MSA(g) and H_2SO_4 agrees very well with a similar value assumed to represent the MSA/non-sea-salt sulfate mixing ratio in background marine aerosol (8). A more detailed discussion of the sulfur measurements made at Cheeka Peak including correlations between sulfur species in the gas and particle phase, and between sulfur species and cloud condensation nuclei (CCN) is given in a recent paper by Berresheim et al. (9).

ACKNOWLEDGEMENT

We thank T. Bates, J. Johnson, and R. Charlson for the opportunity to participate in PSI-3 and for logistical support. This work was funded by the National Science Foundation, grants ATM-9113681, ATM-9021522, and the Office of Exploratory Research, USEPA grant R817121.

REFERENCES

(1) YIN, F., GROSJEAN, D., and SEINFELD, J.H. (1990). Photooxidation of dimethyl sulfide and dimethyl disulfide. I: Mechanism development. J. Atmos. Chem., 11, 309-364.

(2) EISELE, F.L., and BERRESHEIM, H. (1992). High-pressure chemical ionization flow reactor for real-time mass spectrometric analysis of sulfur gases and unsaturated hydrocarbons in air. Anal. Chem., 64, 283-288.

(3) BERRESHEIM, H., EISELE, F.L., and TANNER, D.J. (1992). Real-time measurement of dimethylsulfoxide in ambient air. Anal. Chem., in press.

(4) EISELE, F.L., and TANNER, D.J. (1991). Ion-assisted tropospheric OH measurements. J. Geophys. Res., 96, 9295-9308.

(5) BERRESHEIM H., ANDREAE M.O., IVERSON R.L., and LI S.M. (1991). Seasonal variations of dimethylsulfide emissions and atmospheric sulfur and nitrogen species over the western North Atlantic Ocean. Tellus 43B, 353-372.

(6) WATTS, S.F., and BRIMBLECOMBE, P. (1987). The Henry's law constant of dimethyl sulphoxide. Environ. Technol. Lett., 8, 483-486.

(7) HYNES, A.J., POUNDS, A.J., McKAY, T., BRADSHAW, J.D., and WINE, P.H. (1992). Detailed mechanistic studies of the OH-initiated oxidation of biogenic sulfur compounds under atmospheric conditions. Presented at the 12th International Symposium on Gas Kinetics, Univ. of Reading, UK.

(8) SALTZMAN, E.S., SAVOIE, D.L., ZIKA, R.G., and PROSPERO, J.M. (1983). Methane sulfonic acid in the marine atmosphere. J. Geophys. Res., 88, 10,897-10,902.

(9) BERRESHEIM, H., EISELE, F.L., TANNER, D.J., McINNES, L.M., RAMSEY-BELL, D.C., and COVERT, D.S. (1992). Atmospheric sulfur chemistry and CCN concentrations over the northeastern Pacific coast. J. Geophys. Res., submitted.

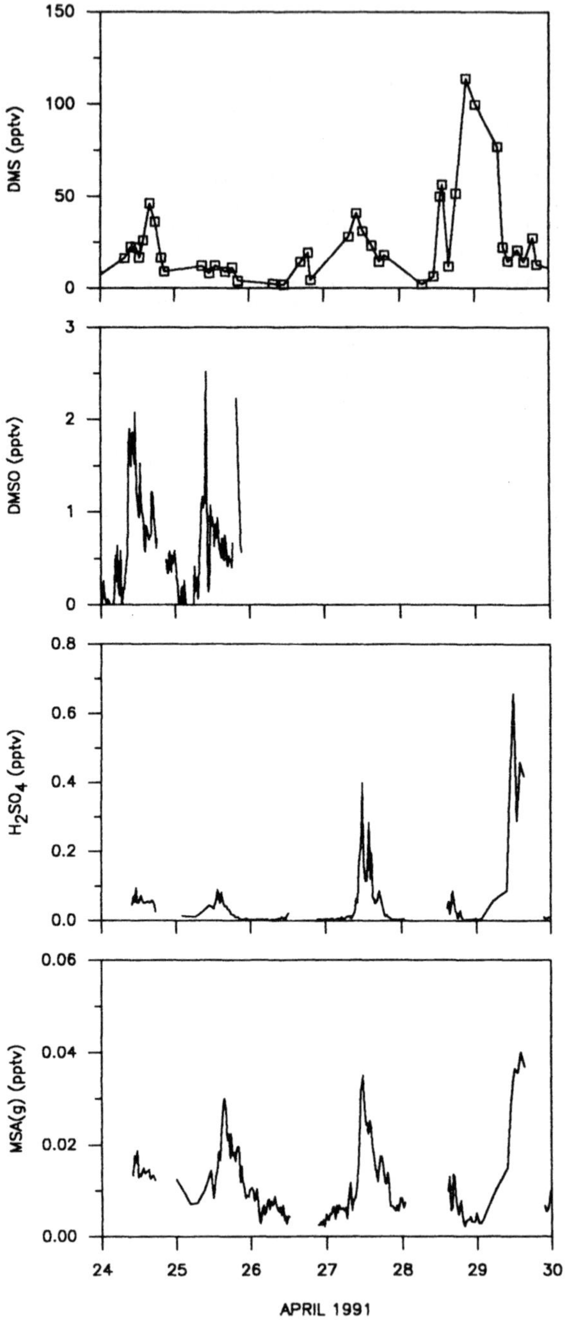

Fig. 1: Results from PSI-3 measurements made at Cheeka Research Station

KINETIC AND MECHANISTIC STUDY OF THE REACTION OF ATOMIC CHLORINE WITH DIMETHYLSULFIDE

R. E. Stickel[a], J. M. Nicovich[a], S. Wang[a], Z. Zhao[b], and P. H. Wine[a,b,c]

Georgia Institute of Technology
Atlanta, Georgia 30332, U.S.A.

a. Physical & Atmospheric Chemistry Branch, EOPSL-GTRI
b. School of Earth & Atmospheric Sciences
c. School of Chemistry & Biochemistry

SUMMARY

The kinetics of the title reaction have been studied over the temperature and pressure ranges 240-421K and 3-700 Torr using time-resolved resonance fluorescence detection of Cl following 266 nm laser flash photolysis of $Cl_2CO/DMS/N_2$ mixtures. The reaction is found to be very fast, occurring on essentially every Cl + DMS encounter. The reaction rate increases with decreasing temperature and shows a significant pressure dependence. In a separate study, time-resolved tunable diode laser spectroscopic detection of HCl has been coupled with 248 nm laser flash photolysis of $Cl_2CO/DMS/CO_2/N_2$ mixtures to measure the HCl product yield from the title reaction as a function of pressure at T = 297K. The HCl yield approaches unity as $P \rightarrow 0$, but decreases with increasing pressure. The mechanism of the title reaction and its potential role in marine atmospheric chemistry are discussed.

1. INTRODUCTION

Dimethylsulfide (CH_3SCH_3, DMS) is a key atmospheric sulfur species. A significant fraction of all sulfur entering the atmosphere does so as DMS volatized from the oceans. Some (1,2), though not all (3), field studies of marine sulfur chemistry have been interpreted as indicating that the DMS removal rate is too rapid to be accounted for entirely by known sink reactions with OH and NO_3 radicals. To explain the apparently short lifetime of DMS in marine air, it was proposed that IO radicals, formed via the reaction of photolytically generated iodine atoms with ozone, could destroy DMS via a rapid reaction producing I + DMSO (4). Subsequent laboratory studies of IO + DMS reaction kinetics (5,6) and field observations of significant concentrations of DMSO in the marine boundary layer (7) seemed to confirm the importance of IO as an initiator of DMS oxidation. However, a direct study of IO + DMS kinetics carried out in our laboratory (8), the results of which have been confirmed (9,10), demonstrates that coupling of the marine iodine and sulfur cycles via the IO + DMS reaction is negligible.

The possibility that reactions occurring on surfaces of sea-salt aerosol particles can lead to significant production of atomic chlorine in the marine boundary layer has received considerable attention in recent years. Production of photochemically labile $Cl_2(g)$ via heterogeneous degradation of O_3 (possibly involving free radical intermediates) is one suggested pathway for generation of gas phase chlorine atoms (11-13); however, recent laboratory (14) and modeling (15) studies suggest that this pathway is not important in the atmosphere. On the other hand, it appears that $ClNO_2$, generated via the heterogeneous reaction of N_2O_5 vapor with moist NaCl(s), may represent a photolytic precursor for atmospherically significant levels of atomic chlorine, even in the remote boundary layer where NO_x levels are very low (14,16,17).

A recent competitive kinetics study of reaction 1 at 298K in one atmosphere of air (18) suggests that this reaction is extremely fast, i.e., $k_1(298K) = 3.2 \times 10^{-10}$ $cm^3molecule^{-1}s^{-1}$.

G. Restelli and G. Angeletti (eds.), Dimethylsulphide: Oceans, Atmosphere, and Climate, 243–249.

$$Cl + CH_3SCH_3 \rightarrow \text{products} \quad (1)$$

Combining the above result with the findings of our study of the kinetics and mechanism of the OH + DMS reaction (19) suggests that $k_1/k_2 \sim 50$ at 298K and one atmosphere of air.

$$OH + CH_3SCH_3 \rightarrow \text{products} \quad (2)$$

Reaction 2 is the dominant known atmospheric sink for DMS under low NO_x conditions, and mean surface level OH concentrations are thought to be around 10^6 molecules per cm^3 (20). Hence, if atomic chlorine concentrations in the marine boundary layer reach levels around 10^4 atoms per cm^3, and if reaction 1 is as fast as the available kinetic data (18) suggest, then reaction 1 could play an important role in marine sulfur chemistry.

In this paper we report the results of a kinetic and mechanistic study of reaction 1. Our results confirm that reaction 1 is extremely fast, and also provide some interesting new mechanistic insights concerning this complex chemical reaction. It should be noted that a more detailed account of this study is being published elsewhere (21).

2. EXPERIMENTAL APPROACH

Two different types of experiments were performed. Laser flash photolysis (LFP) of Cl_2CO (phosgene) at 266 nm was combined with detection of chlorine atoms by time-resolved resonance fluorescence (RF) spectroscopy to investigate the kinetics of reaction 1 as a function of temperature and pressure. To gain insight into the mechanism for reaction 1, a separate set of experiments was carried out where laser flash photolysis of Cl_2CO at 248 nm was combined with detection of HCl by time-resolved tunable diode laser absorption spectroscopy (TDLAS) to measure the HCl product yield at 297K as a function of pressure. Detailed descriptions of the LFP-RF and LFP-TDLAS experiments are given elsewhere (21).

3. RESULTS AND DISCUSSION

All experiments were carried out under pseudo-first order conditions with DMS in large excess over Cl atoms. Observed Cl atom decays were exponential and increased linearly as a function of [DMS], as would be expected if Cl atom removal is dominated by reaction with DMS. The bimolecular rate coefficients of interest, $k_1(P,T)$, are obtained from the slopes of plots of k', the pseudo-first order decay rate, versus DMS concentration. Measured rate coefficients are tabulated in Table 1. Our results confirm that reaction 1 occurs on virtually every Cl-DMS encounter. The reaction rate is found to increase with decreasing temperature as would be expected for a very fast reaction whose rate is determined by the magnitude of long range attractive forces between the reactants. The surprising aspect of the data in Table 1 is our observation of a clear pressure dependence for k_1, at least at temperatures of 297K and below. Reaction 1 appears to occur via both pressure-independent and pressure-dependent pathways; the pressure-dependent pathway must involve collisional stabilization of a $(CH_3)_2S$-Cl adduct.

To obtain the HCl yield from reaction 1 we carried out back-to-back experiments where the photolytically produced Cl atoms reacted with DMS, then with ethane (C_2H_6); the yield of HCl from the Cl + C_2H_6 reaction is known to be unity. About 0.5 Torr of CO_2 was added to the reaction mixture to facilitate rapid relaxation of any HCl which is formed in the v=1 level, and to promote rapid equilibration of the $Cl(^2P_{3/2,1/2})$ spin orbit levels. Measured HCl yields (Φ) as a function of pressure (at 297K) are tabulated in Table 2. The HCl yield approaches unity as $P \rightarrow 0$, but decreases with increasing pressure.

Table 1. Measured bimolecular rate coefficients for the Cl + DMS reaction as a function of temperature and pressure.

T(K)	P(Torr)	$k_1(10^{-10}\ cm^3 molecule^{-1} s^{-1})$*
240	3.1	2.48 ± 0.09
	10	2.70 ± 0.25
	25	2.97 ± 0.16
	25	3.27 ± 0.16
	50	3.38 ± 0.09
	100	3.32 ± 0.15
	150	3.65 ± 0.48
	150	3.64 ± 0.41
	500	3.78 ± 0.28
267	50	2.99 ± 0.18
297	3.1	2.14 ± 0.07
	3.1	1.81 ± 0.29
	6.0	1.95 ± 0.18
	25	2.48 ± 0.07
	26	2.27 ± 0.25
	50	2.56 ± 0.10
	50	2.71 ± 0.09
	150	3.09 ± 0.11
	300	3.03 ± 0.11
	500	3.47 ± 0.17
	700	3.34 ± 0.12
	700	3.05 ± 0.06
356	3.1	1.83 ± 0.06
	25	2.10 ± 0.04
	25	2.01 ± 0.18
	50	2.35 ± 0.07
	150	2.58 ± 0.13
	500	3.56 ± 0.51
385	50	2.33 ± 0.20
421	3.1	1.83 ± 0.07
	25	1.68 ± 0.32
	25	1.77 ± 0.09
	50	1.93 ± 0.18
	150	1.67 ± 0.14
	150	1.82 ± 0.08
	700	2.31 ± 0.10

*Errors are 2σ and represent precision only.

Table 2. Measured HCl yield from the Cl + DMS reaction as a function of pressure at T = 297K.

Buffer Gas	P(Torr)	No. of Expts.	Yield (Φ)
CO_2	0.6	2	0.98
	2.0	19	0.89
	5.0	4	0.79
	10.1	4	0.74
	26	2	0.62
N_2*	5.0	2	0.85
	10.0	2	0.80
	25	2	0.68
	50	2	0.585
	100	2	0.54
	203	2	0.505

* Includes 0.5 Torr CO_2.

Possible channels for reaction 1 include the following:

$$\begin{aligned} Cl + CH_3SCH_3 &\rightarrow CH_3SCH_2 + HCl &\quad (1a)\\ &\rightarrow CH_3SCl + CH_3 &\quad (1b)\\ &\rightarrow CH_3S + CH_3Cl &\quad (1c)\\ &\xrightarrow{M} CH_3S(Cl)CH_3 &\quad (1d) \end{aligned}$$

Based on Benson's estimate of the heat of formation of CH_3SCl(22), channel 1b is endothermic by about 8 kcal $mole^{-1}$.

While quantitative extrapolation of kinetic and HCl yield data to zero pressure is non-trivial, examination of the results in Tables 1 and 2 strongly suggests that the following relationship is obeyed (at T = 297K):

$$\Phi(P) = k_1(P \rightarrow 0) / k_1(P). \qquad (I)$$

Clearly, hydrogen abstraction (channel 1a) is the dominant reaction pathway in the limit of low total pressure. However, channel 1d becomes competitive at higher pressure. Our experiments do not provide definitive information concerning the fate of the collisionally stabilized adduct. The only information we obtain is that adduct decomposition to Cl and/or HCl does not occur on the time scale of the experiments (several milliseconds).

4. IMPLICATIONS FOR ATMOSPHERIC CHEMISTRY

The very fast rate observed in this study for the Cl + DMS reaction, when considered in conjunction with results presented in references 18 and 19, suggests that at 298K and 760 Torr total pressure $k_1 \sim 50\ k_2$. Reaction 2 is generally thought to be the dominant removal mechanism for DMS in the remote marine boundary layer. Since daytime OH concentrations are typically around 10^6 radicals per cm^3 (20), concentrations of Cl atoms would have to reach levels of 10^4 atoms per cm^3 for reaction 1 to be a significant DMS sink. A recent estimate of the formation of atomic chlorine via the gas phase OH + HCl reaction leads to marine tropospheric concentration levels around 10^3 atoms per cm^3, i.e., about one order of magnitude lower than required for reaction 1 to compete with reaction 2 in the marine atmosphere (23). However, recent laboratory work (16,17) suggests that the heterogeneous reaction of gaseous N_2O_5 with moist sea salt aerosol may (via a $ClNO_2$ intermediate) be capable of generating sufficient Cl atoms to make reaction 1 an important atmospheric process. Field observations aimed at evaluating chlorine atom concentrations in the marine boundary layer are needed before a quantitative understanding of the potential coupling of the marine chlorine and sulfur cycles via reaction 1 will be forthcoming.

At 298K and 760 Torr total pressure our results suggest that the H-abstraction pathway accounts for only about 40-50% of overall Cl + DMS reactivity. The other important reaction pathway is formation of the relatively stable adduct $(CH_3)_2SCl$. Based on experimental studies of the related species $(CH_3)_2SOH$ and $(CH_3)_2SBr$ (19,24) and *ab initio* quantum chemical calculations (25), we estimate that the $(CH_3)_2S$-Cl bond dissociation energy is 14±2 kcal $mole^{-1}$. The fate of $(CH_3)_2SCl$ in the atmosphere remains highly uncertain. One interesting possibility is the reaction

$$(CH_3)_2SCl + O_2 \rightarrow (CH_3)_2SO + ClO, \qquad (3)$$

which is exothermic by 18±3 kcal $mole^{-1}$. Reaction 3 could be an unrecognized source of $(CH_3)_2SO$ (DMSO) in the marine boundary layer. Arguing against the occurrence of reaction 3, however, is our recent (unpublished) observation that the analogous reaction with Br replacing Cl is very slow (24). Another energetically feasible pathway for degradation of $(CH_3)_2SCl$ is unimolecular decomposition to $CH_3S + CH_3Cl$; however, since this process requires Cl atom migration from the sulfur atom to a carbon atom, the potential energy barrier may be quite large. If decomposition to $CH_3S + CH_3Cl$ is the dominant removal mechanism for atmospheric $(CH_3)_2SCl$, and if reaction with Cl atoms turns out to be an important removal mechanism for marine DMS, then reaction 1 could be an important source of atmospheric CH_3Cl. Clearly, further research is needed before the atmospheric fate of $(CH_3)_2SCl$ can be assessed with any degree of confidence.

5. ACKNOWLEDGMENTS

This research was supported by the National Science Foundation through grant ATM-9104807 and by the Georgia Tech Research Institute through internal grant E-8904-038. We thank C. A. van Dijk for assistance with some of the experiments.

REFERENCES

(1) NGUYEN, B. C., BONSANG, B. and GAUDRY, A. J. (1983). The role of the ocean in the global atmospheric sulfur cycle, Journal of Geophysical Research, Vol. 88, 10903-10914.

(2) BARNES, I., BECKER, K. H., MARTIN, D., CARLIER, P., MOUVIER, G., JOURDAIN, J. L., LAVERDET, G. and LEBRAS, G. (1989). Impact of halogen oxides on dimethyl sulfide oxidation in the marine atmosphere, in Biogenic Sulfur in the Environment, edited by E. S. Saltzman and W. J. Cooper, ACS Sympsoium Series No. 393, 464-475.

(3) see for example, ANDREAE, M. O., FEREK, R. J., BERMOND, F., BYRD, K. P., ENGSTROM, R. T., HARDIN, S., HOWMERE, P. D., LEMARREC, F., RAEMDONCK, H. and CHATFIELD, R. B. (1985). Dimethyl sulfide in the marine atmosphere, Journal of Geophysical Research, Vol. 90, 12891-12900.

(4) CARLIER, P. (1985). L'ozone serait-il l'oxydant prinicpal du sulfure de dimethyle en milieu océanique?, in Atmospheric Ozone, edited by C. S. Zerefos and A. Ghazi, D. Reidel, Hingham, MA, 815-819.

(5) BARNES, I., BECKER, K. H., CARLIER, P. and MOUVIER, G. (1987). FTIR study of the $DMS/NO_2/I_2/N_2$ photolysis system: The reaction of IO radicals with DMS, International Journal of Chemical Kinetics, Vol. 19, 489-501.

(6) MARTIN, D., JOURDAIN, J. L., LAVERDET, G. and LEBRAS, G. (1987). Kinetic study of the reaction of IO with CH_3SCH_3, International Journal of Chemical Kinetics, Vol. 19, 503-512.

(7) HARVEY, G. R., LANG, R. F. and BROWN, C. (1989). Dimethylsulfoxide and dimethylsulfone over the North Atlantic, Eos Transactions of the American Geophysical Union, Vol. 70, 1044.

(8) DAYKIN, E. P. and WINE, P. H. (1990). Rate of reaction of IO radicals with dimethyl-sulfide, Journal of Geophysical Research, Vol. 95, 18547-18553.

(9) MAGUIN, F., MELLOUKI, A., LAVERDET, G., POULET, G. and LEBRAS, G. (1991). Kinetics of the reactions of the IO radical with dimethyl sulfide, methanethiol, ethylene, and propylene, International Journal of Chemical Kinetics, Vol. 23, 237-245.

(10) BARNES, I., BASTIAN, V., BECKER, K. H. and OVERATH, R. D. (1991). Kinetic studies of the reactions of IO, BrO, and ClO with dimethylsulfide, International Journal of Chemical Kinetics, Vol. 23, 579-591.

(11) ZETZSCH, C., PFAHLER, G. and BEHNKE, W. (1988). Heterogeneous formation of chlorine atoms from NaCl in a photosmog system, Journal of Aerosol Science, Vol. 19, 1203-1206.

(12) BEHNKE, W. and ZETZSCH, C. (1989). Smog chamber investigations of the influence of NaCl aerosol on the concentration of O_3 in a photosmog system, in Ozone in the Atmosphere. Proceedings of the Quadrennial Ozone Symposium 1988 and Tropospheric Ozone Workshop, edited by R. Bojkov and P. Fabian, Deepak, Hampton, VA, 519-523.
(13) KEENE, W. C., PSZENNY, A. A. P., JACOB, D. J., DUCE, R. A., GALLOWAY, J. N., SCHULZ-TOKOS, J. J., SIEVERING, H. and BOATMAN, J. F. (1990). The geochemical cycling of reactive chlorine through the marine troposphere, Global Biogeochemical Cycles, Vol. 4, 407-430.
(14) BEHNKE, W. and ZETZSCH, C. (1990). Heterogeneous photochemical formation of Cl atoms from NaCl aerosol, NO_x and ozone, Journal of Aerosol Science, Vol. 21, Supplement 1, 5229-5232.
(15) CHAMEIDES, W. L. and STELSON, A. W. (1992). Aqueous-phase chemical processes in deliquescent seasalt aerosols, Berichte Bunsengesellschaft für Physikalische Chemie, Vol. 96, 461-470. Journal of Geophysical Research, in press.
(16) ZETZSCH, C. and BEHNKE, W. (1992). Heterogeneous photochemical sources of atomic Cl in the troposphere, Berichte Bunsengesellschaft für Physikalische Chemie, Vol. 96, 488-493.
(17) GANSKE, J. A., BERKO, J. N. and FINLAYSON-PITTS, B. J. (1992). Absorption cross sections for gaseous $ClNO_2$ and Cl_2 at 298K: Potential organic oxidant source in the marine troposphere, Journal of Geophysical Research, Vol. 97, 7651-7656.
(18) NIELSEN, O. J., SIDEBOTTOM, H. W., NELSON, L., RATTIGAN, O., TREACY, J. J. and O'FARRELL, D. J. (1990). Rate constants for the reactions of OH radicals and Cl atoms with diethyl sulfide, di-n-propyl sulfide, and di-n-butyl sulfide, International Journal of Chemical Kinetics, Vol. 22, 603-612.
(19) HYNES, A. J., WINE, P. H. and SEMMES, D. H. (1986). Kinetics and mechanism of OH reactions with organic sulfides, Journal of Physical Chemistry, Vol. 90, 4148-4156.
(20) see for example, LU, Y. and KHALIL, M. A. K. (1991). Tropospheric OH: Model calculations of spatial, temporal, and secular variations, Chemosphere, Vol. 23, 397-444.
(21) STICKEL, R. E., NICOVICH, J. M., WANG, S. ZHAO, Z. and WINE, P. H. (1992). Kinetic and mechanistic study of the reaction of atomic chlorine with dimethylsulfide, Journal of Physical Chemistry, Vol. 96, in press.
(22) BENSON, S. W. (1978). Thermochemistry and kinetics of sulfur-containing molecules and radicals. Chemical Reviews, Vol. 78, 23-35.
(23) SINGH, H. B. and KASTING, J. F. (1988). Chlorine-hydrocarbon photochemistry in the marine troposphere and lower stratosphere, Journal of Atmospheric Chemistry, Vol. 7, 261-285.
(24) NICOVICH, J. M., SHACKELFORD, C. J., KREUTTER, K. D., DAYKIN, E. P., WANG, S. and WINE, P. H. (1993). Chemical kinetic, thermochemical and spectroscopic characterization of the $(CH_3)_2S$-Br adduct, to be published.
(25) McKEE, M. L. (1992). Private communication.

PRODUCTION OF PEROXY RADICALS IN THE DMS OXIDATION DURING NIGHTTIME

G. LE BRAS*, C. GOLZ+ and U. PLATT+

*Laboratoire de Combustion et Systèmes Réactifs-LCSR-CNRS
F-45071-Orléans-Cedex 2-France

+Institut für Umweltphysik, Universität Heidelberg,
INF 366, D-6900 Heidelberg, Germany.

SUMMARY

It is found by box modelling calculations that the NO_3-DMS interaction in semi-polluted coastal areas would produce high levels of peroxy radicals and related photooxidant during nighttime. The calculations use chemical mechanisms based on established or speculative kinetic and mechanistic data of the elementary steps.

1. INTRODUCTION

The atmospheric oxidation of DMS is initiated by reaction with OH during the day and NO_3 during the night. Although the OH initiated oxidation is presently the major process at the global scale, the NO_3 initiated oxidation could become significant as a result of a continuous increase of NO_x in the troposphere. Also, the NO_3 initiated oxidation of DMS can be important in semi-polluted coastal areas where significant levels of NO_3 are produced from reaction of NO_2 with O_3 :

$$NO_2 + O_3 \text{ ---> } NO_3 + O_2$$

The NO_3-DMS interaction had been clearly observed from simultaneous measurements of vertical profiles of NO_3 and DMS in coastal area in Brittany during the CEC OCEANO-NOX campaign in summer 1989, when continental polluted air advected over the site (1).

A box modelling of these field measurements associated with laboratory studies of chemical reactions have shown that high levels of peroxy radicals, RO_2, could be produced (2). The following chain mechanism was considered in the model :

G. Restelli and G. Angeletti (eds.), Dimethylsulphide: Oceans, Atmosphere, and Climate, 251–260.

$$NO_3 + DMS\ (+O_2) \dashrightarrow \ldots \dashrightarrow RO_2$$
$$RO_2 + NO_3 \dashrightarrow RO + NO_2 + O_2$$
$$RO_2 + RO_2 \dashrightarrow RO + RO + O_2$$
$$RO + O_2 \dashrightarrow R'CHO + HO_2$$
$$HO_2 + NO_3 \dashrightarrow OH + NO_2 + O_2$$
$$OH + DMS\ (+O_2) \dashrightarrow \ldots \dashrightarrow RO_2$$

The whole mechanism also includes loss reaction of RO_2 radicals which produces peroxides and formaldehyde by self or cross combination. In the present paper this mechanism is discussed considering the more recent laboratory kinetic data, and results from box modelling calculations are presented.

2. CHEMICAL MECHANISM USED IN THE MODELLING OF THE NO_3 INITIATED DMS OXIDATION.

The schematic diagram of the NO_3-DMS mechanism used in the modelling calculation is represented in figure 1.

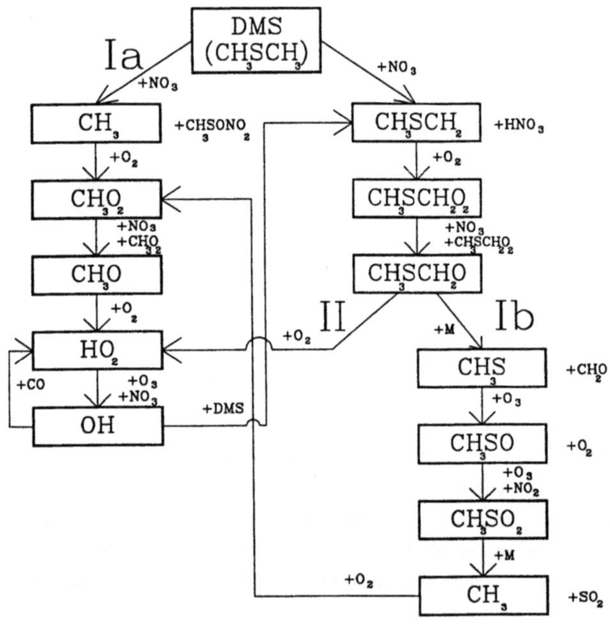

Figure 1 : DMS oxidation by NO_3 radicals :
Mechanism I (DMS oxidation yielding CH_3O_2) is possible through pathway Ia (initiated by reaction (1), see text) or through pathway Ib (initiated by reaction (2), CH_3SCH_2O decomposition to $CH_3S + CH_2O$).
Mechanism II (DMS oxidation yielding HO_2) is also initiated by reaction (2) but assumes CH_3SCH_2O to react with O_2 (reaction (3), see text).

This mechanism is partly based on the kinetic and mechanistic data recently reviewed (3). For the initial step, two channels are possible, considering the end products observed in simulation chamber experiments (4) (5) :

(1) $NO_3 + DMS\ (+O_2)$ ---> $CH_3O_2 + CH_3SONO_2$
(2) ---> $CH_3SCH_2O_2 + HNO_3$

In the reaction sequence converting $CH_3SCH_2O_2$ into CH_3O_2, the peroxy radical, $CH_3SCH_2O_2$, is assumed to be rapidly converted into alkoxy, CH_3SCH_2O by reaction with NO_3 or self combination (see right hand column in figure 1). Another important step of this sequence is the fate of CH_3SCH_2O, which is generally considered to decompose into CH_3S and CH_2O. It can be speculate also that CH_3SCH_2O reacts with O_2 in a similar way as organic peroxy radicals :

(3) $CH_3SCH_2O + O_2$ ---> $CH_3SCHO + HO_2$

In this case $CH_3SCH_2O_2$ is converted into HO_2 instead of CH_3O_2.

The other key steps in the proposed mechanism is the chain propagation through CH_3O_2/CH_3O and HO_2/OH conversion. The CH_3O_2/CH_3O conversion can proceed through:

(4) $CH_3O_2 + NO_3$ ---> $CH_3O + O_2 + NO_2$
(5a) $CH_3O_2 + CH_3O_2$ ---> $2CH_3O + O_2$

Reaction (5a), for which rate constant is now well established ($k_5 = 1.22 \times 10^{-13}$ cm^3 molecule^{-1}s^{-1} at 298 K (6)), can be significant at high CH_3O_2 levels. For reaction (4), a rate constant $k_4 = 2.3 \times 10^{-12}$ cm^3 molecule^{-1}s^{-1} at 298 K, was indirectly determined from computer simulation of modulated photolysis experiments (7) and the mechanism was assumed to be the CH_3O_2/CH_3O conversion. However, a more direct study using the discharge flow mass spectrometric method (8) yields for k_4 a value 10 times lower.

Concerning the HO_2/OH conversion, by the reaction of NO_3 with HO_2 :

(6) $HO_2 + NO_3$ ---> $NO_2 + O_2 + OH$

The relatively high rate constant value firstly obtained ($k_6 = 3.6 \times 10^{-12}$ cm^3 molecule^{-1}s^{-1} at 298 K (9) was further confirmed (10,11), also by direct methods. The k_6 value appears to be now well established although a much lower value has been recently indirectly measured in smog chamber experiments (12).

The simplified mechanism of figure 1 does not include reactions which form peroxynitrates, RO_2NO_2, because these species are unstable under tropospheric conditions. The chain terminating reactions of peroxy radicals considered in the modelling and not included in figure 1 are :

(7) $HO_2 + HO_2$ ---> $H_2O_2 + O_2$
(8) $CH_3O_2 + HO_2$ ---> $CH_3OOH + O_2$

(5b)	$CH_3O_2 + CH_3O_2$	--->	$CH_3OH + CH_2O + O_2$	61 %
(5c)		--->	$CH_3OOCH_3 + O_2$	9 %
(9)	$OH + NO_2 + M$	--->	$HNO_3 + M$	
(10)	$HO_2 + NO_3$	--->	$HNO_3 + O_2$	

3. MODELLING OF PEROXY RADICAL AND PHOTOOXIDANT FORMATION IN THE NO_3 INITIATED OXIDATION OF DMS.

The modelling calculations of the coastal marine atmosphere during the OCEANO-NOX campaigns used the chemistry of the background troposphere conditions (see ref. 2) . For DMS chemistry, the initial NO_3 + DMS reaction ($k = 1.0 \times 10^{-12}$ cm^3 $molecule^{-1}s^{-1}$ at 298 K) was considered to be the limiting step, so that, the DMS chemistry was represented by:

$$NO_3 + DMS \dashrightarrow CH_3O_2 \qquad \text{I}$$

or

$$NO_3 + DMS \dashrightarrow HO_2 \qquad \text{II}$$

depending on the mechanism assumed for CH_3SCH_2O reaction (decomposition or reaction with O_2).

In the calculations, ozone, CO and DMS mixing ratios were held fixed at 40 ppb, 160 ppb and 0.5 ppb respectively. Figures 2 and 3 give the model calculated free radical concentrations as a function of NO_x concentration. The concentration ranges of NO_x encountered during the OCEANO-NOX campaigns in 1988 and 1989 are indicated. Figure 2 and 3 correspond to use of mechanism I (NO_3 + DMS ---> CH_3O_2) and II (NO_3 + DMS ---> HO_2), respectively.

With mechanism I, the calculation of high peroxy (HO_2, RO_2) radical concentrations for NO_2 levels in the ppb range have already been reported (2). However, the lower CH_3O_2/CH_3O conversion rate (8) considered in the present calculation has some effect on the CH_3O_2/HO_2 partitioning (higher CH_3O_2 and lower HO_2 levels). The OH concentration is also slightly reduced. The calculated radical concentrations are very different however, when mechanism II is used instead of mechanism I (figure 3). The HO_2 concentrations are increased by a factor 5 as a result of a fast generation of HO_2 by the NO_3-DMS reaction. The CH_3O_2 concentrations are drastically decreased, the only CH_3O_2 source being in this case the CH_4 oxidation. A significant increase of OH is also observed correlated with the HO_2 increase. Levels as high as 5×10^5 cm^{-3} are reached for c.a. 10 ppb of NO_x. Mechanism II for DMS oxidation together with the $NO_3 + HO_2$ reaction would be an important nighttime source of OH radicals.

The above calculations are also represented in terms of flux diagrams in figures 4 and 5 corresponding to mechanism I and II respectively. The conditions of the calculations are those of figures 2 and 3 with NO_x concentration fixed at 3×10^{11} molecule cm^{-3}. For the intermediates species CH_3O_2, HO_2 and OH, concentrations at equilibrium are indicated (inside box) as well as their fluxes (or reaction rates) in per cent for conversion into other species (inside arrows). The total formation (or loss) rate of these species are indicated outside of the box in molecule $cm^{-3}s^{-1}$. For instance in figure 4, the concentration of HO_2 at equilibrium is 1.1×10^8 molecule cm^{-3} and the formation (or loss) rate is 2.5×10^6 molecule $cm^{-3}s^{-1}$. In the arrows are given the reaction channels. For the stable products i.e. the photooxidants (CH_2O, CH_3OOH, H_2O_2) and

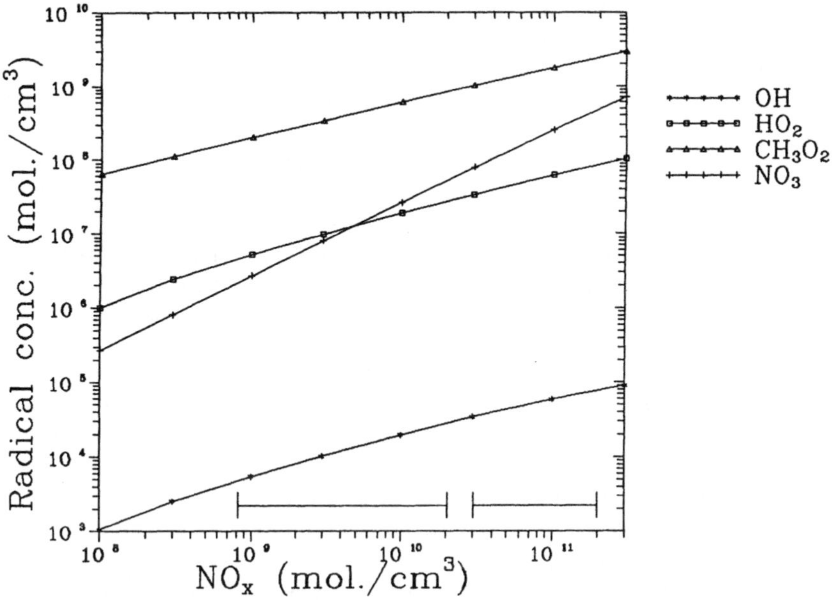

Figure 2 : Results of box model using mechanism I : DMS + NO_3 ---> CH_3O_2.
Radical concentrations are displayed as a function of initial NO_2 concentration (equivalent to the NO_x concentration, which is fixed in the model). O_3, CO and DMS concentrations are fixed to 9.6 x 10^{11}, 4 x 10^{12} and 1.2 x 10^{10} molecule cm^{-3} respectively. Ranges of NO_x encountered during the OCEANO-NOX campaigns in summer of 1988 and 1989 are marked by horizontal bars.

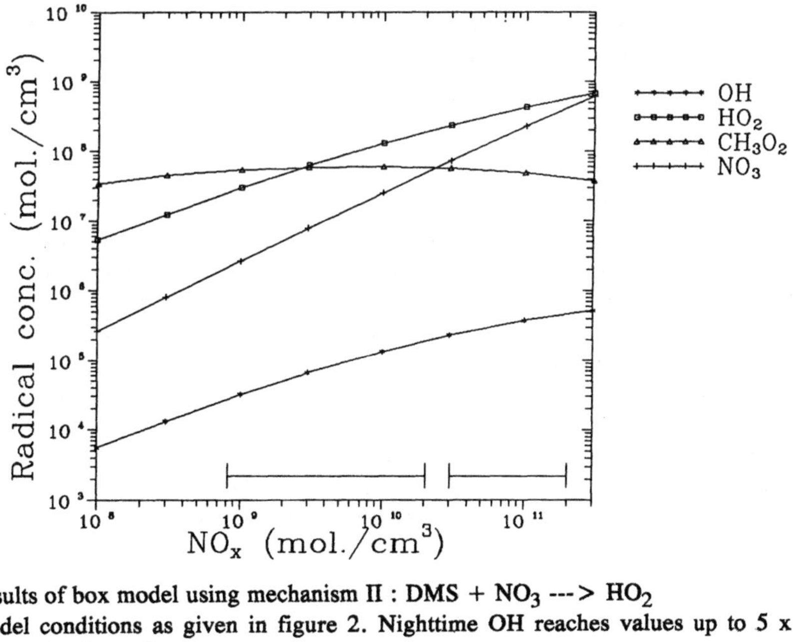

Figure 3 : Results of box model using mechanism II : DMS + NO_3 ---> HO_2
Model conditions as given in figure 2. Nighttime OH reaches values up to 5 x 10^5 molecule cm^{-3}.

HNO_3, the formation rates are also indicated outside of the boxes. Since these compounds accumulate, their concentrations can be calculated after a given time by multiplying the formation rate by time. This applies when equilibrium is reached (typically after 30 min., at $[NO_x] = 3 \times 10^{11}$ molecule cm^{-3}) since the formation rates are given at equilibrium.

There are significant differences in diagrams of figures 4 and 5 due to differences in formation rates of CH_3O_2 and HO_2 depending on whether mechanism I or II is considered. CH_3O_2 reacts mainly with NO_3 and itself to yield CH_3O and subsequently CH_2O (mechanism I), or it mainly reacts with HO_2 to give CH_3OOH (mechanism II). The CH_3O_2/HO_2 conversion occurs mainly by reaction of CH_3O_2 with NO_3 or itself (mechanism I), while the NO_3-DMS reaction is the major source of HO_2 (mechanism II). HO_2 reacts mainly with CH_3O_2 to give CH_3OOH (mechanism I), while the self combination is the major loss channel of HO_2, to give H_2O_2 (mechanism II). The flux of HO_2 through the HO_2/OH conversion (by reaction of HO_2 with NO_3 or O_3), relatively increases when mechanism II is used. Concerning the OH radical, the fluxes increase when mechanism II is used, but the partitioning towards its different reactions remains unchanged. The impact of mechanisms I or II is very important on the formation rate of photooxidants CH_2O, H_2O_2 and CH_3OOH, which are directly related to the concentrations of the precusors CH_3O_2 and/or HO_2. The formation rate of CH_2O decreases roughly by a factor 80 and CH_3OOH by a factor 10 when mechanism II is used. In contrast the H_2O_2 flux increases by a factor 40. Whatever is the mechanism chosen, globally, high amounts of photooxidant are produced but the nature of these oxidants are highly dependent of those mechanisms. The amounts of photooxidant produced after 8 hours are respectively (in molecule cm^{-3}) : 1.2×10^{11} CH_2O, 5.0×10^{10} CH_3OOH and 1.7×10^8 H_2O_2 (mechanism I), and 1.5×10^9 CH_2O, 4.2×10^9 CH_3OOH and 6.8×10^{10} H_2O_2 (mechanism II). Since quilibrium radical concentrations are reached within about 1/2 hour, the difference between these amounts and calculations assuming instantaneous equilibrium is negligible. Note that CH_3OOH and CH_2O destruction via reaction with OH is not included here. Nevertheless the calculated values show that the nighttime oxidation of DMS under semi-polluted conditions can be a significant source of photooxidants which accumulate during the night. Photolysis of these compounds during the day will contribute to the photochemical activity leading to ozone production.

4. CONCLUSION

Box modelling calculations have shown that the NO_3 initiated DMS oxidation can produce high peroxy radical concentrations which would lead to accumulation of photooxidant (CH_2O, H_2O_2 and CH_3OOH) during the night. The calculations are based on both real atmospheric situations encountered in coastal areas and on a more or less speculative mechanism of DMS oxidation. There are in particular needs to get experimental kinetic and/or mechanistic data for some reactions considered in this mechanisms such as the intial step NO_3+ DMS the $CH_3SCH_2O_2/CH_3SCH_2O$ conversion and the fate of CH_3SCH_2O. It will be also important to validate the model calculations by field measurements of peroxy radicals and photooxidant species considered in this paper. In particular, measurements of the CH_3OOH/H_2O_2 ratio should give information about the relative occurence of mechanism I (producing high CH_3OOH/H_2O_2 ratio) and mechanism II (producing low CH_3OOH/H_2O_2 ratio).

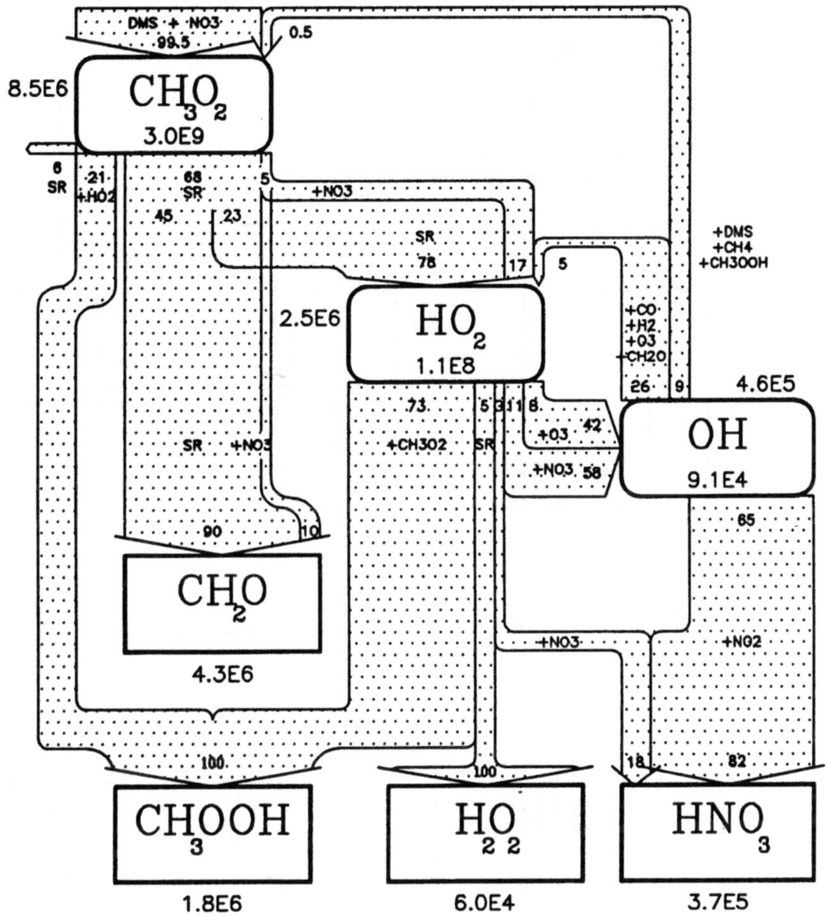

Figure 4 : Radical fluxes in DMS oxidation mechanism I : DMS + NO_3 ---> CH_3O_2. The numbers inside the boxes are concentrations in molecule cm^3. Numbers next to boxes are absolute fluxes in molecule $cm^{-3}s^{-1}$, numbers in the arrows are relative fluxes in percent. In the arrows the reaction partner for the respective channel is given : SR = self reaction. NO_x concentration is fixed to 3 x 10^{11} molecule cm^{-3}, the resulting equilibrium NO_3 concentration is 7.1 x 10^8 molecule cm^{-3}.

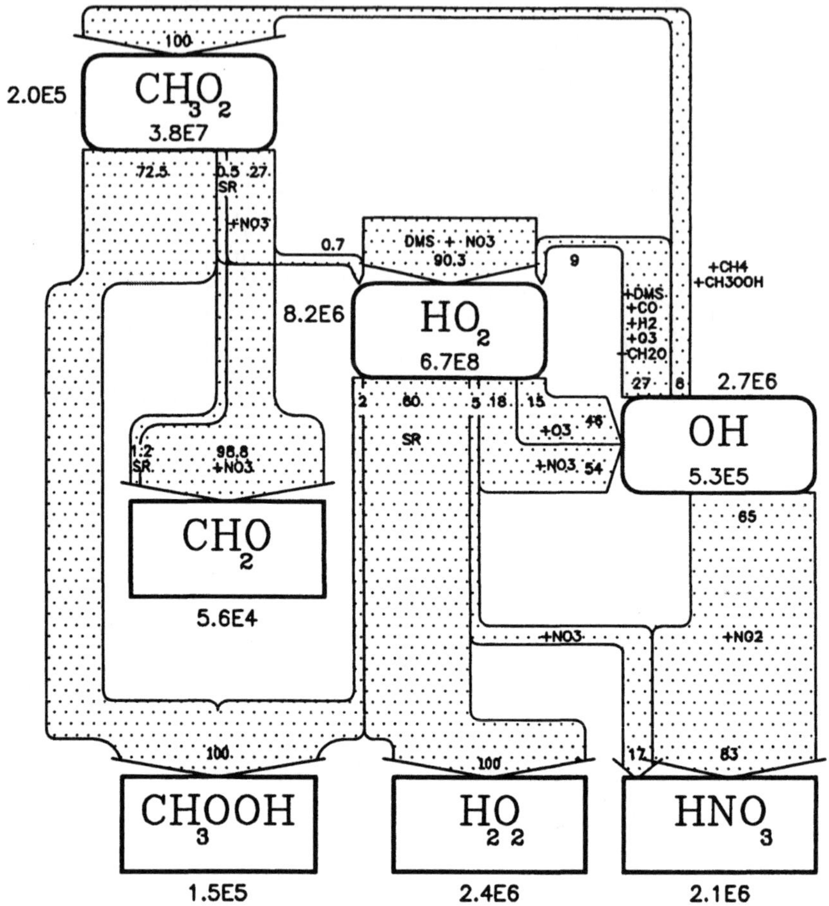

Figure 5 : Radical fluxes in DMS oxidation mechanism II : DMS + NO_3 ---> HO_2.
The numbers in the diagramm have the same meaning as in figure 4. NO_x concentration is fixed to 3 x 10^{11} molecule cm^3, the resulting NO_3 concentration is 6.2 x 10^8 molecule cm^3.

REFERENCES

(1) BARNES, I., BONSANG, B., BRAUERS, T., CARLIER, P., COX, R.A., DORN, H.P., JENKIN, M.E., LE BRAS, G., and PLATT, U., (1991). Laboratory and field studies of oxidation processes occurring in the atmospheric marine boundary layer (OCEANO-NOX CEC Project). CEC Air Pollution Research Report 35.

(2) PLATT, U., LE BRAS, G., POULET, G., BURROWS, J.P., and MOORTGAT, G.K., (1990). Peroxy radicals from nighttime reaction of NO_3 with organic compounds. Nature 348, 147-149.

(3) TYNDALL, G.S., and RAVISHANKARA, A.R., (1991). Atmospheric oxidation of reduced sulfur species. Int. J. Chem. Kinetics 23, 483-527.

(4) TYNDALL, G.S., BURROWS, J.P., SCHNEIDER, W., BINGEMER, H., and MOORTGAT, G.K., (1985). COST 611 (CEC) Workshop on chemistry related to tropospheric ozone, Nov. 1985.

(5) JENSEN, N.R., HJORTH, J., LOHSE, C., SKOV, H., and RESTELLI, G., (1991). Products and mechanism of the reaction between NO_3 and dimethylsulphide in air. Atmos. Environ. 25, 1897.

(6) LIGHTFOOT, P.D., COX, R.A., CROWLEY, J.N., DESTRIAU M., HAYMAN, G.D., JENKIN, M.E., MOORTGAT, G.K., and ZABEL, F. (1992). Organic peroxy radicals : kinetics, spectroscopy and tropospheric chemistry. Atmos. Environ. 26A, 1805-1964.

(7) CROWLEY, J.N., BURROWS, J.P., MOORTGAT, G.K., POULET, G., and LE BRAS, G. (1990). Room temperature rate coefficient for the reaction between CH_3O_2 and NO_3. Int. J. Chem. Kinetics, 22, 673-681.

(8) HELLEIS, F., CROWLEY, J.N., and MOORTGAT, G.K., (1992). The reaction between NO_3 and CH_3O_2 at 298 K. Joint CEC- EUROTRAC (Lactoz Halipp) workshop "Chemical mechanisms describing tropospheric processes" Leuven Sept. 23-25, 1992.

(9) MELLOUKI, A., LE BRAS, G., and POULET, G., (1988). Kinetics of the reaction of NO_3 with OH and HO_2. J. Phys. Chem. 92, 2229-2234.

(10) SCHINDLER, R.N., BECKER, E., and DETHLEFS, J., (1992). The reactions NO_3 + H, OH and HO_2. Determinations of the rate constants and discussion on the basis of calculated energy hypersurfaces. Proceedings of the 12th Int. Symp. in Gas Kinetics, Reading (UK) 19-24 Juil. 1992.

(11) MELLOUKI, A., TALUKDAR, R.K., BOPEGEDERA, A.M., and HOWARD, C.J., (1992). Study of the kinetics of the reactions of NO_3 with HO_2 and OH. Int. J. Chem. Kinetics, in press.

(12) HJORTH, J., CAPPELLANI, F., and RESTELLI, G., (1992). A TDL and FTIR study of the reaction $HO_2 + NO_3 \dashrightarrow OH + NO_2 + O_2$. Joint CEC Eurotrac (Lactoz, Halipp) workshop "Chemical mechanisms describing tropospheric processes". Leuven, Sept. 23-25, 1992.

LABORATORY AND MODELLING STUDIES OF THE FORMATION OF A STABLE INTERMEDIATE IN THE NIGHT–TIME OXIDATION OF DMS

J. HJORTH, N.R. JENSEN*, F. RAES and R. VAN DINGENEN

Commission of the European Communities, Joint Research Centre,

Environment Institute, I–21020 Ispra (VA), Italy

SUMMARY

The formation of gas phase and condensed phase products and the change in aerosol size distribution and number concentration were followed simultaneously during the oxidation of DMS, initiated by NO_3 radicals, in air in large Teflon bags. By application of a combined gas phase chemistry/aerosol physics model, the sticking probability of MSA (gas), α(MSA), on acidic particles was evaluated and found to have a value within the interval between 0.5 and 1.0. A stable peroxynitrate intermediate ($CH_3S(O)O_2NO_2$ or $CH_3S(O)_2O_2NO_2$) was found to play a key role in the kinetics of the system. The thermal stability of this intermediate was investigated and the temperature dependence of the reaction $CH_3S(O)_xO_2NO_2+M \rightarrow CH_3S(O)_xO_2\bullet+M$ was found as $k(T) = 2.15\times10^{18} \exp(-1.39\times10^4\ K/T)\ s^{-1}$.

1. INTRODUCTION

Dimethylsulphide (CH_3SCH_3) from marine phytoplankton is estimated to account for approximately 25% of the total global gaseous sulphur emission (1, 2, 3) and its oxidation products, H_2SO_4 and methanesulphonic acid (MSA), are expected to give an important contribution to the formation of aerosols in the marine troposphere (4). The main sink for DMS in the troposphere is believed to be the reaction with the OH radical, but also the reaction with the NO_3 radical may under some conditions be of importance, in particular in air influenced by anthropogenic pollution (5).

* *Now at NOAA, Aeronomy Laboratory, 325 Broadway, Boulder CO 80303, USA.*

G. Restelli and G. Angeletti (eds.), Dimethylsulphide: Oceans, Atmosphere, and Climate, 261–272.

Although the main products of the oxidation of DMS are well known, the detailed reaction mechanism is only partially known and the reaction steps controlling the end–product distribution have not been identified; neither have the role of possible long–lived intermediates been investigated (6). The main sulphur containing product appears to be SO_2 (which subsequently can be oxidized to sulphate in the marine troposphere) while MSA is normally found at a concentration of below 10% of the non–sea–salt sulphate in marine aerosols. However, the ratio [MSA]/[n.s.s. SO_4^{-2}] is found to be much higher in marine aerosols in arctic areas and apparently the MSA yields can also be considerably higher in a polluted coastal area (5). The latter observation is consistent with results of laboratory experiments, showing that the formation of MSA is favoured by the presence of NO_x.

The present work is an experimental investigation of intermediates, product yields and the growth of aerosols during the oxidation of DMS initiated by NO_3 radicals (7). Two parallel studies were carried out:

1) A series of Teflon bag experiments where the formation of aerosols and SO_2 was followed experimentally and described by combining a chemical model and an aerosol physics model. The tuning of this combined model to fit the experimental data allowed to evaluate the sticking probability, α(MSA), of MSA(gas) on acidic ($MSA/H_2SO_4/H_2O$) particles and to demonstrate the importance of a stable peroxynitrate intermediate, found to be a precursor of MSA and H_2SO_4. The intermediate, which was first observed by Barnes et al. (9), may be either $CH_3S(O)O_2NO_2$ (methylsulphinylperoxynitrate) or $CH_3S(O)_2O_2NO_2$ (methylsulphonylperoxynitrate) and will in the following be named MSPN.

2) In a parallel study (8), the thermal stability of this intermediate was investigated in a laboratory chamber, applying FTIR spectroscopy as analytical technique.

2. EXPERIMENTAL

Aerosol study.

This study was carried out in large (1300 L) 3 meter long pillow–shaped teflon bags, previously treated with O_3 and flushed with dry air. Purified air was used as diluent gas in all experiments. To start an experimental run, the bags were filled with air containing approximately 1500 ppbv ozone (formed by applying a UV–lamp generator) and humidified (by passage over a water bath) to 25% RH. Subsequently DMS was introduced with purified air to reach a concentration of 500 ppbv in the bag. Finally, the reaction was started by introducing an amount of NO_2 with dry air corresponding to a level of 1500 ppbv in the bag.

O_3 was monitored by a UV photometric analyzer (Dasibi), SO_2 by a fluorescence monitor (Environment) and particle size distribution by a differential mobility particle sizer (TSI). The chemical composition of aerosols was measured by ion chromatography after sampling on filters or after bubbling the reaction mixture through scrubbers. The scrubbers applied consisted of two water bubbling flasks connected in series; the efficiency of a flask was found to be 75–80% for aerosols and was thus assumed to be 100% for the soluble gaseous species. Samples on Teflon–filters were collected over time periods of 25–60 min. The reaction was followed over a time period of at least 3 hours. In some runs, an excess of NO was added after 120 minutes in order to remove the peroxynitrate species (MSPN) formed as an intermediate in the reaction, as discussed in the following.

FTIR study

The experiments made to study the thermal stability of MSPN were performed in a 480 L evacuable Teflon–coated chamber equipped with a White–type multiple reflection mirror system connected to a Bruker 113V FTIR spectrometer and adjusted to give a total pathlength of 81 m. N_2O_5 was formed in the chamber by reacting O_3 with NO_2 in purified air. DMS was added subsequently, and the course of the reaction was followed by FTIR analysis. Initial concentrations of N_2O_5, NO_2 and DMS were all in the range of 10–20 ppm. The total pressure in all runs was about 750 Torr while the temperature (measured with a thermocouple placed inside the chamber) was varied within the range of

282–306 K (the experimental set–up did not allow to cover a larger temperature interval). After the initial rapid formation of MSPN, an excess of NO (corresponding to >110 ppm in the chamber) was added and the decay of the peroxynitrate IR bands was followed.

3. MODEL OF AEROSOL DYNAMICS

A model of the aerosol dynamics in the ternary MSA–H_2SO_4–H_2O system (IMAD3, Ref.7) as well as a chemical model to provide the source terms for the condensable chemical species were applied in order to analyse the data obtained in the Teflon bag aerosol experiments. The IMAD3 model describes particle growth by condensation and coagulation, deposition of both particles and gaseous condensable species (MSA and H_2SO_4) and the concentrations of MSA and H_2SO_4 both in gas phase and in particles. Because of the uncertainties involved in measuring and modelling the nucleation burst in the initial phase of the experiment a number of particles at t=0 was prescribed artificially so that modelled and measured aerosol concentrations agreed after 20 minutes. The source terms for MSA and H_2SO_4 were provided by a chemical model and the combined chemistry/aerosol physics model was tuned by comparison with the experimental measurements as discussed in the following.

4. RESULTS

Aerosol study

Results of seven runs are shown in Table 1 and 2. It is seen that the final concentrations of SO_2 generally are close to half of the initially added DMS. The situation for what concerns the MSA is more complicated: before NO is added, a large difference between 'scrubber' and filter measurements is seen, the latter being generally much lower than the former. After addition of NO this discrepancy disappears and most MSA is found in the particle–phase. H_2SO_4 was found to be formed along with MSA in a yield of about 20 % of the MSA yield. The aerosol measurements show a sharp increase in volume upon addition of NO (Fig. 1).

Table 1

SO_2(fin)	MSA(scrubber) (ppbv)			MSA(filter) (ppbv)		
(ppbv)	45 min	120 min	240 min	45 min	120 min	240 min
241	82	64	39	5	15	11
282	106	84	44	1	11	2
262	85	43	19	1	7	4

Measured final concentrations of SO_2 and concentrations of MSA calculated on the basis of analyses of filter samples, collected during time intervals around the elapsed time indicated in the table. Initial conditions always as discussed in text. To facilitate comparison of data, MSA concentrations are given in ppbv, calculated as the volume mixing ratio that MSA would have in the gas phase as an ideal gas.

Table 2

SO_2(fin)	MSA scrubber		MSA filter	
(ppbv)	(ppbv)		(ppbv)	
	60 min	150 min	60 min	150 min
186	236	112	5	129
274	125	55	20	49
223	171	69	5	57
225	176	57	1	31

As Table 1, but in these runs an amount of NO corresponding to 5000 ppbv has been added after 120 min.

Fig. 1 shows the range of the aerosol parameters (total number and volume concentration as well as mean diameter) and the SO_2 concentrations measured in runs of Table 1. The SO_2 concentration is seen to increase rapidly in the beginning and then

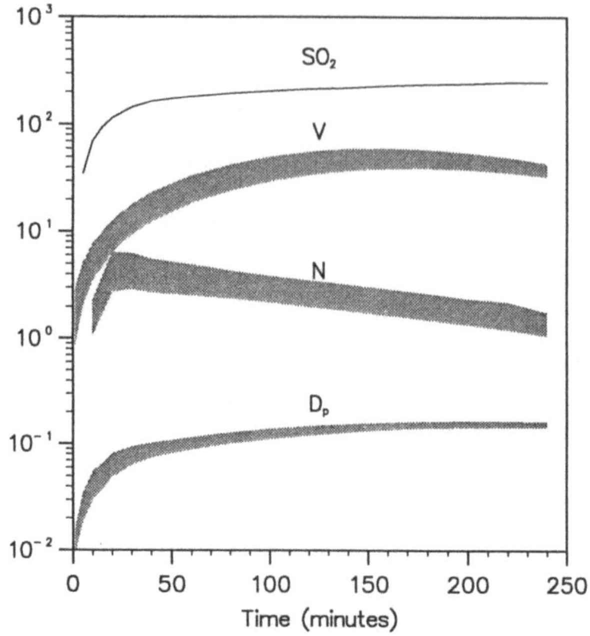

Figure 1. Measured concentrations of SO_2 (ppbv), particle number (N, 10^4 cm^{-3}), particle volume (V, $\mu^3 cm^{-3}$) and particle diameter (D_P, μ) obtained from 3 runs. Initial conditions: 500 ppbv DMS, 1350–1450 ppbv O_3, 1500 ppbv NO_2 and 21–27% relative humidity.

reach a level that remains almost constant. The aerosol volume is more slowly growing to reach a maximum after 2–3 hours reaction time.

FTIR gas–phase study

Two reaction intermediates, both having the characteristic IR bands of peroxynitrate intermediates (RO_2NO_2), could be seen in the course of the reaction between DMS and NO_3; one rapidly formed but shortlived with lifetimes at the order of seconds and IR–bands typical of alkyl–peroxynitrates (790 cm^{-1}, 1300 cm^{-1} and 1720 cm^{-1}) which have tentatively been assigned to $CH_3SCH_2O_2NO_2$ (10). The second intermediate shows also some spectral features typical of peroxynitrate compounds (777 cm^{-1} and 1303 cm^{-1}) with a third band at 1768 cm^{-1}. These bands have been tentatively assigned to $CH_3S(O)O_2NO_2$ (9) but may also be due to $CH_3S(O)_2O_2NO_2$ (10).

For each run, the decay rate of $CH_3S(O)_xO_2NO_2$ in the presence of excess NO, k, was established from the measured decrease of its IR–absorption bands. Under the experimental conditions applied, the reaction leading to the observed peroxynitrate–decay must be

(1, –1) $CH_3S(O)_xO_2NO_2+M \leftrightarrows CH_3S(O)_xO_2\bullet+NO_2+M$

(2) $CH_3S(O)_xO_2\bullet+NO \rightarrow CH_3S(O)_xO\bullet+NO_2$.

As reaction (2) will be much faster than reaction (–1) at high NO concentrations, the decay rate (k) for $CH_3S(O)_xO_2NO_2$ could be considered equal to k_1. Eventually, linear regression on the values of $\ln(k_1)$ vs. $1/T$ allowed to evaluate the parameters of the Arrhenius equation

$k = A \exp(-E/RT)$.

E/R was found as $1.39x10^4$ K (with a standard deviation of $0.13x10^4$) while $\ln(A/s^{-1})$ was found as 42.2 with a standard deviation of 4.5. Thus

$k(T) \mp 2.15x10^{18} \exp(-1.39x10^4\ K/T)\ s^{-1}$.

4. DISCUSSION

Coupled aerosol–chemistry modelling

A simple chemical model, assuming an immediate formation of the SO_2, MSA and H_2SO_4 products following the initial attack by NO_3 radicals on DMS, was first applied. The product distribution was assumed to be 50% SO_2 and 50% MSA and H_2SO_4 with a ratio of 5 to 1 between the latter two. It was found, however, that this simple model could not reproduce the experimental data. Very low values for α(MSA) had to be applied (a best fit of α(MSA) gave a value of $3x10^{-4}$) and even then the variations in the aerosol volume distribution could not be acceptably simulated. The modelling results clearly indicate that there is a time–lag between the initial attack on DMS by the NO_3 radical and the eventual formation of the MSA (and sulfate) products, which can only be explained by the occurrence of a relatively stable intermediate. The rapid formation of MSA (and H_2SO_4) after the addition of NO could well be explained by a reaction between NO and the intermediate. In fact, it is already known from previous work (9, 10, 11) that

MSPN is formed in the reaction between NO_3 and DMS and that this species could be a precursor of MSA. The discrepancy presented by the filter, scrubber and denuder measurements could also easily be understood on the basis of an intermediate that is converted to MSA: what is measured in the scrubbers is then gas phase MSA plus gas phase MSPN, which apparently react to form MSA in liquid water.

Thus a new chemical model was set up (Fig. 2) which included the formation of MSPN and the subsequent transformation of this compound into MSA and some H_2SO_4

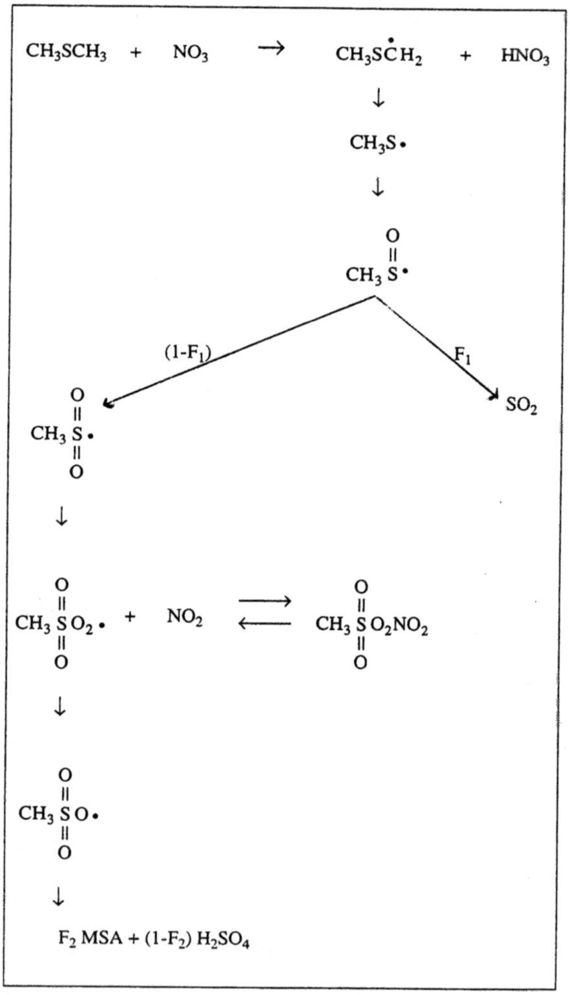

Figure 2. Outline of chemical reaction scheme applied in model. A detailled description of the model is given by Van Dingenen et al., 1993 (7)

as a minor product, introducing a time–lag between the initial reaction between DMS and NO_3 and the formation of MSA and H_2SO_4 while SO_2 is still formed immediately in the first reaction step. The formation of MSA may proceed via the reaction steps

(3, –3) $CH_3S(O)_2O_2NO_2 + M \rightleftharpoons CH_3S(O)_2O_2{\bullet} + NO_2 + M$

(4) $CH_3S(O)_2O_2{\bullet} + NO \rightarrow CH_3SO_3{\bullet} + NO_2$

(5) $CH_3SO_3{\bullet} + RH \rightarrow CH_3SO_3H + R{\bullet}$,

where RH may be any of the H–containing species in the system. It is not clarified whether H_2SO_4 is formed directly from the intermediate, e.g. by the reaction

6) $CH_3SO_3{\bullet} \rightarrow CH_3{\bullet} + SO_3$

7) $SO_3 + H_2O \rightarrow H_2SO_4$,

or whether it is formed by secondary reactions leading to oxidation of SO_2 to H_2SO_4.

The reactions of O_3 with radical intermediates and many reactions of O_2 have been left out in the chemical model. This chemistry is only implicitly treated in the model as it is likely to influence the branching ratio between SO_2 and other products, which is estimated based on the experimental results. The reaction scheme applied in the model is thus much simplified, also leaving out most of the secondary chemistry (e.g. reactions of the $CH_3{\bullet}$ radical). The parameters in the model that are tuned to fit the experimental data are the branching ratio between formation of SO_2 and other products (F_1 in Fig. 2) and that between formation of H_2SO_4 and MSA (F_2 in Fig. 2) as well as loss rates by heterogeneous reactions for N_2O_5 and MSPN. For what concerns the chemical reaction rate constants, k_3 was determined by the FTIR chamber study while k_{-3} was determined by performing a best fit to the data measured in the bag experiments. The remaining rate constants are either known from the literature or values for analogous reactions have been applied.

The chemical model and the aerosol model were finally optimized using an iterative procedure (7), where the chemical model is tuned to reproduce measured concentrations, then used to supply inputs to the aerosol model which is tuned to reproduce the measured aerosol volume measurements by fitting the α(MSA) constant. This eventually allows to obtain a better dataset on time dependent [MSA] and [H_2SO_4], which is used to obtain a better fit of the parameters in the chemistry model which then, in turn, can produce better input to the aerosol model etc. The combined model was found to be optimized by applying α(MSA) values in the range between 0.5 and 1.0 (Fig. 3). This range was calculated taking into account the range of possible values for the wall deposition rate for H_2SO_4 and MSA, β_g, assumed to be from 3.6×10^{-3} s^{-1} to 1.3×10^{-2} s^{-1}. α(MSA) is then higher than the value of the accomodation coefficient of H_2SO_4, found to be within the range 0.02–0.09 (12). Consequently, the main fate of MSA should be that of sticking on already existing particles rather than formation of new particles, as it is already known from previous studies that the nucleation capability of MSA is low (13, 14). Values of other parameters estimated by the fitting procedure, applied to three runs, were 0.6–0.7 for the branching ratio between SO_2 and other products, 0.75 to 0.85 for the branching ratio between H_2SO_4 and MSA formation.

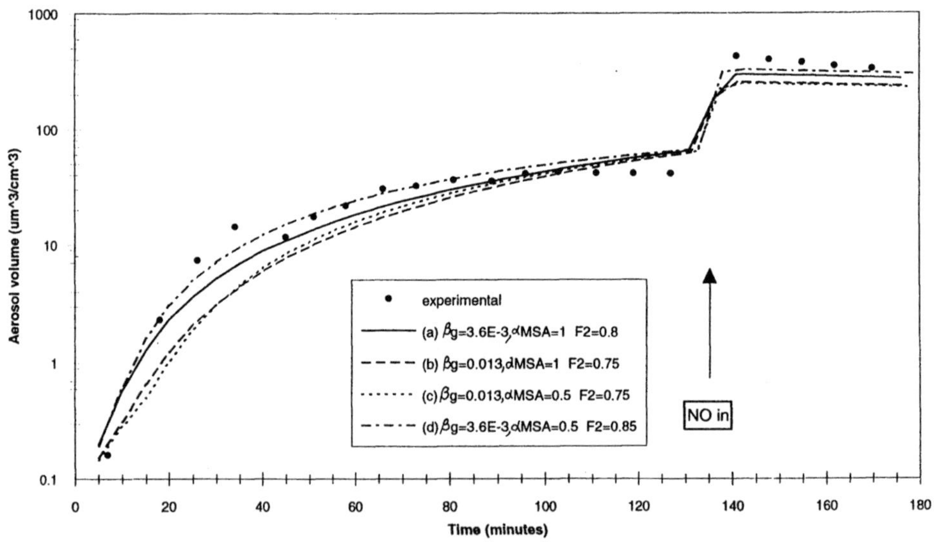

Figure 3. Measured aerosol volume and calculated values of this for four combinations of β_g, α(MSA) and F_2. 'NO in' indicates when NO is added to the reaction mixture.

The model was found to give an adequate simulation also of the sudden increase in particle volume upon addition of NO which confirms the validity of the mechanism applied.

The apparent lifetime of MSPN in this experimental system is particularly long because of the high level of NO_2 which will force the equilibrium (1,–1) towards the unreactive peroxynitrate, away from the reactive peroxyradical. In fact, the FTIR gas–phase study of the thermal stability of MSPN showed that its lifetime with respect to dissociation ($1/k_3$) is only a few minutes at room temperature. However, the lifetime increases dramatically at lower temperatures, e.g. at –10 ^{0}C it is calculated as 11.3 hours.

The formation of the peroxynitrate intermediate is thus expected to be of potential importance at low temperatures and in airmasses influenced by anthropogenic NO_x–sources. Modelling studies are needed to evaluate the contribution of such an intermediate in airborne transport of N and S in the marine troposphere.

Acknowledgements

We thank Dr. G. Restelli for useful discussions and Mr. G. Ottobrini for help in carrying out the laboratory experiments.

References

(1) ANDREAE M.O. (1990) The global biogeochemical sulfur cycle: a review, *Mar.* hem. **30**, 1–29.

(2) BATES T.S., CLINE J.D., GAMMON R.H. and KELLY–HANSEN S.R. (1987) Regional and seasonal variations in the flux of oceanic dimethylsulphide to the atmosphere, *J. Geophys. Res.* **92**, 2930–2938.

(3) SPIRO P.A., JACOB D.J. and LOGAN J.A. (1992) Global inventory of sulfur emissions with 1^0x1^0 resolution, *J. Geophys. Res.* **97**, 6023–6836.

(4) CHARLSON R.J., LOVELOCK J.E., ANDREAE M.O. and WARREN S.G. (1987) Oceanic phytoplankton, atmospheric sulphur, cloud albedo and climate, *Nature* **326**, 655–661.

(5) BARNES I., BONSANG B., BRAUERS T., CARLIER P., COX R.A., DORN H.P., JENKIN M.E., LE BRAS G. and PLATT U. (1991) Laboratory and field studies of oxidation processes occurring in the atmospheric boundary layer, Air Pollution Research Report 35, Commission of the European Communities.

(6) TYNDALL G.S. and RAVISHANKARA A.R. (1991) Atmospheric oxidation of uced sulphur species, *J. Phys. Chem.* **23**, 483–527

(7) VAN DINGENEN R., JENSEN N.R., HJORTH J. and RAES F. (1993) Peroxynitrate formation during the night–time oxidation of dimethylsulphide: its role as a reservoir species for aerosol formation, submitted to J. Atm. Chem.

(8) JENSEN N.R. and HJORTH J. (1992) Evidence for the formation of a stable intermediate in the tropospheric oxidation of DMS, submitted to *Geophys. Res. Lett.*

(9) BARNES I., BASTIAN V., BECKER K.H. and NIKI H. (1987) FTIR spectroscopic studies of the CH_3S+NO_2 reaction under atmospheric conditions, *Chem. Phys. Lett.* **140 (5)**, 451–457.

(10) JENSEN N.R., HJORTH J., LOHSE C., SKOV H. and RESTELLI G. (1991) Products and mechanism of the reaction between NO_3 and dimethylsulphide in air, *Atmos. Environ.*, **9**, 1991.

(11) JENSEN N.R., HJORTH J., LOHSE C., SKOV H. and RESTELLI G. (1992) Products and mechanisms of the gas phase reactions of NO_3 with CH_3SCH_3, CD_3SCD_3, CH_3SH and CH_3SSCH_3, *J. Atm. Chem.* **14**, 95–108.

(12) VAN DINGENEN R. and RAES F. (1991) Determination of the condensation accomodation coefficient of sulfuric acid on water–sulfuric acid aerosol, *Aerosol Sci. Technol.* **15**, 93–106.

(13) HOPPEL W.A. (1987) Nucleation in the MSA–water vapour system, *Atmos. Environ.* **21**, 2703–2709.

(14) VAN DINGENEN R., RAES F. (1993) Ternary nucleation of methane sulphonic acid, sulphuric acid and water vapour, *J. Aerosol. Sci.* **24**, 1–17.

DO DIMETHYL SULPHIDE EMISSIONS FROM THE NORTH ATLANTIC CONTRIBUTE TO RAINWATER ACIDITY AND THE ATMOSPHERIC SULPHUR BURDEN IN THE UK?(*)

C. NICHOLAS HEWITT and B. DAVIDSON

Institute of Environmental and Biological Sciences
Lancaster University
Lancaster LA1 4YQ UK

ABSTRACT

In an attempt to quantify the role played by biogenic emissions of dimethyl sulphide to the UK atmospheric sulphur budget, measurements of DMS and its oxidation products in the atmosphere were made during all seasons of 1988/90 at coastal sites in northwest Scotland and Eire. The annual average air concentration of DMS was estimated to be 58 ng (S) m^{-3} and that of MSA to be 6 ng (S) m^{-3}. Using a steady state approximation, the sea-to-air flux of DMS from the Atlantic north of 45° was estimated to be 0.93 mol m^{-2} d^{-1} or 96 Gg (S) y^{-1}. Three methods were then used to correct the observed air concentrations of the various sulphur species for anthropogenic emissions. The resultant flux of biogenic sulphur into the UK from the North Atlantic, net of any anthropogenic sulphur, was estimated to be 30-96 Gg (S) y^{-1}, or 2-5% of the total anthropogenic emission rate of sulphur from the UK. A simple ten step model was used to describe the equilibria existing between the various sulphur species and CO_2 in cloudwater and the resultant acidity calculated. Values of pH as low as 3.4 were predicted. Biogenic sulphur species from the North Atlantic may therefore make a small but quantifiable contribution to the atmospheric sulphur budget of the UK and may play an important role in determining rainwater acidity in the west of the country.

(*) Only the abstract is available.

G. Restelli and G. Angeletti (eds.), Dimethylsulphide: Oceans, Atmosphere, and Climate, 273.

GAS - TO - PARTICLE CONVERSION AND CCN PRODUCTION

R.J. CHARLSON

Department of Atmospheric Sciences, Department of Chemistry
and Institute for Environmental Studies
University of Washington, Seattle, USA

SUMMARY

There is no longer any doubt that almost all cloud condensation nuclei (CCN) are produced by chemical reactions and gas-to-particle conversion within the atmosphere. Oxidation reactions involving both natural and anthropogenic reduced sulfur gases are of particular importance because the resulting sulfates often are a dominant chemical component and because these sulfate particles have nearly optimal size, solubility and hygroscopic properties to allow them to act as CCN. However, in spite of the clear identification of the composition, chemical and physical pathway by which CCN are created, knowledge is meager regarding the rate of CCN production and resulting concentrations as functions of the gas-phase chemical inputs. Both empirical evidence and fundamental theory suggest that the rate of new CCN production must have a highly non-linear dependence on the concentrations of both the input gaseous reactant and the pre-existing aerosol. Numerous processes are involved, making it difficult to develop *a priori* models or to make enough simultaneous measurements to allow a sufficient description of the entire system. Among the consequences of the inherent non-linearity is the paradoxical possibility that at low levels, variations of dimethyl sulfide concentration can cause variations in CCN concentration while at much higher concentration in polluted air, SO_2 may have little effect on CCN number concentration. Because the sulfur gas budget of the Northern Hemisphere is now totally dominated by anthropogenic SO_2, and because CCN can influence cloud and global albedo, a more quantitative understanding of the dependence of CCN production rate on input reactant gases is needed.

1. INTRODUCTION

Twomey [1-3] proposed the notion that aerosol particles, acting as cloud condensation nuclei (CCN), are an important factor governing the albedo of clouds. Here, CCN are the subset of the total aerosol particle population (often called

G. Restelli and G. Angeletti (eds.), Dimethylsulphide: Oceans, Atmosphere, and Climate, 275–286.

condensation nuclei, or CN) that become droplets at the low supersaturations that occur within real clouds. (In actuality, CN is an operational term -- the number of particles detected with a CN counter.) The sensitivity of cloud albedo to CCN number concentration was calculated to be sufficiently high to be of importance to the global heat balance and, therefore, climate. A change of CCN population of ±30% only over the oceans and only influencing marine stratiform clouds is calculated [4] to yield a change of heat balance, $\Delta F \simeq \pm 1$ watt/meter squared (Wm^{-2}). Compared to the forcing from the increase in CO_2, which yields $\Delta F \simeq +1.2$ Wm^{-2}, this sensitivity cannot be ignored. Because of its long lifetime in air, CO_2 changes only slowly; however, CCN populations are known to vary greatly, often a factor of 10 diurnally. Thus a mere 30% change amounts to a tiny shift of the average of a rapidly varying quantity, and understanding such noisy variations is difficult indeed.

While there is as yet no complete or controlled experiment with real clouds that demonstrates the cloud brightening theory of Twomey, increased cloud reflectances at visible wavelengths have been observed to be associated with ships underneath stratus clouds acting as aerosol sources [5,6]. The observed enhancement of cloud albedo is in quantitative agreement with the Twomey theory; however, simultaneous changes in liquid water content in one case study, possibly caused by cessation of removal of water by drizzle, confounded the interpretation of results [6].

Numerous treatments of the CCN/climate connection have addressed this point of the sensitivity of cloud albedo and climate to the CCN number concentration [7,8,9]. However, very few papers exist which attempt to consider the factors that actually control the CCN *population*. Still fewer papers have been done on factors controlling CCN *properties*, e.g. the so-called supersaturation spectrum or number of particles activated as a function of supersaturation.

The earliest view, probably based on observations that cloud droplet populations were always much smaller then the total aerosol particle (or Aitken particle) population, was that atmospheric dynamical processes governed the number of droplets that activate. That is, the rate of ascent of an air parcel (upward vertical velocity) initially governs the rate of increase of supersaturation and the rate of activation or nucleation of CCN. Then, shortly after cloud droplets form near cloud base, these activated CCN consume water vapour and return the supersaturation to near zero. Thus the conclusion was made that "only a very small fraction of the (Aitken) nuclei can be involved in their (cloud droplet) formation"[10]. Nonetheless, substantial interest was placed on what the CCN were composed of and where they came from. Some attempts were made to ascertain source strengths of CCN, e.g. over the ocean[11].

Somewhat more recently, the notion of the relative source strengths of different types of CCN was emphasized. Included as one possibility was that industrial activity produced CCN and led to the elevated concentrations observed over the

Eastern U.S. and Europe [12-14]. Indeed, these treatments suggested that there was an enormous potential for production of CCN by industrial activity. While no global budgets of CCN number (or mass) production were given, it was implied that industrial aerosol production was likely to be larger than natural production. The other implication of these early approaches was that the CCN number population was an initial condition imposed upon the formation of a cloud. The CCN population was one of the independent variables leading up to the activation and growth of droplets. Little consideration seemed to be given to the notion that cloud processes also were the main modifiers and removers of CCN.

Still more recently two new ideas have emerged: (i) that most CCN are produced in the atmosphere from gaseous precursors and (ii) that the CCN number concentration is itself a dependent variable operating within a set of interdependent variables [15].

Against this background of lore relating to the population of CCN and its potential effects on climate, a major connection to the atmospheric chemistry of sulfur compounds has emerged. Two key gaseous species are important as precursors of sulfate aerosol particles: natural dimethyl sulfide (DMS) from oceanic phytoplankton and anthropogenic sulfur dioxide from the combustion of fossil sulfur in coal and oil. Figure 1 summarizes the chemical reaction pathways that are involved [16]; however, it does not reflect at all any of the factors governing the physically important quantity of CCN *number* concentration.

Recognizing this chemical link between DMS emissions and CCN, Charlson et al.[4] proposed that CCN in unpolluted marine air should be mainly sulfate compounds. They further hypothesized that a feedback system exists whereby changes in sea-surface temperature and/or solar irradiance control the DMS emission. While considerable progress has been made via qualitative demonstration that DMS indeed is a source of CCN [17,18] the feedback hypothesis is still a largely open problem (not the least because of the necessity of understanding microbial ecology).

One substantial test of the CLAW (Charlson, Lovelock, Andreae and Warren) hypotheses [4] was posed by Schwartz [8], suggesting that (a) the aerosol of the Northern Hemisphere (NH) was significantly enhanced by anthropogenic sulfate, (b) that there was no evidence for any "excess midlatitude cloud component of albedo in the NH relative to the SH" or "any reduced warming in the NH relative to the SH", and (c) "that these variables are not controlled by anthropogenic sulfates, and by extension, are not controlled by marine DMS emissions". While this argument was extremely cogent, it fails in face of the evidence above that DMS *does* influence the concentration of CCN, at least at Cape Grim, Tasmania, as in Fig.2 [18]. Thus, Schwartz's test of CLAW generates a paradox. How is it possible that DMS influences CCN at low levels but anthropogenic SO_2 does not (or only has a small effect) at higher overall concentrations of sulfur species? The dramatic enhancement of reactive sulfur gas input to the NH and the globe makes this paradox even more amazing (Fig.3).

2. SOURCE STRENGTHS OF CLOUD CONDENSATION NUCLEI

Because atmospheric CCN have a short residence time in the air, it is convenient to think of their concentration, N(cm^{-3}), as a function of both source and sink processes. The rate of change of CCN is:

$$dN/dt = Q - S$$

where Q and S represent the source and sink strengths respectively ($cm^{-3}\ sec^{-1}$). In the simplest case of a steady state, dN/dt = 0 and $Q \simeq S$. While there is no way to measure Q or S directly, they can be estimated from a variety of different approaches.

If we suppose that the natural steady-state CCN population in marine air (e.g. at a supersaturation of 1%) is 200 cm^{-3} and that they have a lifetime as individual particles of ca. one day or 10^5 sec, then $Q_{DMS} \simeq 2x10^{-3}\ cm^{-3}sec^{-1}$. (Here, 10^5 sec is used for the lifetime of an individual CCN before it consolidates with another particle or is removed from the atmosphere, rather than the lifetime of the mass of aerosol substance which is longer, perhaps $5x10^5$ sec. The reason for this difference is that there are more processes to reduce number concentration than to reduce mass concentration. This number life-time is highly uncertain -- perhaps to a half an order of magnitude).

For comparison the actual flux of anthropogenic CCN from industrial regions has been estimated from aircraft observations of N ($10^3 \simeq cm^{-3}$ at 1% supersaturation [14]). If there is a flow from the entire Eastern U.S. of 10 m sec^{-1}, a planetary boundary layer (PBL) depth of 10^3 m and a coastline length of 10^6 m, a volume of ca. $10^{10}\ m^3\ sec^{-1}$ or $10^{16}\ cm^3\ sec^{-1}$ emanates from this industrial region. We assume that most anthropogenic CCN are in the PBL. Thus if $N \simeq 10^3\ cm^{-3}$ as measured, this region produces ca. 10^{19} CCN sec^{-1}. Given other regions in the NH, e.g. Asia, $2\text{-}8x10^{19}\ sec^{-1}$ are produced globally. Now the volume of the PBL of the NH is ca. $2.5x10^{23}\ cm^3$ so the volume average for $Q_{anthropogenic}$ is roughly $10^{-4}\ cm^{-3}sec^{-1}$, which is smaller than the estimate for Q_{DMS}. Of course, this quantity, too, is uncertain; however, it seems clearly possible that $Q_{anthropogenic}$ is not necessarily much larger than Q_{DMS}, and in fact may be much smaller when globally averaged.

However, the potential magnitude of $Q_{anthropogenic}$ is very large; indeed even the amount that could be made from fossil-fuel SO_2 is large. If the anthropogenic SO_2-sulfur source strength is 70 Tg yr^{-1} ($70x10^{12}$ g yr^{-1}), with half being converted to $SO_4^{=}$ aerosol particles of number mean radius 0.02 μm (critical supersaturation, $S \simeq 1\%$), $Q_{anthropogenic}$ would be ca. $10^{-1}cm^{-3}sec^{-1}$ averaged over a 1 km deep PBL for the NH.

To summarize these sketchy estimates;

$$Q_{DMS} \simeq 2x10^{-3}cm^{-3}sec^{-1}$$

$$Q_{anthropogenic} \simeq 10^{-4}cm^{-3}sec^{-1}$$

$$\text{Potential } Q_{anthropogenic} \simeq 10^{-1}cm^{-3}sec^{-1}$$

illustrating that it is possible that very little enhancement of CCN may occur due to anthropogenic aerosol, or at least, due to an anthropogenic sulfate. This, taken along with the results of Ayers and Gras [17] and Figs. 2 and 3 suggests that Q_{CCN} is a very non-linear function of the source strength of aerosol-genic gases, e.g. DMS and SO_2. This leads to the question: what controls this non-linearity?

3. FACTORS INFLUENCING THE PRODUCTION AND REMOVAL OF CCN

Numerous factors contribute to the above result that even large increases of the source strength of SO_2 might not make a substantial increase in CCN. As yet, these factors have not been thoroughly described in a first-principle mathematical model. While the following qualitative discussion obviously lacks the formality of mathematical description, it is a necessary precedent to the formulation of any more quantitative approach. Five main factors can be recognized:

(i) The portion of SO_2 that makes new particles is a very small fraction of the anthropogenic emission.
(ii) New particle production decreases as the surface area of preexisting aerosol increases.
(iii) Cloud droplets and preexisting aerosol are both sinks for newly formed molecular clusters or the smallest aerosol particles.
(iv) As the total number of particles increases (e.g. as the total aerosol mass increases) the fraction that are active CCN decreases.
(v) The lifetime of any individual CCN is short compared to the lifetime of mass concentration in the air [19].

At this juncture, it is not possible to estimate which of these factors are dominant and it also is not possible to indicate the expected mathematical form of any model algorithm. However, some features of this system are clear. At very low levels of aerosol surface area concentration (a few square micrometers per cm^3) much of the oxidation product of oxidation of DMS or SO_2 can and does make new particles. Indeed, if the initial particle content were zero, almost all the product would make new particles (that is, all except that lost to the earth's surface).

This is not a simple system to depict, e.g. as chemists do with chemical reactions. In large part this is due to the simultaneity of chemical reactions and physical processes that have similar characteristic times. The time for oxidation of SO_2 to SO_3 is perhaps of order days, while the time for removal of aerosol by precipitation also is days. Another complication arises from (ii) above in that surface area goes up as pollution increases; however, the amount of SO_2 also increases so the concentration of SO_3 or $H_2SO_4 \cdot xH_2O$ gas also increases. Which goes up more rapidly[20]?

Item (v) is not really separated from (i)-(iv) in that it simply summarizes that there are many more and more rapid processes that destroy the integrity and uniqueness of single particles than there are processes that remove the mass of those particles from the atmosphere. The latter is essentially dominated by removal by precipitation which, of course, also removes particles by number as well. However, in addition, Brownian coagulation, in-cloud scavenging of unactivated particles and coalescence all join particles together and reduce N without reducing mass concentration.

4. A SYSTEM FRAMEWORK FOR THE MULTIPHASE ATMOSPHERIC PROCESSES OF SULFUR

Figure 4 introduces a flow scheme depicting the existence of competing physical pathways for anthropogenic and natural SO_2 (the latter e.g. from oxidation of DMS or from volcanoes). Approximate percentages of initial partitioning of SO_2 removal are given per the estimates of Langner et al.[21]. What is immediately evident is that if the boxes labelled CN and CCN have a low surface area concentration, new particles have a chance to form, first as CN and then as CCN. On the other hand, if CCN are put into clouds repeatedly and frequently, the loop (CCN --> cloud droplets --> fewer but larger cloud droplets --> CCN) could be effective in clamping the magnitude of the CCN population.
Clearly, the percentages indicated in Fig. 4 indicate that the response of the CCN population to anthropogenic SO_2 would be small even if the system were linear. The additional non-linearity implicated through the multiphase processing between aerosol CCN and clouds must result in a weak response indeed.

Quantifying and understanding this complex picture will be difficult. To make matters still worse, means must be found for parametric representation of this microscale processes on the spatial scale inherent in global models (e.g. a few degrees of latitude and longitude and small altitude (pressure range). Item (v) above offers hope of understanding the regional to global scale effects of the complexities of Fig. 4. For example, if the lifetime of a $SO_4^{=}$-CCN as an individual particle is ca. one fourth of the lifetime of the mass concentration of $SO_4^{=}$, the geographical area under influence of the enhancement of CCN would be ca. one sixteenth (i.e. $(1/4)^2$) of the area of enhanced mass concentration (assuming radial diffusion as a first crude approximation). Since we now have studies of H_2SO_4 deposition showing that the mass has a typical horizontal scale of 1000 km enhanced CCN would be expected only very near the SO_2 sources, perhaps $1000/4 \simeq 250$ km.

5. CONCLUSION

The consequences of this expected, very strong non-linearity could be that the CCN population returns essentially to normal within a day of transport from the immediate source region. Thus, the impact of enhanced cloud albedo from anthropogenic-SO_2-derived CCN would be expected to be small and perhaps inconsequential. It certainly is not expected to be a linear function of Q_{SO2}. If it were, CCN would have increased many fold over the NH, with radiative forcing of $10+Wm^{-2}$, which clearly has not happened.

Another consequence of this putative non linearity is that the direct forcing due to reflection of sunlight by anthropogenic sulfate aerosol itself may be much more important than previously thought [22]. Indeed that forcing appears to be a linear function of Q_{SO2}. Thus, while direct forcing by aerosol from Q_{DMS} is demonstrably small, e.g. based on measurements of aerosol optical depth of Cape Grim, Tasmania, direct forcing by aerosol from Q_{SO2} might very well be a globally important quantity [23,24,9].

Returning to Schwartz's paradox, it seems that one possible explanation is clear; e.g. the system is very non-linear. Testing this hypothesis would be a useful step in understanding the links between aerosol and climate.

REFERENCES

(1) TWOMEY, S. (1971), Inadvertent Climate Modification, C.L.Wilson and W.H.Mottben (Directors), MIT Press, Cambridge, MA, Figure 8.9.
(2) TWOMEY, S. (1976), "The Role of Aerosols in Influencing the Radiative Properties of Clouds", in Proceedings Symposium on Radiation in the Atmosphere, Garmisch-Partenkirchen, FRG. Science Press, Princeton, USA, 171-176.
(3) TWOMEY, S. (1977), Atmospheric Aerosols, Elsevier Scientific Publishing Co., Amsterdam, Figure 12.6.
(4) CHARLSON, R.J., LOVELOCK, J.E., ANDREAE, M.O. and WARREN, S.G. (1987), "Oceanic Phytoplankton, Atmospheric Sulphur, Cloud Albedo and Climate", Nature, **326**, 655-661.
(5) COAKLEY, J.A. Jr., BERNSTEIN, R.L. and DURKEE, P.A. (1987), "Effect of Ship-Stack Effluents on Cloud Reflectivity", Science, **237**, 1020-1022.
(6) RADKE, L.F., COAKLEY, J.A. Jr. and KING, M.D. (1989), "Direct and Remote Sensing Observations on the Effects of Ships on Clouds", Science, **246**, 1146-1149
(7) COAKLEY, J.A. Jr., BERNSTEIN, R.L. and DURKEE, P.A. (1988), "Effect of Ship-Stack Effluents on the Radiative properties of Marine Stratocumulus: Implications for Man's Impact on Climate", in "Aerosols and Climate", P.V. Hobbs and M.P. McCormick, Eds., A.Deepak Publishing, 253-260.
(8) SCHWARTZ, S.E. (1988), "Are Global Cloud Albedo and Climate Controlled by Marine Phytoplankton?", Nature, **336**, 441-445.
(9) CHARLSON, R.J., SCHWARTZ, S.E., HALES, J.M., CESS, R.D.,

COAKLEY, J.A. Jr., HANSEN, J.E. and HOFMANN, D.J. (1992), "Climate Forcing by Anthropogenic Aerosols", Science, **255**, 423-430.

(10) FLETCHER, N.H. (1966), The Physics of Rainclouds, Cambridge University Press, p. 98.

(11) MASON, B.J. (1957), The Physics of Clouds, Clarendon Press, Oxford, p. 73.

(12) SQUIRES, P. (1966), "An Estimate of the Anthropogenic Production of Cloud Nuclei", J. Recherche Atmospherique, II, No.2-3, 297-308.

(13) HOBBS, P.V., RADKE, L.F. and SHUMWAY, S.E. (1970), "Cloud Condensation Nuclei from Industrial Sources and Their Apparent Influences on Precipitation in Washington State", J. Atmos. Sci., **27**, 81.

(14) RADKE, L.F. and HOBBS, P.V. (1976), "Cloud Condensation Nuclei on the Atlantic Seaboard of the United States", Science, **193**, 999-1002.

(15) BAKER, M.B. and CHARLSON, R.J. (1990), "Bistability of CCN Concentration and Thermodynamics of the Cloud-Topped Boundary Layer", Nature, **345**, 142-145.

(16) TYNDALL, G.S. and RAVISHANKARA, A.R. (1991), "Atmospheric Oxidation of Reduced Sulfur Species", Int. J. Chem. Kinetics, **23**, 483-527.

(17) AYERS, G.P. and GRAS, J.L. (1991), "Mean Seasonal Relationship between CCN and Aerosol MSA in Marine Air at Cape Grim", Nature, **353**, 834-835.

(18) CHARLSON, R.J. (1993), in Proceedings CHEMRAWN VII Dec. 2-6, 1991; Baltimore, MD.

(19) ACKERMAN, A. (1992) Personal communication.

(20) SCHWARTZ, S.E. (1992), Personal communication.

(21) LANGNER, J.H., CRUTZEN, P.J. and ZIMMERMANN, P. (1992), "Anthropogenic Influence on the Distribution of Sulphate Aerosol", Nature, **359**, 712-716.

(22) BOLIN, B. and CHARLSON, R.J. (1976), "On the Role of the Tropospheric Sulfur Cycle in the Shortwave Radiative Climate of the earth", Ambio, **5**, 47-54.

(23) CHARLSON, R.J., LANGNER, J. and RODHE, H. (1990), "Sulphate Aerosol and Climate", Nature, **348**, 22.

(24) CHARLSON, R.J., LANGNER, J., RODHE, H., LEOVY, C.B. and WARREN S.G. (1991), "Perturbation of the Northern Hemisphere Radiative Balance by Backscattering from Anthropogenic Sulfate Aerosols", Tellus, **43AB**, 152-163.

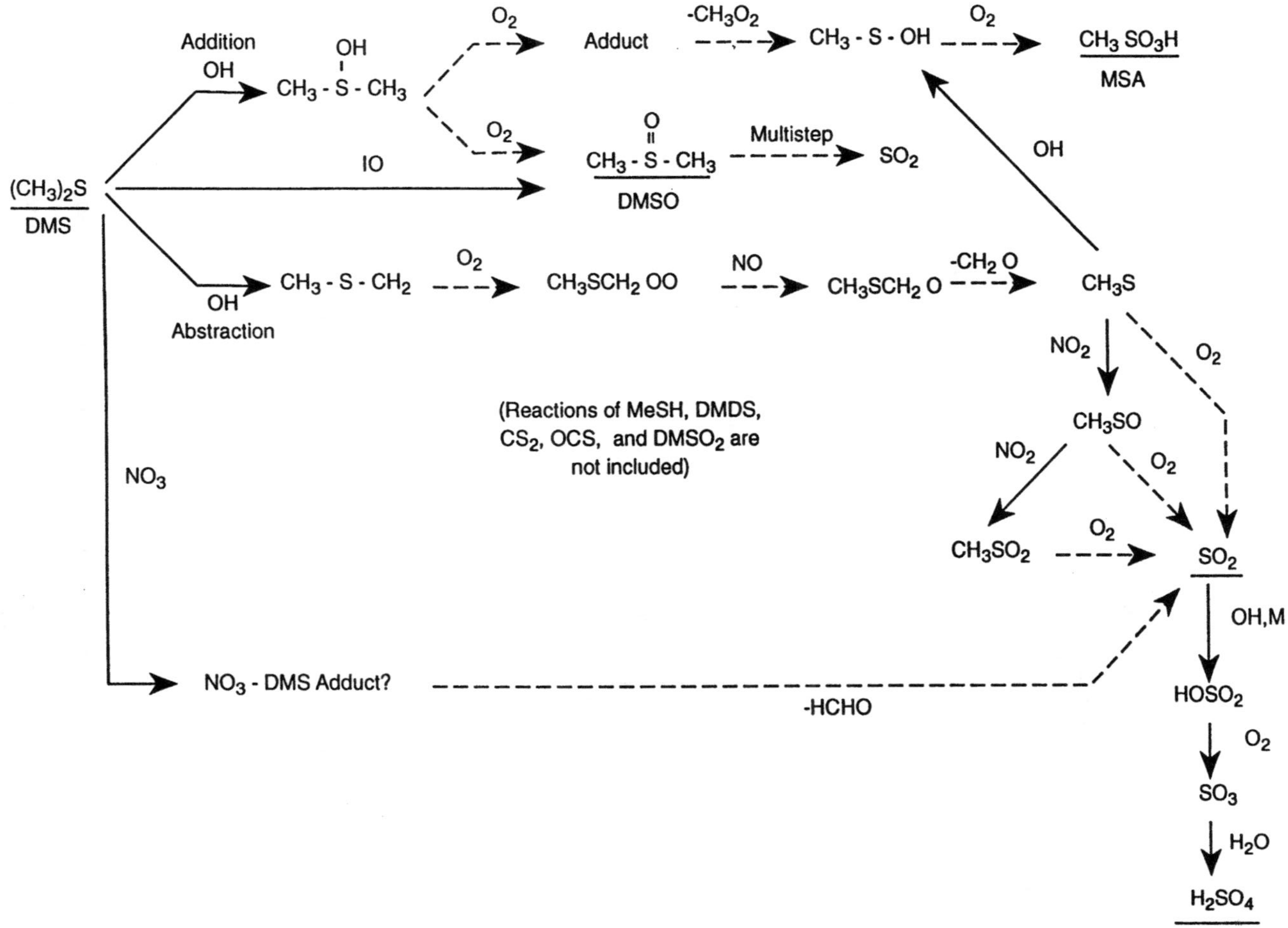

Figure 1. Atmospheric chemical reaction pathways for sulfur (from Ref.[16]).

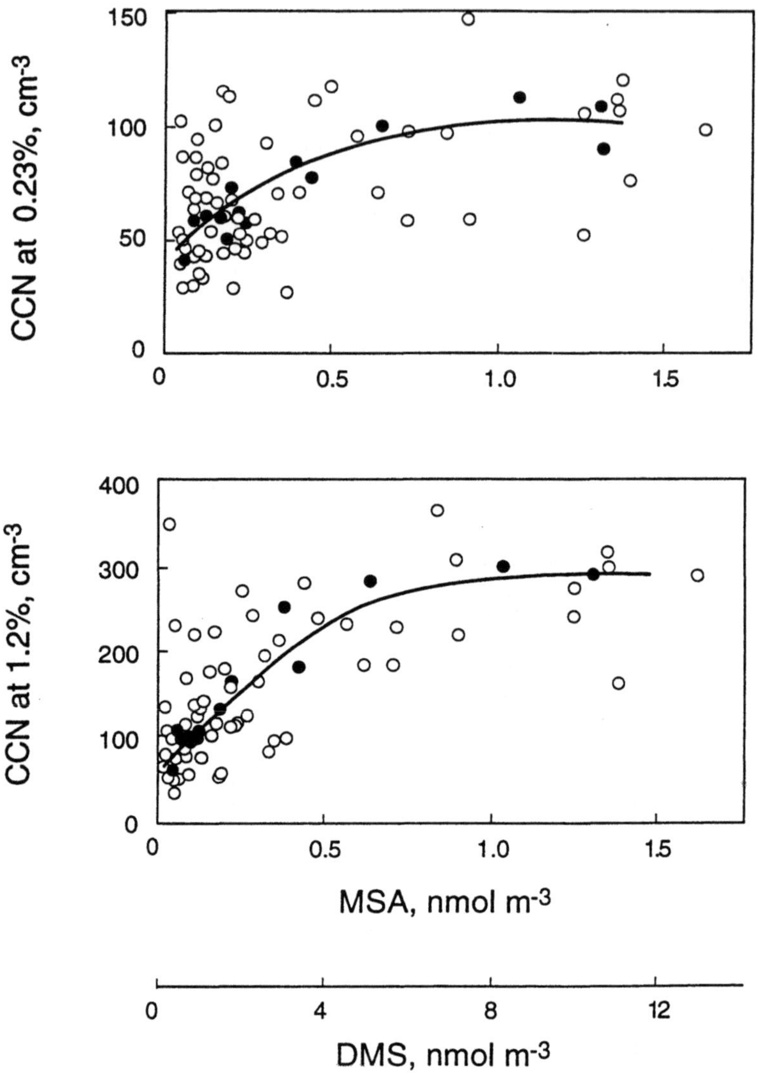

Figure 2. Measured cloud condensation nucleus concentration (cm^{-3}) at 0.23 and 1.2% supersaturation versus near simultaneously measured aersosol methane sulphonate (MSA). Surrogate axis for dimethylsulphide (DMS) is from other simultaneous observations of DMS and MSA (adapted from Ref.[17]).

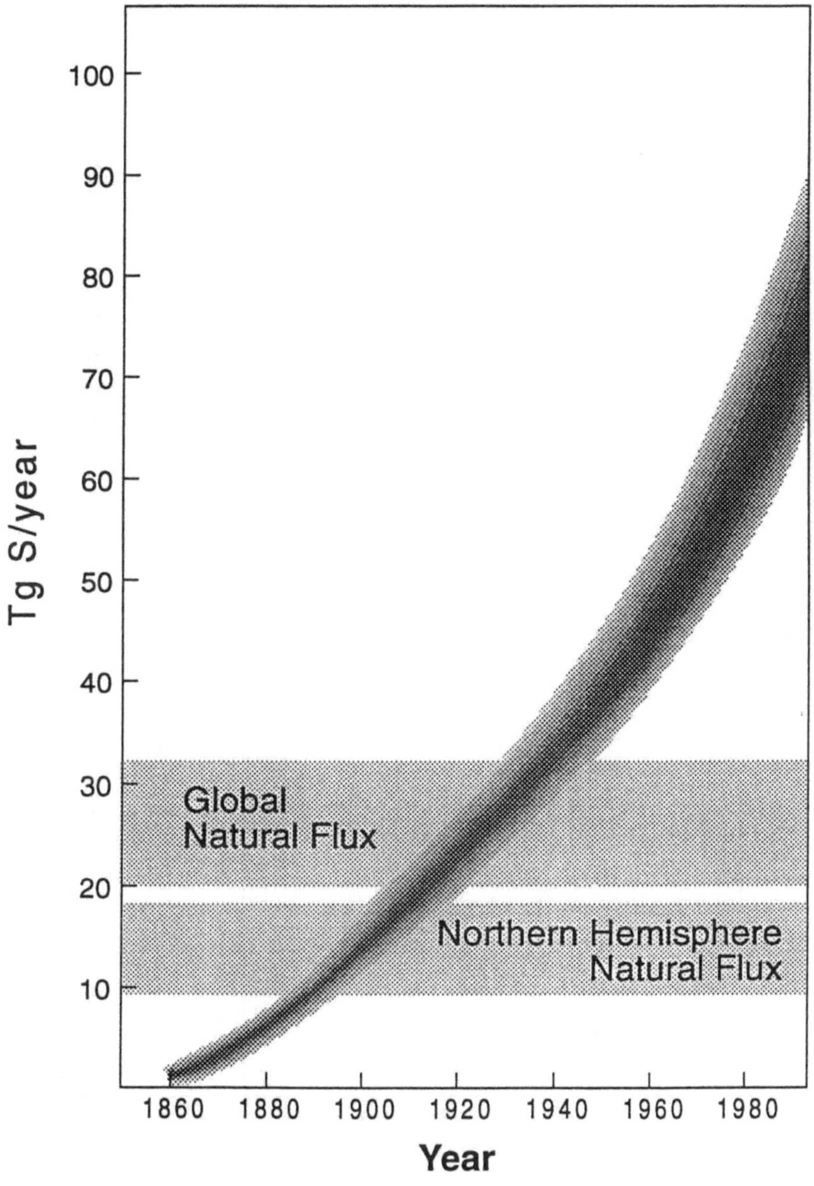

Figure 3. Growth of the flux of anthropogenic SO_2-sulfur to the atmosphere since 1850 A.D. (after Ref.[9]). The range of estimates of the global and Northern hemisphere (NH) natural fluxes are given as the crosshatched range of values. Because anthropogenic sulfur is not efficiently mixed into the Southern Hemisphere, and because 90% of the anthropogenic emission is in the NH the relevant comparison is between the anthropogenic and NH natural fluxes.

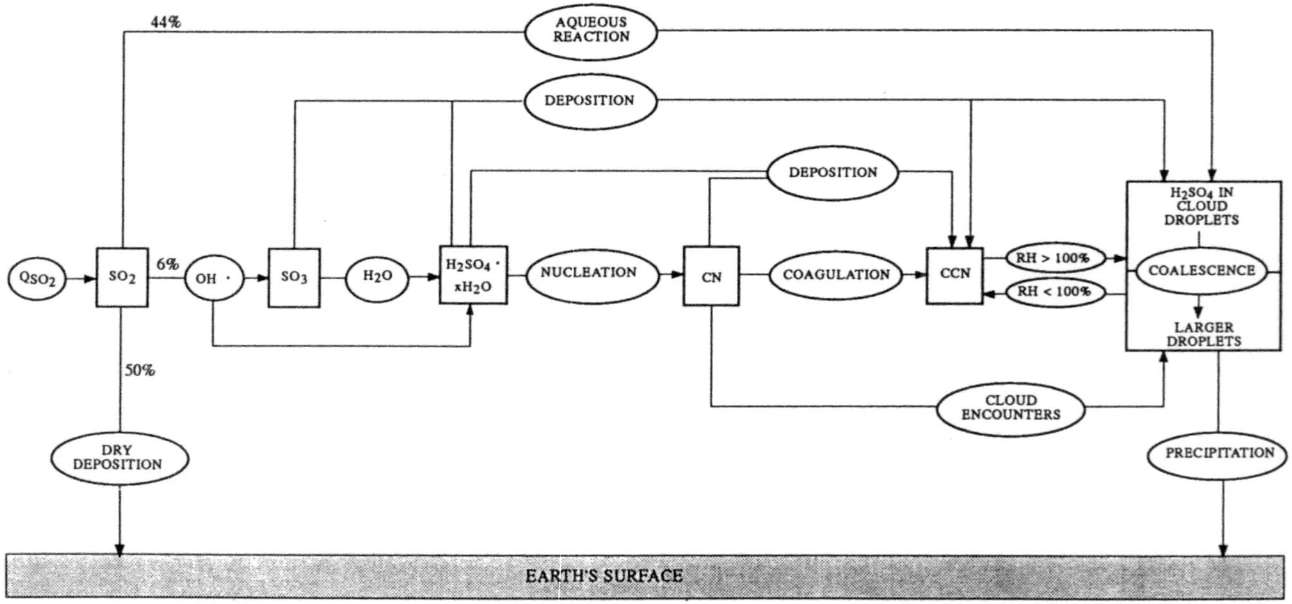

Figure 4. Flow diagram for production and control of CCN population. Approximate percentage of SO_2 pathways per Langner et al.[21].

PACIFIC MARINE AEROSOL: EQUATORIAL GRADIENTS IN SULFATE, AMMONIUM AND CHLOROPHYLL

A. D. CLARKE and J. N. PORTER

School of Ocean and Earth Science and Technology, University of Hawaii, Honolulu, HI 96822

SUMMARY

Measurements of aerosol physicochemistry were made during five transects across the equator between 15°N and 10°S. Marked gradients in both the aerosol $NH_4^+/SO_4^=$ ratio and strong surface water chlorophyll were associated with boundaries separating oligotrophic waters and regions of equatorial upwelling. Highest aerosol ammonium concentrations corresponded to regions of highest chlorophyll concentrations. Favorable aerosol chemistry, wind directions and cruise tracks in conjunction with rapid aerosol sampling made it possible to estimate the flux of ammonia from the surface in these transition regions at about 10 μmol $m^{-2}day^{-1}$ with possibly greater values in regions where higher chlorophyll concentrations exceeded about 0.25 mg m^{-3}. Aerosol sulfate concentrations were also generally elevated over the upwelling region but showed a less pronounced relationship to chlorophyll abundance, suggesting independent source mechanisms for ammonium and sulfate. Rapid variability in aerosol mass and number was evident in convective regions. Both depletion of larger aerosol (mass) through precipitation scavenging and an increase in the number of smaller aerosol in these regions indicate the complex relationship between CN, CCN and aerosol mass in the remote marine boundary layer.

1. INTRODUCTION

During February and March of 1990, we participated in our third Soviet American Gas and Aerosol Experiment (SAGA-3) in order to study aerosol processes over remote marine regions and to investigate previously encountered interhemispheric gradients of aerosol properties in the equatorial Pacific (1,2). The five sequential equatorial transects are shown in Figure 1. Each box indicates a location where the aerosol size distribution was measured. Indicated winds are mostly northeasterly north of 5°N and easterly between 5°N and 5°S.

A laser optical particle counter (OPC) was used to obtain the aerosol size distribution averaged over about 3 hour periods. The aerosol was heated to various temperatures in order to drive off volatile components related to sulfuric acid (150C), ammonium sulfate/bisulfate (300C) and leaving refractory sea-salt, dust or soot (3). When combined with laboratory calibrations this volatility was used to estimate of the sulfate mass, the $NH_4^+/SO_4^=$ ratio, and the sulfuric acid fraction (3).

G. Restelli and G. Angeletti (eds.), Dimethylsulphide: Oceans, Atmosphere, and Climate, 287–296.

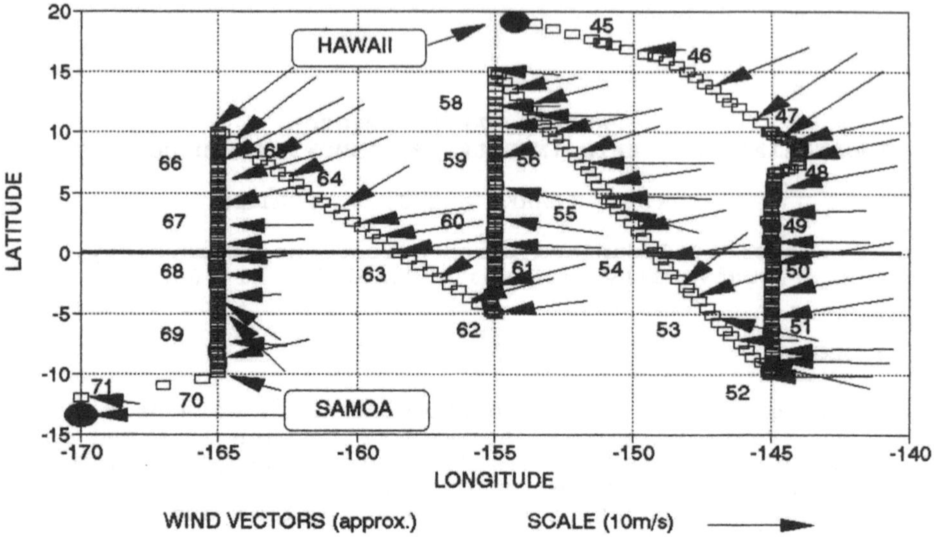

Figure 1. The cruise track for SAGA3 showing the approximate mean winds and the locations where size distributions were measured (boxes).

2. RESULTS

Figure 2 shows the resulting total non sea-salt sulfate mass (solid line) and sulfuric acid (dotted line) for each of the five transects. The triangles are independent corroborative concentrations determined using a differential mobility analyzer and will not be discussed further. The NH_4^+/SO_4^- molar ratio (open rectangles) derived from aerosol volatility provides a more revealing assessment of the aerosol chemistry in Figure 3. On each leg the ratio undergoes a dramatic transition from values near zero (i.e. sulfuric acid) to values near 1 (i.e., ammonium bisulfate) somewhere between 5° and 8°N, the region where chlorophyll (solid rectangle) undergoes a marked transition.

Figure 4 shows the result of plotting aerosol ammonium and surface chlorophyll (less frequently measured) values for all legs as a function of latitude. In spite of differences in advective scales, relative motion between air and water, atmospheric lifetimes (eg. removal of aerosol by precipitation etc.), a relationship between the two is evident. Although no chlorophyll values were measured below 0.05 mg/m³ the lowest values were found between 10 and 15°N, where a sulfuric acid aerosol was most common.

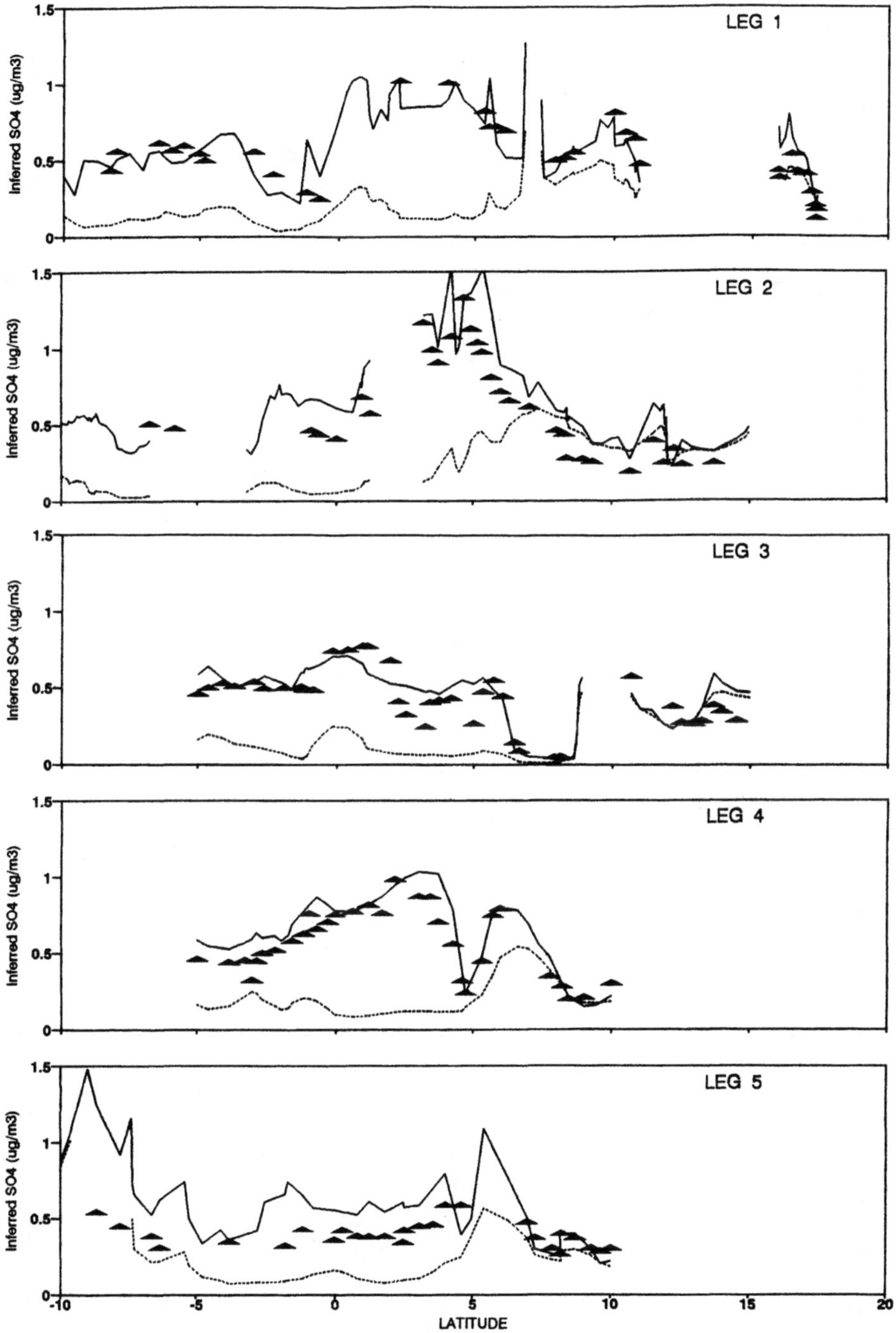

Figure 2. OPC estimated values for each leg of sulfate (solid line), sulfuric acid (dashed line). Total mass less than 0.6um seen by a differential mobility analyzer shown for comparison.

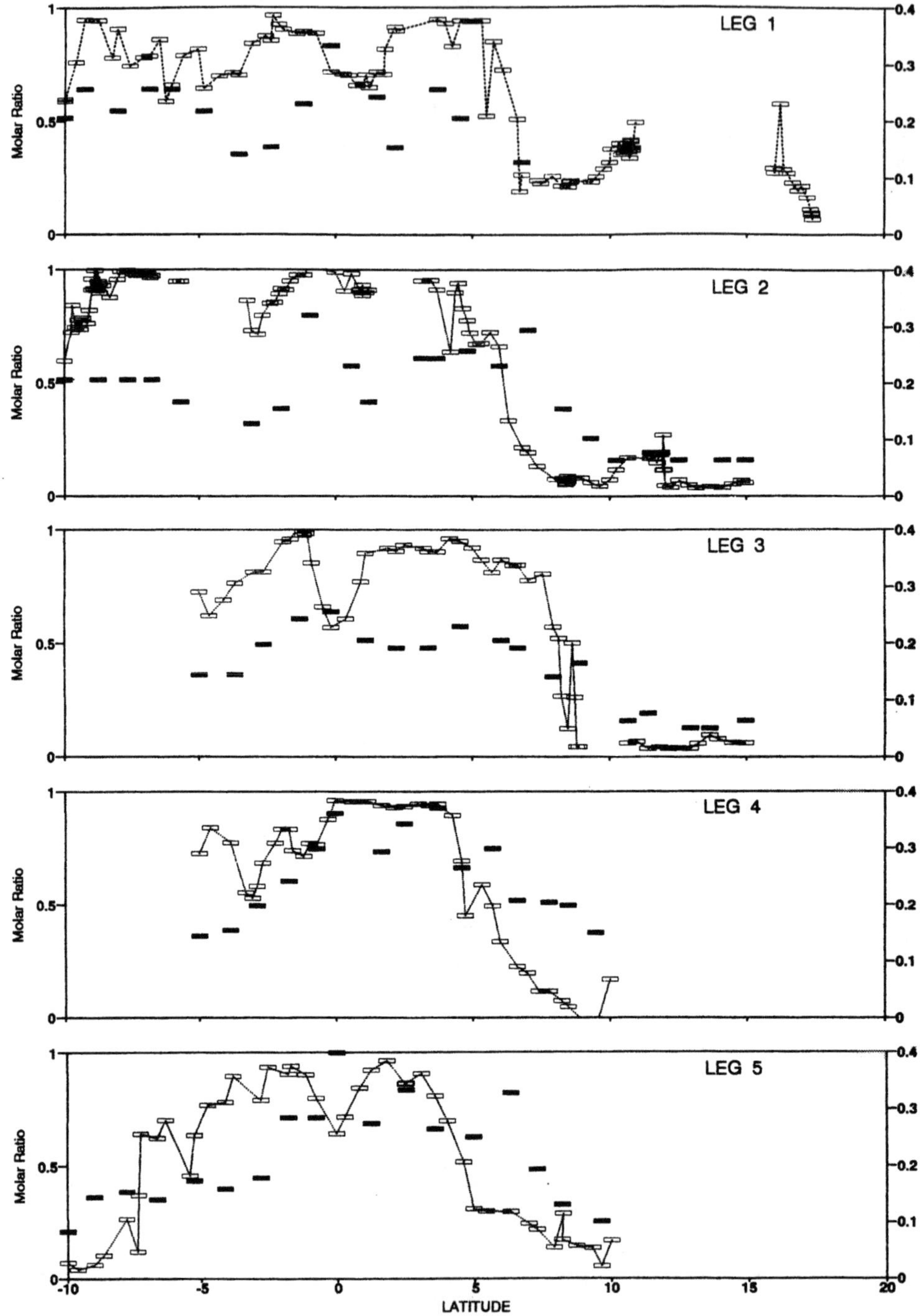

Figure 3. The NH_4^+/SO_4^- molar ratio (open boxes) and surface chlorophyll (solid boxes) for each leg of SAGA3.

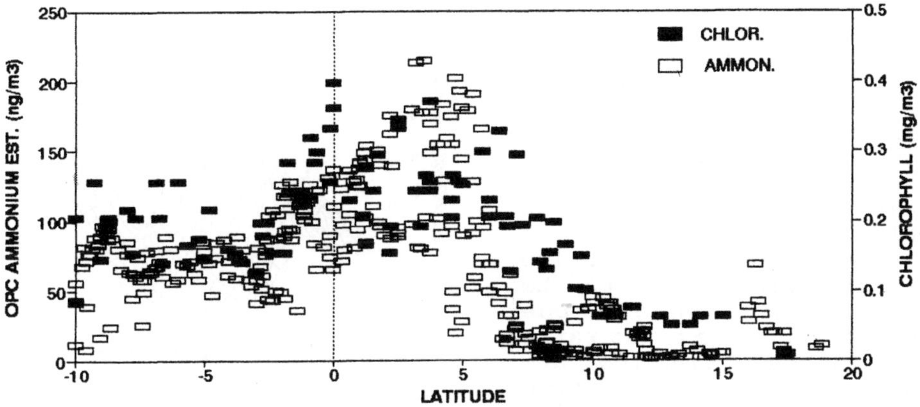

Figure 4. The latitudinal distribution for all measurements of aerosol ammonium and surface chlorophyll for all five legs.

Figure 5 is a plot limited to only concurrently measured values of aerosol ammonium against surface chlorophyll (open boxes). A direct correspondence between individual measurements is not expected for reasons mentioned above. However, if a dependency is present then it should show up in an average over all data for all legs. Clearly considerable scatter exists in the data. However, the scatter is markedly reduced if we average both ammonium and chlorophyll values for each chlorophyll concentration interval of 0.05 mg m^{-3} resulting in what appears as a near linear relationship (triangle). This suggests that higher surface chlorophyll concentrations are

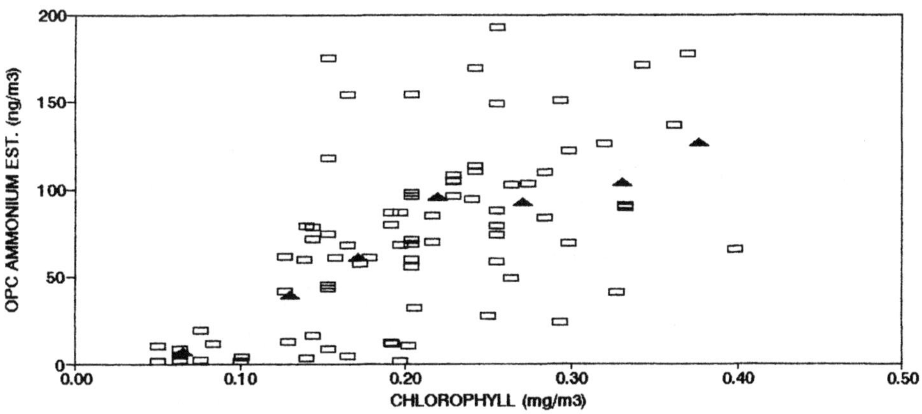

Figure 5. Aerosol estimated ammonium vs. surface chlorophyll for coincident measurements only (open box) and for values averaged over intervals of 0.05 mg m^{-3} in chlorophyll (triangle).

associated with higher aerosol ammonium. Here we emphasize that although a relationship to chlorophyll is suggested, the actual origin of the ammonium probably results from excretion by zooplankton that graze upon the chlorophyll.

A regional perspective of this relationship can be visualized when the chlorophyll, the molar ratio and aerosol ammonium are each projected as three dimensional surfaces in Fig. 6. Surfaces are fit to data collected for the locations shown in Fig. 1 and for values displayed in Figs. 2 and 3. All plots show a pronounced equatorial maximum that is broader to the east and more narrow to the west. Because the molar ratio is not directly affected by changes in the aerosol concentration, the variability in this ratio appears similar to that seen in the chlorophyll. Peaks in aerosol ammonium concentrations are also seen to be centered over the region of high chlorophyll. However, these values show a more peaked structure than either chlorophyll or the molar ratio. This is to be expected since atmospheric concentrations are rapidly modulated in both space and time by removal via precipitation.

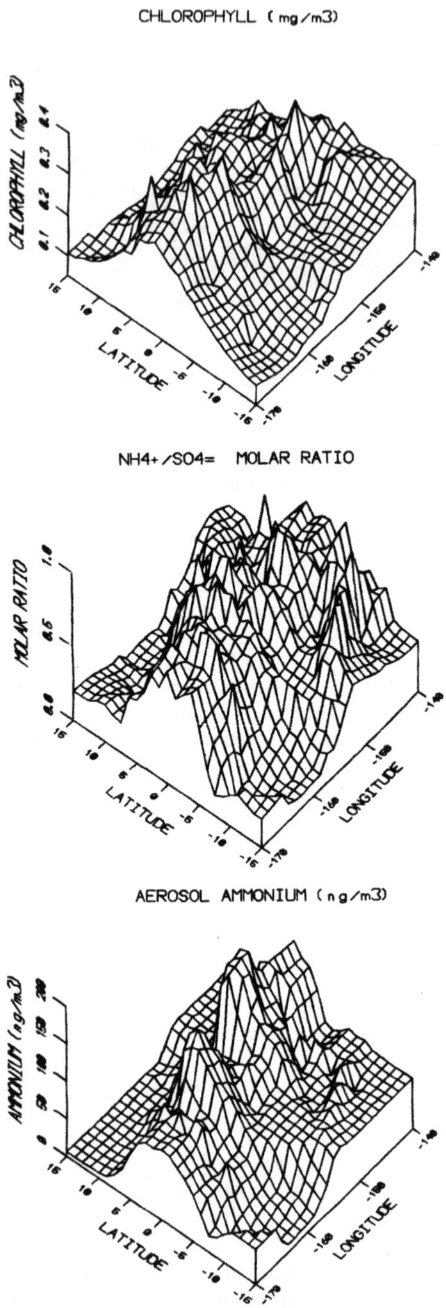

Figure 6. Surface plots of chlorophyll, $NH_4^+/SO_4^=$ ratio and aerosol ammonium (see text for details).

3. AN EQUATORIAL AMMONIA FLUX?

If we presume that the equatorial surface waters are a source of the ammonia that appears as ammonium in the aerosol, then the unique conditions afforded by SAGA3 provide an opportunity to estimate the ammonia gas flux from the surface. In order to estimate this we make the following observations:

1) A net northerly component of the wind is advecting aerosol across the latitude range exhibiting the chlorophyll and NH_4^+/SO_4^- gradient.

2) The aerosol north of this region behaves like sulfuric acid and appears to be transformed to mostly ammonium bisulfate during this advection (Fig. 4).

3) Total sulfate concentrations increase by only about 50% during passage through this transition region that varies over 1-5° of latitude (Fig. 2) for these legs.

Other related assumptions are:

1) We can obtain values that are representative of the mean conditions for the region by averaging relevant properties over the variability present in all legs (Fig 3).

2) The change in NH_4^+/SO_4^- from about 0.2 to 0.8 can be used to characterize the transition latitudes. The changes in the associated sulfate concentration, wind speed and latitude range are defined accordingly for each leg.

3) The aerosol is well mixed in the boundary layer and ammonia uptake on the aerosol occurs on time scales that are short compared to passage across the transition region.

4) The observed increase in sulfate concentration and NH_4^+/SO_4^- values for all legs crossing the transition region could be due to either a) gas to particle conversion in the region or b) mixing with the more easterly air with higher sulfate concentrations that were already partially neutralized. The former possibility would require sufficient ammonia emissions to neutralize both advected and locally derived sulfuric acid. On the other hand, the alternative possibility of mixing would require less ammonia emissions to explain the observed increase of NH_4^+/SO_4^- values because these would reflect mixing with aerosol already partially neutralized. Since both processes are likely to occur and since they tend have opposite implications for the required ammonia flux, we will assume that the flux can be estimated by allowing that only the advected sulfuric acid is fully neutralized to ammonium bisulfate.

TABLE 1. CHARACTERISTICS OF THE TRANSITION REGION

LEG	$SO_4^=$ @MR=0.2 [μg m^{-3}]	$SO_4^=$ @MR=0.8 [μg m^{-3}]	Wind North (m s^{-1})	Wind East (m s^{-1})	Gradient Region	Bdy. Layer Height(m)
1	0.4	0.6	6	11	2°	1,700
2	0.6	1.0	3	9	3°	1,600
3	0.5	0.5	1	8	1°	900
4	0.5	0.8	7	8	4°	?
5	0.3	0.5	7	7	5°	1,800
REGIONAL AVG.	0.46	0.68	4.4	8.6	3°	1,500

MR = $NH_4^+/SO_4^=$ molar ratio

The relevant characteristics for each leg required to estimate a flux are summarized in Table 1 along with the average "regional" values for all five legs. Here we will use the regional average characteristics to obtain a regional flux. The mean north-south advection is 4.4 m s^{-1} such that passage over the mean 3° transition zone is about 18 hours. The corresponding easterly component is 8.6 m s^{-1} and the resulting easterly advection during this time is about 6° of longitude. The presence of a similar transition region for all five legs spanning 20° of longitude suggests that this easterly component should not significantly impact the transition characteristics described here. At the "transition" latitudes the height of the marine boundary layer, MBL, was determined from radiosonde data to be about 1500 m and for the mixed layer about 500m. The mixing times are estimated to be 10-20 minutes for the mixed layer and about 3-12 hr for the MBL. These are both shorter than the average 18 hour advection time and will be expected to add only a few hours to the time response of atmospheric column MR to a surface source of ammonia.

Hence, we will assume that aerosol MR values and sulfate concentrations measured at the surface are representative of the entire MBL up to about 1500 m. The estimated lifetime of ammonia gas in this type of region of 1-6 hr (4) is also shorter than this advection time, such that the assumption of both complete mixing and scavenging of ammonia gas to the aerosol during this advection is reasonable. Using the mean boundary layer column burden of advected sulfuric acid of about 750 μg m^{-2} (approx. 0.5μg m^{-3} X 1500 m), this suggests that about 7.5 μmol m^{-2} $SO_4^=$ was approximately neutralized to NH_4HSO_4 in about 18 hrs. The surface ammonia flux required for this neutralization during advection over the transition region is therefore about 10 μmol m^{-2} day^{-1} with an estimated uncertainty of about a factor of two. This is consistent with the only other known point estimate for this region (14 μmol m^{-2} day^{-1} (4) for a similar chlorophyll concentration of about 0.23 mg m^{-3}.

It is important to emphasize that this estimated flux is based upon five transects averaged over the entire region. The gradient behavior evident in Fig. 3 and 4 suggests that this is a persistent regional feature since it is repeated spatially over 20° of longitude (2000km) and temporally over a three week period. A composite plot of "regional" aerosol sulfate and ammonium (nmol m^{-1}) along with surface chlorophyll is shown above where the data are combined for all five legs and have been averaged over 2 degrees of latitude. Although all three peak over the region of equatorial upwelling, the ammonium and chlorophyll show the most similar behavior. Sulfate in remote regions is expected to originate from oxidation of dimethylsulfide, also produced by surface plankton. If both ammonium and sulfate are dominated in this region by an oceanic source, this suggests that the processes resulting in their emissions may be different with sulfate showing less of a relationship to chlorophyll.

4. CONCLUSION

Our data for SAGA-3 indicate that the transition in the aerosol NH_4^+/SO_4^- molar ratio consistently occurred in regions associated with transitions to higher concentrations of surface sea-water chlorophyll. Consideration of the rate of neutralization of the aerosol during advection across these gradients by the northerly wind component yielded an estimate for a surface water ammonia flux of about 10 μmol $m^{-2}day^{-1}$ in the transition region. This flux is small (a few percent) compared to the daily cycling of ammonia/ammonium in the ocean mixed layer(5) but the potential influence on the global atmospheric ammonia cycle may be large. We estimate a global oceanic ammonia source to the atmosphere based on a factor of two uncertainty in above derived flux and a similar uncertainty in how it might scale with chlorophyll concentrations. Based upon global average chlorophyll of 0.5 mg m^{-3} (W. Esiasias, NASA Goddard), the resulting global oceanic ammonia flux in the range of 10-80x10^{12} g yr^{-1} with an intermediate value most probable. This is large and comparable to recent assessments of emissions from land of about 40x10^{12}g yr^{-1} (5,6).

5. ACKNOWLEDGEMENTS

We wish to extend our gratitude to the scientists and crew aboard the R/V Akademik Korolev for their interest and efforts that made this work possible. We also appreciate the data made available to us for chlorophyll (Dan Lee, PMEL), meteorology (J. Johnson, PMEL), and soot carbon (A.D.A. Hansen, McGee Scientific). This work was supported by NSF grant ATM-9001434. This is the School of Ocean and Earth Science and Technology contribution 3187.

REFERENCES

(1) CLARKE, A. D., Aerosol physical chemistry in remote marine regions, Jour. Aerosol Sci., 19, 1195, 1988.

(2) CLARKE, A. D., N. C. ALQUIST, and D. S. COVERT, The Pacific marine aerosol: Evidence for natural acid sulfates, J. Geophys. Res., 92, D4, 4179, 1987.

(3) CLARKE, A. D., Thermo-optic technique for in situ analysis of size resolved aerosol physicochemistry, Atmos. Env., 25A, 635-644, 1991.

(4) QUINN, P. K., T. S. BATES, J. E. JOHNSON, D. S. COVERT, and R. J. CHARLSON, Interactions between the sulfur and reduced nitrogen cycles over the central Pacific Ocean, J. Geophys. Res., 95, 16,405-16,416, 1990.

(5) WHEELER, P. A., D. L. KIRCHMAN, M. R. LANDRY and S. A. KOKKINAKIS, Diel periodicity in ammonium uptake in the oceanic subarctic Pacific: Implications for interactions in microbial food webs, Limnology Oceanography, 34, 1025-1033, 1989.

(6) DAWSON, G. A., Atmospheric ammonia from undisturbed land, Jour. Geophys. Res., 21, 3125-3133, 1977.

(7) MACKENZIE, F. T., Biogases and biotic feedbacks on global climate change, Workshop Report on Biotic Feedbacks in the Global Climate System, Woods Hole Research Center, Oct. 25-29, 1992.

FORMATION AND DISTRIBUTION OF CLOUD CONDENSATION NUCLEI IN THE MARINE ENVIRONMENT

B. SCHÄFER, H.-W. GEORGII and R. STAUBES

Institute for Meteorology and Geophysics, University of Frankfurt,
Feldbergstr. 47, 6000 Frankfurt/Main, Germany

Measurements of condensation nuclei (CN) over the remote Atlantic Ocean averaged to concentrations of 100 - 400 cm^{-3}. Over the tropical ocean the activation rate of CN as CCN at 1.5 % supersaturation was up to 64% (mean value). In addition bulk particle samples were taken and analyzed for sulfate and methanesulfonate (MSA). The concentration of non-sea-salt sulfate was in the range of 40 - 840 ng S(nss-$SO_4^{=}$)/m^3 and 1 - 28 ng S(MSA)/m^3. Regression analysis on CCN and nss-sulfate showed no significant linear correlation. A significant linear correlation exists between daily means of CN and sulfate ($R^2 = 0.78$, n = 17). For daily means of CCN at 1% supersaturation and dimethyl sulfide (DMS) a linear correlation was found ($R^2 = 0.63$, n = 8) but only for a small region.

INTRODUCTION

Aerosol particles, especially condensation nuclei (CN) and cloud condensation nuclei (CCN) are affecting the physical properties of the atmosphere by contributing to the backscattering of incident solar radiation (Twomey 1974, 1977). A possible change in CCN concentrations due to an increased emission of its potential biogenic precursor (DMS) from the ocean is of particular interest (Nguyen 1983, Charlson *et al.* 1987). In the marine environment most of the CN will be activated and lead to cloud droplets (Bigg *et al.* 1984, Georgii 1990). A change in CCN concentration will lead to a change of the cloud droplet spectra altering the albedo much more effective than the CCN itself.

EXPERIMENTAL

CN measurements were performed using a condensation nuclei counter TSI 3020. CCN were investigated by means of a Static Thermal Gradient Diffusion Cloud Chamber. The chamber consists of two parallel oriented brass cylinders covered by synthetic material to minimize thermal effects. The surfaces facing to the inside are humidified and kept at different temperatures. The upper surface is maintained at environmental temperatures (~ 20°C) while the lower one is maintained at lower temperature. Supersaturation in the range 0.1 to 2% are adjustable - depending on the temperature difference.

The chamber is 80 mm in diameter and 14 mm high. The light of a Hg-lamp is passed through a lense system and thereby focused to the middle of the chamber with a depth of 1.3 mm. The activated CCN become visible in the light beam and can be registrated directly by camera and automatically by using the effect of lightscattering. The lightscattering signal was calibrated by evaluating the photographic measurments.

G. Restelli and G. Angeletti (eds.), Dimethylsulphide: Oceans, Atmosphere, and Climate, 297–302.

MEASUREMENTS

Concentrations of condensation nuclei (CN) and cloud condensation nuclei (CCN) were measured over the Atlantic Ocean between 50°N and 45°S (Fig. 1) aboard the german research vessel 'Polarstern' (Oct.-Nov. 1990).

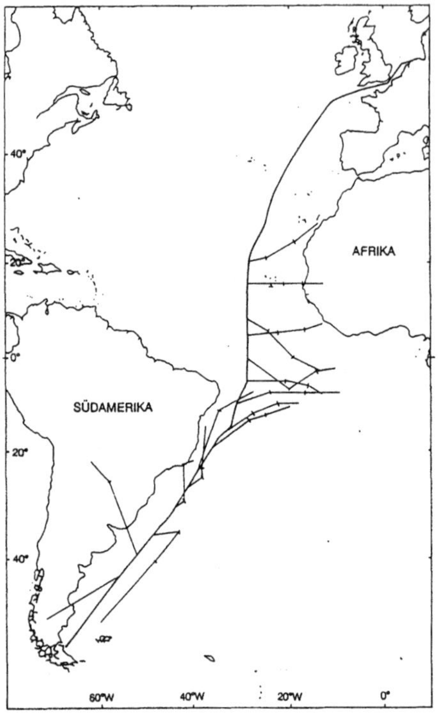

Fig. 1: 'Polarstern' research cruise ANT IX/I (1990). The lines show 48 hours back-trajectories airmasses calculated from isobaric wind fields at 850 hPa (Deutscher Wetterdienst).

Fig. 2 shows the latitudinal distribution of the CN and CCN at 1.5% supersaturation derived from continuous measurements. The highest concentrations at all were detected passing the English Channel. After leaving the Channel the particle concentration decreased to 500 CN/cm^3. The CN concentration over the remote Atlantic ocean were in the range of 100 - 400 cm^{-3}. Southwards from 20°S the concentrations increased again significantly. The trajectories showed that after the passage of a frontal system the current wind changed from easterly to westerly direction. Therefore a possible influence of particles originating from South-America can not be neglected. The number concentration of CN slowly increased up to 1500 CN/cm^3 at 40°S.

The variation of CN and CCN concentrations in general is similar. Over the tropical ocean - between 20°N and 20°S - the ratio of CCN to CN and therefore the activation rate is highest with 64±2%. Compared to this, the activation rate between 40 and 20° was 56±4% on the northern hemisphere and 31±5% on the southern hemisphere.

A comparison to individual measurements (by camera) shows the same general trend - highest activation rates over the tropical ocean of 59±6%. Outside the tropic region the activation at the northern hemisphere dropped to 44±11% compared to the southern hemisphere with 31±12%.

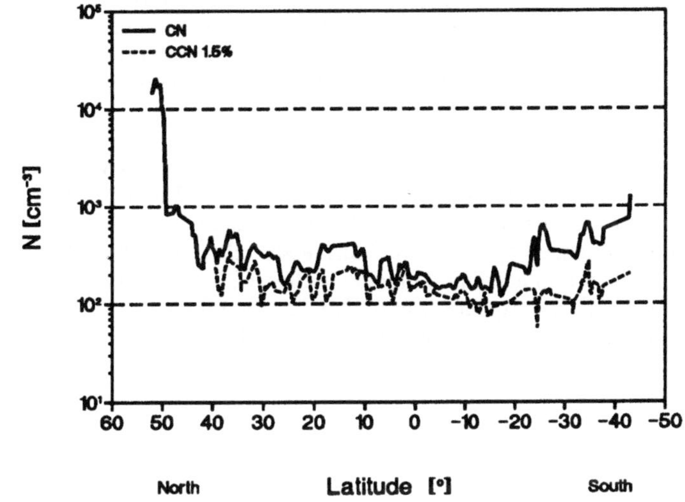

Fig. 2: Continuous measurements of CN and CCN at 1.5% supersaturation (3h average).

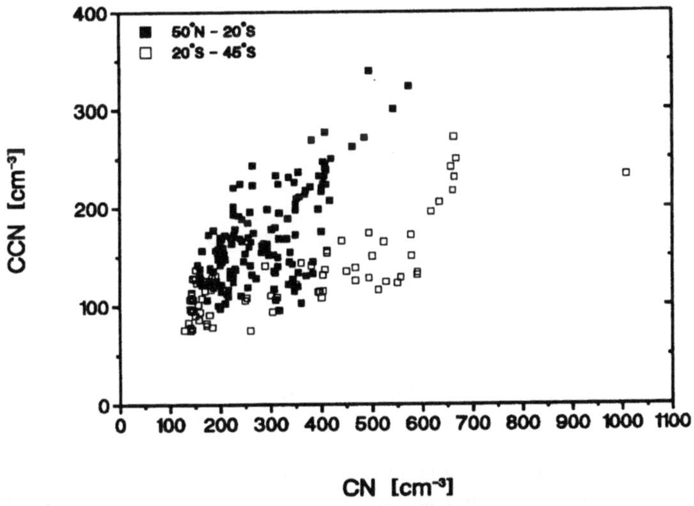

Fig. 3: CCN as a function of CN (data set from Fig. 2).

Fig.3 shows the influence of different airmasses on the activation of CN as CCN and explains the different activation ratios. The closed squares stand for the activation between 50°N and 20°S while the open squares show the activation after the passage of the frontal system at 20°S. - It is likley that these different ratios are due to different particle size distributions. A lower activation rate along with higher concentrations of CN found for the airmasses represented by the open squares can be caused by a larger amount of small particles that were not activated.

For the remote tropical ocean a calculated supersaturation spectrum following the fundamental relationship $N = C \cdot S^k$ (Twomey 1959) yields values of 142 cm^{-3} for the constant C and 0.5 for the exponent k (N = number of CCN [cm^{-3}], S = supersaturation [%]). This result of 0.5 for k corresponds very well to mean values frequently reported by other authors (Laube and Höller 1987).

The latitudinal variation of the concentration of sulfate (Fig. 4) as well as of non-sea-salt sulfate (nss-$SO_4^{=}$) was similar to that of the aerosol particles. The highest values for sulfate (1825 ng S/m^3) as for nss-sulfate (840 ng S/m^3) were measured at 46°N and 42°S. In both cases the regions were influenced by continental airmasses. However, at 42°S high biogenic activity also has to be taken into account. Dimethylsulfide (DMS) was high concentrated in the seawater as well as in the atmosphere (Fig.6). Methane sulfonic acid (MSA) - a reaction product of DMS oxidation - was high concentrated as well (28 ng S/m^3).

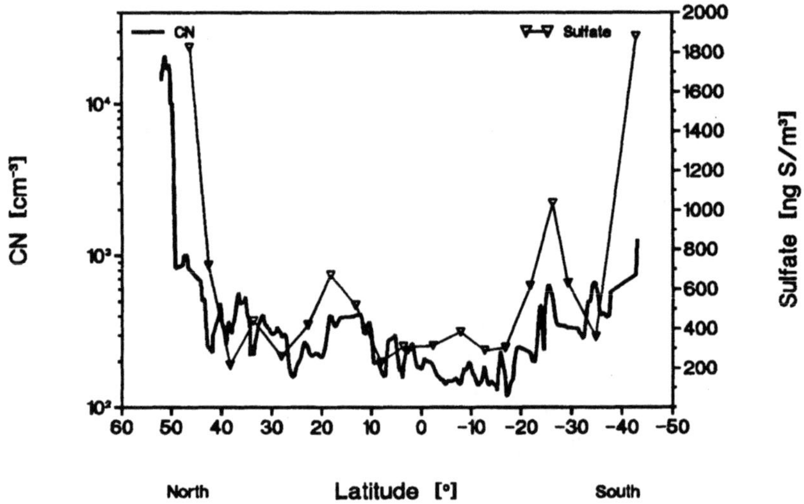

Fig 4: Latitudinal concentration variation of CN and sulfate.

A regression analysis between CN and sulfate (Fig. 6a) shows a significant linear correlation between daily means of CN and sulfate with a determination coefficient of 0.78, significant at the 99% level (n = 17). The resulting regression equation is

$$[CN] = 0.45\ (\pm 0.06)\ [SO_4^{=}] + 97\ (\pm 40) \qquad (1)$$

Unfortunatley the data did not provide more information in the high concentration range which of course is somehow critical. Nevertheless, a positive relationship between CN and sulfate is obvious. Despite of a similar concentration variation of CCN and nss-sulfate, no significant linear correlation for the present data set was found.

The latitudinal concentration variation of CCN at 1% supersaturation and DMS (Fig. 5) showed for most of the geographical regions a similar behaviour, except the contradictory course from 5 to 15°S. Based on daily mean concentrations it was possible to derive a linear relationship between CCN and DMS only for the region from 40 to 5°N (Fig. 6b). The regression equation is

$$[CCN] = 1.4\,(\pm 0.4)\,[DMS] + 78\,(\pm 24) \qquad (2)$$

The determination coefficient is 0.63, significant at the 98% level (n = 8).

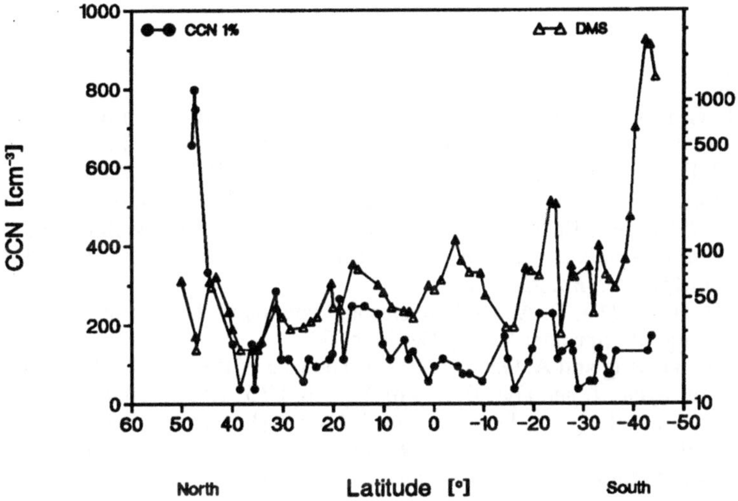

Fig. 5: Latitudinal concentration variation of CCN at 1% supersaturation and DMS in the atmosphere.

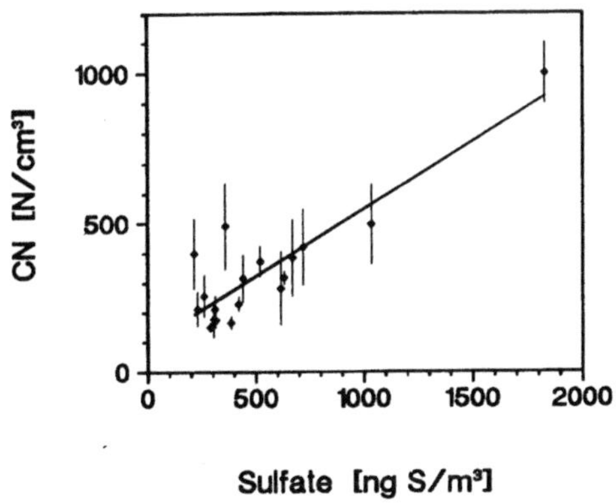

Fig. 6a: CN as a function of sulfate (daily means) ($R^2 = 0.78$, n = 17, signif. level 99%).

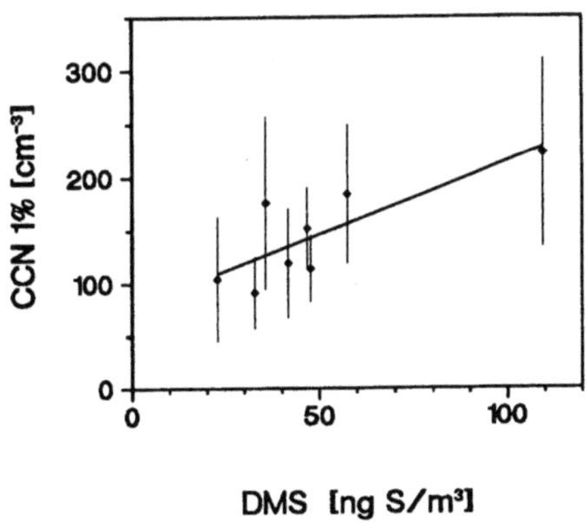

Fig. 6b: CCN at 1% supersaturation as a function of DMS (daily means) ($R^2 = 0.63$, n = 8, signif. level 98%).

CONCLUSIONS

(1) Simultaneous measurements of DMS by Staubes (1992) support an interrelation between DMS and CCN.

(2) A simple and significant linear relationship between DMS and CCN over the complete latitudinal range was not found. Nevertheless, the chemical analysis showed that most particles contained sulfate (mainly caused by oxidation of DMS).

(3) The present data set only shows for CN a significant linear relationship with sulfate while CCN and nss-sulfate were not linear correlated.

REFERENCES

Bigg E. K., J. L. Gras, Ch. Evans (1984). Origin of the aitken particles in remote regions of the southern hemisphere. *J. Atmospheric Chemistry,* Vol. 1, 203-214.

Charlson R. J., J. E. Lovelock, M. O. Andreae, S. G. Warren (1978). Oceanic phytoplankton, atmospheric sulphur, cloud albedo and climate. *Nature,* No. 6114, 655-661.

Georgii H.-W. (1990). Beeinflussen biogene atmosphärische Schwefelverbindungen das Klima? *Sitzungsbericht der wissenschaftlichen Gesellschaft an der J. W. Goethe Universität Frankfurt/M., Bd. XXVI,* Nr. 1.

Laube, M. and H. Höller (1987). Aerosol physics and chemistry. G. Fischer (ed.), *Landolt Börnstein* Band 4b, Springer Verlag Berlin, p. 57.

Nguyen D. C., B. Bonsang and A. Gaudry (1983). The role of the ocean in the global atmospheric sulfur cycle. *J. geophys. Res.*, Vol. 88, No. C15, 10,903-10,914.

Staubes R. (1992). Verteilung von Dimethylsulfid, Carbonylsulfid und Schwefelkohlenstoff in Ozean und mariner Atmosphäre. *PhD-Thesis, University of Frankfurt.*

Twomey S. (1974). Pollution and planetary albedo. *Atmospheric Environment,* Vol. 8, 1251-1256.

Twomey S. (1977). *Atmospheric Aersols.* Elsevier, Amsterdam.

Twomey S. (1959). The nuclei of natural cloud formation - Part II: The supersaturation in natural clouds and the variation of cloud droplet concentration. *Geofis. Pura Appl.,* 43, 243-249.

ARE DIMETHYLSULFIDE AND CONDENSATION NUCLEI CONNECTED OVER THE TROPICAL NORTHEASTERN ATLANTIC OCEAN ?

J.P. PUTAUD, B.C. NGUYEN, S. BELVISO and N. MIHALOPOULOS
Centre des Faibles Radioactivités, Laboratoire Mixte CNRS-CEA,
F-91198 Gif-sur-Yvette Cedex, France

SUMMARY

Atmospheric dimethylsulfide (DMS), size segregated aerosol and condensation nuclei (CN) were simultaneously sampled in September-October 1991 during the oceanic cruise EUMELI 3 in the 20°-30°N, 20°-30°W area. ^{222}Radon activity was monitored to distinguish air masses with oceanic and continental characters. Atmospheric DMS concentrations presented marked variations (range 1-6 nmol.m^{-3}), mainly due to variations in DMS sea-air flux. MSA and nss-SO_4 in submicrometer aerosols ranged from 0.06 to 0.4 and from 0.4 to 33 nmol.m^{-3}, respectively. DMS and nss-SO_4 were not directly correlated over the whole sampling period, partially because of the influence of continental and/or anthropic sources. CN number concentrations ranged from 150 to 1600 cm^{-3}. A correlation ($R=0.81$, $n=23$) appears between daily mean DMS and CN concentrations, although the highest CN concentrations were observed in continental air masses, suggesting a strong contribution of anthropic sources to the CN population close to the continent.

1. INTRODUCTION

It has been recognized that tropospheric submicrometer aerosols and mainly sulfate particles can affect the Earth's radiative budget either through a direct effect of backscattering the incident solar radiation (1), or through their role of precursors of cloud condensation nuclei (2). Over the oceans, the major source of sulfur appears to be the emission of dimethylsulfide (DMS) from the ocean-atmosphere interface (3). Once in the atmosphere, DMS is oxidized by radicals such as OH or NO_3 to give mainly methanesulfonic acid (MSA), sulfur dioxide (SO_2), and sulfuric acid (H_2SO_4)(ref. 4) Globally, the marine source of sulfur is estimated to be about 5 times smaller than the anthropogenic source (5). One can therefore wonder if the process involving oceanic DMS emission, its particulate oxidation products and their effect on the Earth's albedo (6) is still relevant for climate control in an oceanic area downind of a continent. Indeed, seasonal covariations between atmospheric DMS and cloud condensation nuclei (CCN) concentrations have only been pointed out at Cape Grim, Tasmania (7, 8). In the northern hemisphere, the extent of the oceanic areas where the condensation nuclei (CN) population is not totally dominated by continental and/or anthropogenic aerosol, but where biogenic DMS could play a distinguishable role as a CN precursor is still unknown.

In this work, we report simultaneous measurements of atmospheric DMS, MSA and non-sea-salt sulfate (nss-SO_4^{2-}) in submicrometer aerosols and CN number concentrations obtained during an oceanic cruise in the northeastern tropical Atlantic Ocean. ^{222}Radon activity was used as a tracer of continental air masses. Variations of these parameters are compared in order to discuss the influence of the marine sulfur source on the CN population over this oceanic area.

2. EXPERIMENTAL

Samples were collected on board R/V *L'Atalante* along the track of the oceanographic cruise EUMELI 3 during the 17-28 September and 5-22 October 1991 periods between 3 points, Canaria Islands, (27°45'N, 18°02'W), "M" point (18°29'N, 21°05'W), and "O" point (21°02'N, 31°08'W) as depicted on Figure 1.

Atmospheric samples were collected from the top of the bow mast of the ship at 16

G. Restelli and G. Angeletti (eds.), Dimethylsulphide: Oceans, Atmosphere, and Climate, 303–310.

meters above sealevel. For DMS analysis, air samples were collected every 6 hours and analysed within 1.5 hr by cryogenic trapping at -90°C and subsequent measurement by GC/FPD (9). The detection limit was about 0.8 nmol.m $^{-3}$ for this experiment. Aerosols were collected using a 6-stage cascade impactor (Sierra) at a flow rate of 70 m^3/hr over 24 hr periods, from sunrise to sunrise (about 8.00 a.m. local time). A wind vane and an anemometer controlled the pump for the sampling to be performed only when the relative wind was forward of the ship's beam (±50°) and greater than 1 m.s^{-1}. Impaction substrates and final filters were made of Whatman-41 filter paper previously washed in 0.1 M HNO_3 and rinsed 3 times in deionized water (R=18.2 MOhm). The cut-off diameters were 9.8, 4.2, 2.5, 1.2, and 0.6 µm for stages 1 to 5 respectively. The last stage (n°6) was a "total" Whatman-41 filter paper, the efficiency of which to trap particles in the radii range 0.01-0.2 µm has been proved to be better than 80% under high-volume regime (10). Soluble species were extracted in 20 ml of deionized water. Measurements were performed by ionic chromatography, using Dionex columns AS4 for MSA and SO_4^{2-}, and CS10 for Na^+ which was used as a sea-salt tracer. Our detection limit corresponds to concentrations below 2pmol.m^{-3}.

The number concentration of condensation nuclei was measured in real time with a Condensation Particle Counter (CPC) TSI 3022 designed to detect nearly 100% of particles greater than 0.02 µm. It was positioned at the top of the bow mast and connected to a very short inlet tube (30 cm) to limit impaction of particles within the tube.

^{222}Radon activity was monitored to document the transport of continental air masses according to a method already descibed (11).

3. RESULTS AND DISCUSSION

Meteorological conditions and air mass characterization

Except during the 13-15 and 18-20 October periods, northeastern to eastern trade winds occured, so that sampling points were most of the time downwind of the African continent. ^{222}Rn activities were above 7 and 10 pCi.m^{-3} for 17-20 September and 5-10 October periods, respectively, denoting the continental character of the air masses sampled during these periods. ^{222}Rn was close to the oceanic background on the other days (Fig. 2a).

Sulfur species atmospheric concentrations

Atmospheric DMS concentations ranged from <0.8 to 12.3 nmol.m^{-3} (mean 3.1, figure 3a), with the highest values observed between Canaria Islands and M point, i.e. close to the west-African upwelling area.

We focused our attention on submicrometer aerosols and present here the concentrations of MSA and SO_4^{2-} in particles in the diameter range < 0.6 µm only. The size distribution of these compounds studied on some of the 6-stage series collected during this cruise show that 49±12% of the total MSA and 60 ± 18% of the total nss-SO_4^{2-} were in particles with diameter < 0.6 µm (12).

MSA concentrations in submicrometer particles ranged from 0.06 to 0.41 nmol.m^{-3} (average 0.15 nmol.m^{-3}, Figure 2b). MSA seems to roughly covary with DMS during the first leg, whereas during the second one, no relationship appears between DMS and MSA daily mean concentrations, certainly because of advection of sulfur-enriched air masses from the upwelling area, or to changes in the kinetics and the branching ratio of DMS oxidation in polluted air masses.

Nss-sulfate concentrations in particles with diameter < 0.6µm ranged from 0.4 to 33 nmol.m^{-3} (mean 7.4, Figure 2c). No significant correlation was observed between nss-SO_4^{2-} and DMS concentrations from the entire data set, but, excluding the values obtained in continental airmasses, nss-SO_4^{2-} concentrations correlate with DMS atmospheric concentrations (R=0.66, n=21).

Condensation Nuclei

Over 402 hr of measurements, the CN number concentration ranged from 155 to 1890 cm^{-3} (mean 586, figure 3b). Considering only values obtained under "pure" oceanic influence, the mean CN concentration decreases to 354±108 cm^{-3}, close to the oceanic background (about 380 cm^{-3}) measured at 14°S and 41°S in the Pacific Ocean (13). A correlation appears between the daily mean CN concentrations and the submicrometer fraction of nss-SO_4^{2-} (R = 0.91, n=17), suggesting an important role of H_2SO_4 in CN production over this oceanic area. The ratio between the mean CN number concentrations obtained under continental and "pure" oceanic influence (1055/354), indicates that the CN population can be strongly affected by continental inputs over the oceanic area situated less than 500 km downind of the continent. Therefore, the correlation observed between atmospheric DMS and CN number concentration (R=0.81, n=23, Figure 4) must be carefully considered since the highest DMS concentrations were always observed close to the M point, with strong northeastern winds, and therefore advection of air masses affected by continental inputs. The atmospheric circulation patterns encountered during EUMELI 3 experiment therefore generally precludes determining any connection between DMS and CN over the study area.

However, indications of continental influences were never detected at O point and in "pure" oceanic conditions, the increase of the daily mean atmospheric DMS concentration (from 1.3 to 3.6 nmol.m^{-3}, figure 3a) was followed by an increase in the daily mean CN number concentration (220 to 434 cm^{-3} on 15 October and 609 cm^{-3} on 16 October, figure 3b). As a strong influence of sea-salts or of another volatile compound emitted from the ocean seems unlikely, our observations indicate that DMS could have in this case directly influenced the CN number concentration over the ocean, despite the vicinity of a continent (less than 1500 km upwind).

4. CONCLUSION

During the oceanographic cruise EUMELI 3, no obvious correlation appeared between DMS and its oxidation products (MSA and H_2SO_4) over the whole period of experiment, probably due to differences in lifetimes between reactive gases and aerosols and/or continental inputs of nss-SO_4 and/or oxydizing species changing the DMS lifetime. However, the selection of data from the continental tracer ^{222}Rn activity allowed us to demonstrate that the submicrometer nss-SO_4 concentration could be connected to the DMS concentration in unpolluted air masses. The correlation between the daily mean CN concentrations and the submicrometer fraction of nss-SO_4^{2-} (R = 0.91, n=17) suggests an important contribution of sulfur compounds in CN production over the tropical northeastern Atlantic ocean. However, CN number concentration was found to be 3 times higher in continental air masses than in maritime ones. Consequently, the connection between DMS and CN cannot be concluded from the correlation between their daily mean value (R=0.81, n=23), since ^{222}Radon monitoring indicate continental influence during the periods when the highest CN and DMS atmospheric concentrations were simultaneously observed. However, further than 1500 km from the African continent and under pure oceanic influence, a tripling in atmospheric DMS concentration was followed by a tripling in CN concentration number. This shows that DMS could in some cases influence the CN population over oceans even closer than 1500 km downwind of a continent.

Acknoledgments. We thank the officers and the crew of the IFREMER ship *L'Atalante* for their assistance and cooperation. This work was supported by the Centre National de la Recherche Scientifique, the Commissariat à l'Energie Atomique and the Ministère de la Recherche et de la Technologie.

REFERENCES

(1) CHARLSON R.J., J. LANGNER, H. RHODE (1990). Sulfate aerosol and climate, *Nature, 348*, 22.

(2) CHARLSON R.J., S.E. SCHWARTZ, J.M. HALES, R.D. CESS, J.A. COACKLEY Jr, J.E. HANSEN, D.J.HOFFMAN (1992). Climate forcing by anthropogenic aerosols. *Science, 255*, 423-430.

(3) ANDREAE, M.O. (1986). The ocean as a source of atmospheric sulfur compounds, in *The Role of Air-Sea Exchange in Geochemical Cycling*, edited by P. Buat-Ménard, pp. 331-362, D. Reidel, Horwell, Mass.

(4) YIN F., D. GROSJEAN, J.H. SEINFELD (1990). Photooxidation of dimethylsulfide and

dimethyldisulfide, I: mechanism development. *J. Atmos. Chem., 11*, 309-344.
(5) BATES, T.S., B.K. LAMB, A.B. GUENTHER, J. DIGNON, E.S. STOIBER (1992). Sulfur emissions to the atmosphere from natural sources, *J. Atmos. Chem., 14*, 315-337.
(6) CHARLSON R.J., J.E. LOVELOCK, M.O. ANDREAE, S.G. WARREN (1987). Oceanic phytoplankton, atmospheric sulphur, cloud albedo and climate, *Nature, 326*, 655-661.
(7) AYERS, G.P., J.P. IVEY, R.W. GILLETT (1991). Coherence between seasonal cycles of dimethyl sulphide, methanesulphonate and sulphate in marine air, *Nature, 349*, 404-406.
(8) GRAS, J.L. (1990). Cloud concensation nuclei over the southern ocean, *Geophys. Res. Lett., 17*, 1565-1567.
(9) NGUYEN, B.C., N. MIHALOPOULOS, and S. BELVISO (1990). Seasonal variation of atmospheric dimethyl sulfide at Amsterdam Island in the southern Indian Ocean, *J. Atmos. Chem., 11*, 123-143.
(10) WATTS, S.F., R. YAAQUB and T. DAVIES (1986). Comments on "The use of Whatman 41 filter papers for high volume aerosol sampling", *Atmos. Environ., 21*, 2731-2732, 1987.
(11) POLIAN, G., G. LAMBERT, B. ARDOUIN, A. JÉGOU (1986). Long range transport of continental radon in subantarctic areas, *Tellus, 38B*, 178-189.
(12) PUTAUD, J.P., S. BELVISO, B.C. NGUYEN, N. MIHALOPOULOS. Dimethylesulfide, aerosols and condensation nuclei over the tropical northeastern Atlantic Ocean, submitted to *J. Geophys. Res.*
(13) BATES, T.S., R.J. CHARLSON and R.H. GAMMON (1987), Evidence for the climatic role of marine biogenic sulphur, *Nature, 329*, 319-321.

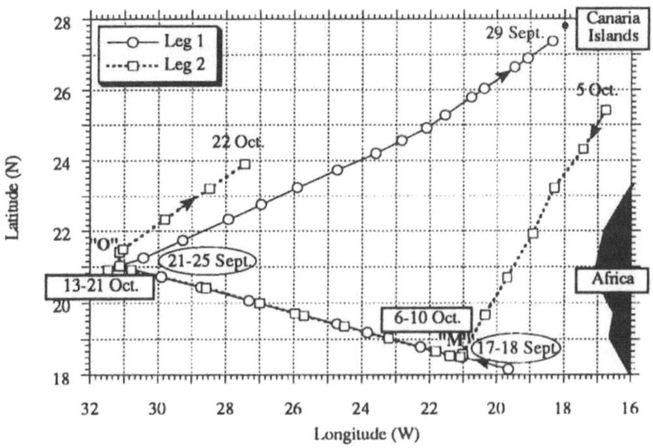

Figure 1: Ship's track and DMS sampling points during the cruise Eumeli 3 (17 Sept.-22 Oct. 1991). Circled and framed dates indicates periods of stations during legs 1 and 2, respectively.

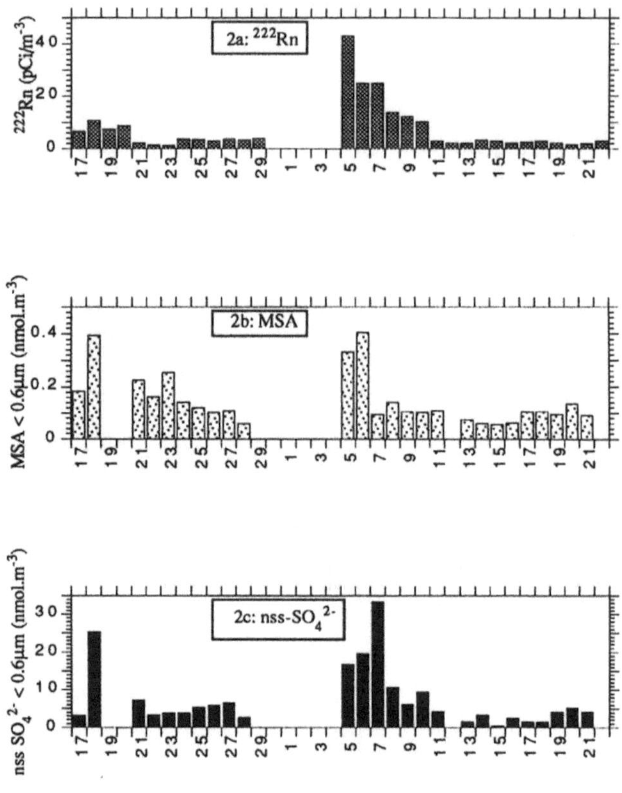

Figure 2: Variations of ^{222}Rn activity (2a) and concentrations of MSA (2b) and nss-SO_4^{2-} (2c) in submicrometer aerosol.

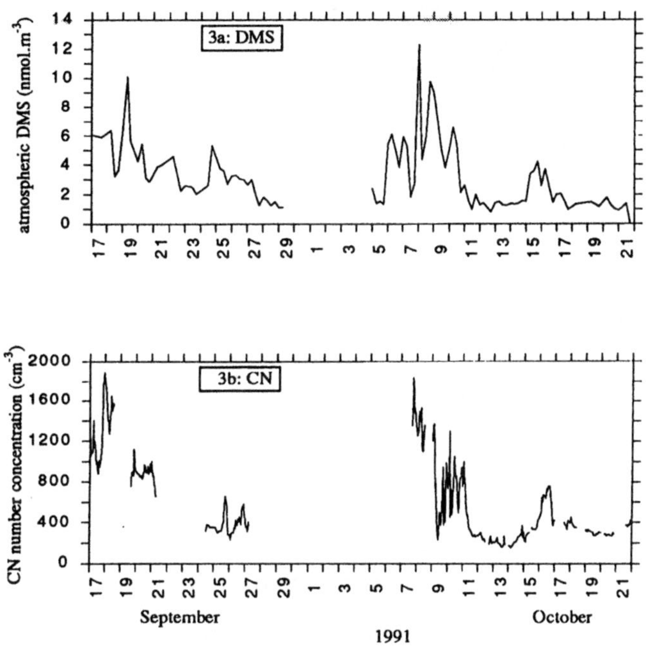

Figure 3: Variations of atmospheric DMS (3a) and Condensation Nuclei number (3b) concentrations.

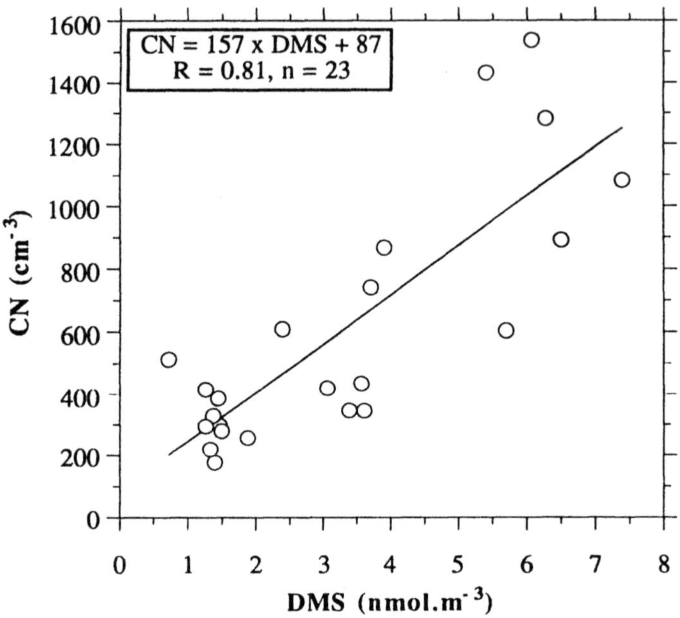

Figure 4: Correlation between daily mean atmospheric DMS and CN number concentrations.

CLOUD CONDENSATION NUCLEI FROM DIMETHYL SULPHIDE IN THE NATURAL MARINE BOUNDARY LAYER: REMOTE VS. IN-SITU PRODUCTION

F. RAES, R. VAN DINGENEN, J. WILSON and A. SALTELLI

Commission of the European Communities
Environment Institute
I-21020 Ispra (VA) Italy

SUMMARY

In this modelling study, we have compared two scenario's for the production of cloud condensation nuclei from DMS. The first assumes that DMS emission, chemical reaction and subsequent aerosol formation and growth all take place within the marine boundary layer. The second assumes that DMS is transported to the free troposphere, where it is transformed into aerosol particles that eventually subside back to the boundary layer. Only the second scenario predicts aerosol concentrations, cloud condensation nuclei in particular, that correspond with the observations. The important role of coagulation in this second scenario also offers an explanation for the observed constancy of the marine boundary layer aerosol.

1. INTRODUCTION

Large scale subsidence of air may create a distinct temperature inversion layer, under which extensive stratus cloud decks can form. This is true, in particular for the sub-tropical and polar oceanic regions, as is shown by weather satellite pictures [1]. The persistence and reflectivity of these clouds are sensitive to variations in the cloud droplet population, consequently they may be important actors in climate regulation and climate change [2,3,4].

It has been suggested that, away from continental influences, these clouds form primarily on sulphate particles derived from dimethyl sulphide (DMS, see Ref. 3 for a review). This is supported by long term monitoring programs that reveal correlations between DMS, its oxidation products and CCN [5,6,7], and by sulphur isotope measurements of the marine sub-micron NSS-sulphate and MSA aerosol [8].

The link between DMS molecules in air, and aerosol particles large enough to act as CCN for stratus clouds implies:

1) chemical transformation of DMS into sulphuric and methane sulphonic acid (H_2SO_4, MSA)
2) nucleation of those acids into sub-nanometer sized droplets and
3) growth of these droplets by condensation and coagulation to radii larger than 0.05 µm.

Whether this sequence will be accomplished in practice, depends on the overall time scale of these processes compared to the time that is available for them in the atmosphere.

In an earlier paper [9], we have assumed that all processes mentioned above occur within the marine boundary layer (MBL), and we have calculated that their overall times exceed 3 days. Comparing this with the 0.5 to 4 days an air parcel on average spends in between clouds, we concluded that the above processes alone do not explain CCN production. We therefore proposed additional processes, among which

G. Restelli and G. Angeletti (eds.), Dimethylsulphide: Oceans, Atmosphere, and Climate, 311–322.

1) heterogeneous transformation of SO_2 on sub-micron sulphate aerosol
2) nucleation of particles in the free troposphere with subsequent transport to the MBL.

Based on an extended version of the model, we will reconsider in-situ formation of CCN from DMS within the MBL and expand on the significance of nucleation outside the MBL. In doing so we will adopt boundary conditions which are relevant to remote sub-tropical oceanic regions during summertime.

2. DESCRIPTION OF THE MODEL

The model describes aerosol formation in an air parcel, that follows a prescribed trajectory which defines the external conditions (temperature, humidity, presence of clouds,....). At each point along the trajectory, the model predicts the aerosol size distribution, from which one can infer the total number concentration (N_{tot}), the fraction of particles that is observable with a regular condensation nucleus counter (CN), as well as CCN concentrations at various supersaturations. The core of the model is the aerosol dynamics model AERO2, which was originally written for analysing smog chamber experiments[10], but later adapted for describing aerosol formation in the ambient atmosphere[9]. The main differences with the model used in Ref. 9 is that we now start from DMS emissions, rather than from prescribed SO_2 concentrations, and that we also treat cloud processing, albeit in a highly parametrised way. A short description is given below.

emission

Air parcels in contact with the sea surface experience a flux of DMS , for which we take values between 10^9 and 4×10^9 molecules/cm^2/s, typical for summer conditions[11]. Measurements of Bates et al.[12] show a light dependence of the DMS flux, from which we first inferred that DMS is only emitted during daytime. Recent measurements[13], however, show that atmospheric DMS actually builds up during night-time, indicating DMS emissions also during night-time. In the 'Hadley cell scenario' (see below) all model runs, except run 60, were done with a daytime-only DMS flux, in the 'MBL scenario' all runs were done with day- and night-time DMS flux. As we will see below, this will not affect the comparability of the two scenario's. We further assume a mixing height of 1000 m.

homogeneous chemistry

The photo-oxidation of DMS is described by the following simplified scheme:

$$DMS + OH \longrightarrow SO_2 \qquad k_1 = 8\times10^{-12}\,cm^3molec^{-1}s^{-1} \qquad (1)$$

$$SO_2 + OH \longrightarrow H_2SO_4 \qquad k_2 = 1\times10^{-12}cm^3molec^{-1}s^{-1} \qquad (2)$$

This scheme neglects MSA as the second product of reaction (1). At sub-tropical temperatures however, the MSA/H_2SO_4 ratio is $< 10\%$ [14] , hence the size of the aerosol particles (which determines whether they can be activated in clouds) will be underestimated by at most and only a factor $(1.1)^{1/3} = 1.03$.

nucleation, coagulation , condensation and dry deposition.

H_2SO_4 formed by reaction 2 will nucleate together with H_2O molecules and form stable droplets of about 1nm. These droplets will further grow by coagulation among themselves, and by additional condensation of H_2SO_4 and H_2O molecules. The description of these processes and of dry deposition, is done with AERO2, which also provides the numerical framework for describing the following processes.

heterogeneous chemistry

Formation of condensable material by heterogeneous reactions in wet sub-micron aerosol particles has been observed[15] . Recently, Luria and Sievering showed that SO_2 can be oxidised on marine aerosol particles[16]. From their measurements we derived that SO_2 sticks to sub-micron particles with an efficiency of 1%, followed by immediate oxidation. We model this process as an additional condensation process [8].

cloud processing

When air becomes supersaturated, particles larger than a critical radius become activated and quickly grow to cloud droplets. Within those droplets SO_2 can be absorbed and oxidised to sulphate, the droplets also scavenge the non activated (interstitial) aerosol particles. These two processes increase the sulphate content of the droplets, which, upon evaporation, results in larger particles. We have treated these processes in the following way. At each cloud encounter, the integration of AERO2 is stopped, and a time step is taken equal to the time spent in the cloud. A supersaturation of 2 % for cumulus clouds and 0.2 % for stratus clouds is prescribed, and all particles larger than the respective critical radii form cloud droplets of 10 μm size. If the resulting droplet concentrations is larger than 3 cm^{-3}, a fraction of the available gas phase SO_2 is distributed as H_2SO_4 over the droplets, while the amount of interstitial H_2SO_4 scavenged during the clouds lifetime is calculated analytically. After this cloud-time step, we re-equilibrate the aerosol with the ambient relative humidity, and AERO2 continues with this new size distribution.

This separation of aerosol dynamics and cloud processing, by introducing a prescribed supersaturation is artificial. In fact the size distribution of the aerosol determines (together with the cooling rate of the air) the maximum supersaturation that will be reached in the air parcel, and hence the minimum particle radius that will be activated. This activation and subsequent cloud processing changes the size distribution, etc.. The modelling of this interaction between aerosol dynamics and cloud microphysics is attempted by Saltelli et al.[17]. Note however that the prescribed supersaturation for stratus clouds of 0.2% is high; this will make our estimates of CCN activating in stratus clouds optimistic.

precipitation scavenging

Some clouds rain out. In such a cloud we oxidise the available SO_2 and scavenge the interstitial aerosol as above, after which we reduce the number of activated droplets by a certain fraction.

3. MODEL PREDICTIONS

A notable feature of the remote MBL is the constancy of the aerosol concentrations ; in summer [CN] varies around 300-500 cm^{-3}, with [CCN] being a constant fraction of [CN] . For CCN active in MBL Stratus clouds (S < 0.2%), [CCN(0.2%)]/[CN] > 0.2 [7,18,19].

The question is whether the combination of processes considered in the model does reproduce these features within the time available. Defining this 'available time' depends on the premises. In a first scenario, will first presume that CCN for MBL St clouds are formed 'in-situ', i.e. all processes (DMS emission, chemical transformation, nucleation, ...) happen within the subtropical MBL. The maximum 'available time' then equals to the transport time through the subtropics, i.e. 12 days [20]. In a second scenario, we will consider the connection between the subtropical MBL with the tropics and mid-latitudes, through large scale atmospheric circulations (e.g. Hadley cell circulation), and that processes outside the MBL might eventually be responsible for the MBL CCN observed. In this case, the 'available time' is that spent within the MBL plus the transport time between tropics (mid-latitudes) and sub-tropics.

In-situ formation of MBL CCN (the MBL scenario)

During the 12 days an air parcel travels through the sub-tropical MBL, it will be processed many times by Stratus clouds (St), which will affect the particle formation process. According to Lelieveld et al. [22], the average time spent in-between sub-tropical St cloud is 12 hours, and the average time spent inside those clouds 2 hours. We further assume that every tenth cloud is a precipitating one, which removes CCN(0.2%) with a certain efficiency. We thus calculate the formation of H_2SO_4-H_2O droplets in an air parcel which is initially free of particles and follows a history shown in Fig.2; the input data are summarised in Table I.

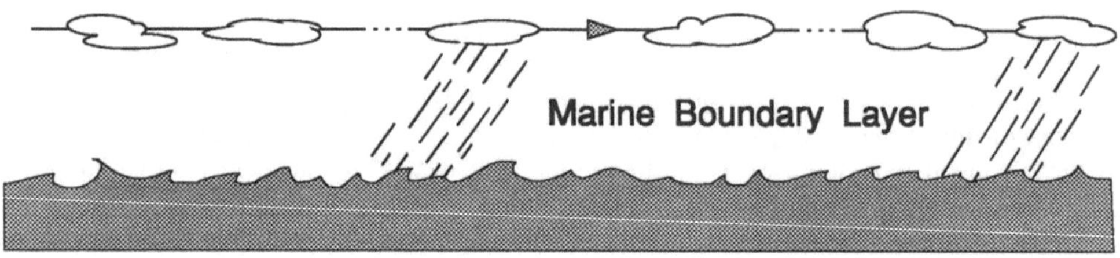

FIG. 1 *Schematic of the MBL scenario. The air parcel starts free of particles. Outside clouds the parcel experiences a DMS flux and sulphate aerosol forms by photochemical reactions followed by nucleation and growth. Inside clouds, a supersaturation of 0.2 % is assumed, and the activated particles acquire more sulphate by in-cloud SO_2 oxidation and scavenging of the interstitial aerosol. The air parcel goes through a number of such cloud cycles before it encounters a precipitating cloud.*

Fig 2 shows the resulting [N_{tot}], [CN] and [CCN(0.2%)] for a DMS flux of 2×10^9 molec/cm^3/s . The model predicts the right number of CN, however fails to produce enough CCN(0,2%) even after 12 days. This is also clear from the calculated size distributions; after 12 days, a bimodal aerosol develops which is typical for the remote MBL [23,24] however the largest mode is far too small compared with the observations. We have performed a simple sensitivity analysis, varying the input parameters, one at the time, as indicated in Table I. The results of this analysis are summarised in Fig 3a and 3b, which show [CN] and [CCN(0.2%)] after 12 day, just before the last precipitation event.

	reference run 3	sensitivity analysis	run	units
DMS flux	10^9 - $4x10^9$			molec/cm^2/s^1
also during night-time	yes			
time spent in MBL	12			days
cloud cycle times	2 (in cloud) + 12 (in between)			hours
# cloud cycles	20			
# precipitating clouds	2	4	7	
precipitation efficiency	0.9	0.5	9	
temperature MBL	20	25	11	°C
relative humidity MBL	80	90	12	%
pressure	950			mbar
OH concentration	$2x10^6$	$5x10^5$	13	molec/cm^3
fraction of SO_2 oxidised in cloud	0.9	0.5	14	
heterogeneous SO_2 oxid.	no	yes	4	

TABLE I Input data for MBL scenario. The second column shows the data for the reference case. The third column shows variations applied, one at the time, in the various sensitivity runs.

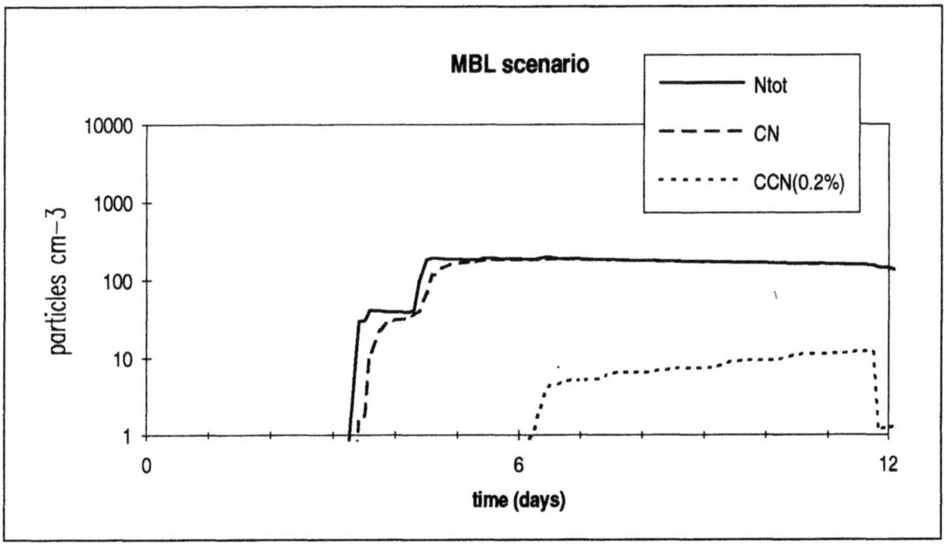

FIG. 2 Concentrations of N_{tot}, CN and CCN(0.2%) in an air parcel travelling through the marine boundary layer, and experiencing a flux of DMS of $2x10^9$ molec cm^{-2} s^{-1}. Reference run 3.

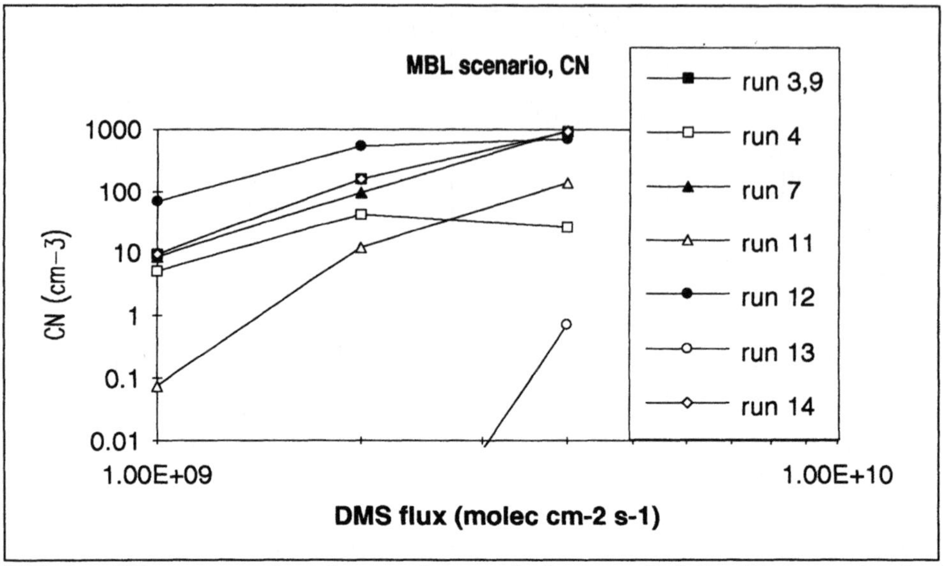

FIG. 3a *CN concentrations in an air parcel that travelled for 12 hours in the MB, calculated with a variety of input conditions (see Table I).*

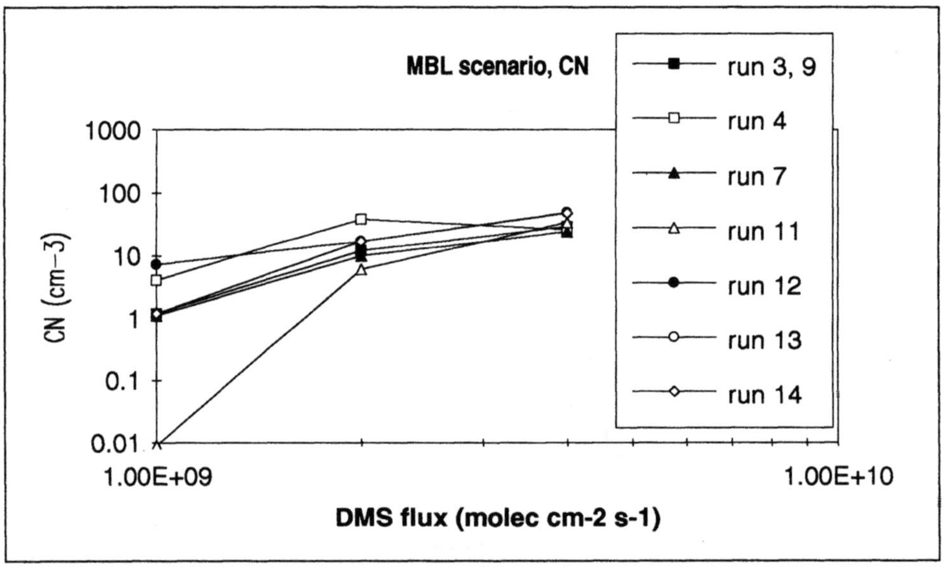

FIG. 3b *CCN(0.2%) concentrations in an air parcel that travelled for 12 hours in the MBL, calculated with a variety of input conditions (see Table I).*

Based on the criteria that $300 < [CN] < 500 cm^{-3}$ and $[CCN(0.2\%)]/[CN] > 0.2$ within 12 days, we conclude that MBL in-situ processes alone do not readily explain the presence of the necessary CCN for sub-tropical MBL clouds. Another feature of Fig. 3 deserves attention, namely that in the MBL scenario, [CN] are very sensitive to variations in the environmental conditions. We will discuss this later.

MBL CCN and large scale atmospheric circulation (the Hadley cell scenario)

The sub-tropical MBL is a product of large scale subsidence which brings air from the upper troposphere down to sea level. This transport takes about 12 days[21], which has to be added to the time spent in the MBL. The air in the upper troposphere is pumped up both by tropical and mid-latitude storm systems. However, to probe for possible CCN formation in the upper troposphere and during subsidence, we will consider only the Hadley cell circulation (see Fig. 4) which links tropics with sub-tropics, and we will model CCN formation in an air parcel that is cycled through this cell. In doing so, we extend the ideas of Chatfield and Crutzen[25], who already showed that importance of convective clouds for the marine sulphur cycle. The parameters and environmental conditions used in the calculations are shown in Table II.

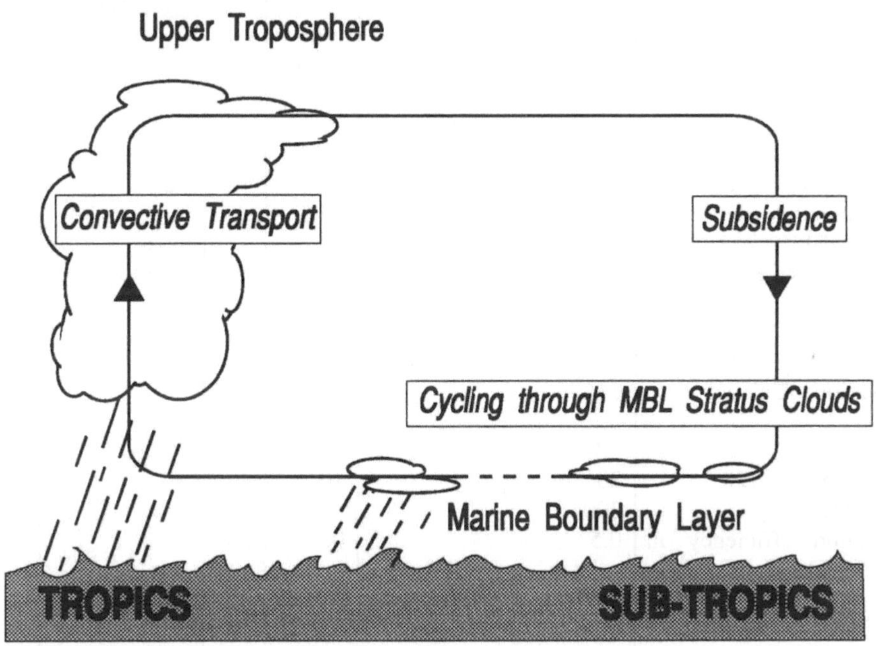

FIG. 4 The Hadley cell scenario. The air parcel starts in the tropical upper troposphere and is transported to the sub-tropical MBL, during which it does not encounter any clouds. In MBL it experiences a DMS flux and cycles through a number of Stratus clouds, before it encounters a precipitating one. It further travels to the tropics where it is pumped up by a convective cloud. During this convective transport, DMS will remain undisturbed, but a fraction of the SO_2 and the particles will be rained out.

The parcel is cycled through the Hadley cell until a stationary state is established. Figure 5 shows [N_{tot}], [CN] and [CCN(0.2%)], for a DMS flux of $2x10^9$ molecules/cm^2/s. The model predicts new aerosol production in the upper troposphere, favoured by both the low temperature, and the low concentration of background aerosol that remains after cleansing by the convective cloud. By the time the sub-tropical MBL is reached, the model predicts concentrations of 380 and 150 cm^{-3} for CN and CCN(0.2%) respectively. Despite the high relative humidity and the source of DMS, the model does not predict nucleation within the MBL; all H_2SO_4 formed condenses on the aerosol already present or deposits in the sea. The calculated aerosol size distribution in the MBL shows the typical bimodal structure, produced by cloud processing. Figures 6a and 6b again compile the results of a sensitivity analysis, showing [CN] and [CCN(0.2%)] averaged over the MBL.

	reference run 51	sensitivity analysis	run	units
DMS flux	10^9 - $4x10^9$			molec/cm^2/s
also during night-time	no	yes	60	
Hadley cell cycle times	7 (in UT) +	3 +	54	days
	14 (subsidence) +	6 +		
	7 (in MBL) +	3 +		
	0.04 (convective cloud)	0.04		
temperature MBL	25			^oC
relative humidity MBL	80	90		%
pressure MBL	950			mbar
cloud cycle time MBL	2 (in cloud) + 12 (in between)			hours
# cloud cycles MBL	10	7	54	
# precipitating clouds MBL	1			
precipitation efficiency MBL	0.9	0.5	56	
temperature UT	- 34			^oC
relative humidity UT	30	10	45	%
pressure UT	300			mbar
precipitation efficiency in convective cloud	0.5	0.9	53	
OH concentration	$2x10^6$	5 10^5	55	molec/cm^3
fraction of SO_2 oxidised in clouds	0.9	0.5	52	
heterogeneous SO_2 oxid.	no	yes	59	

TABLE II Input data for Hadley cell scenario. The second column shows the data for the reference case. The third column shows variations applied, one at the time, in the various sensitivity runs

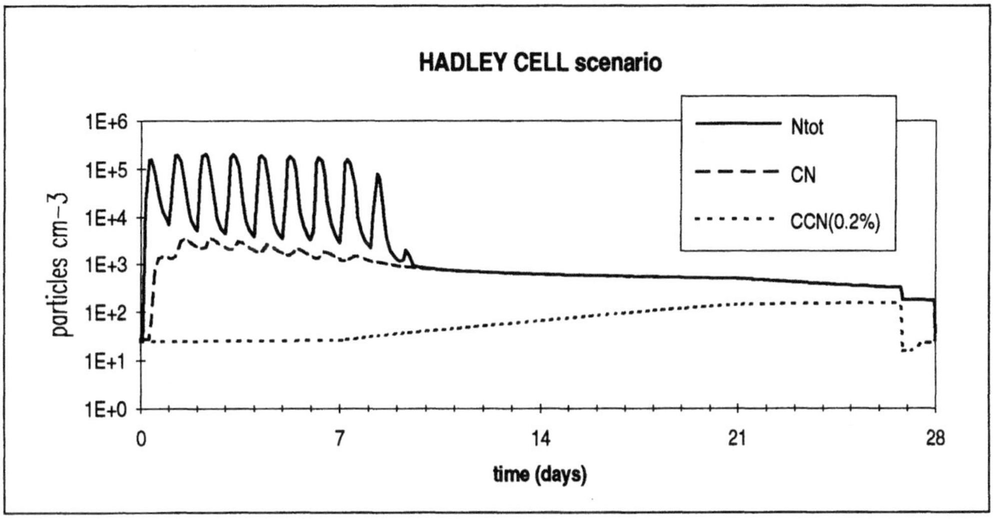

FIG. 5 *Concentrations of N_{tot}, CN and CCN(0.2%) in an air parcel that is moved through a Hadley cell. At time zero, the parcel starts in the free troposphere. In the MBL, it experiences a DMS flux of 2 x10^9 molec/cm^2/s. Reference run 51.*

Under all conditions considered, the model reproduces the right values for [CN] and [CCN(0.2%)]. Note that, contrary to the MBL scenario above, [CN] is now most insensitive to external parameters, whereas [CCN(0.2%)] primarily depends on the DMS flux. We explain this lack of sensitivity by coagulation theory where:

$$\frac{dN}{dt} = -KN^2 \text{ hence } N(t) = \frac{1}{1/N_0 + Kt}$$

N is the total particle number concentration, N_0 the initial particle concentration, K the coagulation coefficient (= 10^{-8} $cm^{-3}s^{-1}$), and t the time. It follows that, when nucleation in the upper troposphere creates at least 10^4 particles per cm^3 (which it does easily given the favourable conditions), N($t > 10^6$ s) is independent of N_o, and hence of the conditions under which nucleation takes place. Furthermore, after 10^6 s (12 days, typically the time needed for subsidence) N = 1/(K.t) = 100 cm^{-3}, and fairly constant over a period of days. (Note that in Fig. 5, $[CN]_{MBL}$ is larger than 100 cm^{-3} since the air parcel is eventually compressed from 300 mbar to 1000 mbar).

3. CONCLUSION AND DISCUSSION

From this modelling exercise, we conclude that it is not likely that CCN for sub-tropical MBL Stratus clouds form within the MBL, but rather that they are formed in the free troposphere and subside into the boundary layer as part of the Hadley cell circulation.

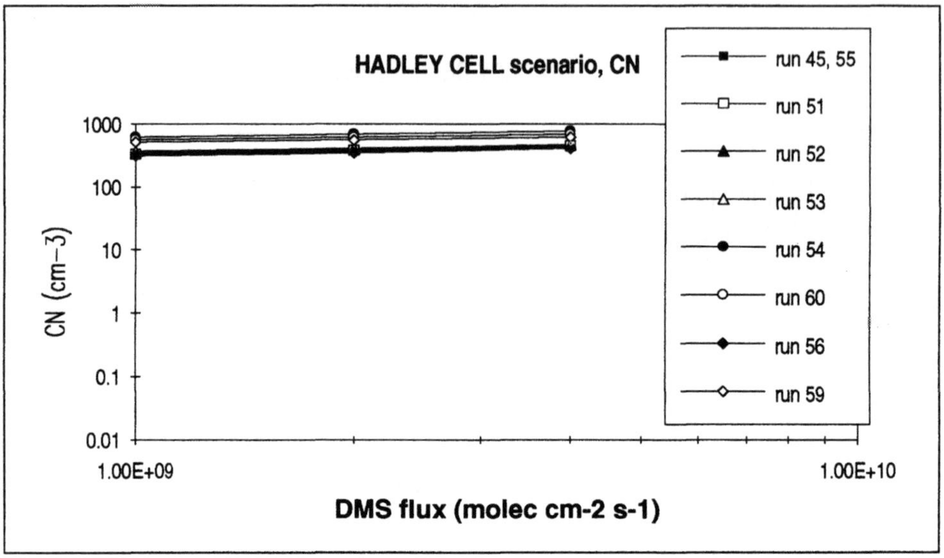

FIG. 6a Average CN concentrations in the MBL, in air parcel that is constantly moving through the Hadley cell, calculated with a variety of input conditions (see Table II).

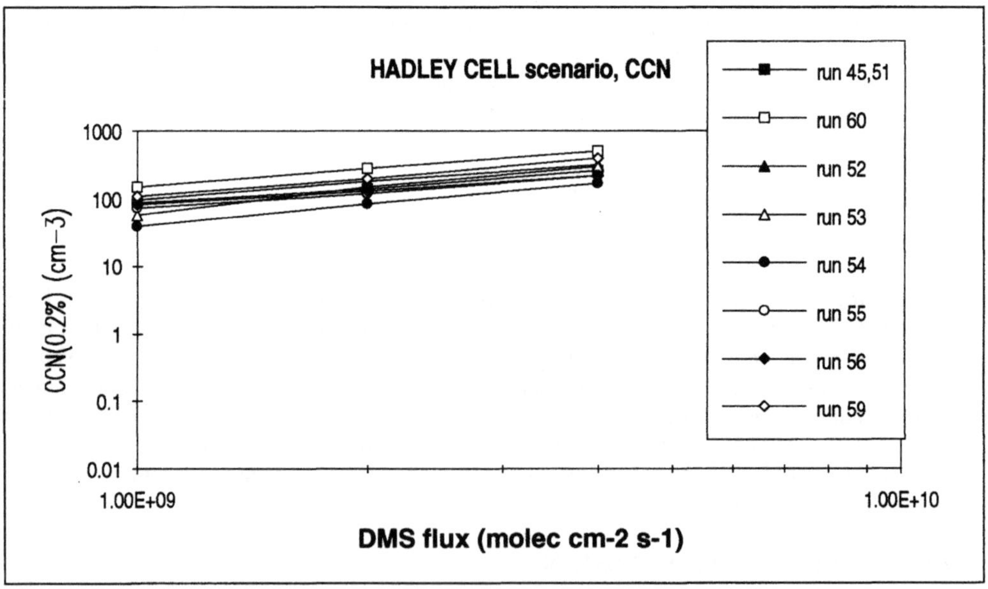

FIG. 6b Average CCN(0.2%) concentrations in the MBL, in air parcel that is constantly moving through the Hadley cell, calculated with a variety of input conditions (see Table II).

A fundamental feature of the latter mechanism is that the nucleation process is not very important for defining the aerosol number concentration. This gives us a clue for explaining the lack of sensitivity of the MBL CCN system. Indeed, whereas in-situ nucleation cannot be reconciled with the constancy of the MBL aerosol, the Hadley cell scenario lets nucleation happen far away from the MBL so that the exact way in which it occurs is 'forgotten' in the coagulation process by the time the particles reach the MBL. Although this conclusion was reached from a study on the sub-tropical MBL, we believe that the mechanism can be at work, each time air is pumped up in the troposphere, and slowly subsides back to the surface.

The Hadley cell scenario described above, is supported by extensive observations of Clarke [26, 27] who observed new particle formation from the gas phase in the tropical upper troposphere and absence of new particle formation in the tropical and subtropical lower atmosphere [28].

One should keep in mind , however, that our model only treats CCN derived from DMS. Although this is considered to be the main mechanism (see Introduction), measurements, both in the remote MBL and the free troposphere, have revealed the presence of aerosols that are not derived from the gas phase. Such aerosols, like sea salt, will not only affect the CCN population but can also interfere with the coagulation process, as was shown by Clarke [27].

The results of our model calculations anyway point at and quantify the importance of the free troposphere, and the exchange between free troposphere and boundary layer in producing boundary layer CCN and clouds. An aerosol characterisation experiment within a well localised large scale circulation cell is needed to give further support to this piece of understanding.

REFERENCES

1. CHAHINE M.T. (1992) The hydrological cycle and its influence on climate. Nature, 359, 373-380.
2. TWOMEY S. (1977) The influence of pollution on the short-wave albedo of clouds. J. Atmos. Sci., 34, 1149-1152.
3. CHARLSON R.J., LOVELOCK J.E. ANDREAE M.O. and WARREN S.G. (1987) Oceanic phytoplankton, atmospheric sulphur, cloud albedo and climate. Nature, 326, 655-661.
4. WIGLEY T.M. (1989) Possible climate change due to SO_2 derived cloud condensation nuclei. Nature, 339, 365-367.
5. AYERS G.P. and GRAS J.L. (1991) Seasonal relationship between cloud condensation nuclei and aerosol methanesulphonate in marine air. Nature, 353, 834-835
6. NGUYEN B.C., MIHALOPOULOS N., PUTAUD J.P., GAUDRY A., KEENE W.C. and GALLOWAY J.N. (1992) Covariations in oceanic dimethyl sulphide, its oxidation products and rain acidity at Amsterdam Island in the Southern Indian Ocean. J. Atmos. Chem. 15, 39-53.
7. GRAS J.L. (1990) Cloud condensation nuclei over the southern ocean. Geophys. Res.Lett. 17, 1565-1567
8. CALHOUN J.A., BATES T.S. and CHARLSON R.J. (1991) Sulfur isotope measurements of submicrometer sulfate aerosol particles over the pacific ocean.. Geophys. Res. Lett, 18, 1877-1880
9. RAES F. and VAN DINGENEN R. (1992) Simulations of condensation and cloud condensation nuclei from biogenic SO_2 in the remote marine boundary layer. J. Geophys. Res., 97, 12901-12912.

10. RAES F., SALTELLI A, and VAN DINGENEN R. (1992) Modelling formation and growth of H_2SO_4-H_2O aerosols: uncertainty analysis and experimental evaluation. J. Aerosol Sci., 23, 759-771.
11. NGUYEN B.C., MIHALOPOULOS N. and BELVISO S.(1990) Seasonal variation of atmospheric dimethyl sulphide at Amsterdam Island in the Southern Indian Ocean. J. Atmos. Chem. 11, 123-141
12. BATES T.S., CHARLSON R.J. and GAMMON R.H. (1987) Evidence for the climatic role of marine biogenic sulphur. Nature, 329, 319-321.
13. ANDREAE M.O. (1992) DMS discoveries, Gaian Science, 3, 10.
14. BATES T.S, CALHOUN J.A. and QUINN P.K. (1992) Variations in the methanesulphonate to sulfate molar ratio in submicrometer marine aerosol particles over the south pacific ocean. J. Geophys. Res. 97, 9859-9865
15. McMURRY P.H. and WILSON J.C. (1983) Droplet phase (heterogeneous) and gas phase (homogeneous) contributions to secondary ambient aerosol formation as functions of relative humidity. J. Geophys. Res., 88, 5101.
16. LURIA M. and SIEVERING H.(1991) Heterogeneous and homogeneous oxidation of SO_2 in the remote marine atmosphere. Atmos. Environ., 25A, 1489-1496.
17. SALTELLI A, RAES F., VAN DINGENEN R. and WILSON J. (1993) Modelling of the sulphur cycle; modules development and analysis of uncertainties. This volume.
18. TWOMEY S. and WOJCIECHOWSKI T.A. (1969) Observations of geographical variations of cloud nuclei. J. Atmos. Sci., 26, 684-688.
19. PUTAUD J-P. (1992) Personal Communication
20. The upper estimate for the time available in the sub-tropical MBL can be set equal to the transport time through the sub-tropics. For this we taken the local average wind velocity at ground level (2.5 m/s)[21] , and a linear dimension of 2000 km, derived from the areas of low level clouds in the sub-tropics [1]. This yields 12 days.
21. NEWELL R.E., KIDSON J.W., VINCENT D.G. and BOER G.J. (1972,1974) The general circulation of the tropical atmosphere. Vol. 1, 2. MIT Press, Cambridge, Massachusetts.
22. LELIEVELD J., CRUTZEN P.J. and RODHE H. (1989) Zonal average cloud characteristics for global atmospheric chemistry modelling. Rep. CM-76, University of Stockholm, Dept of Meteorology.
23. HOPPEL W.A., FITZGERALD J.W., FRICK G.M., LARSON R.E. and MACK E.J., (1990) Aerosol size distributions and optical properties found in the marine boundary layer over the Atlantic Ocean. J. Geophys. Res., 95, 3659-3686.
24. VAN DINGENEN R., JENSEN N.R., GROENE T., HJORTH J., MURRAY N. and RAES F. (1992) Physical and chemical aerosol characterization across the North Atlantic. EGS XVIII General Assembly, Wiesbaden, 3-7 May 1993.
25. CHATFIELD R.B. and CRUTZEN P.J. (1984) Sulfur dioxide in remote oceanic air: cloud transport of reactive precursors. J. Geophys. Res., 89, 7111-7132
26. CLARKE A.D. (1992) Atmospheric nuclei in the remote free-troposphere J. Atmos. Chem., 14, 479-488
27. CLARKE A.D. (1992) Atmospheric nuclei in the Pacific mid-troposphere: their nature, concentration and evolution. J. Geophys. Res. (in press).
28. Note that nucleation has been observed in the MBL[29], however not as a regularly occurring process, but as a result of peculiar conditions.
29. HEGG D. A., COVERT D. S. and KAPUSTIN V.N. (1992) Modeling case of particle nucleation in the marine boundary layer. J. Geophys. Res., 92, 9851-9857.

EQUATORIAL CONVECTION AS A SOURCE OF TROPOSPHERIC NUCLEI OVER THE REMOTE PACIFIC

ANTONY CLARKE

Department of Oceanography
University of Hawaii
Honolulu, HI, USA 96822

SUMMARY

Continuous production of new atmospheric nuclei is required to balance the depletion of nuclei scavenged by cloud processes and dry deposition to the earth's surface. These new nuclei, their origin and potential influence on climate are a focus of current interest since they may achieve sizes that can result in a significant influence on atmospheric radiative transfer (1) and/or a contribution to the population of cloud condensation nuclei (CCN) that help to determine cloud albedo (2) and ultimately global albedo (3). The relatively low and stable concentrations of nuclei common to the marine boundary layer (MBL), in spite of active cloud removal processes, indicates that compensating production of new nuclei must occur. However, unambiguous evidence for in-situ particle production in the clean MBL has been difficult to observe (4) and rarely documented (5). Measurements of surface nuclei measurements described here for the equatorial region suggest that conversion of DMS to sulfate is likely to occur heterogeneously upon existing nuclei. However, above cloud top over these regions and in clean air scavenged of boundary layer aerosol due to deep equatorial convection, prolific new nuclei are present in spite of very low mass concentrations. It is suggested that the source for these nuclei may be due to cloud pumping of reactive gases (6), such as DMS, into an environment aloft that favors homogeneous nucleation. The measurements indicate production of new particles occurs in the mid and upper troposphere and that these may evolve and subside to eventually provide a source for nuclei in the boundary layer.

1. INTRODUCTION

Evidence for photochemical particle production from gaseous precursors above exposed tidal flats, in polluted air and volcanic plumes are well known. However, concentrations of gaseous precursors in these locations are generally far higher than those common over the open ocean. Nonetheless, seasonal changes in condensation nuclei (CN) and CCN observed in the remote South Pacific (7, 8) have been argued to be due to the gas to particle (GPC) conversion of gases originating from the oxidation of dimethylsulfide (DMS) produced by plankton. This source is considered responsible for the excess (non-sea salt) sulfate mass found in most clean marine regions (9) and has been suggested to result increased numbers of particles being formed (3). However, recent models that attempt to describe nucleation from GPC under conditions present in the MBL indicate that the formation of new nuclei is often unfavorable unless the surface area of existing aerosol is lower than typically present in the MBL (10, 11, 12). Most previous studies looking into the relationship of DMS to particle formation have focussed upon experiments in marine boundary layer. Here I present data indicating that certain regions the free troposphere may be important as source regions for the homogeneous nucleation of new nuclei. Emphasis is placed upon the importance of equatorial deep-convection and the role of these clouds in the scavenging of aerosol mass from marine boundary layer air and their creation of conditions favorable for homogeneous nucleation of aerosol aloft.

G. Restelli and G. Angeletti (eds.), Dimethylsulphide: Oceans, Atmosphere, and Climate, 323–330.

2. NUCLEI IN MARINE BOUNDARY LAYER

If significant production of new nuclei does occur in the clean MBL and if DMS is a common precursor we might expect that regions of elevated DMS should frequently be associated with increased relative concentrations of new and smaller nuclei. In general this relationship has not been observed (4). The region of upwelling in the equatorial Pacific exhibiting moderate concentrations of DMS (13,9) associated with elevated nutrients has also been found to be associated with fine particle sulfate mass (14). However, the low relative and absolute concentration of smaller nuclei in this region (Figure 1) is not consistent with the notion that they are being produced via DMS oxidation. The fraction of all measured nuclei with diameters larger than 150nm are seen to reach a maximum over the upwelling region while the total number of smaller nuclei are at a minimum. The lack of evidence for new particles forming while aerosol mass is increasing suggests that existing particles are accumulating mass through heterogeneous GPC and growing larger during advection toward the intertropical convergence zone (ITCZ).

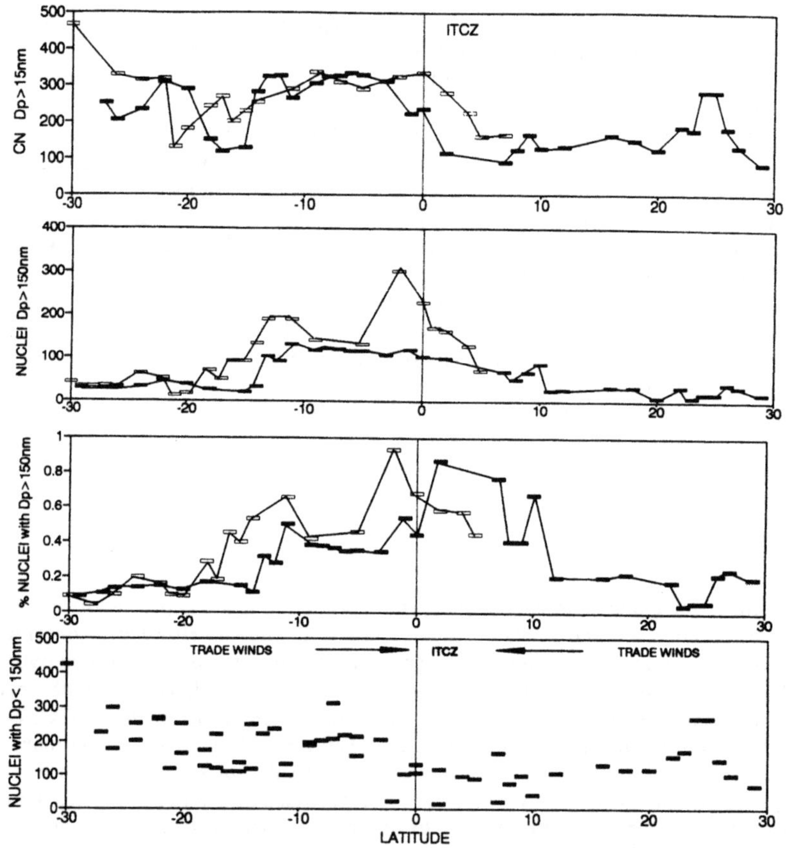

Figure 1. Surface measurements of nuclei for a north-south transect along 155W (solid box) and along 135W (open box) during SAGA1 (Soviet-American Gas and Aerosol) in Oct.-Dec., 1983 aboard the Akademik Korolev (14). From top to bottom the panels indicate the behaviour of total CN, large CN (Dp > 150nm), the ratio of larger to total CN and the small CN (Dp < 150nm). See text for discussion.

3. NUCLEI IN THE FREE TROPOSPHERE

This behavior of nuclei in the MBL stands in marked contrast to observations made aloft over this region (15,16). Nuclei with diameters larger than 3nm were measured near 8-12 km altitude above the North and South Pacific during the Global Backscatter Experiment (GLOBE) aboard the NASA DC-8 during May-June 1990. A comparison of nuclei averaged over ca. 200km intervals for most of the 35,000 km encompassed by the GLOBE transects (15) generally show that highest number concentrations were observed in air having lowest aerosol mass and aerosol surface area as measured with a laser optical particle counter (Figure 2).

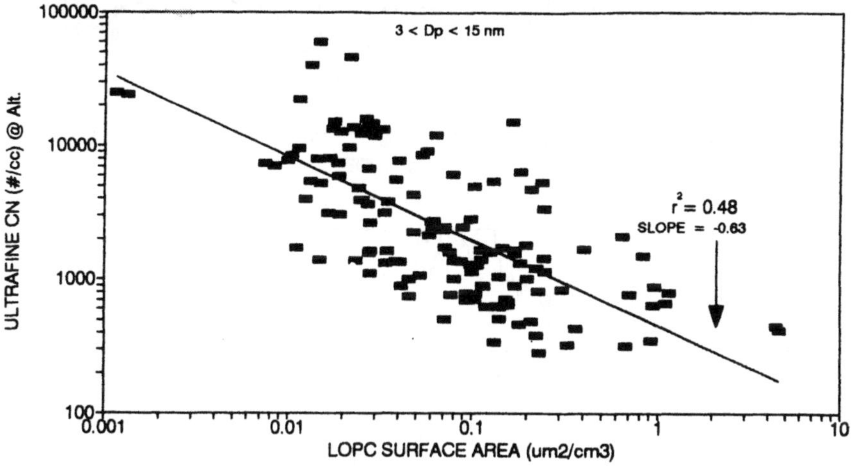

Figure 2. The relationship of Ultrafine nuclei concentrations to the surface area of larger aerosol (Dp > 150nm) for about 35,000 km at 8-12km altitude over the Pacific.

Conditions of low aerosol mass and high nuclei counts at altitudes of 8-12km were very apparent for flights above the ITCZ (Figure 3a,b) and SPCZ (South Pacific Convergence Zone) and near or above other convective regions where surface area of larger aerosol were also at a minimum. This figure reveals the behavior of the various classes of measured CN that, based upon instrument performance (16), I operationally define as Ultrafine (UCN, 3nm < Dp < 15nm); Volatile (VCN-volatile at 300C, 15nm < Dp < 3000nm) and Refractory (RCN-nonvolatile at 300C, 15nm < Dp < 3000nm). UCN, those nuclei most recently formed from gas to particle conversion, are frequently highest in the vicinity of deep convection. VCN, frequently found to be volatile sulfate originating from gas to particle conversion, are also generally higher in these regions while the RCN derived from the surface as either sea-salt, dust or soot tend to be lowest. The similar variability apparent in UCN and VCN evident here and in other GLOBE data suggests that the evolve from the UCN in regions least influenced by surface sources as reflected by the presence of RCN (15,16).

Elevated relative humidity at 8km between 12N and 4S above the convective ITCZ can be seen in Figure 3. Vertical profiles of humidity fields (not shown) generated along the flight path [D. Bowdle, NASA-Marshall Space Flight Center] indicate that the air aloft over this region had its origin near the ocean surface. Nuclei concentrations in dense cloud with 100% RH have been deleted due to spurious particle counts caused by cloud droplet shatter on the inlet probes. Otherwise, during most periods of high (low) RCN both the VCN and UCN are relatively low (high). Since the flight was typically just above the altitude of convective cloud top, the variability in these nuclei may reflect passage through alternating regions of nearly complete and incomplete cloud scavenging of marine boundary layer aerosol mass.

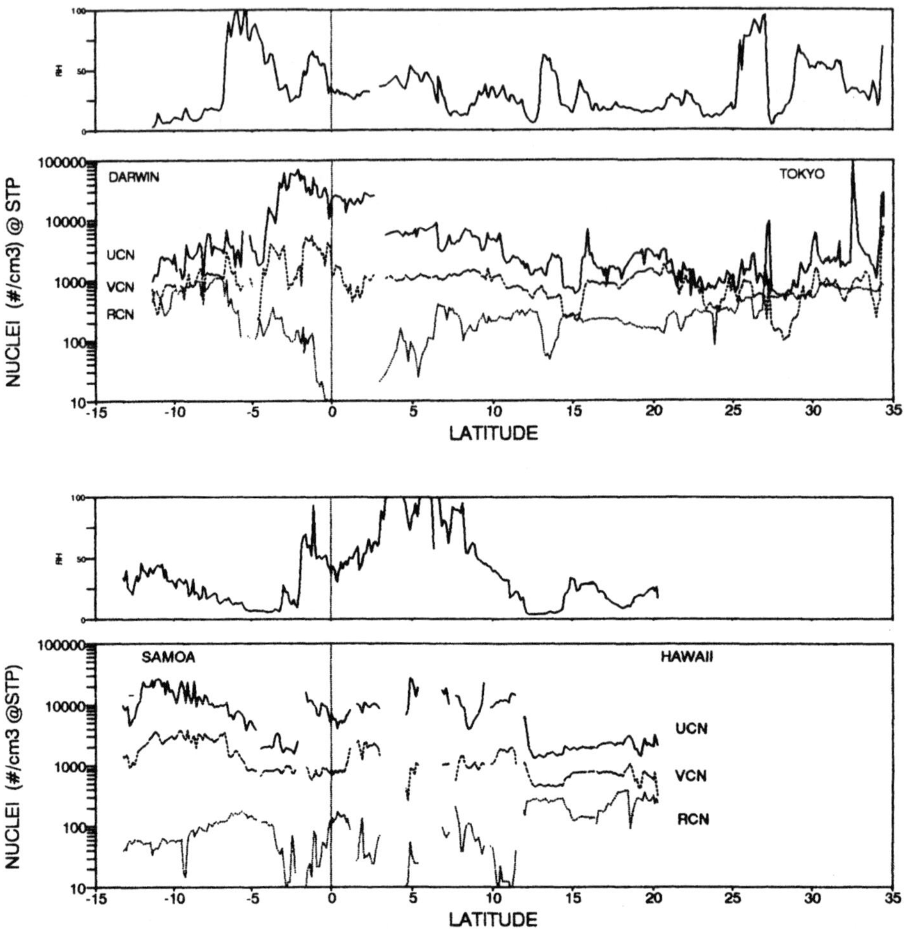

Figure 3 a,b. Relative humidity and the nuclei variability (UCN, VCN and RCN @ STP-see text) for horizontal transects from Darwin to Tokyo (a) and from Hawaii to Samoa (b) at altitudes of 8-11 km in the free troposphere.

Previous comparisons of aerosol mass and RCN for the flight from Darwin to Tokyo (Figure 3a) showed that RCN are a good indicator of aerosol mass (15, 16). Increasing RCN and the low variability seen for nuclei on this flight north of 12N and south of 7S coincide with over an order of magnitude higher aerosol mass measured by the laser optical particle counter (not shown), indicative of air masses quite different from that of the moist equatorial region. Both of the equatorial transects (Fig. 3) suggest that when air originates from the marine boundary layer through deep convection then it is both cleansed of existing aerosol and enhanced in water vapor and reactive gases (6) that favor nucleation of new particles. This notion is in accordance with modeling results described elsewhere in these proceedings that show similar particle formation aloft and subsequent return to the boundary layer via the Hadley cell circulation (11).

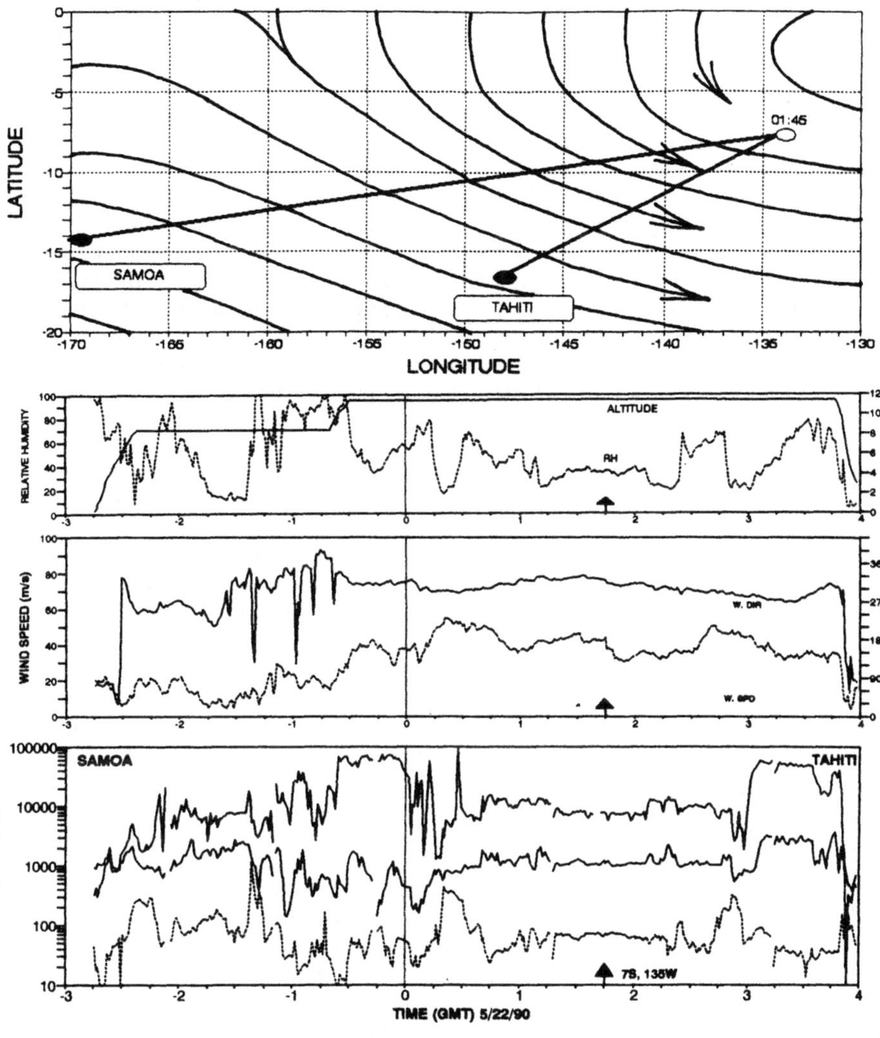

Figure 4. Panels showing data from the GLOBE flight from Samoa to Tahiti including an excursion to 7S, 134W. From top to bottom the panels depict the flight path superimposed upon the 250mb wind field, the altitude and relative humidity variability, aircraft measurements of wind direction and wind speed, and the behaviour of UCN, VCN and RCN as in Figure 2.

An opportunity to explore characteristics of the convective regions near the equator was afforded by the flight from Samoa to Tahiti that included the excursion to 7S and 135W shown in Figure 4. The waypoint time at the turn is indicated as point of reference. The top panel shows the 250mb flow field for 12:00 GMT superimposed on the flight path (ca. 220mb). The several hour time difference will result in the actual flow field at our altitude being slightly different. The second panel shows relative humidity and the flight altitude which was increased about half an hour before 00:00GMT in order to climb above towering cumulus. Wind direction and speed measured on aboard the aircraft is shown in the middle panel and is in reasonable agreement with the 250 mb analysis. Wind speeds are lower in the more westerly flow near Samoa but intensify and become more northerly near about 155W, about 1 hour before 00:00 GMT. At this point we were passing through the cores of large cumulus towers responsible for relative humidity values near 100% and the rapid excursions in wind direction affected by cloud outflow. Variations in the nuclei shown in the lower panel are driven by whether we are in or out of cloud. In general both UCN and VCN were lower and RCN higher in cloud and the reverse was observed between cloud tops.

The climb above the clouds (-00:30GMT) to about 11 km is accompanied by a lowering of RH to around 50% (Fig. 4). This is also accompanied by a dramatic increase in UCN and lower but variable RCN. Shortly after 00:00 GMT there is a temporary transition to drier air, higher wind speed, lower UCN and higher RCN. This 120 km wide region with highest winds is close to the boundary between the easterly flow coming recently from the south Pacific and the northerly flow coming recently from above the ITCZ just north of the equator. From here to the turn point and back the UCN and CN change slowly with number concentrations about a factor of two greater in the regions of higher RH. Note the pronounced similarity in the aerosol, humidity, wind speed, wind direction etc. during the approach and return from the turn point indicating a spatial coherence and relationship of the aerosol and meteorological fields present on horizontal scales up to 500km.

For the indicated flight leg at 11 km the UCN are generally 8,000 cm^{-1}, VCN about 1,000 cm^{-1} and RCN about 80 cm^{-1}. The corresponding low aerosol mass concentrations measured by the OPC were on the order of about 0.005 ıg m^{-1} and about twice that in the "dry" regions. Closer to Tahiti the narrow region of low RH, low UCN, low VCN but high RCN is encountered again (02:50GMT) suggesting it may be an extended belt of air with these properties. Nuclei volatility and concentration in this transitional dry belt are not unlike that shown near 10S on the Darwin-Tokyo flight (Figure 2). This location, just north of Australia, is also the likely position of this dry air about 24 hours before based analysis of the 250mb flow field for the previous day. The dry belt (03:00 GMT) is followed by marked increases in RH, UCN, and CN but a decrease in RCN prior to our descent. Here we flew between extensive cirrus layers that appear to be an extension of the region of deep convection noted on the outbound portion just before 00:00 GMT.

The descent profile into Tahiti for various nuclei from 7km to the surface is shown in Fig. 5. Some thin cirrus aloft gave way to clear air until about 5 km altitude where isolated cumulus convection was common down to the boundary layer. Below 2 km the air was contaminated by anthropogenic sources and is

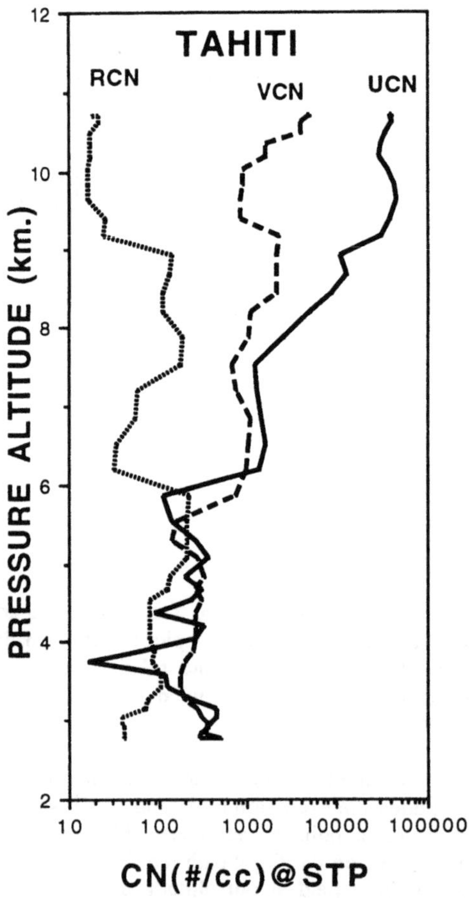

Figure 5. Profile of nuclei during descent into Tahiti.

not shown. In the upper troposphere the winds were predominately westerly and at the surface easterly, typical of mean conditions. Climatologically, this air is often convected aloft both by deep convection over the "warm pool" in the western Pacific (17) and convective cumulus towers in the ITCZ and SPCZ, not unlike the conditions present for this flight. A high degree of scavenging is expected from this convection and is consistent with our observation of low aerosol mass and relatively low surface derived nuclei. This cloud pumping over the equatorial upwelling also has the potential for making the air relatively rich in DMS (6).

The Tahiti profile has the highest concentrations of ultrafine nuclei and the most conspicuous gradient in concentration toward the surface of all GLOBE profiles (16). Very low RCN indicative of low aerosol mass and effective cloud scavenging of surface aerosol is also evident. Between 11 and 5 km the STP concentrations of UCN decrease by two orders of magnitude, the VCN decrease by one order of magnitude and the RCN increase by about one order of magnitude. Below 5 km concentrations are variable but show little gradient probably reflecting the cumulus convection present in the area up to that altitude. Since this region has been identified both climatologically and at this time of year as a region of subsidence (17) the gradient from aloft toward the surface for VCN and UCN is indicative of a the mid troposphere being a source of nuclei for the lower troposphere in this region. The role of equatorial convection in driving the Hadley circulation implies that these measurements in these regions may have global significance to our understanding of the source of tropospheric nuclei and their relationship to global circulation. They are also consistent with model results of the formation and evolution of such nuclei in the Hadley cell and the implication for supplying and maintaining the marine boundary layer CN and cloud condensation nuclei concentrations through subsidence (11).

4. SUMMARY

Regions of deep convection appear to play a dual role in the cycling of atmospheric aerosol. In-situ nucleation of cloud droplets, their coagulation and the subsequent removal through precipitation continuously depletes the mass, surface area and number of atmospheric aerosol brought up from the surface. However, the same convective process can pump reactive gasses aloft into regions with higher oxidant, higher insolation and much lower temperatures all of which markedly favor supersaturations required for homogeneous nucleation. Through formation and aging in clean air aloft these new particles have sufficient time to coagulate with themselves and evolve to sizes more resistant to diffusional losses before they are mixed into the boundary layer.

Although these particles would be smaller than MBL aerosol during subsidence, once mixed into the MBL they could continue to grow by heterogeneous GPC if a source of precursors was present. This suggests a coupling may exist between dynamically and thermally driven global processes and the cycling and renewal of tropospheric aerosol. As such, the argument presented above indicates a link between large scale dynamically/thermally forced convection and the source/sink regions, mechanisms, lifetimes and microphysics of new particle production in the troposphere.

Most other GLOBE profiles tend to show a layer of similar clean air separated by layers with greater surface influence (15), indicating that horizontal advection and superposition of different air masses is a common feature that must be considered in the application of the model described above. Nevertheless, the new nuclei formed aloft over convective regions must ultimately be returned to the surface in subsiding air that will contribute to nuclei populations over large regions. The evolution of these nuclei will inevitable grow some of them into sizes that can act as CCN in the MBL. Ultimately these too will nucleate cloud drops that will be scavenged through precipitation and provide a region of clean air aloft where the process of homogeneous nucleation and replenishment may begin anew.

5. ACKNOWLEDGEMENT

I would like to thank the scientists and crew members of the Soviet R/V Akademic Korolev and the NASA DC-8 for all of their efforts in making these measurements possible. This work was supported under NSF-ATM 8513776 and NASA-NAG8-848. This is SOEST contribution number 3188.

REFERENCES

(1) CHARLSON, R.J. et al. (1992), Climate Forcing by Anthropogenic Aerosols, Science, 255, 423-430.

(2) TWOMEY, S., M. PIEPGRASS, and T.L. WOLF (1984), An assessment of the impact of pollution on global cloud albedo, Tellus, 36B, 356-366.

(3) CHARLSON, R.J., J.E. LOVELOCK, M.O. ANDREAE, and S.G. WARREN (1987), Oceanic Phytoplankton, atmospheric sulphur, cloud albedo and climate, Nature, 326, 655-661

(4) BATES, T. S., A. D. CLARKE, V. N. KAPUSTIN, J. E. JOHNSON and R. J. CHARLSON (1989), Oceanic dimethyl sulfide and marine aerosol: difficulties associated with assessing their covariance, Global Biogeochem. Cycles, 3, 299-304, 1989.

(5) COVERT D.S. (1992), New Particle Formation in the Marine Boundary Layer, Jour. Geophys. Res., 20,581-20589.

(6) CHATFIELD, R. B. and P. J. CRUTZEN (1984), Sulfur dioxide in remote oceanic air: cloud transport of reactive precursors, J. Geophys. Res., D: Atmos, 89, 7111-7132.

(7) AYERS, G. P., J. P. IVEY and R. W. GILLETE (1991), Coherence Between Seasonal Cycles of Dimethyl Sulfide, Methanesulfonate, and Sulfate in Marine Air, Nature, 349, 404-406.

(8) BIGG, E.K., J.L. GRAS, and C.J. EVANS (1984), Origin of Aitken particles in remote regions of the Southern Hemisphere, J. Atmos. Chem., 1, 203-214.

(9) BATES, T. S. et al., (1987), Regional and Seasonal Variations in the Flux of Dimethylsulfide to the Atmosphere, J. Geophys. Res., 92, 2930-2938.

(10) KREIDENWEISS S. M., J. E. PENNER, F. YIN and J. H. SEINFELD (1991). The effects of Dimethyl Sulfide on Marine Aerosol Concentrations, Atmos. Environ., 25A, 2501-2512, .

(11) RAES, F. et al., (1993) Cloud condensation nuclei from dimethylsulfde in the natural marine boundary layer, in these proceedings.

(12) LIN, X. ET AL., (1992), A model study of the formation of Cloud Condensation Nuclei in remote marine areas, Jour. Geophys. Res., 18,161-18,71.

(13) ANDREAE, M. O. and H. RAEMDONCK (1983) Dimethyl sulfide in the surface ocean and the marine atmosphere: A global view, Science, 221, 744-747.

(14) CLARKE, A.D., N.C. AHLQUIST, and D.S. COVERT (1987), The Pacific marine aerosol: evidence for natural acid sulfates, J. Geophys. Res., 92, 4179-4190.

(15) CLARKE, A.D., (1992), Atmospheric Nuclei in the Remote Free Troposphere, Jour. Atmos. Chem., 479-488.

(16) CLARKE, A.D., (1993), Atmospheric Nuclei in the Pacific mid-troposphere: their nature, concentration and evolution. (in press)

(17) MURAKAMI, T. and B. WANG,(1993) Annual Cycle of East-West circulation over the Indian and Pacific Oceans, in press Jour. of Climate.

RELATIONSHIPS BETWEEN DMS-DERIVED PARTICULATE MASS, PARTICLE SURFACE AREA, AND CN AND CCN NUMBER CONCENTRATIONS(*)

P.K. QUINN, T.S. BATES, D.S. COVERT, D.C. RAMSEY-BELL and L. McINNES

NOAA/PMEL and University of Washington, Seattle, WA, USA

ABSTRACT

The oxidation of atmospheric dimethylsulfide (DMS) yields water soluble aerosol particles which contain non-seasalt sulfate (nss $SO_4^=$) and methanesulfonic acid (MSA). These particles can act as cloud condensation nuclei (CCN) whose number concentration may influence cloud droplet concentrations and, hence, cloud albedo. Quantification of this indirect effect of nss $SO_4^=$ aerosol on the earth's radiation budget depends on knowing: 1) what controls the DMS oxidation pathway and, hence, ratio of end products; 2) if these end products result in new particle production or particle growth; 3) what the quantitative relationship is between DMS-derived particulate mass and condensation nuclei (CN) number concentration, and 4) what fraction of the DMS-derived CN becomes CCN.

These questions were addressed during the Pacific Sulfur/Stratus Investigation (PSI) in April of 1991 in a region dominated by marine air and stratus clouds. Simultaneous measurements included gas phase DMS and SO_2, particulate phase mass size distributions of nss $SO_4^=$ and MSA, number size distributions of particles having diameters between 0.02 and 0.6 μm, total number concentration of particles larger than 0.003 μm diameter, CCN concentrations at a supersaturation of 0.3%, relative humidity, atmospheric temperature, and degree of cloudiness.

The amount of nss $SO_4^=$ and MSA found in particles with diameters greater than 1 μm ranged between 8 and 44% and 8 and 53%, respectively. Total available particle surface area, estimated from number and mass size distributions, ranged between 8 and 200 $\mu m^2\ cm^{-3}$. Statistically significant correlations were found between both nss $SO_4^=$ and MSA mass and particle surface area. These correlations as well as the magnitude of the total available particle surface area suggest that newly formed H_2SO_4 and MSA condensed onto pre-existing aerosol instead of forming new particles.

No correlations were found between mass and CN number concentrations for particles smaller than 0.16 μm in diameter. However, both nss $SO_4^=$ and MSA mass and CN concentrations did correlate for two larger particle size ranges. Between 0.16 and 0.25 μm, a doubling of $SO_4^=$ and MSA mass corresponded to a 50% increase in number concentration. Between 0.25 and 0.5 μm, a doubling of $SO_4^=$ and MSA mass corresponded to a 70 and 60% increase in number concentration, respectively. The lack of a correlation between mass and number in the smallest size ranges confirms that H_2SO_4 and MSA were involved in particle growth and not new particle production.

CCN measurements indicated that 60% of CN with diameters greater than 0.07 μm were activated at a supersaturation of 0.3%. This suggests an aerosol composed of both soluble and insoluble material such that not all DMS-derived particulate mass was incorporated into CCN at ambient supersaturations.

(*) The full paper has been submitted to: J. Geophysical Research.

G. Restelli and G. Angeletti (eds.), Dimethylsulphide: Oceans, Atmosphere, and Climate, 331.

ATMOSPHERIC CONCENTRATION OF DMS AND ITS OXIDATION PRODUCTS ESTIMATED IN A GLOBAL 3-D MODEL

H. RODHE
Department of Meteorology,
Stockholm University
S-10691 Stockholm
Sweden

J. LANGNER
Swedish Meteorological and Hydrological Institute,
S-60176 Norrköping
Sweden

SUMMARY

A global three-dimensional transport model (MOGUNTIA), originally developed at Max-Planck Institute for Chemistry in Mainz, has been used to simulate the atmopheric distribution of sulfur species (DMS, SO_2 and aerosol sulfate) from natural and anthropogenic sources. The simulations are broadly consistent with observations of sulfur concentrations in surface air and precipitation, most of them obtained in and around polluted regions in Europe and North America. Here we focus on the contribution to the atmospheric sulfur burden of DMS emissions from the oceans. The total oceanic emission of DMS was assumed to be 16 Tg S/yr. Such emissions are shown to be reasonably consistent with observed concentrations of DMS and SO_2 at remote marine locations. We present maps of estimated concentrations in surface air of DMS and SO_2 originating from the oxidation of DMS. We also compare calculated and observed seasonal cycles of the concentration of DMS and SO_2 at two remote oceanic sites. DMS concentrations are consistently overpredicted at high latitudes in winter, presumably due to overestimated emissions or underestimated removal. Finally, we discuss the potential effect on radiation fluxes and climate of aerosol sulfate derived from oxidation of DMS.

1. INTRODUCTION

Emissions of DMS from the oceans have been hypothesized to be of climatic significance through a feedback loop involving the global climate (6). Atmospheric oxidation of DMS is assumed to lead mainly to sulfate. This sulfate may either form new, or attach to preexisting aerosol particles to form particles that can act as cloud condensation nuclei (CCN) (8). Due to lack of understanding of the processes regulating the number concentration of CCN this indirect effect of sulfate aerosols is presently difficult to evaluate quantitatively. Recently, the direct scattering of shortwave radiation by sulfate aerosols has received increased attention. The direct scattering by anthropogenic sulfate aerosols is estimated to cause a cooling of ~1 W m^{-2} averaged over the Northern Hemisphere (9, 10). Here we focus on the contribution to the atmospheric sulfur burden of DMS emissions from the oceans. We present maps of estimated emissions of DMS to the atmosphere along with calculated concentrations in surface air of DMS and SO_2 at a few selected remote sites. Finally we discuss the potential effect on radiation fluxes and climate of aerosol sulfate derived from oxidation of DMS.

G. Restelli and G. Angeletti (eds.), Dimethylsulphide: Oceans, Atmosphere, and Climate, 333–343.

2. MODEL

A three-dimensional Eulerian transport model, MOGUNTIA (31), has been used in the calculations. The model is based on climatological monthly mean meteorological data and covers the global troposphere with a horizontal resolution of 10° longitude by 10° latitude and with ten layers in the vertical between the surface and 100 hPa. The present version simulates the tropospheric distribution of dimethyl sulfide (DMS), sulfur dioxide (SO_2) and non-seasalt (nss) sulfate aerosol (SO_4^{2-}) (16). The major part of the emissions of volatile sulfur compounds occurs as DMS and SO_2. SO_4^{2-} is assumed to be the end product. COS is the most abundant sulfur species in the troposphere and is an important compound in the stratospheric sulfur budget, but its contribution to the fluxes of sulfur through the troposphere is small and it is therefore not included in the model. Other reduced sulfur compounds are estimated to account for less than 5% of the total natural emission of sulfur gases (1) and are excluded from the model calculations.

Gas phase oxidation of DMS to SO_2 and of SO_2 to sulfate is proportional to the concentration of hydroxyl radicals (OH). Three-dimensional time dependent OH fields are derived from a version of the MOGUNTIA model that contains full CH_4-CO-NO_x-OH-O_3 chemistry (11). Aqueous phase oxidation of SO_2 to sulfate is assumed to be proportional to cloudiness, wet deposition of SO_2 and sulfate to precipitation, and dry deposition of SO_2 and sulfate to their surface concentrations. Further details regarding the model formulation are given in Langner and Rodhe (16).

3. EMISSIONS

Global and hemispheric sulfur emissions are summarized in Table 1. Anthropogenic emissions and natural emissions of SO_2 are given for purposes of comparison. Three types of natural emissions are considered in the calculations: Oceanic emissions, emissions from plants and emissions from soils. The distribution of DMS emissions from the oceans was based on studies by Bates et al. (4) and Leck and Rodhe (18). Two seasons are distinguished, May - October and November - April. Within each season the flux from the ocean varies with latitude and in the equatorial region also with longitude based on measured concentrations of DMS in ocean surface waters and estimates of the air-sea exchange based on climatological wind speed. Considerable uncertainty still pertains to both the distribution of (in time and space) and magnitude of emission of DMS from the oceans. Bates et al. (4) estimate a global DMS emission from the oceans of 16 Tg S yr^{-1} with an uncertainty of a factor of two, while Andreae (1) gives a range of 19-51 Tg S yr^{-1}. The difference is due to different choices of transfer coefficients for the flux of DMS between ocean and atmosphere, the inclusion of seasonality in the Bates et al. study (4) and different extrapolations of seawater concentration data.

The emission from soils and vegetation was calculated using the temperature dependent emission functions for different soil and vegetation types together with global soil type and land use data bases presented by Henderson-Sellers et al. (14). A global emission of H_2S and DMS from inland soils and plants of 1 Tg S yr^{-1} was estimated and included in the model (as DMS). The emission from coastal wetlands may be of similar magnitude as the emission from inland soils and plants (1) but it has been neglected in the calculations.

The uncertainty in the estimate of the natural emission is of the order of a factor of two (1), mainly due to the uncertainties in the emission of DMS from the oceans.

TABLE 1. Emissions of sulfur inTg S yr^{-1}. NH and SH data were used in the model. The range of values corresponds to recent estimates given in the indicated references.

Emission	Global	Range	NH	SH	References *
Industrial (SO_2)‡	70	(62-80)	64	6	a,b,c,d,e
Biomass burning (SO_2)	2.5	(2.3-2.8)	1.1	1.4	a,e,f
Oceans (DMS)	16	(8-51)	6.9	9.1	e,f,g
Volcanoes (SO_2)	8.5†	(7.4-9.4)	5.8	2.7	h,f
Soil and plants (DMS, H_2S)	1	(0.2-4.3)	0.6	0.4	a,e,f,i
Total:	107.0		82.3	24.7	

†The fraction emitted into the troposphere.
‡5% is emitted as sulfate.
*a is (16), b (13), c (20), d (29), e (26), f (1), g (4), h (28) and i (6).

The geographical distribution of the DMS emissions used in the calculations for January and July is shown in Figures 1a and 1b. The oceanic emission exhibits a pronounced seasonal variation, especially at mid and high latitudes. The biogenic emissions from the continents are estimated to be relatively small in comparison with the other preindustrial emissions of sulfur (Table 1), resulting in low emissions over the continents.

4. RESULTS

4.1. Concentration of DMS

Figures 2a and 2b show the calculated surface concentrations of DMS for December-February and June-August, respectively. Long-term measurements from marine environments are also included. The modelled DMS distribution exhibits a strong seasonal behaviour with highest concentrations over the high latitude oceans in the summer hemisphere. High concentrations are also found over parts of the equatorial oceans all year around. Modelled concentrations are low over land surfaces due to lower emission rates (Table 1). The measurements confirm this general picture although the spatial coverage is poor. The fair agreement between observed and calculated concentration provides some support for the estimated global flux. The winter season concentrations at high latitudes in both hemispheres seem to be overpredicted by the model. This is likely to be due to overestimated emissions, but underestimated removal (due to the neglect of reaction with the NO_3 radical?) could also contribute. Figures 3a and 3b show the observed and model calculated seasonal cycles of DMS at Cape Grim (41°S, 145°E) and Amsterdam Island (38°S,77°E). The phase of the seasonal cycle in the model is shifted two to three months compared to the observations and the amplitude is too small.

4.2. Concentration of SO_2

Figure 4a shows the calculated annual average surface concentrations of SO_2 resulting from both natural and anthropogenic emissions. A few available measurements are also included. The most prominent feature is the high concentration over the industrialised regions in Europe, North America and East Asia. The SO_2 concentrations in Figure 4b are those resulting from the DMS emissions alone. A comparison between Figures 4a and 4b shows that the DMS source has a dominating influence on the surface SO_2 concentration over the world's oceans. It is only in coastal regions and in the North

Atlantic that other sources are important. The limited number of observations is in fair agreement with the calculations. The low SO_2 concentrations at Cape Grim, Tasmania (around 10 pptv as an annual average) are partly due to the fact that these measurements have been selected to represent baseline conditions only; i.e. situations with a continental influence have been excluded from the records. Thus, the observed SO_2 concentrations, albeit few, do not contradict the assumption of a DMS flux of the magnitude given in Table 1. Despite the reasonable agreement between modelled and observed concentrations of both DMS and SO_2 a larger flux of DMS cannot be excluded provided that the rates of removal of DMS and SO_2 are larger than what we have assumed in the model. However, the flux of DMS cannot be much larger in view of the observed concentrations of nss sulfate (c.f. 14). On the other hand, the DMS flux can hardly be much lower than what we have assumed since it would then be difficult to explain the observed concentrations of DMS and SO_2.

Figure 5 shows the observed and model calculated seasonal cycles of SO_2 at Amsterdam Island. Except for a certain overprediction during the winter season, both phase and amplitude are in reasonable agreement with observations.

4.3 Direct radiative effect

Figure 6 shows the annual average direct sulfate aerosol forcing due to emissions of DMS. This forcing was calculated using the same procedures as in Charlson et al. (10). The maximum forcing is found over the equatorial ocean areas where the annual emissions of DMS are strongest (Figure 1). The maximum calculated forcing is about 1 W m^{-2} in the equatorial Pacific of South America. An extended area from the equatorial Atlantic across northern Africa and into the northern parts of the Indian Ocean is also substantially (larger than 0.5 Wm^{-2}) affected. The large values over this land area are due to the low cloudiness which increases the lifetime of the sulfate particles and also makes their impact on the radiation balance more pronounced. The calculated values shown in Figure 6 can be compared with the corresponding values for the anthropogenic sulfate given in Charlson et al. (10) - 1.1 Wm^{-2} as an average for the Northern Hemisphere - and with the current global average of the anthropogenic greenhouse forcing - about 2.5 Wm^{-2} (positive forcing; (15)). The DMS induced climate forcing due to the direct radiative effect is thus relatively small but far from insignificant, especially in the tropics.No attempt is made here to estimate the possible climate forcing of DMS due to its indirect effect, via changes in cloud condensation nuclei and cloud albedo (8).

5. CONCLUDING REMARKS

We have shown results from model calculations of the global distribution DMS and SO_2 based on a three-dimensional climatological tracer model. A comparison with observations of DMS and SO_2 at remote oceanic sites gives a certain support to the assumptions made in the model, in particular the general magnitude of the total emisssions of DMS. However, the number of observations is very limited and many uncertainties remain. A better understanding of the role of DMS in the global sulfur cycle and its potential impact on climate requires further investigations of several key processes including:

- The DMS flux across the air/sea interface.
- The oxidation pathways of DMS in the atmosphere.
- The oxidation of SO_2 in cloud droplets and in aerosol particles.
- The precipitation scavenging of sulfate particles.

What is needed to achieve this is both more high quality monitoring data from remote sites and comprehensive field campaigns including measurements of the vertical distribution of sulfur compounds, H_2O_2, clouds etc. under well characterised meteorological conditions.

ACKNOWLEDGEMENT

This work has been carried out within the GLOMAC subproject of EUROTRAC with financial support from the Swedish National Environment Protection Board under contracts No. 5620341-7 and No. 34203.

REFERENCES

(1) ANDREAE, M.O. (1990). Ocean-atmosphere interactions in the global biogeochemical sulfur cycle. *Mar. Chem.* 30, 1-29.

(2) ANDREAE, M.O., FEREK, R.J., BERMOND, F., BYRD, K.P., ENGSTROM, R.T., HARDIN, S., HOUMERE, P.D., LeMARREC, F. and RAEMDONCK, H. (1985). Dimethyl sulfide in the marine atmosphere. J. Geophys. Res. 90, 12,891-12,900.

(3) AYERS, G.P., GILLETT, R.W., IVEY, J.P. and GRAS, J.L. (1993). Measurement of dimethyl sulfide, sulfur dioxide, methane sulphonic acid and seasalt sulfate at three sites in the southern hemisphere. This book.

(4) BATES, T.S., CLINE, J.D., GAMMON, R.H. amd KELLY-HANSEN, S.R. (1987). Regional and seasonal variations in the flux of oceanic dimethylsulfide to the atmosphere. J. Geophys. Res. Vol. 92, 2930-2938.

(5) BATES, T.S., JOHNSON, J.E., QUINN, P.K., GOLDAN, P.D., KUSTER, W.C., COVERT, D.D. and HAHN, C.J. (1990). The biogeochemical sulfur cycle in the marine boundary layer over the Northeast Pacific Ocean. J. Atmos. Chem. Vol. 10, 59-81.

(6) BATES, T. S., LAMB, B. K., GUENTHER A., DIGNON, J. and STOIBER, R.E. (1992). Sulfur emissions to the atmosphere from natural sources. J. Atmos. Chem. Vol. 14, 315-337.

(7) BERRESHEIM, H., ANDREAE, M.O., IVERSON, R.L. and LI, S.M. (1991). Seasonal variations of dimethyl sulphide emissions and atmospheric sulfur and nitrogen species over the western North Atlantic Ocean. Tellus 43B, 353-372.

(8) CHARLSON, R.J., LOVELOCK, J.E., ANDREAE, M.O. and WARREN, S.G. (1987). Oceanic phytoplancton, atmospheric sulfur, cloud albedo and climate. Nature. Vol. 326, 655-661.

(9) CHARLSON, R.J., LANGNER, J. AND RODHE, H. (1990). Sulphate aerosol and climate. Nature. Vol. 348, 22.

(10) CHARLSON, R.J., LANGNER, J., RODHE, H., LEOVY, C.B. and WARREN, S.G. (1991). Perturbation of the northern hemisphere radiative balance by backscattering from anthropogenic sulfate aerosol. Tellus Vol.43 AB, 152-163.

(11) CRUTZEN, P. J. and ZIMMERMANN, P. H. (1991). The changing photochemistry of the troposphere. Tellus Vol. 43 AB, 136-151.

(12) DAVISON, B and HEWITT, C.N. (1992). Natural sulphur species from the North Atlantic and their contribution to the United Kingdom sulphur budget. J. Geophys. Res. Vol. 87, 2475-2488.

(13) HAMEED, S. and DIGNON, J. (1988). Changes in the geographical distributions of global emissions of NO_x and SO_x from fossil fuel combustion between 1966 and 1980. Atmos. Environ.Vol. 22, 441- 449.

(14) HENDERSON-SELLERS, S., WILSON, M.F., THOMAS, G., DICKINSON, R.E. (1986). Current global land-surface data sets for use in climate related studies. NCAR/TN-272 + STR. Atmospheric analysis and prediction division, National Center for Atmospheric Research, Boulder, CO.

(15) IPCC (1990). Climate Change; The IPCC Scientific Assessment, World Meteorological Organization and United Nations Environment Programme, Cambridge University Press, Cambridge.

(16) LANGNER, J., and RODHE, H. (1991). A global three-dimensional model of the tropospheric sulfur cycle. J. Atmos. Chem. Vol. 13, 225-263 .

(17) LECK, C., NILSSON, D., PERSSON, C., BIGG, K and COVERT, D.C. (1993). Biogenic sulfur in the marine boundary layer of the Arctic; The International Arctic Ocean Expedition 1991. This book.

(18) LECK, C. AND RODHE, H. (1991). Emissions of marine biogenic sulfur to the atmosphere of Northern Europe. J. Atmos. Chem. Vol. 12, 63-86

(19) LEVY, H., GALLOWAY, J.N., ELIASSEN, A., FISCHER, B.E.A., GORZELSKA, K., HASTIE, D.R., MOODY, J.L., RYABOSHAPKO, A.G. SAVOIE, D.and WHELPDALE, D.M. (1990). The long-range transport of sulfur and nitrogen compounds, in A.H. Knap (ed.) The Long-Range Atmospheric Transport of Natural and Contaminant Substances. Kluwer Academic Publishers, Dordrecht, 231-257.

(20) MÖLLER, D. (1984). Estimation of the global man-made sulphur emission. Atmos. Environ. Vol. 18, 19-27.

(21) NGUYEN, B.C., BONSANG, B. and GAUDRY, A. (1983). The role of the ocean in the global atmospheric sulfur cycle. J. Geophys. Res. Vol. 88, 10903-10914.

(22) NGUYEN, B.C., MIHALOPOULOS, N. and BELVISO, S. (1990). Seasonal variation of atmospheric dimethylsulfide at Amsterdam Island in the Southern Indian Ocean. J. Atmos. Chem. Vol. 11, 123-142.

(23) PUTAUD, J.P., MIHALOPOULOS, N., NGUYEN, B.C., CAMPIN, J.M. and BELVISO, S., (1992). Seasonal variations of atmospheric sulfur dioxide and dimethylsulfide concentrations at Amsterdam Island in the Southern Indian Ocean. J. Atmos. Chem. Vol. 15, 117-132.

(24) QUINN P. K., BATES, T.S., JOHNSON, J.E., COVERT, D.S and CHARLSON, R.J. (1990). Interactions between the sulfur and reduced nitrogen cycles over the central Pacific Ocean. J. Geophys. Res. Vol. 95, 16405-16416.

(25) SALTZMAN, E.S. and COOPER, D.J. (1988). Shipboard measurements of atmospheric dimethylsulfide and hydrogen sulfide in the Caribbean and Gulf of Mexico. J. Atm. Chem. 7, 191-209.

(26) SPIRO, P. A, JACOB, D. J. and LOGAN, J. A. (1992). Global inventory of sulfur emissions with 1°x1° resolution. Submitted to J. Geophys. Res.

(27) STAUBES, R. and GEORGII, H.-W. (1993). Measurements of atmospheric and seawater DMS concentrations in the Atlantic, the Arctic and Antarctic region. This book.

(28) STOIBER, R. E., WILLIAMS, S. N., and HUEBERT, B. (1987). Annual contribution of sulfur dioxide to the atmosphere by volcanoes. J. Volcanol. Geotherm. Res. Vol. 33, 1-8.

(29) VARHELYI, G. (1985). Continental and global sulfur budgets-I Anthropogenic SO_2 emissions. Atmos. Environ. Vol. 19, 1029-1040.

(30) WYLIE D.J., HARVEY, M.J., DE MORA, S.J., BOUYD, I.S. and LILEY, J.B. (1993). Dimethylsulfide and aerosol measurements on Ross Island, Antarctica. This book.

(31) ZIMMERMANN, P. H. (1987). MOGUNTIA: A handy global tracer model. Proceedings of the sixteenth NATO/CCMS International Technical Meeting on air pollution modeling and its application. Lindau, Germany, Apr. 6-10, Reidel.

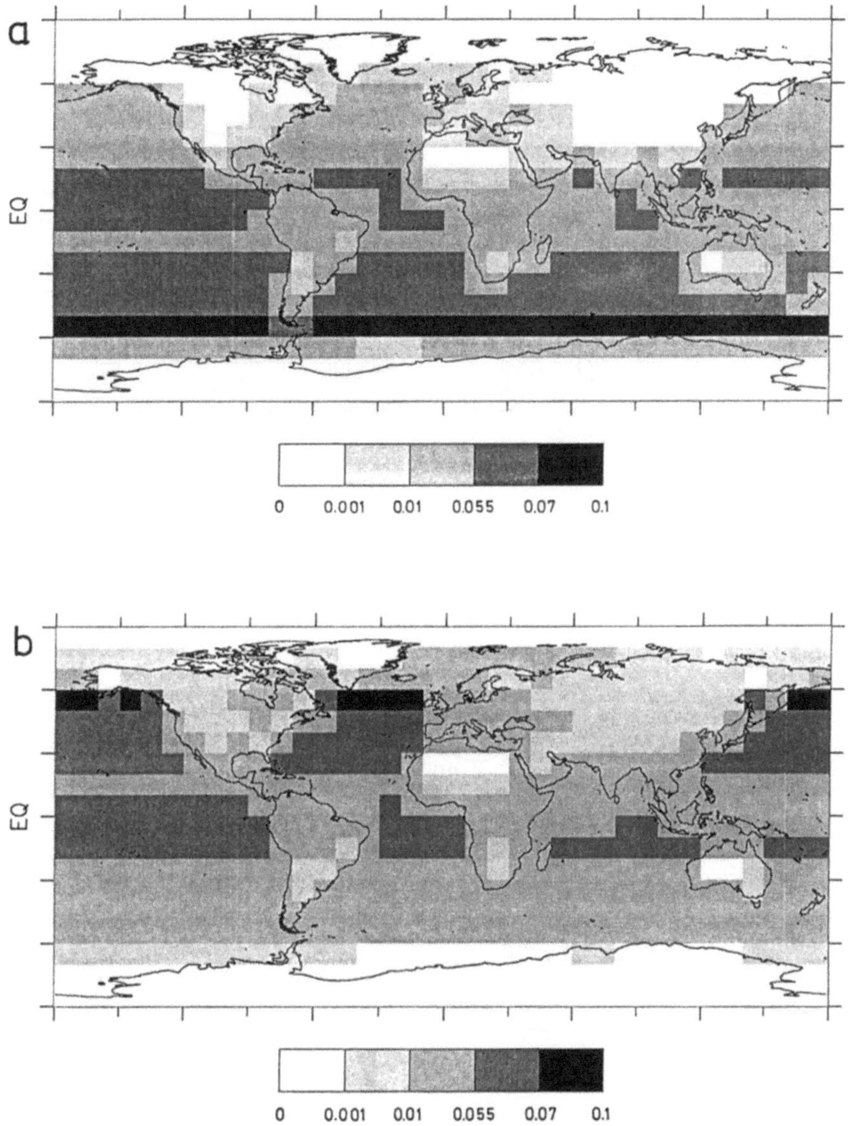

Figure 1. DMS emissions used in the model calculations for: a) January, b) July. Unit: g S m^{-2} yr^{-1}.

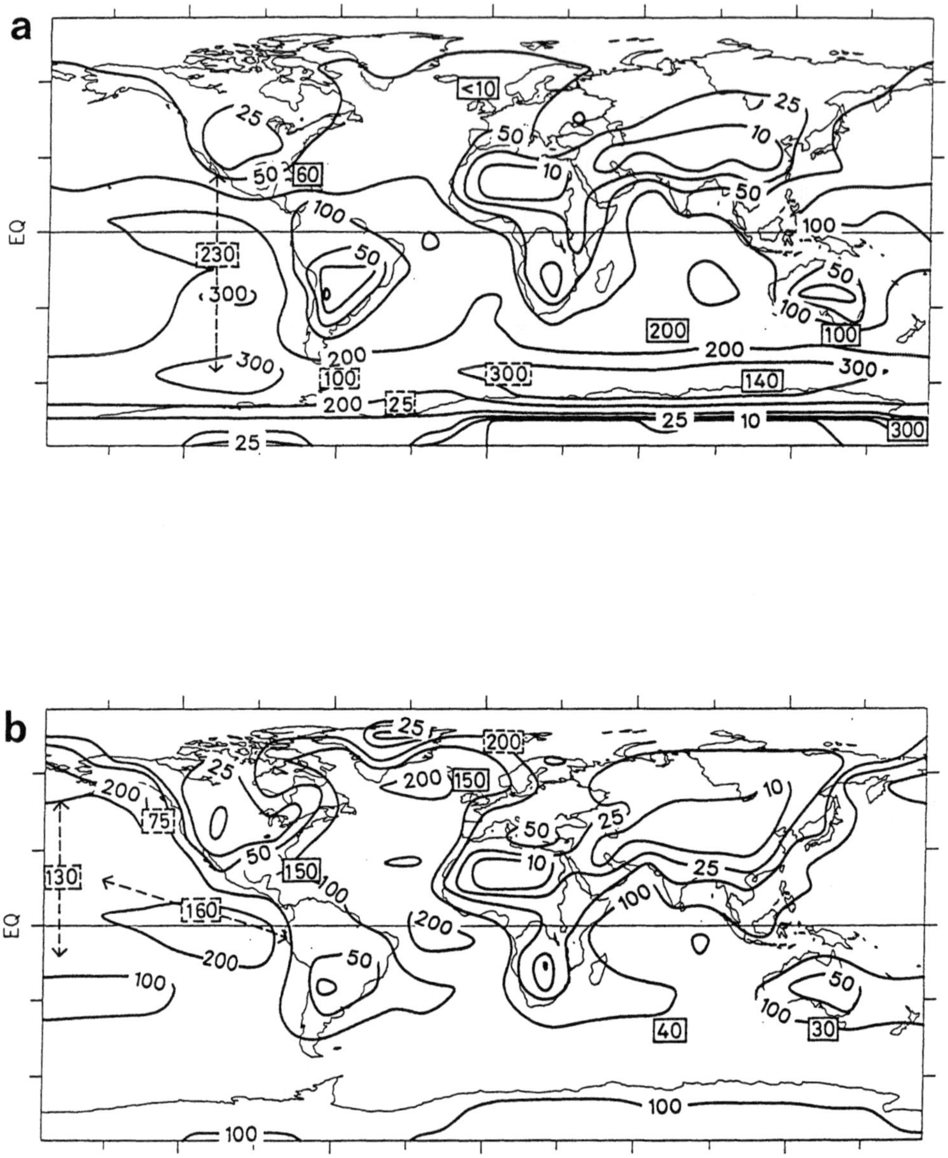

Figure 2. Model calculated DMS concentrations in oceanic surface air for: a) December-February, b) June-August. Numbers in squares indicate measured concentration values taken from references: (2), (3), (5), (7), (12), (17), (22), (24), (25), (27) and (30). Dashed squares and arrows indicate observations from ship cruises during limited periods. Unit: pptv.

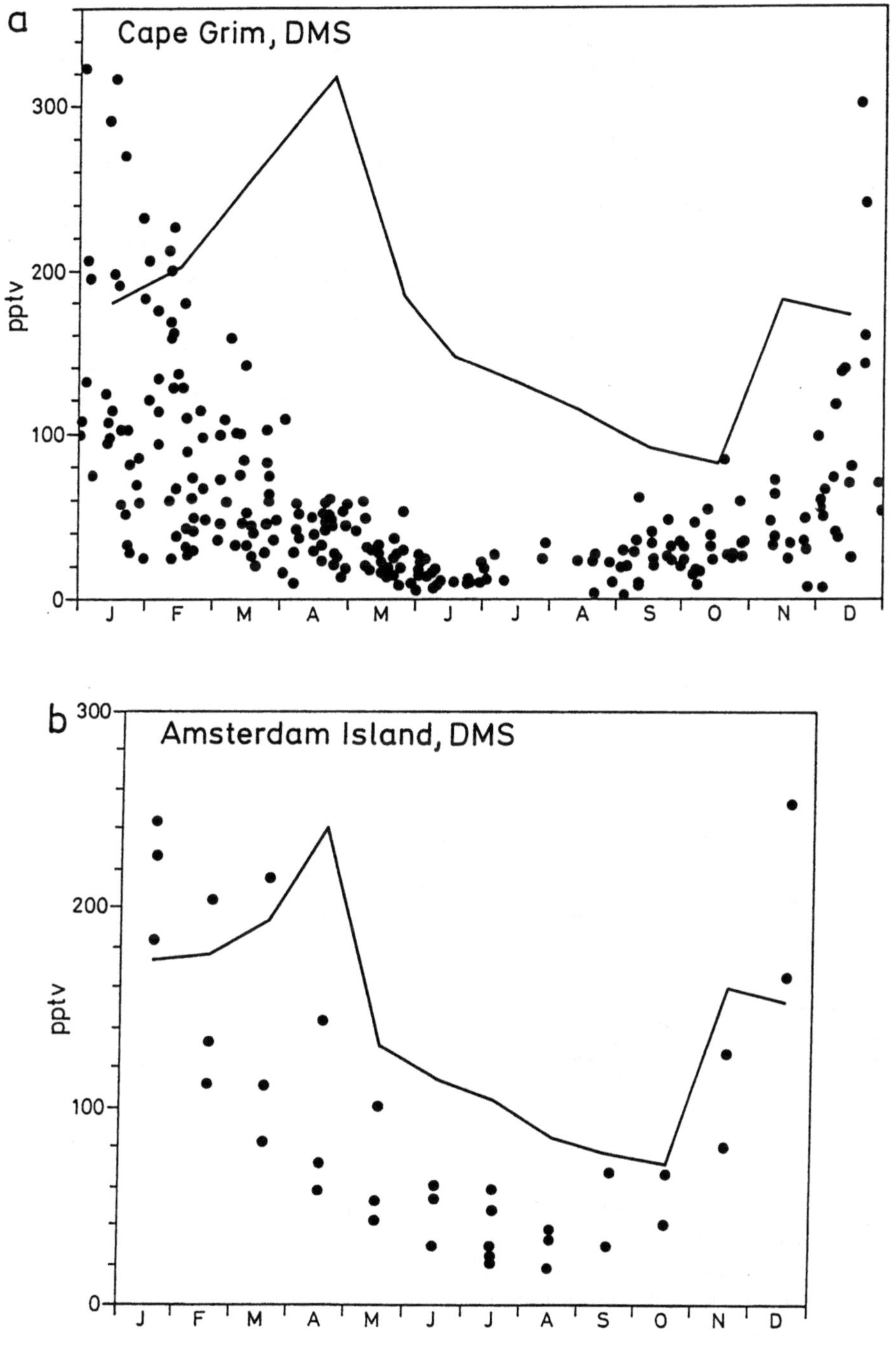

Figure 3. Model calculated and observed seasonal cycle of DMS at: a) Cape Grimm (3), b) Amsterdam Island (22). Unit: pptv.

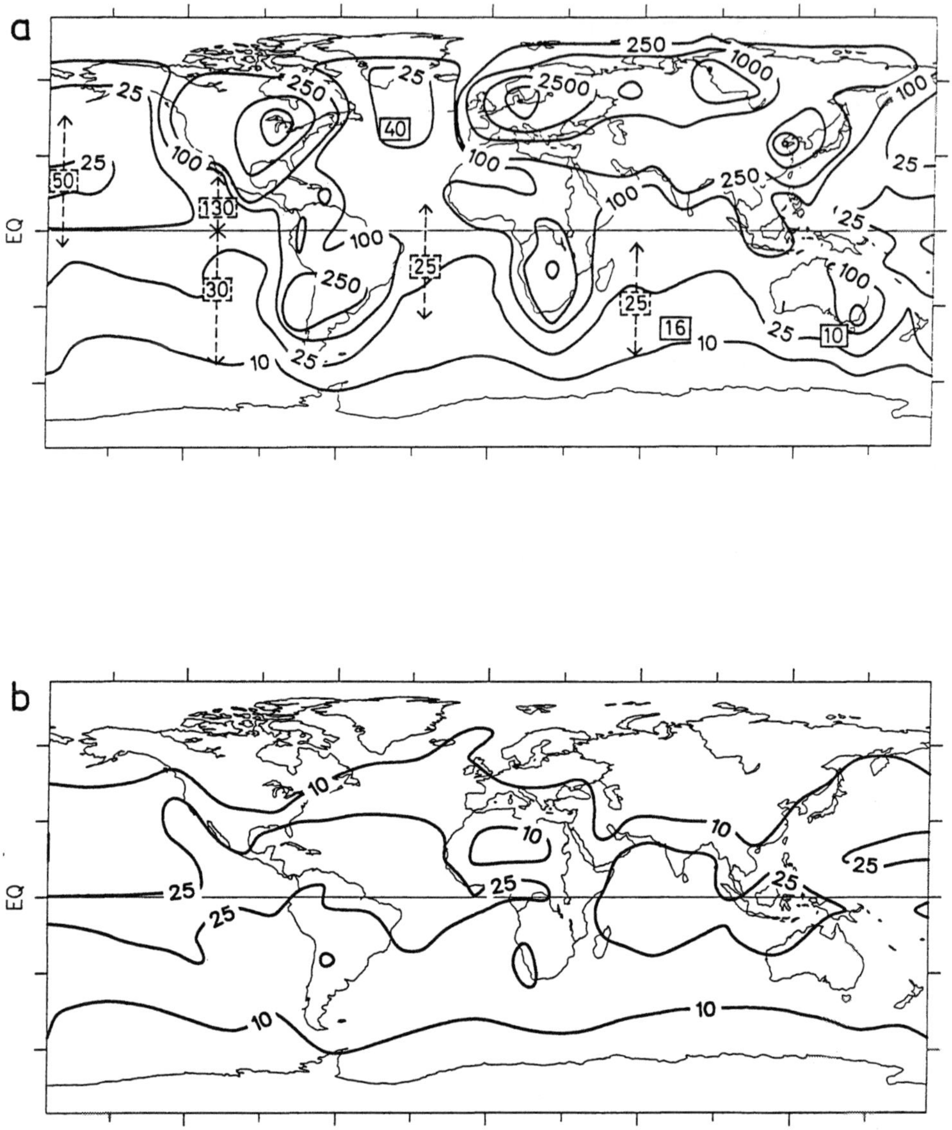

Figure 4. Model calculated SO_2 concentrations in surface air for: a) all sources, b) DMS sources only. Numbers in squares indicate measured concentration values taken from references: (3), (19), (21), (23) and (24). Dashed squares and arrows indicate observations from ship cruises during limited periods.
Unit: pptv

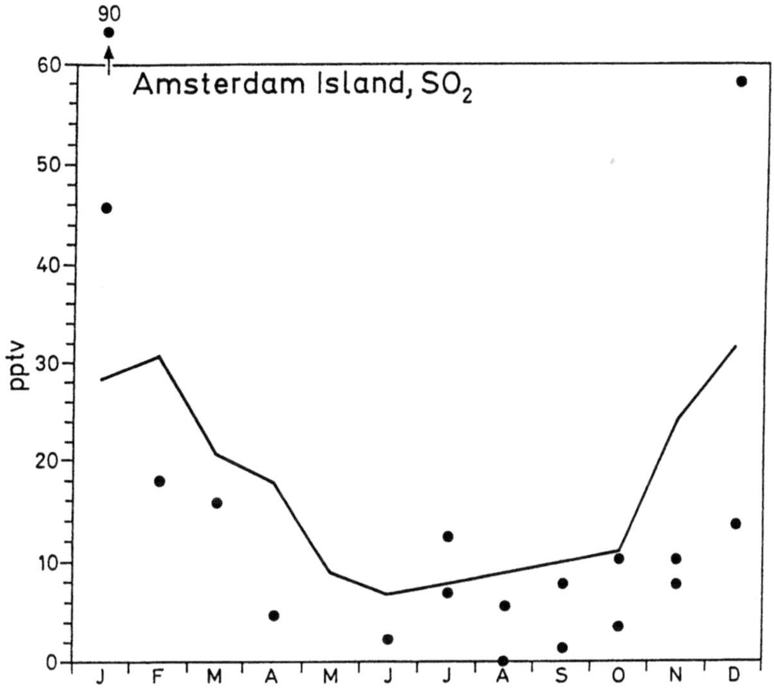

Figure 5. Model calculated and observed seasonal cycle of SO_2 at: Amsterdam Island (23). Unit pptv.

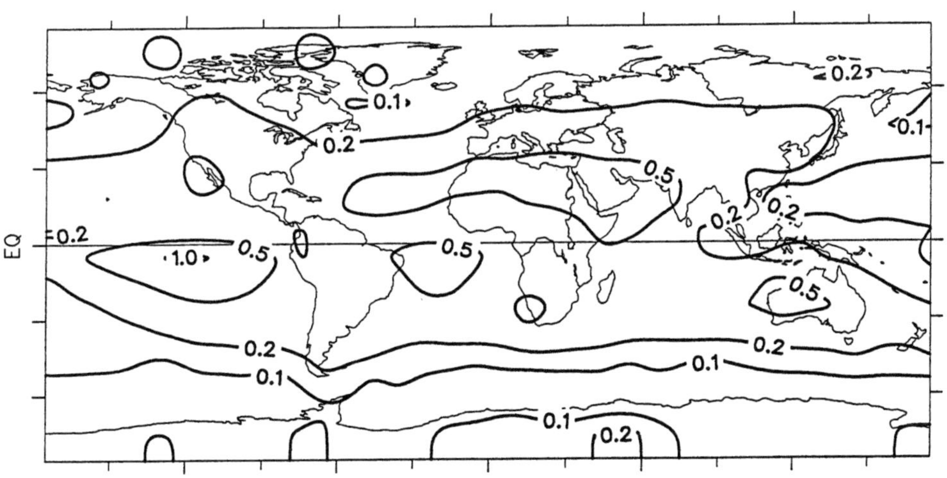

Figure 6. Annual average sulfate aerosol forcing due to DMS emissions. Unit: W m^{-2}.

Figure 6.[illegible] Annual average sulfate aerosol forcing due to LMS emissions. Unit: W m^{-2}.

RECENT FIELD STUDIES OF SULFUR GASES, PARTICLES AND CLOUDS IN CLEAN MARINE AIR AND THEIR SIGNIFICANCE WITH RESPECT TO THE DMS-CLOUD-CLIMATE HYPOTHESIS

PETER V. HOBBS, DEAN A. HEGG AND RONALD J. FEREK

Cloud and Aerosol Research Group,
Atmospheric Sciences Department, AK-40
University of Washington,
Seattle, Washington 98195, USA

SUMMARY

Airborne measurements have been made of the vertical profiles of dimethylsulfide (DMS), SO_2 (in the sub-100 pptv range), particle concentrations and size spectra, non-sea-salt (NSS) sulfate, and cloud condensation nucleus (CCN) spectra during conditions of clean marine airflow over the northeastern Pacific Ocean. More limited measurements have also been obtained over the Arctic Ocean.

DMS concentrations in the marine boundary layer over the Pacific Ocean varied seasonally from as little as a few to several hundred pptv and tended to decrease from the surface up to the top of the marine boundary layer. DMS concentrations over the Arctic Ocean ranged from a few pptv in early June to ≈ 300 pptv later in the month; the latter concentration is among the highest ever measured. The concentrations of DMS were large enough to account for most of the measured SO_2 and NSS sulfate.

DMS and CCN are correlated, but not by the same linear relation at all supersaturations. Cases of new particle production, in and around clouds, have been documented. We attribute this to the homogeneous-bimolecular oxidation of SO_2.

The significance of these new data to the DMS-cloud-climate hypothesis is discussed.

1. INTRODUCTION

The dimethylsulfide-cloud-climate hypothesis, first proposed by Nguyen et al. (1) and Shaw (2), and elaborated on by Charlson et al. (3), depends on the following assumed linkages:

a) Phytoplankton in the ocean excrete dimethylsulfide (DMS), some of which passes into the air.

b) DMS (a gas) is converted into particles (via either SO_2 and H_2SO_4 or methansulfonic acid). In clean marine air this is the major source of non-sea-salt (NSS) particles which, in turn, account for most of the particles that serve as the centers on which water vapor condenses to form cloud drops (i.e. cloud condensation nuclei-CCN).

c) The number concentrations of drops in marine stratiform clouds are controlled by the number concentrations of CCN, and therefore, in clean marine air, by the concentration of DMS.

d) The amount of solar radiation reflected by marine stratiform clouds increases as the number concentrations of cloud drops increases (and therefore, in clean marine air, as the DMS concentration increases).

e) The radiation balance of the earth is sensitive to the reflectivity of marine stratiform clouds (and therefore to the concentration of DMS in marine air).

f) Feedbacks may occur, whereby changes in global temperatures affect DMS emissions from phytoplankton and therefore the radiation balance of the earth. These feedbacks could be positive or negative, depending on whether the DMS emissions decrease or increase with increasing temperature.

Over the past several years we have carried out field and theoretical studies with the goal of evaluating some of the proposed linkages listed above and, where appropriate, quantifying them. The results of these studies are summarized briefly in this paper; for

G. Restelli and G. Angeletti (eds.), Dimethylsulphide: Oceans, Atmosphere, and Climate, 345–353.

more details, the reader is referred to the references listed herein.

Unless mentioned otherwise, all of the field measurements to be described were obtained in clean marine air over the northeastern Pacific Ocean (off the coast of Washington State) aboard the University of Washington's Convair C-131 research aircraft. The instrumentation aboard this aircraft has been described by Hobbs et al. (4).

2. VERTICAL PROFILES OF DMS AND SO_2 IN CLEAN MARINE AIR

The starting point for our studies was to measure the vertical profiles of the concentrations of DMS and SO_2 in clean marine air. Since these gases are present in low concentrations this is not an easy task, in fact our measurements of the vertical profiles of SO_2 are probably the first reliable measurements to be obtained in clean marine air. The technique used to measure SO_2 has been described by Ferek et al. (5), and the technique for DMS by Ferek and Hegg (6).

Figure 1 shows some typical profiles of DMS and SO_2 in the marine atmospheric boundary layer. These measurements were obtained about 80 km off the Washington coast in a strong airflow from the west with no signs of any continental influences. The airmass was postfrontal, with good vertical mixing. The concentrations of DMS are higher near the ocean surface (≈ 30 – 40 pptv); between about 0.5 – 3 km the concentrations are lower and uniform at ≈ 20 pptv. By contrast the SO_2 concentrations are lower (≈ 40 pptv) below about 1.5 km and higher (≈ 55 pptv) between 1.5 and 3.0 km. These measurements are consistent with the view that SO_2 in clean marine air derives from DMS that originates in the ocean and is transported upwards in the atmosphere. For more information on such measurements see Ferek et al. (5).

We have also obtained measurements of DMS in the arctic atmosphere (7). The concentrations ranged from a few pptv in early June to nearly 300 pptv in late June. The latter concentration is among the highest measured in any environment. It could be due to the increase of exposed seawater as sea ice melts.

3. VERTICAL PROFILES OF PARTICLES IN CLEAN MARINE AIR

Shown in Fig. 2 are some typical measurements of total particle (CN) concentrations and cloud condensation nucleus (CCN) concentrations at 1% supersaturation in the clean atmospheric marine boundary layer. CN concentrations are a few hundred per cubic centimeter in the lower part of the boundary layer, but they increase sharply to ≈ 500 – 600 cm^{-3} near the top of the stratocumulus cloud layer. (We will suggest an explanation for this increase in Section 6.)

The CCN concentrations decreased from ≈ 60 cm^{-3} just below cloud base to ≈ 5 cm^{-3} inside the clouds where they were present interstitially with the cloud drops. Note that in contrast to the CN, the CCN did not increase near cloud top. This is because the freshly nucleated CN were too small to act as CCN (even at 1% supersaturation). Similar increases in the concentrations of CN near cloud top, but generally peaking just above cloud top, were seen in all of our spring and summer flights, but in only one out of five winter flights.

The same case shown in Fig. 2 is depicted in a different manner in Fig. 3. Here we have fitted the CCN data to the expression:

$$[CCN] = c(SS)^k \tag{1}$$

where, [CCN] is the number concentration (cm^{-3}) of CCN active at a water supersaturation SS (in %), and the values of c and k have been derived empirically from measurements of the CCN spectra (i.e. measurements of [CCN] at various supersaturations). We can see from (1) that c is the concentration of CCN at 1% supersaturation. Plotted in Fig. 3 are the ratio of c/CN, and k. Comparison of Figs. 2 and 3 indicates that k is a measure of the number of small (CN) particles. We also see from Fig. 3 that the number concentration of CCN, even at the relatively high supersaturation of 1% (only CCN active at less than about

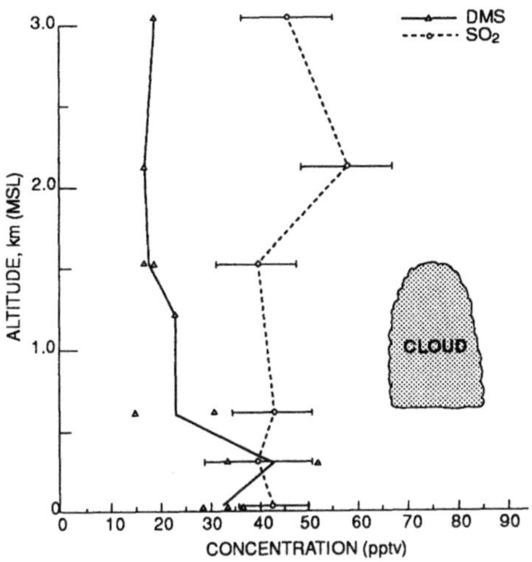

Fig. 1. Vertical profiles of DMS and SO_2 in the marine boundary layer over the NE Pacific Ocean.

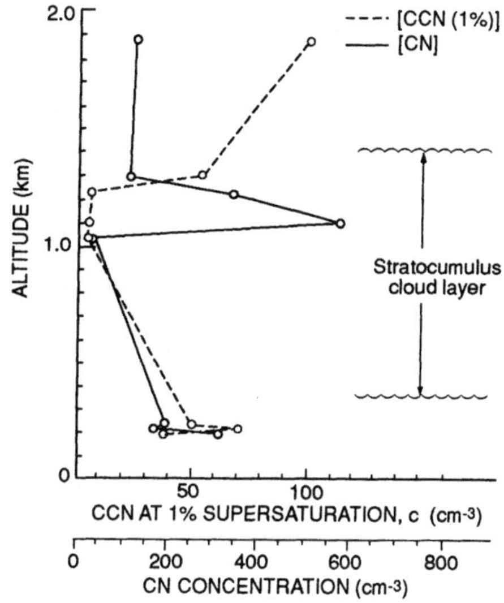

Fig. 2. Vertical profiles of total particle concentrations (CN) and cloud condensation nucleus concentrations active at 1% supersaturation [CCN(1%)] in the marine boundary layer over the NE Pacific Ocean.

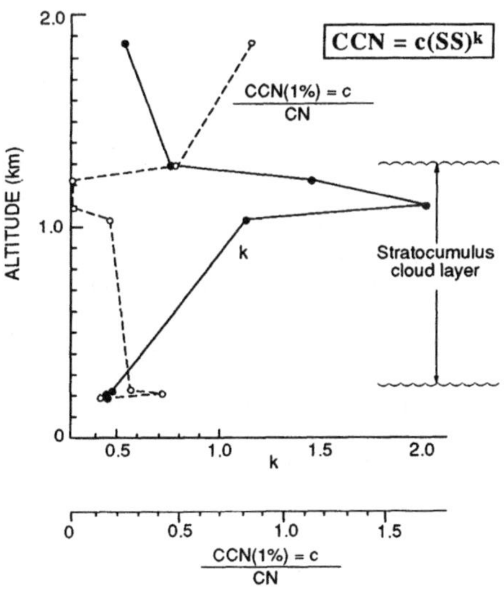

Fig. 3. As for Fig. 2 but for CCN (1%)/CN and the empirical constant k in Eqn. (1).

0.5% are involved in the formation of stratiform clouds), is only about 50% of the total particle concentration just below cloud base. This has proven to be a widespread phenomenon (8). This is particularly important with respect to the DMS-cloud-climate hypothesis because it means that not all CN are CCN even in clean marine air, so that an increase in CN (due, for example, to an increase in DMS) does not necessarily produce an increase in effective CCN and cloud droplets.

Before leaving this topic, it is important to point out that it can be shown theoretically that the concentrations of cloud drops that form on CCN in marine stratus clouds depends primarily on the value of c in (1) rather than on k (8, 9, 10). The relationship between drop concentrations in marine stratus clouds and c for our data set is shown in Fig. 4.

For more details on the topics discussed in this section see References (8), (11) and (12).

4. RELATIONSHIPS BETWEEN CCN AND DMS

Ayers et al. (13) have presented field data from Australia that show strong correlations, on a seasonal basis, between the sea-level concentrations of DMS and NSS sulfate in marine air. A *non-concurrent* CCN data set (14) showed similar seasonal variations. Another important study (15) has revealed a high correlation between methanesulfuric acid (MSA) – a product of DMS oxidation – and CCN concentration.

We have obtained the first *concurrent* data set of DMS and CCN concentrations, from airborne measurements in marine air over the northeastern Pacific Ocean. As can be seen from Fig. 5, our measurements also show a good correlation (r = 0.72) between (flight) mean CCN concentrations measured at 1% supersaturation and (flight) mean DMS concentrations measured in the boundary layer of clean marine air.

Using the data shown in Fig. 5, and a similar data set for DMS versus CCN concentrations at 0.3% supersaturation, empirical relations can be derived between DMS and CCN. These are listed in Table 1, together with the corresponding relations derived from the Ayers et al. data sets. In the latter case, the relations are based on just four data points (one for each season). Therefore, although the correlative coefficients (r) are high they are relatively uncertain. Our empirical relations are based on thirteen independent data points, therefore the r values are lower but more certain.

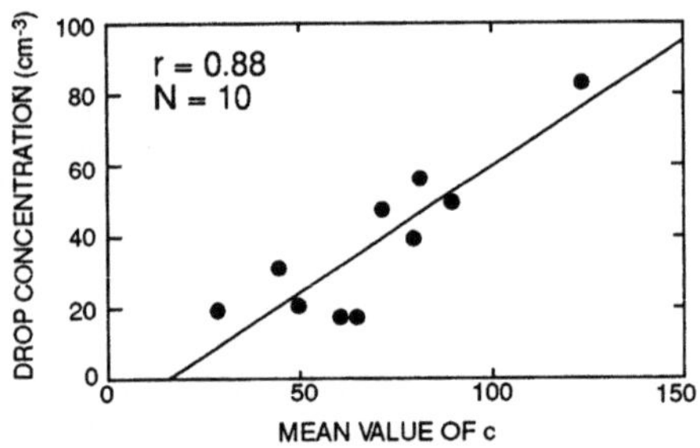

Fig. 4. Mean concentration of cloud drops in marine stratus clouds versus the empirical constant c in Eqn. (1).

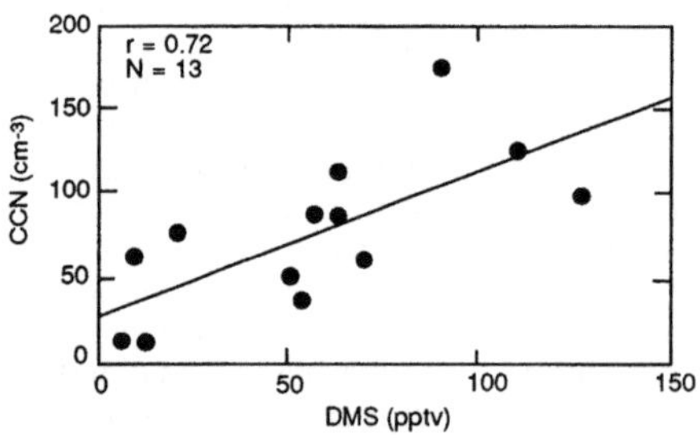

Fig. 5. Mean concentration of CCN at 1% supersaturation versus mean DMS concentration in the marine boundary layer over the NE Pacific Ocean.

Within the ranges of the uncertainties in the values of the empirical coefficients, our relationships between the CCN concentrations at 1% and 0.3% supersaturation and DMS concentrations for the NE Pacific Ocean are essentially identical to those of Ayers et al. for Cape Grim, Tasmania (13). Note that both data sets show a large change in the CCN-DMS relationship in going from 1% to 0.3% supersaturation, with the CCN concentration at 1% being much more sensitive to changes in DMS concentration than those at 0.3% (as indicated by the values of the "slopes" in the CCN-DMS relationship in Table 1 (e.g. 0.86 ± 0.25 in the case of our relationship at 1% supersaturation). This is important, since only those CCN active at supersaturations less than about 0.5% are involved in the formation of stratus clouds. It appears, therefore, that in addition to DMS there is another important natural source of the CCN that are active in clean marine air at low supersaturations. We will return to this point in Section 6.

TABLE 1. EMPIRICAL RELATIONS BETWEEN DMS AND CCN CONCENTRATIONS DERIVED FROM FIELD MEASUREMENTS IN MARINE AIR AT CAPE GRIM, TASMANIA (13, 14) AND OVER THE NE PACIFIC OCEAN (8). [CCN(X%)] IS THE CONCENTRATION OF CCN AT X% SUPERSATURATION IN CM^{-3} AND [DMS] THE CONCENTRATION OF DMS IN PPTV.

EMPIRICAL RELATION	CORRELATION COEFFICIENT (r) AND NUMBER OF DATA POINTS (n)	REFERENCE
[CCN(1%)] = (0.86 ± 0.25)[DMS] + (23 ± 17)	r = 0.72 n = 13	8
[CCN(1%)] = (0.99 ± 0.14)[DMS] + (56 ± 13)	r = 0.98 n = 4	13, 14
[CCN(0.3%)] = (0.28 ± 0.17)[DMS] + (13 ± 11)	r = 0.46 n = 13	8
[CCN(0.3%)] = (0.40 ± 0.06)[DMS] + (50 ± 5)	r = 0.98 n = 4	13, 14

The "intercept" values in the CCN-DMS relations (e.g. 23 ± 17 in the case of our relationship at 1% supersaturation) are the non-DMS contributions to the CCN concentrations. These could derive from both natural and anthropogenic sources. Interestingly, the non-DMS contributions appear to be greater for the Australian data than for our data, and significantly so at 0.3% supersaturation. However, the small number of points in the regression for the Australian data preclude any definite conclusions on this point.

5. RELATIONSHIP BETWEEN CCN AND NON-SEA-SALT (NSS) SULFATE

One of the tenuous links in the DMS-cloud-climate hypothesis is that between particulate sulfate mass generated by biogenic gas emissions (i.e, NSS sulfate) and CCN number concentrations. Our fifty-five simultaneous pairs of measurements of CCN (1%) and NSS sulfate show no significant correlation. We interpret this as indicating that NSS sulfate and CCN are not in equilibrium, which is not surprising considering the relatively large particles required to serve as CCN even at 1% supersaturation. However, if we look at the average values of CCN (1%) and NSS measured on each of our sixteen flights and plot these sixteen pairs of measurements, a reasonably good correlation (r = 0.52) is obtained (Fig. 6).

6. OTHER FACTORS AFFECTING PARTICLES IN MARINE AIR

The DMS-cloud-climate hypothesis assumes that DMS is the dominant source of the particles large enough to serve as CCN in marine air. However, there are other sources of particles, although they have not been well quantified. For example, it is apparent that increasingly large areas of the marine atmosphere are being invaded by anthropogenic pollution that is relatively rich in particles. However, even in the pristine marine boundary layer, there are other sources of particles including organics from the ocean (about which little is known), sea salt particles (these can serve as excellent CCN, however early measurements, which need to be verified, indicate that their concentrations are low (16)), sedimentation from the free troposphere, and clouds (Fig. 7). Since the latter source is potentially important, and has been studied more extensively than the other sources, we will discuss it briefly below.

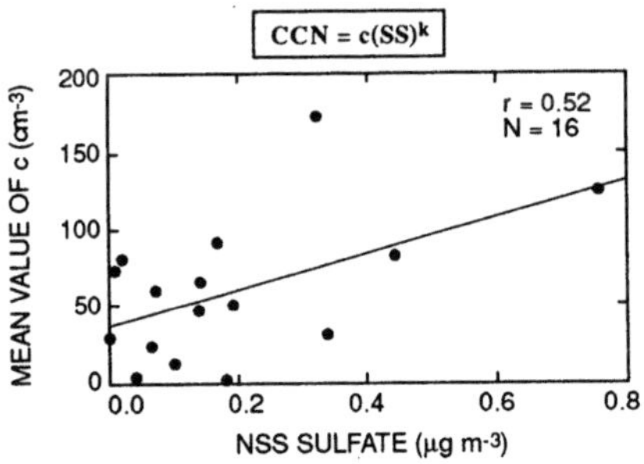

Fig. 6. Flight-average values of c in Eqn. 1 versus non-sea-salt sulfate in the marine boundary layer over the NE Pacific Ocean.

Clouds can be involved in the production of sulfate mass in two ways: through chemical reactions in cloud drops, and through gas-to-particle conversion in and around clouds. The role of chemical reactions in cloud drops in enhancing the activity of CCN was identified in field studies by our group over twenty years ago (17), subsequently verified by further field studies (18, 19, 20), and modeled numerically (21, 22). Since then, prompted mainly by the interest in acid rain, many laboratory, field and theoretical studies of sulfate production in cloud drops have been carried out. However, for the present discussion, the main point is that chemical reaction in cloud drops can produce sulfate mass relatively rapidly. When a cloud drop evaporates, this sulfate is deposited onto the CCN on which the drop first formed. Consequently, the

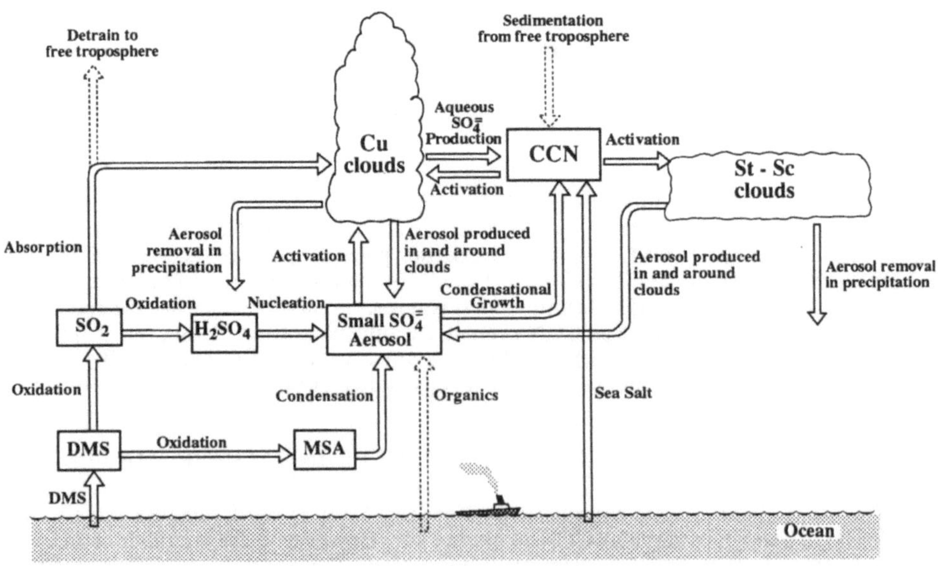

Fig. 7. Schematic showing the sources and sinks of aerosol in the atmospheric marine boundary layer.

resulting CCN is larger, and therefore can be activated at a lower supersaturation, than the original CCN. Recent calculations suggest that newly created particles in the marine atmosphere (e.g. from the oxidation of DMS) can, in general, only achieve sizes, by means of gas-to-particle processes, sufficient to serve as CCN in the relatively high super-saturations achieved in cumuliform clouds, but not at the lower supersaturations present in stratiform clouds (23). However, the addition of sulfate to these particles, through aqueous phase chemical reactions in cumulus clouds, can increase their sizes sufficiently to serve as CCN in stratiform clouds.gas-to-particle processes, sufficient to serve as CCN in the relatively high super-saturations achieved in cumuliform clouds, but not at the lower supersaturations present in stratiform clouds (23). However, the addition of sulfate to these particles, through aqueous phase chemical reactions in cumulus clouds, can increase their sizes sufficiently to serve as CCN in stratiform clouds.

The production of new particles by gas-to-particle conversion in and around clouds is a new discovery by our group (11, 24). We have already presented field observations illustrating this phenomenon in Section 3, where it can be seen as an increase in CN near the top of the stratiform cloud layer in Fig. 2. Similar increases in particle concentrations are seen in "halo" regions surrounding cumuliform clouds (24). We have explained these large concentrations of small particles in terms of homogeneous-bimolecular nucleation of H_2SO_4 – H_2O droplets, which is promoted by the high relative humidities and actinic radiation fluxes in and around clouds (11, 12). Shown in Fig. 8 are the results of numerical modeling calculations, based on this theory, of the vertical profile of CN for the case shown in Fig. 2. The modeling results show the same trends as the measurements, with a pronounced peak in CN near the top of the cloud.

The CN produced in and around clouds by gas-to-particle conversion are (like the particles produced by the gas-phase oxidation of DMS) initially too small to serve as CCN. However, given time, they might grow sufficiently to act as CCN in cumuliform clouds, and then, through the aqueous-phase chemical reactions discussed above, become large enough to serve as CCN in stratiform clouds.

The sources of particles in marine air discussed in this section have another potentially important influence on the DMS-cloud-climate hypothesis. There are two possible pathways for the conversion of DMS gas into sulfate: the production of new particles, and deposition onto existing particles. A necessary condition for the DMS-cloud-climate hypothesis to be viable is that the former pathway create at least the majority of the new particles in the marine boundary layer. This is most likely to be the case when there are relatively few pre-existing particles in the air on which the DMS can deposit and with which it must compete as a particle source. Consequently, the presence of other significant sources of particles in the marine atmosphere reduces the likelihood of the DMS-cloud-climate hypothesis being valid.

7. SUMMARY

In this paper we have summarized results from airborne measurements over the NE Pacific Ocean that are relevant to the DMS-cloud-climate hypothesis. The following results provide support for some of the linkages involved in this hypothesis:

- Measurements of the vertical profiles of DMS and SO_2 in marine air (Fig. 1) are consistent with the view that the SO_2 derives from DMS that originates in the ocean and is transported upward.
- CCN concentrations (even at 1% supersaturation) in the atmospheric boundary layer of clean marine air are low ($\approx 10 - 100$ cm^{-3}) (Fig. 2).
- Drop concentrations in marine stratus clouds correlate with CCN concentrations (Fig. 4).
- CCN concentrations correlate with DMS concentrations (Fig. 5 and Table 1).
- CCN concentrations correlate with non-sea-salt sulfate (Fig. 6).

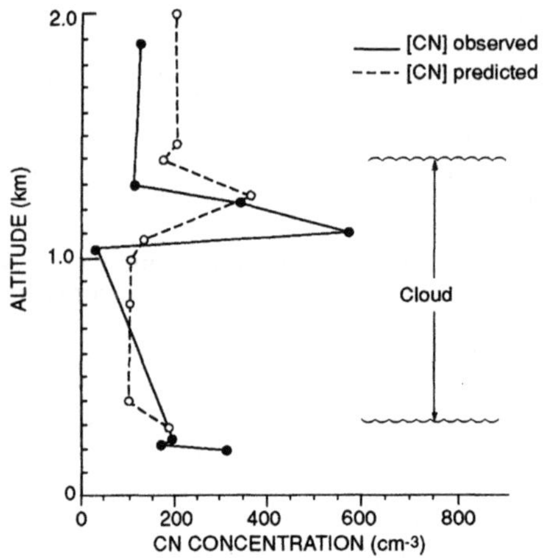

Fig. 8. A case study of the concentrations of aerosol (CN) for the atmospheric marine boundary layer predicted by the homogeneous-bimolecular theory compared to observations.

The following results, on the other hand, indicate that some of the required linkages in the DMS-cloud-climate hypothesis may be weak:

- Not all of the particles in the boundary layer of clean marine air act as CCN (even at 1% supersaturation) (Fig. 3).
- CCN concentrations at low supersaturations (typical of those active in stratus clouds) are not as sensitive to changes in DMS concentrations as those at higher supersaturations. (Table 1).
- The particles produced from DMS are initially too small to serve as CCN in stratus clouds. However, after suitable processing (e.g. by clouds or after long-range transport) they may grow large enough to act as CCN in stratus clouds.
- In addition to DMS, there are other sources of particles in marine air (Fig. 7). These sources need to be better quantified.

The results presented here are a first step toward obtaining the types of field data required to evaluate the DMS-cloud-climate hypothesis. In parallel with this effort, we are developing a model of the cloud-topped marine boundary layer that will be used to study the roles and interactions of many of the physical and chemical processes involved in the hypothesis (25). Eventually, such studies should lead to the types of parameterizations required to include the DMS-cloud effect in global climate models.

Acknowledgements. This research was supported by a series of grants from the Atmospheric Chemistry Program, Division of Atmospheric Sciences, National Science Foundation, the latest of which is NSF grant ATM 90-08639. The paper was written while one of us (P.V.H.) was on study leave at FISBAT/CNR, Bologna, Italy.

REFERENCES

(1) NGUYEN, B. C., BONSANG, B. and GAUDRY, A. (1983). The role of the oceans in the global atmospheric sulfur cycle. J. Geophys. Res. Vol. 88 10,903–10,914.
(2) SHAW, G. (1983). Bio-controlled thermostasis involving the sulfur cycle. Clim. Change. Vol. 5 297–303.
(3) CHARLSON, R. J., et al. (1987). Oceanic phytoplankton, atmospheric sulphur, cloud albedo and climate. Nature. Vol. 326 655–668.
(4) HOBBS, P. V. et al. (1991). Airborne measurements of particle and gas emissions from the 1990 volcanic eruption of Mount Redoubt. J. Geophys. Res. Vol 96 18,735–18,752.
(5) FEREK, R. J., et al. (1991). An improved filter pack technique for airborne measurements of low concentrations of SO_2. J. Geophys. Res. Vol. 96 22,373–22,378.

(6) FEREK, R. J. and HEGG, D. A. (1992). Measurements of dimethyl sulfide and SO_2 during GTE/CITE-3. J. Geophys. Res. (In press).
(7) FEREK, R. J., HOBBS, P. V. and RADKE, L. F. Dimethyl sulfide in the Arctic atmosphere. J. Atmos. Chem. (Submitted).
(8) HEGG, D. A., RADKE, L. F. and HOBBS, P. V. (1991). Measurements of Aitken nuclei and cloud condensation nuclei in the marine atmosphere and their relation to the DMS-cloud-climate hypothesis. J. Geophys. Res. Vol. 96 18,727–18,733.
(9) BIGG, E. K. (1990). Aerosols over the Southern Ocean. Atmos. Res. Vol. 25 583–600.
(10) HEGG, D. A. and LARSON, T. V. (1990). The effects of microphysical parameterization on model predictions of sulfate production in clouds. Tellus. Vol. 42B 272–284.
(11) HEGG, D. A., RADKE, L. F. and HOBBS, P. V. (1990). Particle production associated with marine clouds. J. Geophys. Res. Vol. 95 12,917–13,926.
(12) HEGG, D. A. (1991). Particle production in clouds. Geophys. Res. Lett. Vol. 18 995–998.
(13) AYERS, G. P., IVEY, J. P. and GILLETT, R. W. (1991). Coherence between seasonal cycles of dimethyl sulphide, methanesulphonate and sulphate in marine air. Nature Vol. 349 404–406.
(14) GRAS, J. L. (1990). Cloud condensation nuclei over the Southern Ocean. Geophys. Res. Lett. Vol. 17 1565–1567.
(15) AYERS, G. P. and GRAS, J. L. (1991). Seasonal relationship between cloud condensation nuclei and aerosol methanesulphonate in marine air. Nature. Vol. 353 834–835.
(16) HOBBS, P. V. (1971). Simultaneous airborne measurements of cloud condensation nuclei and sodium-containing particles over the ocean. Quart. J. Roy. Meteor. Soc. Vol. 97 263–271.
(17) RADKE, L. F. and HOBBS, P. V. (1969). Measurements of cloud condensation nuclei, light scattering coefficient, sodium containing particles, and Aitken nuclei in the Olympic Mountains of Washington. J. Atmos. Sci. Vol. 26 281–288.
(18) HEGG, D. A., HOBBS, P. V. and RADKE, L. F. (1980). Observations of the modification of cloud condensation nuclei in wave clouds. J. Rech. Atmos. Vol. 14 217–222.
(19) HEGG, D. A. and HOBBS, P. V. (1981). Cloud water chemistry and the production of sulfates in clouds. Atmos. Environ. Vol. 15 1597–1604.
(20) HEGG, D. A. and HOBBS, P. V. (1982). Measurements of sulfate production in natural clouds. Atmos. Environ. Vol. 16 2663–2668.
(21) EASTER, R. C. and HOBBS, P. V. (1974). The formation of sulfates and the enhancement of cloud condensation nuclei in clouds. J. Atmos. Sci. Vol. 31 1586–1594.
(22) HEGG, D. A. and HOBBS, P. V. (1979). The homogeneous oxidation of sulfur dioxide in cloud droplets. Atmos. Environ. Vol. 13 981–988.
(23) HEGG, D. A. (1990). Heterogeneous production of cloud condensation nuclei in the marine atmosphere. Geophys. Res. Lett. Vol. 17 2165–2168.
(24) RADKE, L. F. and HOBBS, P. V. (1991). Humidity and particle fields around some small cumulus clouds. J. Atmos. Sci. Vol. 48 1190–1193.
(25) ACKERMAN, A. S., TOON, O. B. and HOBBS, P. V. (1992). Numerical modeling of the stratocumulus-topped marine boundary layer. Proceedings 11th International Conference on Clouds and Precipitation, Montreal, Canada, 17–21 August 1992, 378–380.

MODELLING OF THE SULPHUR CYCLE. FROM DMS TO CLOUD PARTICLES

A. SALTELLI, T. HOMMA, F. RAES, J. WILSON AND R. VAN DINGENEN
Commission of the European Communities
Environment Institute,
I-21020 Ispra (VA) Italy

SUMMARY

A model of the sulphur cycle is being assembled to assist in the investigation of the roles of DMS and of anthropogenic sulphur in climate change.

At present the model consists of two modules: AERO2, a model for aerosol dynamics and LPMR, a simplified model for cloud particle growth in cumulus clouds.

AERO2 and LPMR are embedded into a set of subroutines which act as a statistical driver and enable sensitivity and uncertainty analysis of the output from the codes.

In the present paper AERO2 is briefly described, with reference to the existing literature. For the newest module LPMR a description of the relevant equations is given together with the results of a verification exercise conducted on 4 test cases available in the literature. The performances of the statistical driver are exemplified with references to the model AERO2 in its "reaction chamber" and "field" versions.

1. INTRODUCTION

A model for the sulphur cycle is being assembled as part of the JRC activities in the area of global change. This model will assist in the investigation of the role of the DMS cycle (Figure 1) and of the anthropogenic sulphur on global climate change. The ensemble of codes (the physical models plus the statistical subroutines) has been temporarily named OCCAM[1]. The first model to be developed was the AERO2 code for aerosol dynamics. In its upgraded version AERO2 can model the process of aerosol particles formation and growth starting from a DMS source term in the ocean. The second module of the cycle is the LPMR code, a preliminary model of water droplet formation in a cumulus cloud.

It is peculiar of this system that the input for the cloud micro physics model, ie the

1 Originally OBCAAM, acronym for Ocean Biology Chemical Atmosphere Aerosol Model, then the more euphonic OCCAM, which is the Latin name of the philosopher William of Ockham, known for his "razor". Occam's razor is "the philosophic rule that entities should not be multiplied unnecessarily" /3/. Such a rule can be relevant to the construction of an integrated model of the sulphur cycle.

G. Restelli and G. Angeletti (eds.), Dimethylsulphide: Oceans, Atmosphere, and Climate, 355–373.

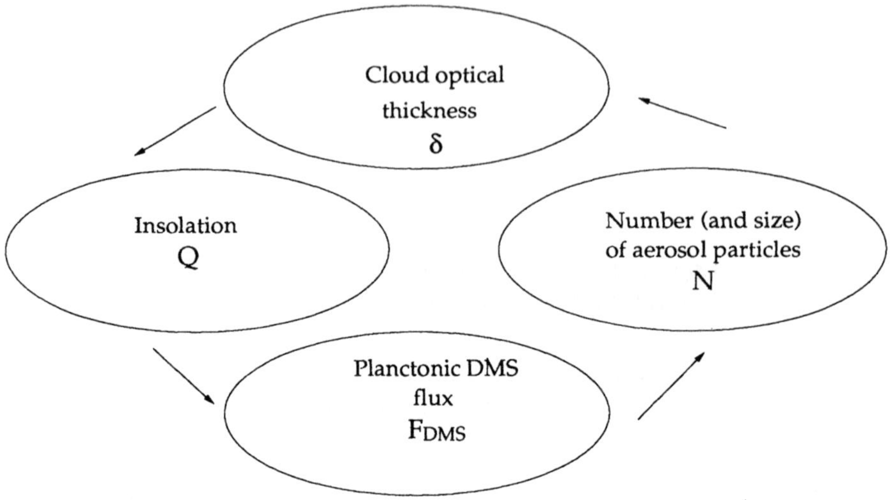

Figure 1. DMS cycle after Bates /1/ and Foley /2/. F_{DMS} is sensitive to Q and influences N. N affects δ which governs Q.

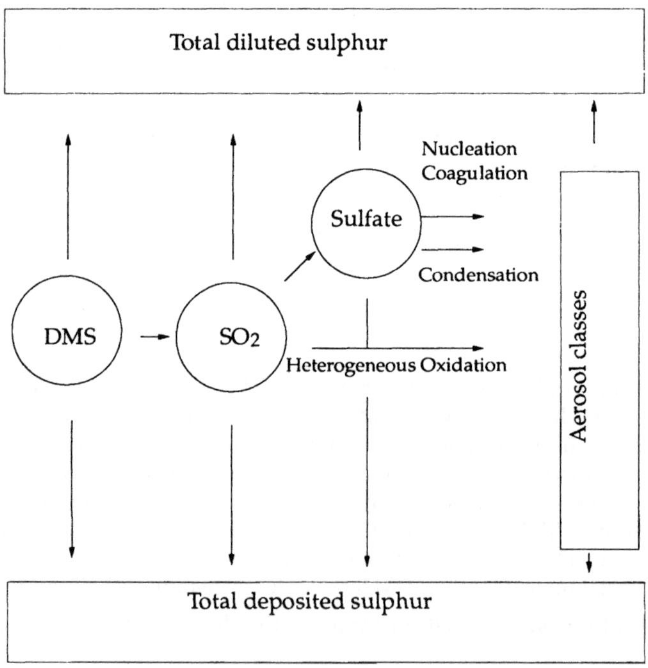

Figure 2. Processes considered in the DMS kinetics - aerosol formation model AERO2. The arrows indicate deposition, dilution, nucleation, coagulation, condensation, homogeneous and heterogeneous oxidation of SO_2

dry radii particle distribution at cloud base, can be generated from an aerosol dynamics model starting from the conditions at the sea surface (DMS fluxes, sea temperature ...), rather than being assigned based on experimental data of aerosol populations.

Another characteristic of the model is that both AERO2 and LPMR are embedded into a statistical driver which allow parameters to be input in the form of distribution, thus enabling uncertainty analysis on the model predictions. This also implies that the impact of any given input parameter on the predicted output can be immediately quantified by simply acting on the appropriate execution switches, once the uncertainty in that parameter is known.

2. MODEL DESCRIPTION

AERO2 is a model of aerosol dynamics. Initially developed at the University of Gent (B) /4/ to describe the formation of H_2SO_4-H_2O aerosol in a smog chamber, AERO2 was upgraded at the JRC in order to model the entire process of CCN formation starting from a DMS source term in the ocean /5/. AERO2 includes the following processes (Figure 2):

- Spatially averaged (constant or time dependent) DMS source term from the ocean
- DMS to SO_2 reaction kinetics
- Photolytic SO_2 to H_2SO_4 reaction kinetics
- Aerosol particle nucleation
- Aerosol growth by particle coagulation and H_2SO_4 condensation on existing particles
- Heterogeneous SO_2 to H_2SO_4 oxidation (by SO_2 diffusion in existing particles)
- Aerosol deposition and - for the smog chamber version - dilution

A description of AERO2 can be found in /4,5/.

The other physical module of OCCAM presently available is the LPMR code for cumulus cloud water droplet formation. This is in fact a re-coding - with some modifications - of a model from Lee and Pruppacher described in /6,7/. The model was selected for didactic purposes; it is well documented in a textbook, where a number of bench-marks are also presented, and the description of the test cases is accurate enough to allow a verification exercise.

LPMR is a Lagrangian plume (or ascending thermal) model, where the velocity and the radius of the rising parcel as well as the water drop spectra (in size, concentration and salt content) inside the parcel can be predicted. The aerosol initial conditions for LPMR can be generated by AERO2 (initial aerosol size distribution at cloud base). LPMR uses interpolated sounding values for the external temperature and relative humidity, and handles entrainment of water vapour as well as entrainment of external aerosol particles.

LPMR computational procedure can be summarised as follows: first the initial conditions are assigned; an aerosol population as function of radius (and salt content) is either input from AERO2 or assigned as dry radii distribution. For the purpose of the verification exercise discussed further on the dry radii were assigned through a Junge law distribution of particles calibrated on the test case at hand. For the general use of the model the initial H_2SO_4 aerosol distribution generated by AERO2 is stoichiometrically converted to $(NH_4)_2SO_4$ aerosol. Then the conditions at the cloud base are imposed: based on the initial parcel relative humidity (assumed to be 99%) the radii are re-computed by the solution of the Koeler equation at that relative humidity /7/.

For each class (identified by its radius) the radius itself, the population and the class "total" salt content are integrated, including entrainment of both air and aerosol. Here "total" means total salt for all the particles in that class, which is used to compute the average particle salt content for that class. This is an approximation, as for the same radius there will be - in the aerosol entrainment case - particles with different salt content. It is not clear from /6,7/ how this problem was treated in the original model.

The growth rate equation for the radii is:

$$a\frac{\partial a}{\partial t} = \frac{\left[\dfrac{e_a}{e_{sat,w}} - e^{y}\right]}{\left[\dfrac{\rho_s'' R T}{e_{sat,w} D'_v M_w}\right] + \left[\dfrac{e^{y} L_{e,0} \rho_s''}{k'_a T}\left(\dfrac{L_{e,0} M_w}{R T} - 1\right)\right]} \tag{1}$$

where:

$$y = \frac{2 M_w \sigma_{w/a}}{R T a \rho_w} + \frac{6 M_w B'_{(NH_4)_2SO_4} m_s}{4 R \rho_w^2 \pi T a^4} - \frac{m_s \Phi_s \nu\, M_w/M_s}{\left[\dfrac{4 \pi a^3 \rho_s''}{3} - m_s\right]}$$

See Table 1 for the symbols legend. The first two terms in "y" are commonly referred to as the curvature (Kelvin) term, while the third term represents the solute effect. Note that the Koeler equation is equal to the root of the numerator in (1), ie $e_a/e_{sat,w} - e^{y} = 0$.

For the entrainment case the class population changes according to:

$$\frac{\partial n_i}{\partial t} = -\mu_j\left[n_i - n_i'\right] \tag{2}$$

where $\mu_j = \dfrac{C}{R_j}$. The quantities n_i are then adjusted for the expansion by assuming that:

$$\frac{1}{n_i}\frac{\partial n_i}{\partial t} = \frac{1}{\rho}\frac{\partial \rho}{\partial t} \tag{3}$$

The total amount (g) of salt in the class (sum over all the droplets) changes according to:

Table 1. Input data for the cloud micro-physics model

VARIABLE DEFINITION AND UNITS	SYMBOL
drop radius, cm	a
term to compute solution surface tension (dyne/cm/g)	$B'(NH_4)_2SO_4$
entrainment parameter	C
specific heat of dry air, cal/(g °K)	C_{pa}
water vapour modified diffusivity, cm^2/s	D_v'
water vapour pressure, mbar	e_a
water vapour pressure at saturation, mbar	$e_{sat,w}$
acceleration of gravity, cm/s^2	g
moist air thermal conductivity,cal/(cm s °K)	k_a'
water latent heat of evaporation, cal/g	$L_e \approx L_{e,0}$
molecular weight of water	M_w
molecular weight of salt	M_s
molecular weight of dry air	M_a
mass of salt in drop, g	m_s
total mass of salt in all particles of radius a_i	m_i^T
population of particles with radius a_i	n_i
air pressure, dyne, atm, mbar	p
gas state eq. constant, erg/(°K mol)	R
gas constant for 1 g of dry air, erg/(°K g)	R_a
jet radius (m)	R_j
water vapour supersaturation	$s_{v,w}$
temperature inside rising parcel, °K	T
temperature outside rising parcel, °K	T'
parcel vertical velocity m/s	W
liquid water mixing ratio in rising parcel	w_L
water vapour mixing ratio in rising parcel	w_v
water vapour mixing ratio outside rising parcel	w'_v
saturation water vapour mixing ratio	$w_{v,sat}$
constant for induced mass acceleration term in Equation 12	γ
ratio M_w/M_a	ε
number of ions salt molecule dissociates into	ν
density of salt containing drop, g/cm^3	ρ_s''
density of dry air, g/cm^3	ρ
pure water density, g/cm^3	ρ_w
surface tension above drop of radius a, pure water, dyne / cm	$\sigma_{w/a}$
surface tension above drop of radius a, function of salt content dyne / cm	$\sigma_{s/a}$
molal (or practical) osmotic coefficient	Φ_s

$$\frac{\partial m_i^T}{\partial t} = -\mu_j \, W \, [\, n_i \, m_{s,i} - n'_i m'_{s,i} \,] \tag{4}$$

where primed quantities refer to the external aerosol classes. The external temperature and relative humidity are computed based upon soundings. The external aerosol population and salt content for each class is constant, while the class radius is re-computed based on the Koeler equation and the external relative humidity. The "parcel" quantities also integrated are : liquid water mixing ratio, water vapour mixing ratio, temperature, pressure, parcel (plume) rising rate, plume radius, plume height: the relevant equations are:

$$\frac{\partial w'_v}{\partial t} = interpolation - soundings \tag{5}$$

$$\frac{\partial T'}{\partial t} = idem \tag{6}$$

$$\frac{\partial w_L}{\partial t} = \frac{4\pi\rho_w}{3\rho}\int_0^\infty [3a^2 n(a)\frac{\partial a}{\partial t} + a^3\frac{\partial n}{\partial t}]da \; - \; \frac{4\pi\rho_w}{3\rho^2}\frac{\partial \rho}{\partial t}\int_0^\infty a^3 n(a)\, da \tag{7}$$

$$\frac{\partial w_v}{\partial t} = -\frac{\partial w_L}{\partial t} - \mu_j \left(w_v - w'_v + w_L\right) W \tag{8}$$

$$\frac{\partial R_j}{\partial t} = \frac{R_j}{2}\left[\mu_j \, W - \frac{\partial \ln \rho}{\partial t} - \frac{\partial \ln W}{\partial t}\right] \tag{9}$$

$$\frac{\partial T}{\partial t} = -\frac{g\, W}{C_{pa}} - \frac{L_e}{C_{pa}}\frac{\partial w_v}{\partial t} - \mu_j \, [\, \frac{L_e}{C_{pa}}(w_v - w'_v) + (T - T')] \, W \tag{10}$$

$$\frac{\partial p}{\partial t} = -\frac{g p W}{R_a T} \tag{11}$$

$$\frac{\partial W}{\partial t} = \frac{g}{1+\gamma}\left(\frac{T - T'}{T'} - w_L\right) - \frac{\mu_j}{1+\gamma} W^2 \tag{12}$$

$$\frac{\partial z}{\partial t} = W \tag{13}$$

The plume supersaturation is computed from $w_v$and$w_{v,sat}$. $w_{v,sat}$ is computed from pressure and the $e_{sat,w}$ (saturation water vapour pressure). $e_{sat,w}$ is given by the formula (Lowe and Ficke, 1974, quoted in /7/):

$$e_{sat,w} = a_0 + T\,(a_1 + T\,(a_2 + T\,(a_3 + T\,(a_4 + T\,(a_5 + a_6\, T\,))))) \tag{14}$$

valid in the range -50 °C , 50 °C for water, where the a_i's are constants. The derivative of the air density is computed from the equation of state of dry air:

$$\frac{\partial \rho}{\partial t} = \frac{1}{R_a\, T} \frac{\partial p}{\partial t} - \frac{p}{R_a\, T^2} \frac{\partial T}{\partial t} \tag{15}$$

LPMR uses to a large extent the system equations of the original model of Lee and Pruppacher. The results obtained by those Authors are used for the verification part of the study. Some important differences are nevertheless worth being mentioned. In the original model from Lee and Pruppacher the parcel supersaturation was explicitly integrated. As far as the water vapour saturation pressure is concerned the Authors refer to the Clausius Clapeyron equation in the form:

$$\frac{\partial \ln(e_{SAT,w})}{\partial T} = \frac{L_{e0}\, \varepsilon}{R_a\, T^2} \tag{16}$$

in /7/, and to the Smithsonian Tables in /6/. The Lowe formula is used in LPMR. The original formulation for the supersaturation was also tried in the present model development, but was later abandoned as it could lead to inconsistencies; supersaturation, water vapour mixing ratio, temperature and pressure are interrelated ie if three of them are known the fourth is known as well if an expression for $e_{SAT,w}$ is available. These quantities cannot be integrated independently. The effect of the difference between the two formulations on the model results was seen to be tiny. Both references /6/ and /7/ give scarce data on the technicalities of incorporating aerosol entrainment in the treatment. Model assumptions for the two models cannot be compared. The effect of expansion on the aerosol volume concentration is very briefly discussed /7/; our Equation 3 represents our understanding of the process; furthermore in the original model the equation for $\frac{\partial w_L}{\partial t}$ does not include the $\frac{\partial \rho}{\partial t}$ term.

3. STATISTICAL SUBROUTINES

The PREP preprocessor is a FORTRAN utility to assist in the implementation of a Monte Carlo analysis /8/. This analysis implies that a given model, coded into a computer program, is run for several combinations of input parameter value, in order to gain insight both on the distribution of value of the model output (prediction) and on the relationship between output uncertainty and uncertainty on the individual parameter. In order to perform such an analysis two main tasks are to be tackled:

a) Preparation of the input sample

b) Modification of the user model to allow the stochastic calculation (eg to introduce an iterative loop in the code to compute the output for each selected combination of input parameters)

The PREP preprocessor assists in performing tasks a) and b) by

a) generating the input sample for a Monte Carlo simulation. This sample can be thought of as a matrix which contains, for each of the N runs which compose the Monte

Carlo analysis, the sampled values for each of the K variables under examination. The sample is taken in such a way as to respect the individual distributions of the variables. The sampling strategy can either be purely random sampling or Latin hypercube sampling.

b) preparing a number of FORTRAN program subsections which can be automatically included by the user program, thus minimising the degree of program modification which falls on the user.

The SPOP (Statistical POst Processor) code is a tool to perform Uncertainty and Sensitivity Analyses on the output of a User implemented model /9/. SPOP performs Sensitivity Analysis (SA) and Uncertainty Analysis(UA) on a sample output from a Monte Carlo simulation. The sample is generated by the User and contains values of the output variable (in the form of a time series) and values of the input variables for a set of different simulations (runs), which are realised by varying the model input parameters. The User may generate the Monte Carlo sample with the PREP pre-processor. Among the tasks performed by the code are the computation of Tchebycheff and Kolmogorov confidence bounds on the output variable (UA), and the use of effective non-parametric statistics to rank the influence of model input parameters (SA).

Techniques for uncertainty and sensitivity analysis implemented in PREP and SPOP are further illustrated in /10-12/.

As mentioned before the two physical models (AERO2 and LPMR) are both concatenated with this set of statistical subroutines to enable sensitivity and uncertainty analysis. Regression and parametrization studies are also made easier by having the physical models linked with PREP and SPOP.

4. COMPUTATIONS

The execution of the package is driven by shell scripts which take care of executing the appropriate makefiles and executable modules, routing intermediate and output files. The package can be run stochastically or deterministically by simply changing the value of the appropriate switch. All the input values for AERO2 and LPMR are in fact assigned by default in a deterministic fashion at the beginning of the execution. When the switch for Monte Carlo execution is activated those values are overridden by the stochastic values of the input sample matrix generated by PREP.

The integration is performed for both AERO2 and LPMR using a Runge Kutta 4th order algorithm with adaptive step control /13 /. At the beginning the integration subroutines are called to integrate the following equations:

- NCLASS equations describing the variation in the aerosol population (particles/cm^3) for the different size classes, where NCLASS is the number of classes.
- Seven more equations describing respectively the H_2SO_4, SO_2 and DMS concentrations in the gas phase, the total amount of $(NH_4)_2SO_4$ molecules generated by the SO_2 oxidation, the total number of sulphur atoms deposited and diluted and the total number of DMS molecules generated by the ocean. Some of those equations are needed to check mass conservation during the integration.

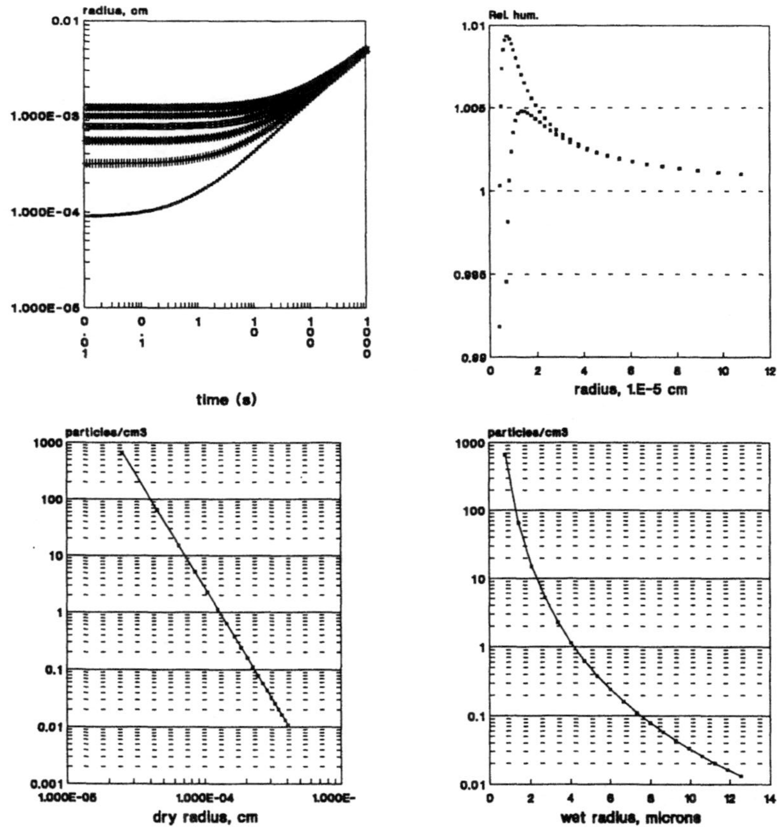

Figure 3. Testing of LPMR. Upper left: growth kinetics (Equation 1) for different ammonium sulphate content in the drop, ranging from 1.1 10^{-13} to 2.5 10^{-10} g. Upper right: solution of the Koeler equation for pure ammonium sulphate (ms = 1.0 10^{-16}) and a 40% volume mixture of salt and insoluble component (silica). Lower left: input dry radius spectrum for the four test cases. Lower right: wet (at 99% relative humidity) particle spectrum

Once this integration is completed the integration subroutines are called to integrate:

- NCLASS equations to describe the rate of change of class radius (Equation 1).
- NCLASS equations to describe the rate of change of class population (Equations 2 and 3).
- NCLASS equations to describe the rate of change of class salt content (Equation 4).
- The equations describing the plume liquid water mixing ratio, the water vapour mixing ratio, the temperature and pressure in the plume, the parcel rising rate, radius and height (Equations 7 through 13).

The Equations 2 for the class population includes an entrainment term, ie external aerosol particles are incorporated in the parcel during its ascent. These particles are previously equilibrated with the external relative humidity. Entrainment introduces a complication in the integration: in AERO2 in fact the particle histogram, defined by the particle radius, is fixed, ie the histogram abscissa does not move with time as the aerosol population grows; in LPMR instead the radii are changing with time and the particles histogram is moving both in the abscissa (the radii) and in the ordinate (the populations). The entrainment also alters the salt content of each class so that the particle histogram is also moving in a third dimension defined by the (average) salt content of each class. In LPMR the moving histogram is re-computed at each time step on the original coordinates (radii), redistributing populations and salt content.

5. CODE VERIFICATION

The verification exercise has been conducted on the four continental aerosol test cases described in /6/. Figure 3 shows some results from the preliminary testing of LPMR, ie the testing of the particle growth formula and of the Koeler implicit equation for droplets containing a mixture of ammonium sulphate and an insoluble component (silica).

The initial dry aerosol size distribution for the parcel has been obtained by calibrating a Junge power law to the data in /6/. The results for the dry radii are also shown in Figure 3 together with the corresponding radii at 99% relative humidity. The results for test cases one to four are shown in Figures 4 to 7; LPMR results can be compared with those of the original model given on the same figures.

Case 1. Continental aerosol, adiabatic parcel ascent, W computed (Figure 4). Parcel velocity, liquid water mixing ratio and plume radius are in good agreement. There is disagreement on the supersaturation (by a factor of two or more).

Case 2. Continental aerosol, adiabatic parcel ascent, W fixed (Figure 5). Liquid water mixing ratio and plume radius are in good agreement. There is again disagreement on the supersaturation.

Case 3. Continental aerosol, entrainment of air (no aerosol) (Figure 6). Same as above. Good agreement on all the predicted quantities but the supersaturation.

Case 4. Continental aerosol, entrainment of air and aerosol (Figure 7). Good agreement on all the quantities, including - this time - supersaturation.

The good agreement on most of the predicted quantities is encouraging, although the differences in the supersaturation results for cases one to three could not be eliminated. It seems that for the four test cases considered the initial water content of the parcel (99% relative humidity) is so large as to guarantee positive supersaturation of the parcel all the way up. Water vapour mixing ratio stays always larger than the saturation value (even in test case 3). Explicitly integrating the supersaturation does not change this trend. It is inevitable that there are differences between LPMR and the original model; these could be due to the differences in the modelling and in the numerical calculation mentioned in Sections 2 and 4, as well as to differences in the number of classes used to discretize the continuum particle radii spectrum, but it is interesting to note that supersaturation seems to be very sensitive to these differences in test cases 1 to 3. As a general trend supersaturation tends to be higher in our results than in those of the original model.

6. COUPLING AERO2 WITH LPMR

Figure 8 shows the results of a simulation where LPMR input (the aerosol histogram distribution, n(a) and $m_s(a)$) is generated by AERO2. The input size distribution has been obtained by assuming a three days aerosol generation phase, at an average temperature of 10 °C, with a DMS flux at the ocean surface assumed constant at the value $5.\ 10^{18}$ molecules/m^2/day and a mixing height of 1,000 m for the DMS concentration in the gas phase. Such rather unrealistic constant values are due to the preliminary nature of these simulations and to the lack of reliable data; both modules could easily process time dependent flux and temperature. The test case of Figure 8 includes entrainment of wet air (same LPMR setting as for test case 3 above, Figure 6). The aerosol is assumed to contain ammonium sulphate alone. Sea salt and insoluble components are neglected.

It is interesting that in spite of the constant low temperature allowed for a three days particle growth period the resulting particle histogram is positioned at much lower particle sizes than those normally assumed for ocean conditions. LPMR computations show that only about 50 particles are activated to cloud drops in these conditions. In a maritime cumulus cloud test case described in /6,7/ about 150 particles are taken as input to the model, with dry radii ranging from 0.1 to 5 microns. As a result supersaturation rises to unrealistically large values above 6%. The apparent inability of aerosol dynamics to nucleate and grow a sufficient number of particles is discussed elsewhere in these proceedings /14/.

7. WORKED EXAMPLE OF STOCHASTIC CALCULATION

In so far a stochastic analysis has been only performed for the AERO2 code in its "smog chamber" /15/ and "field" /5/ conditions. As mentioned above all what it takes to run the entire package (AERO2 plus LPMR) stochastically is to activate the appropriate switch in the input file; this would also require distribution of parameter value for the cloud parcel model, which are not available at present, ie distribution of input parameter values have only been assessed for the older module of the package.

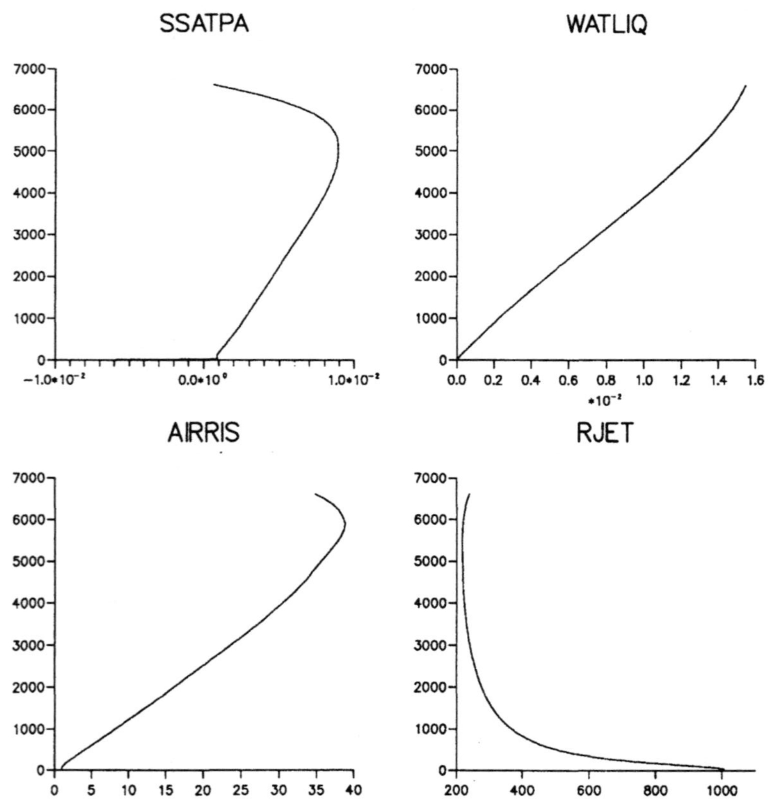

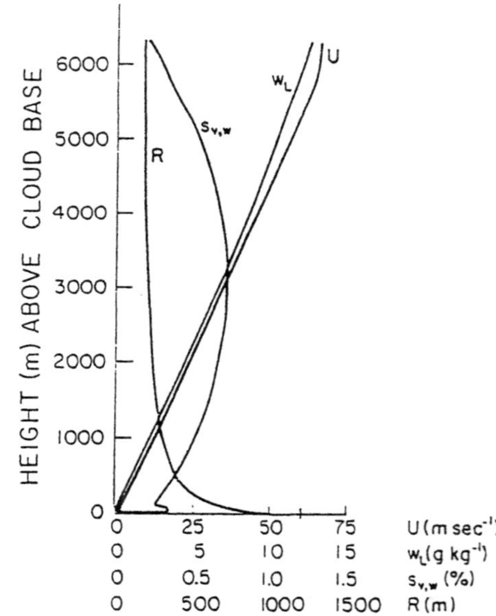

Figure 4. Test Case 1.

Left. LPMR prediction for the variation with height above cloud base of the cloud parameters sv,w (SSATPA, [0]), wL (WATLIQ, [0]), W (AIRRIS, [m/s]) and R (RJET, [m]). Adiabatic cloud model (no entrainment), continental aerosol, W computed from Equation 12.

Above. Same quantities for the original model /6,7/ (Courtesy of the Authors and of the Editors).

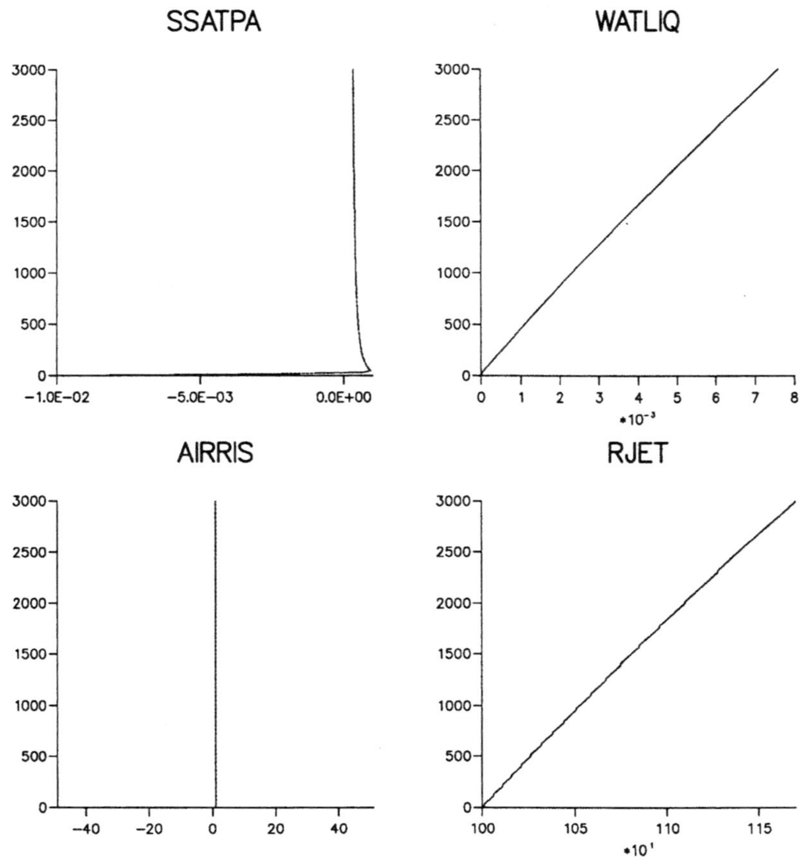

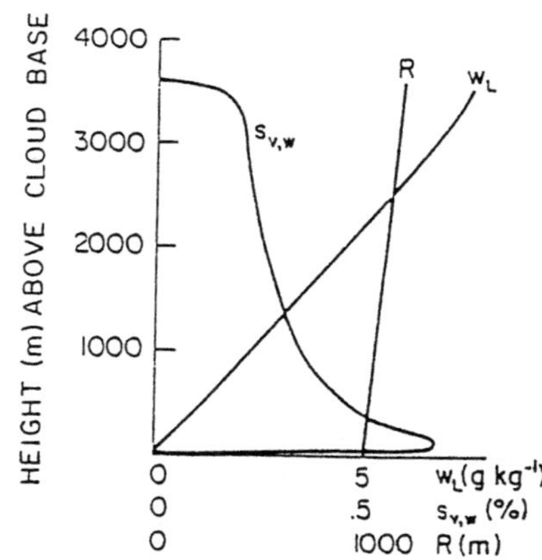

Figure 5. Test Case 2.
Left. LPMR prediction for the variation with height above cloud base of the cloud parameters sv,w (SSATPA, [0]), wL (WATLIQ, [0]), W (AIRRIS, [m/s]) and R (RJET, [m]). Adiabatic cloud model (no entrainment), continental aerosol, W fixed to 1 m/s.
Above. Same quantities for the original model /6,7/ (Courtesy of the Authors and of the Editors).

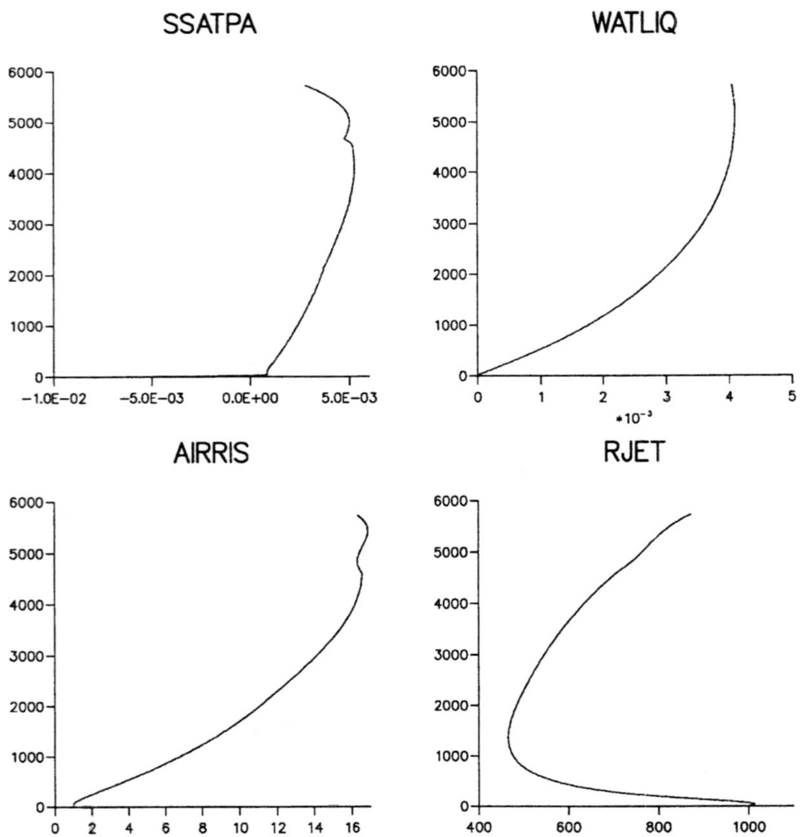

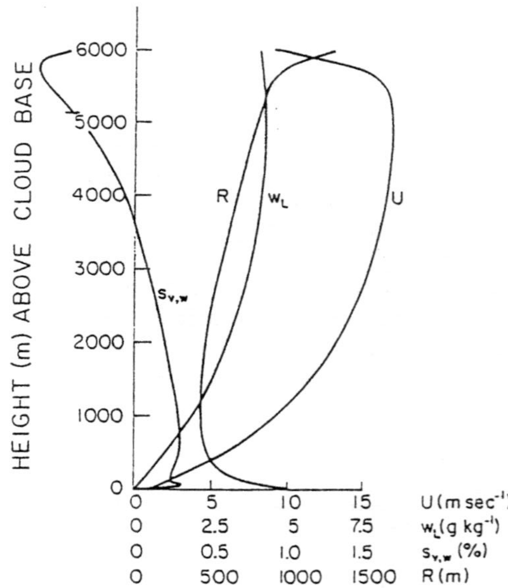

Figure 6. Test Case 3.
Left. LPMR prediction for the variation with height above cloud base of the cloud parameters sv,w (SSATPA, [0]), wL (WATLIQ, [0]), W (AIRRIS, [m/s]) and R (RJET, [m]). Entrainment model, continental aerosol, W computed, no aerosol particles entrained.
Above. Same quantities for the original model /6,7/ (Courtesy of the Authors and of the Editors).

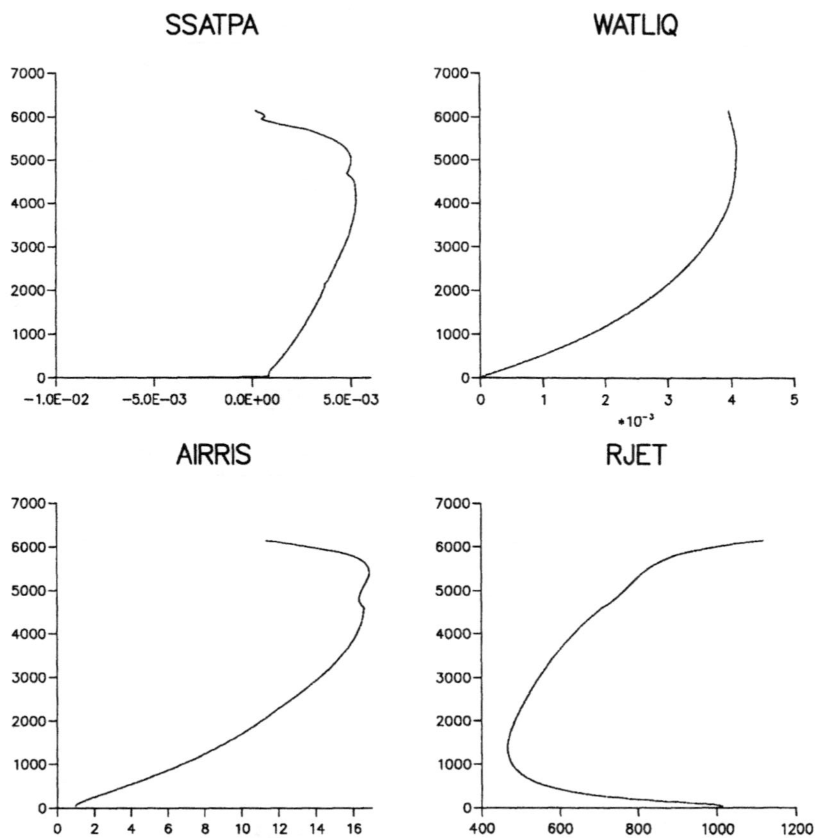

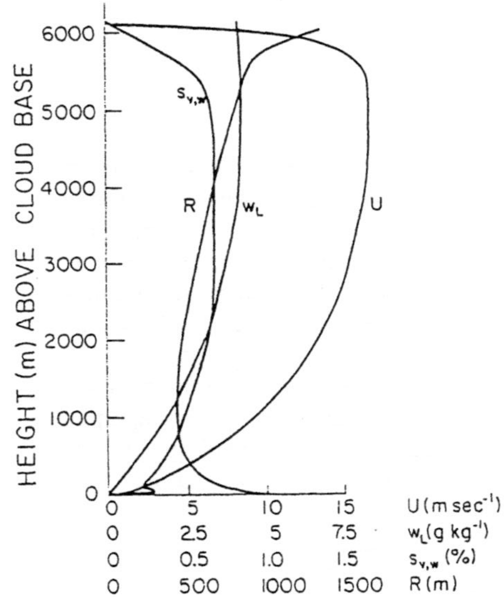

Figure 7. Test Case 4.

Left. LPMR prediction for the variation with height above cloud base of the cloud parameters sv,w (SSATPA, [0]), wL (WATLIQ, [0]), W (AIRRIS, [m/s]) and R (RJET, [m]). Entrainment model, continental aerosol, W computed, aerosol particles entrained.

Above. Same quantities for the original model /6,7/ (Courtesy of the Authors and of the Editors).

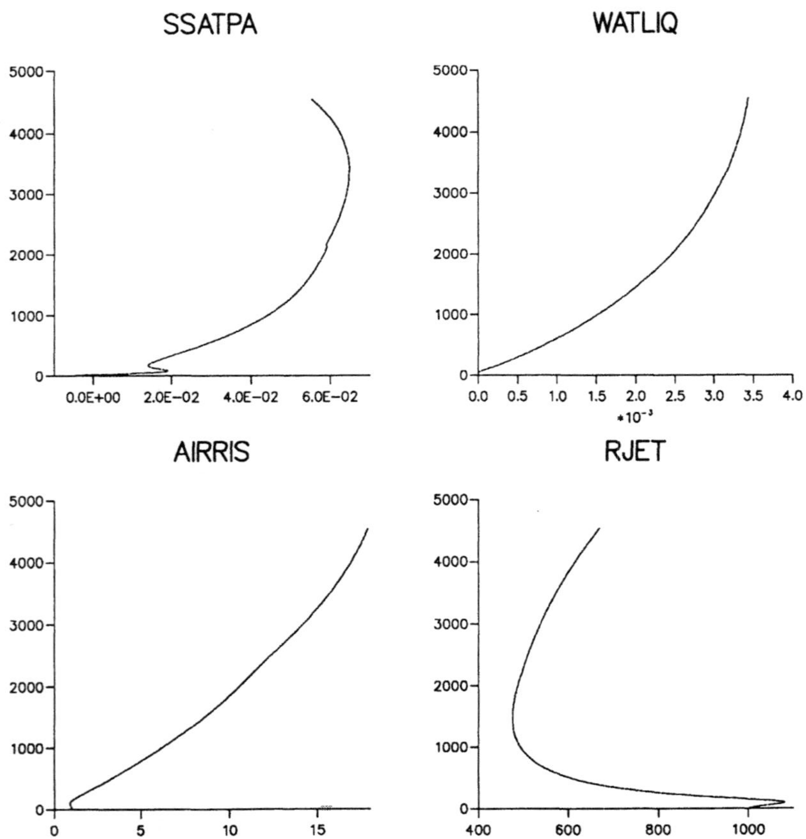

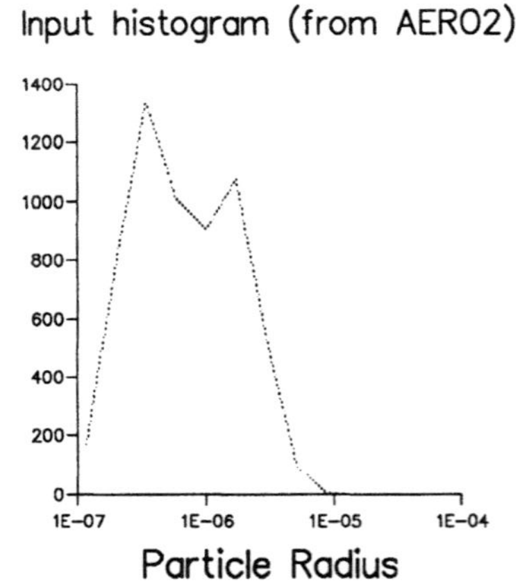

Figure 8. AERO2 + LPMR

Left. LPMR prediction for the variation with height above cloud base of the cloud parameters $s_{v,w}$ (SSATPA, [0]), w_L (WATLIQ, [0]), W (AIRRIS, [m/s]) and R (RJET, [m]). Entrainment model, continental aerosol, W computed, no aerosol particles entrained. Input from AERO2

Above. Particle histogram input distribution (from AERO2).

These parameter distributions allowed a validation exercise to be conducted on AERO2, to ascertain the adequacy of this model to describe the aerosol dynamics as observed in a set of laboratory controlled experiments. Both model prediction and experimental results are affected by uncertainties. As a result the validation requires that the comparison between model and experiments be done over the entire space of the uncertainties. A four step procedure was adopted. First the model prediction uncertainty was evaluated by propagating through the model the uncertainties in the input parameters as obtained from the literature. Second, a regression analysis was performed on the uncertain parameters using the mean experimental results. Third, model uncertainties were propagated again using reduced uncertainty range for the input parameters around the regressed values. The reduced ranges were finally compared with the original literature value intervals. The exercise allowed to confirm and reduce the range of uncertainty found in literature for the parameter K_1, the empirical reaction constant for the the formation of H_2SO_4 from SO_2 and OH radicals, and to criticise the accepted literature values for the nucleation reaction parameters /15/. Figure 9 shows the results of an uncertainty analysis on the "field" version of AERO2 /5/. The plot shows the median, 25th and 75th percentile of the distribution of condensation nuclei (CN) for a typical MBL condition. The uncertain input parameters responsible for the uncertainty in CN are the nucleation tuner (an empiric parameter used to adjust the thermodynamic value of the nucleation reaction rate), the accommodation coefficient for H_2SO_4 molecules on the H_2SO_4 aerosol, and the kinetic constant for the SO_2 oxidation reaction.

8. CONCLUSIONS AND FUTURE WORK

A description has been given of the on-going activities in the development of a model for the sulphur cycle. Such a model will allow the feedback mechanism of the sulphur cycle on the forcing due to greenhouse gases to be studied. At present only two modules of such a model are developed. These roughly describe some of the processes which link the DMS flux at the ocean floor to the cloud droplets in the atmosphere. Those modules are embedded into a statistical Monte Carlo driver.

It is evident from this note that the analysis and the quantification of the uncertainty will play an important role in the study. Uncertainty propagation and sensitivity analysis will be played at all the stages of the development, as shown in the two worked examples discussed in this note. One of the next exercises on schedule is an "aggregation analysis", ie a study of the impact of the number of classes used in both AERO2 and LPMR on the model results done in a Monte Carlo fashion (see /16/). At a later stage a simplified model for stratus cloud will be implemented in Occam in order to allow simple parametric studies of the Hadley cell with uncertainty and sensitivity analyses.

Acknowledgements

The authors wish to thank Pekka Korhonen and Markku Kumala of the University of Helsinki (SF) for their useful suggestions and advice.

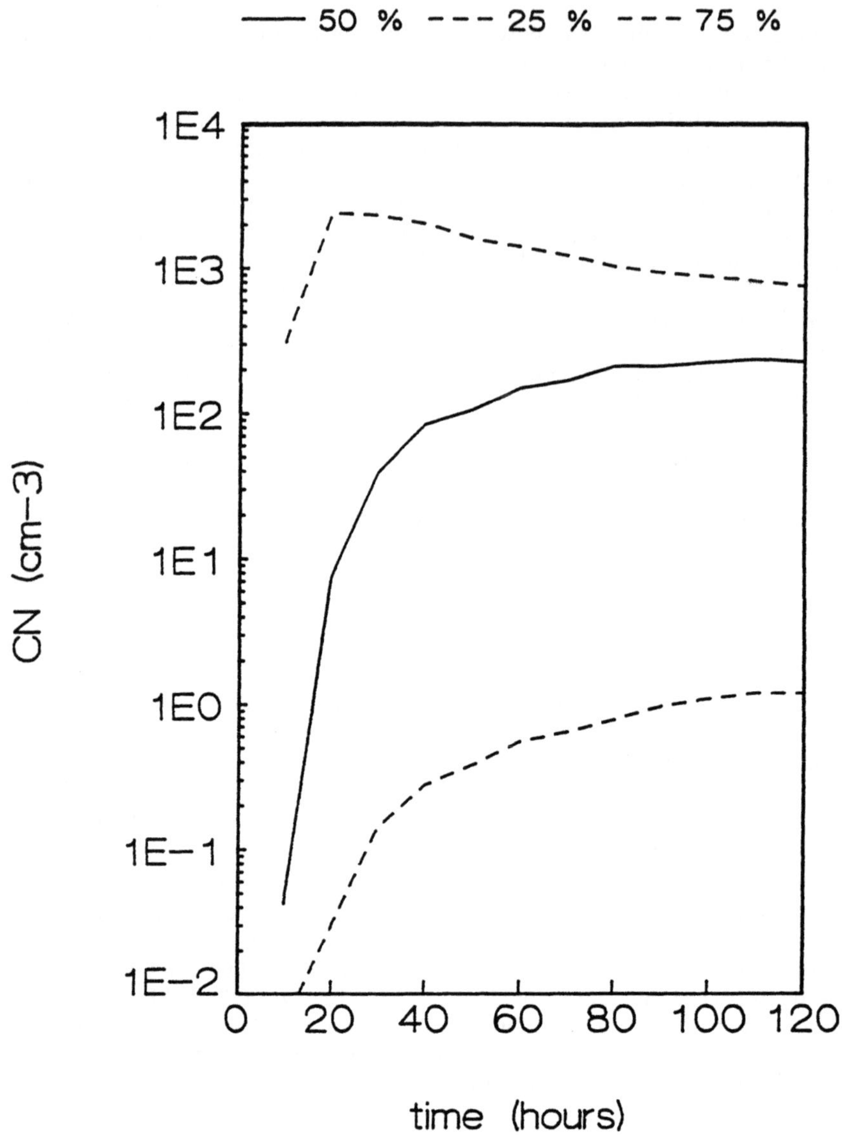

Figure 9 (from Reference 5). Uncertainty in CN resulting from uncertainty on AERO2 model input parameters. Curves are the median, the 25th and the 75th percentile as function of time for conditions typical of a MBL.

REFERENCES:

1. BATES, T. S., CHARLSON, R. J. and GAMMON, R. H. (1987). Evidence for the climatic role of marine biogenic sulphur. *Nature*, 329.
2. FOLEY, J. A., TAYLOR, K. E. and GHAN, S. J. (1991). Planctonic dimethylsulfide and cloud albedo: an estimate of the feedback response. *Climatic Change*, 1-15.
3. WEBSTER's new third international dictionary (unabridged). Merriam-Webster Inc. Publishers, Springfield (1886).
4. RAES, F. and JANSSENS, A. (1986). Ion-induced aerosol formation in a H_2O-H_2SO_4 system - II. Numerical calculations and conclusions. *Journal of Aerosol Science* **17**.
5. RAES, F. and VAN DINGENEN, R. (1992). Simulation of condensation cloud and condensation nuclei from biogenic SO_2 in the remote marine boundary layer. *Journal of Geophysical Research*, **97**, 12901-12912.
6. LEE, I.Y. and PRUPPACHER, H. R. (1977). A comparative study on the growth of cloud drops by condensation using an air parcel model with and without entrainment. *Pure and appl. geophysics*, **115**, 523-545.
7. PRUPPACHER, H. R. and KLETT, J. D. (1980). Microphysics of clouds and precipitation. D. Reidel Publishing Co. Dordrecht NL.
8. HOMMA, T. and SALTELLI, A. (1991). LISA package user guide. Part I. PREP (Statistical Pre-Processor); Preparation of input sample for Monte Carlo Simulation; Program description and user guide. CEC/JRC Scientific & Technical Report EUR 13922 EN, Luxembourg.
9. SALTELLI, A. and HOMMA, T. (1991). LISA package user guide. Part III. SPOP; Uncertainty and sensitivity analysis for model output. Program description and user guide. CEC/JRC Scientific & Technical Report EUR 13924 EN, Luxembourg.
10. SALTELLI, A. and MARIVOET, J. (1990). Nonparametric statistics in sensitivity analysis for model output; a comparison of selected techniques. *Reliability Engineering and System Safety* **28**, 229.
11. SALTELLI, A. and HOMMA, T. (1992). Sensitivity analysis for model output. Performance of black-box techniques on three international benchmark exercises. *Computational Statistics and Data Analysis* **13**, 73.
12. SALTELLI, A. ANDRES, T. and HOMMA, T. (1993). Some new techniques in sensitivity analysis of model output. *Computational Statistics and Data Analysis*, **15**, 211.
13. PRESS, W. H. et al. (1988). Numerical recipes. Cambridge University Press.
14. RAES, F. , VAN DINGENEN, R., WILSON, J. and SALTELLI, A. (1993). Cloud condensation nuclei from dimethyl sulphide in the natural marine boundary layer: remote vs. in-situ production. These proceedings.
15. RAES, F., SALTELLI, A. and VAN DINGENEN, R. (1992). Modelling the dynamics of H_2SO_4-H_2O aerosols. Uncertainty analysis and experimental evaluation. *Journal of Aerosol Science* **23**, (7) .
16. HELTON, J. C., IMAN, R. L., JOHNSON, J. D. and LEIGH. C. D. (1989). Uncertainty and sensitivity analysis of a dry containment test problem for the MAEROS aerosol model. *Nucl.Sci. Eng.*, **102**, 22-42.

MODEL STUDY OF THE RATIO BETWEEN METHANESULPHONIC ACID (MSA) AND NONSEASALT-SULPHATE IN COASTAL AIR

O. HERTEL[1], J. CHRISTENSEN[1], and Ø. HOV[2]

(1) National Environmental Research Institute, Department of Emissions and Air Pollution, 4000 Roskilde, Denmark
(2) University of Bergen, Geophysical Institute, 5007 Bergen, Norway

SUMMARY

Model results for the ratio between Methane Sulphonic Acid (MSA) and Non-Sea-Salt-Sulphate (NSSS) from a Lagrangian trajectory model are compared to measurements from the Oceano-NO_x project in Brittany, France. The development of Sulphate and MSA are followed along 96 h backtrajectories to the receptor point.

1. INTRODUCTION

In the present paper a model study of emissions, transport and chemistry of natural sulphur compounds is presented and results are compared to measurements from a campaign in Brittany, France (Barnes et al, 1991).

The model (1-dim) used for the computations is a Lagrangian boundary layer model. A one-dimensional column is advected along 96 hours back-trajectories. Vertical transport is described by eddy-diffusion within a log-linear grid of 20 grid points. The grid has a fine mesh close to the ground and is coarser as the free troposphere is approached.

The goal of the work presented is to quantify the distribution of natural sulphur compounds, based on the knowledge of their chemistry and emission. It is the hope to explain the relationship between DMS, MSA and nss SO_4 found in the Oceano-NO_x project.

2. MODEL DESCRIPTION

The model will only briefly be described in the following. A more detailed description can be found in Hertel et al (1993).

2.1 Chemistry

The chemical scheme for the natural sulphur compounds used in the model is primarily based on Yin et al (1990a,b). Rate constants are when possible adobted from more recent literature, as many of the reaction rate coefficients described in Yin et al (1990a,b) are estimates and very uncertain. Reaction rate coeffients for the natural sulphur compounds used in the model are shown in Appendix 1.

One problem in the chemical scheme is the rate coefficient of the thermal loss of CH_3SO_2. Yin et al (1990a,b) estimated a value of 11/s. A loss rate of that size gave nearly no MSA production in the scheme. Good and Thynne (1967) measured a rate constant of $2.7\cdot10^{-3}$/s, a value nearly 5000 times smaller than the estimated value of Yin et al (1990a,b). Ravishankara (1992) claims that CH_3SO_2 is highly stable, and that the lower value of Good and Thynne (1967) is most realistic. The low value means that thermal loss of CH_3SO_2 is of no significance.

The total chemical scheme used in the model contains about 80 species and 200 reactions. The numerical method used for solving the chemical reactions is the well known Quasi

G. Restelli and G. Angeletti (eds.), Dimethylsulphide: Oceans, Atmosphere, and Climate, 375–383.

Steady State Approximation (QSSA) Hesstvedt et al (1978). A time step of 90 seconds is used for the chemistry.

2.2 Transport

925 hPa winds from DNMI (The Norwegian Meteorological Institute) on the EMEP grid system are used to calculate 96 hours back-trajectories to the measurement site in Brittany. We assume that 925 hPa is representable for the boundary layer, disregarding wind turning with height. Positions are computed for each 2 hours. The one dimensional column is advected along the trajectory with a time step of 90 seconds, interpolating between the computed positions. For each time step chemistry and atmospheric diffusion in the vertical are performed. The atmospheric diffusion are represented by eddy diffusion coefficients. The numerical method used for the atmospheric diffusion is a semiimplicit method. For details see Hertel et al. (1993).

2.3 Dry and wet deposition

The dry deposition is described by the resistance method, $V_d = (r_a + r_b + r_s)^{-1}$, where V_d is the dry deposition velocity, r_a is the aerodynamic resistance, r_b is the laminar sublayer resistance and r_s is the bulk surface resistance. r_a and r_b are computed in 2 m height for each time step using DNMI meteorological data on EMEP grid. r_s is given as a mean value for land and a mean value for sea for each species in the model.

Wet deposition is modelled by scavenging coefficients. Cloud base is assumed to be situated at the mixing height at 1200 h the former day. Scavenging coefficients of $7 \cdot 10^5$ and $2 \cdot 10^5$ /s are used for quick and slow scavenged species respectively.

For both dry and wet deposition it is planned to make a better description in future. Especially dry deposition on sea is of interest here.

2.4 Heterogeneous processes

It is assumed that HNO_3 is transformed through heterogeneous processes to organic nitrates at a constant rate of $1 \cdot 10^{-5}$ /s. The organic nitrates are assumed to evaporate back to HNO_3 by a rate of $5 \cdot 10^{-6}$ /s. Barnes et al (1991) estimated heterogeneous loss rates of NO_3 radical to organic nitrates in the range 10^{-4} to 10^{-3} /s. The higher value is used in the following computations.

3. MODEL RESULTS

In this section model results are presented and compared to measurements.

3.1 Emissions

Emissions of SO_2, NO_x, VOC and NH_3 species for the grid shown in Fig. 1 are used in the computations.

The fluxes of natural sulphur compounds are modelled as a function of wind speed and sea concentration, $F = k \cdot U_{10}^2 \cdot (C_{sea} - C_{air})$, where F is the flux, k an empirical constant, U_{10} is the wind speed in 10 meters height, C_{sea} is the concentration in sea surface and C_{air} is the concentrations in air. The air concentration is assumed to play neglegible role for the flux, as the concentration in the water phase is considerably higher than the concentration in air. The empirical constants are found from the flux data presented in Leck and Rohde (1991). The transfer velocities and sea concentrations are given in Table I.

The DMS concentration in Table I is observed by Turner et al (1988). For the other species it is assumed that the ratios of DMS and the other sulpher compounds in the English

Table I. Transfer velocities and sea concentrations used in the model.

Species	K_w [cm/h]	Sea conc. [ng(S)/l]
DMS	$0.16 \cdot U_{10}^2$	220.0
DMDS	$0.15 \cdot U_{10}^2$	1.3
CH_3SH	$0.17 \cdot U_{10}^2$	11.0
CS_2	$0.15 \cdot U_{10}^2$	1.5

Channel is similar to the ratios observed in Kattegat/Skagerrak by Leck and Rohde (1991).

3.2 Results of computations

In the Oceano-NO_x project measuring campaigns were performed in July 1988 and June 1989. Fig. 1 shows 4 trajectories for 20 June 1989 representing air parcels arriving at 0000, 0600, 1200 and 1800 h.

The concentrations arriving at the receptor point can be compared to the measurements performed in the campaign in Brittany. First we will focus on the trajectory arriving at 20 June 1200 h. Fig. 2 shows the emissions of NO_x, SO_2 and biogenic sulphur in molecules cm^{-2} h^{-1} along the trajectory. 96 h prior to arrival the trajectory starts in Scotland, where the anthropogenic emissions are relatively low. When the coast of the Netherlands is approached, the anthropogenic emissions increase considerably.

Fig. 3 shows the concentrations at 2 m height along the trajectory for selected species. In Fig. 3a the SO_2 and DMS concentrations are shown. DMS increase slowly until the end of the second day. At this time the NO_x emissions are increasing, giving high NO_3 radical concentrations. High NO_3 concentrations lead to the degradation of DMS, and this causes an increase in SO_2. In Appendix 1 it is seen that the products from the DMS reactions are $DMSO_2$ and DMSO. $DMSO_2$ transforms either to CH_3SO_2 or CH_3SO, and these species go to SO_2 or MSA. DMSO leads to MSEA or CH_3SO_2. Fig. 3b shows the MSA and MSEA concentrations along the trajectory. It is seen that the MSEA concentrations are almost as high as the MSA concentrations. This is not seen in the measurements, and it is an open question whether this is true. The key to this question is the reaction of DMS and OH leading to DMSO, and the fate of DMSO. DMSO has a very high sticking coefficient (Ravishankara, 1992), and this makes it very difficult to measure its concentration and reaction rate coeffients. Fig. 3c shows ammonium sulphate and the sum of MSA and MSEA along the trajectory. On the second day the ammonium sulphate concentration increase both due to SO_2 emissions and the break up of DMS. On the third day concentrations increase strongly primarily due to SO_2 emissions. The ratio of the sum of MSA and MSEA to ammonium sulphate are shown in Fig. 3d. It is seen to vary between 0 and 0.2 along the trajectory and is 0.14 at the receptor point.

Fig. 4 shows the results of a calculation similar to Fig. 3, but in this case all the SO_2 emissions are shut off. This means that the SO_2 concentrations seen in Fig. 4a come from destruction of DMS. It is seen from Fig. 4a, that the SO_2 concentrations are relatively constant along the trajectory. The concentration of DMS increase until the middle of second day, when NO_x increases. When DMS reacts, the MSA, MSEA and ammonium sulphate concentrations increase. At the start-point the ratio of MSA/MSEA contra ammonium sulphate is 1 to 10 and in the receptor point it is 1 to 3.

In Table II concentrations computed by the model are compared to concentrations obtained from the measuring campaign.

It is seen from Table II that the calculated MSA and ammonium sulphate fits fairly well with the observed concentrations. It must, however, be remembered that these observations are 6 hours means. The DMS concentrations does not fit with the observed concentrations. There may be

Tabel II Computed and measured concentrations at Point de Penmarc'h. Observed ammonium sulphate and MSA mixing ratios are 6 hours means. All other data are 1 hour means.

0000 h 20 June 1989

Species	Observed	Computed
DMS	0.1 ppb	1.0 ppb
MSA	0.2 ppb	0.2 ppb
SO_2	1.0 ppb	3.6 ppb
ASO_4	2.3 ppb	2.4 ppb

1200 h 20 June 1989

Species	Observed	Computed
DMS	1.0 ppb	0.2 ppb
MSA	0.2 ppb	0.3 ppb
SO_2	2.1 ppb	3.4 ppb
ASO_4	5.9 ppb	2.9 ppb

0000 h 21 June 1989

Species	Observed	Computed
DMS	1.6 ppb	0.9 ppb
MSA	0.2 ppb	0.4 ppb
SO_2	1.0 ppb	5.0 ppb
ASO_4	2.3 ppb	2.4 ppb

more than one explanation to this, but local emissions are probably the prime reason. At the coast sea algae dies at low tide. This local effect is seen also in other locations, (Kirst, 1992), and causes the pronounced tidal variation in DMS concentrations which is seen in the Oceano-NO_x project. This local effect is, however, difficult to model, as the emissions used in the model are mean emissions for a grid of $150 \cdot 150$ km^2. It can be seen from Table III the calculated mixing ratio for SO_2 differ from the observations, but not as much as DMS.

4. CONCLUSIONS

A Lagrangian model for transport and chemistry in the PBL has been presented. The chemistry of the model has a detailed description of the biogenic sulphur compounds. Model results are compared to measured concentrations from the Oceano-NO_x project. Concentrations of MSA, the prime end product of the biogenic sulphur compounds from the sea, seems to be in fair agreement with observed concentrations. The DMS concentrations are, however, not in agreement with observed values. This may be due to a local variation in DMS emissions. In the Oceano-NO_x project a pronounced variation with the tide was observed. This effect has also been observed in other locations, and is caust of algae when they dry out at low tide.

The description of the emissions can be improved as more information becomes available on sea concentrations and the coastal meteorology. Also the description of dry deposition to sea can be improved.

5. ACKNOWLEDGEMENTS

J. Hjort and N.R. Jensen from JRC, Ispra are gratfully acknowledged for their help with the chemical scheme. Ruwim Berkowicz is acknowledged for discussions of the model results.

6. REFERENCES

Atkinson, R., 1986. Kinetics and mechanisms of the gas-phase reactions of the hydroxyl radical with organic compounds under atmospheric conditions. **Chem. Rev., 86, 69-201.**

Barnes, I., Bonsang, B., Brauers, T., Carlier, P., Cox, R.A., Dorn, H.P., Jenking, M.E., Le Bras, and Platt, U., 1991. Laboratory and field studies of oxidation processes occurring in the atmospheric marine boundary layer. Oceano-NO_x, CEC project. Report No 35 in the air pollution series.

Good, A., and Thynne, J.C.J., 1967. Reaction of free radicals with sulphur dioxide. Part 1 and 2. **J. Chem. Soc. Faraday. Trans., 63, 2708-2727.**

Hertel, O., Hov, Ø., Christensen, J., and Asman, W. 1993. One dimensional vertical model for gases in the atmospheric boundary layer. In preparation.

Hesstvedt, E., Hov, Ø., and Isaksen, I.S.A, 1978. Quasi-steady state approximation in air pollution modelling: Comparison of two numerical schemes for oxidant prediction. **Int. J. Chem. Kinet., 10, 971-994.**

Jensen, N.R., 1992. A laboratory study of the gas phase reactions of nitrate NO_3, with reduced organic sulphur compounds. Ph.D. thesis, University of Odense, Denmark, June 1992.

Kirst, G.O., 1992. Personal communication.

Leck, C., and Rohde, H., 1991. Emissions of marine biogenic sulfur to the atmosphere of Northern Europe. **J. Atm. Chem., 12, 63-86.**

Ravishankara, A.R., 1992. Personal communication.

Turner, S.M., Malin, G., Liss, P.S., Habour, D.S., and Holligan, P.M., 1988. The seasonal variation of dimethyl sulfide and dimethylfulfioniopropionate concentrations in nearshore waters. **Limnol. Oceanogr., 30, 59-70.**

Tyndal, G.S., and Ravishankara, A.R., 1989. Kinetics and mechanisms of the reactions of CH_3S with O_2 and NO_2 at 298 K. **J. Phys. Chem., 93, 2426.**

Tyndal, G.S., and Ravishankara, A.R., 1991. Atmospheric oxidation of reduced sulphur species. **Int. J. Chem. Kin., 23, 483-527.**

Yin, F., Grosjean, D., and Seinfeld, J.H., 1990a. Photooxidation of dimethyl sulfide. I: Mechanism development. **J. Atm. Chem., 11, 309-364.**

Yin, F., Grosjean, D., and Seinfeld, J.H., 1990b. Photooxidation of dimethyl sulfide. II: Mechanism Evaluation. **J. Atm. Chem., 11, 365-389.**

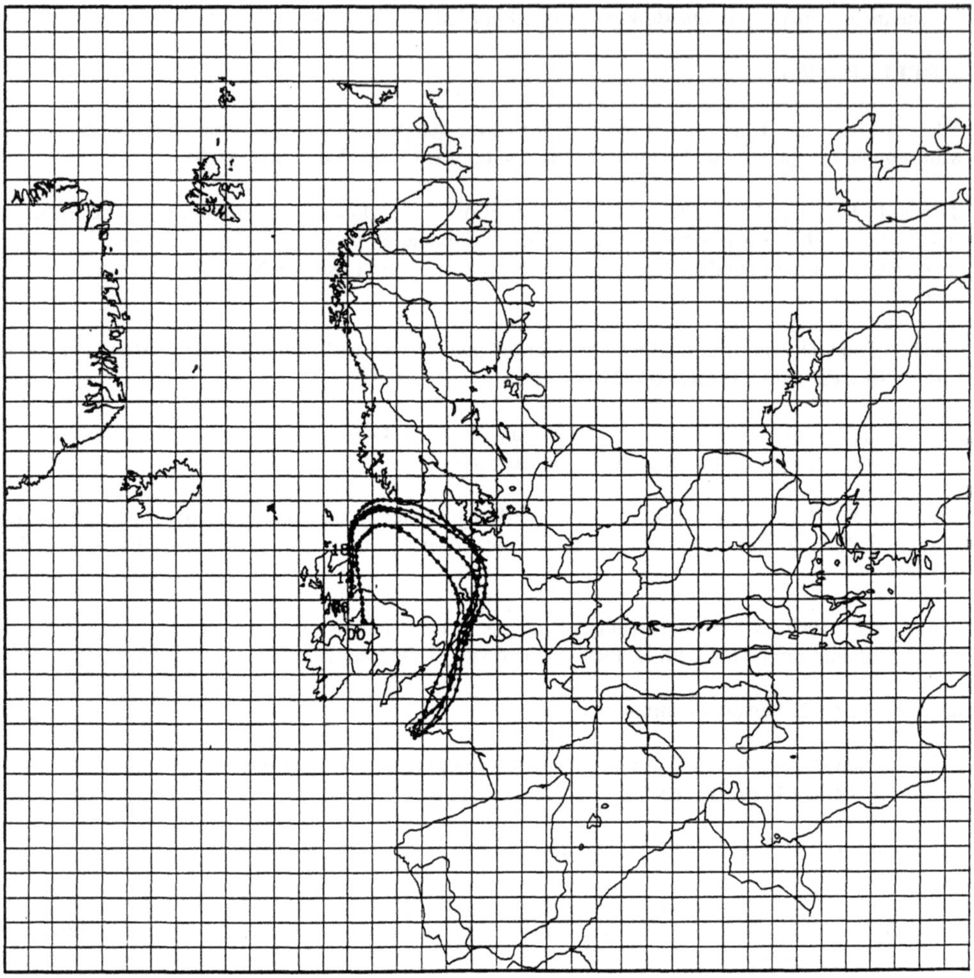

Fig. 1 925 hPa backtrajectories for 0000, 0600, 1200 and 1800 h for 20 June 1989 for the receptor point at Pointe de Penmarc'h.

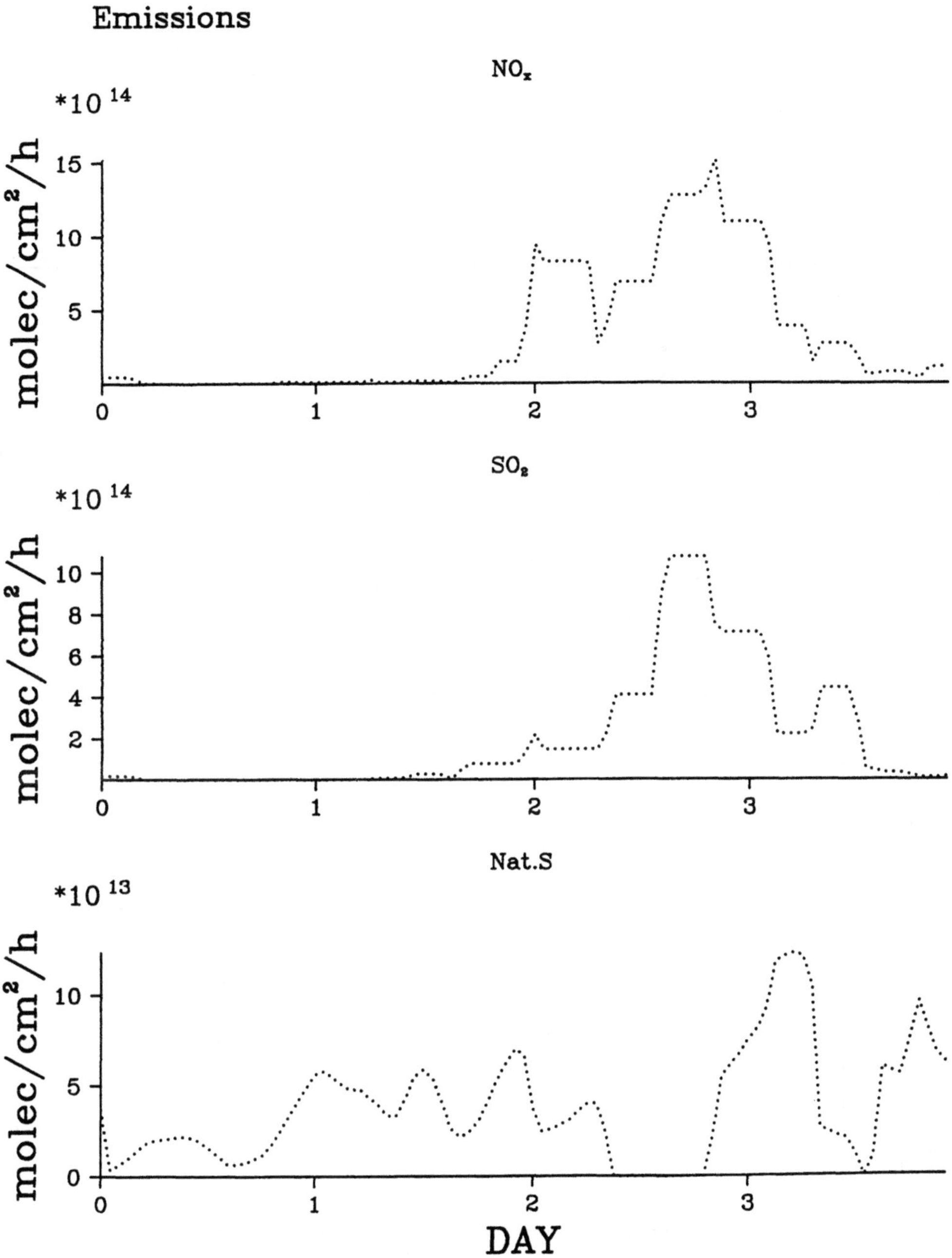

Fig. 2 Emissions of NO_x, SO_2 and biogenic sulphur along the backtrajectory for 1200 h 20 June 1989.

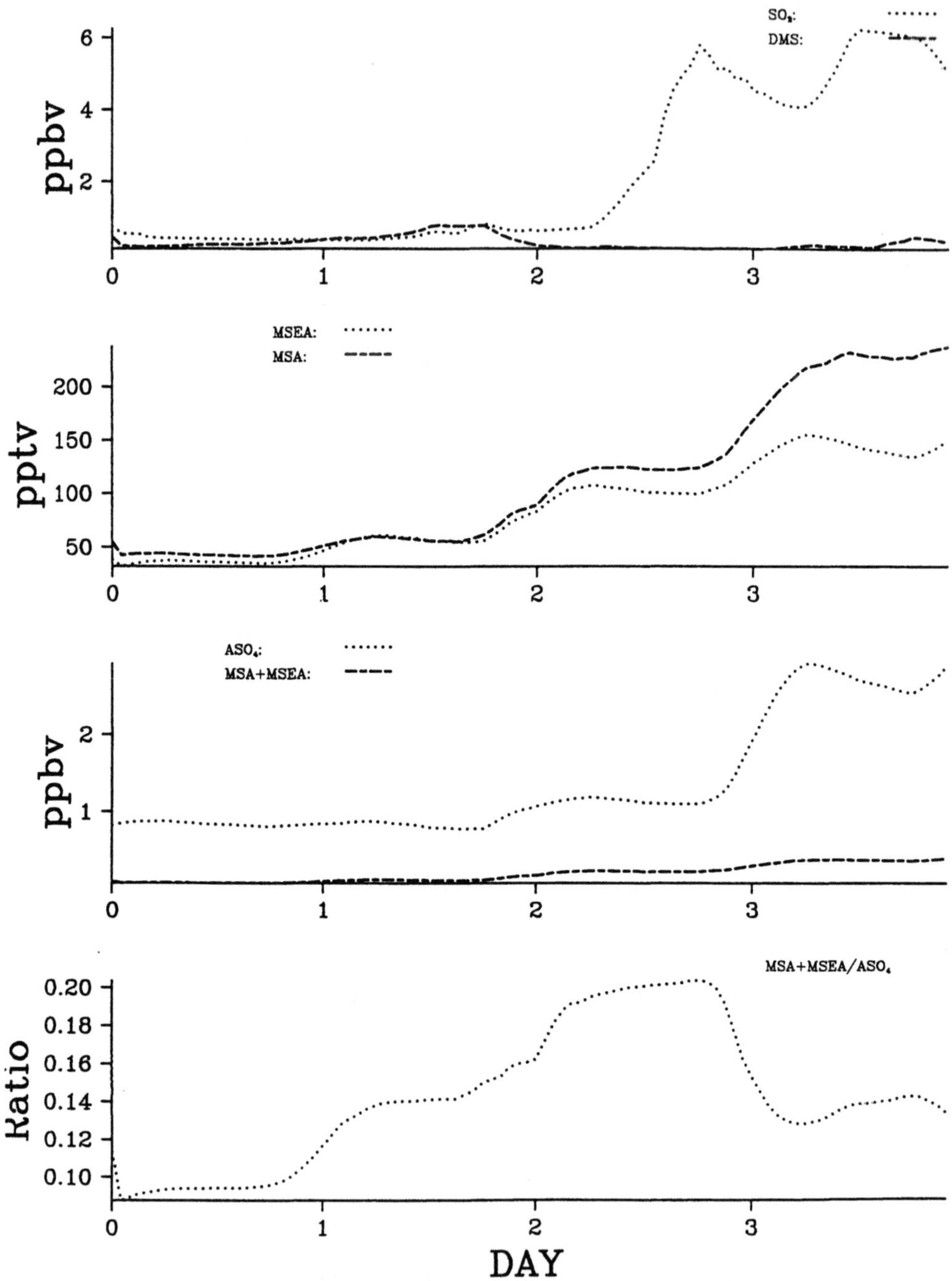

Fig. 3 Mixing ratios along the 925 hPa backtrajectory 1200 h 20 June 1989, of SO_2, DMS, MSEA, MSA, ASO_4 and the ratio (MSA+MSEA)/ASO_4, at 2 m height.

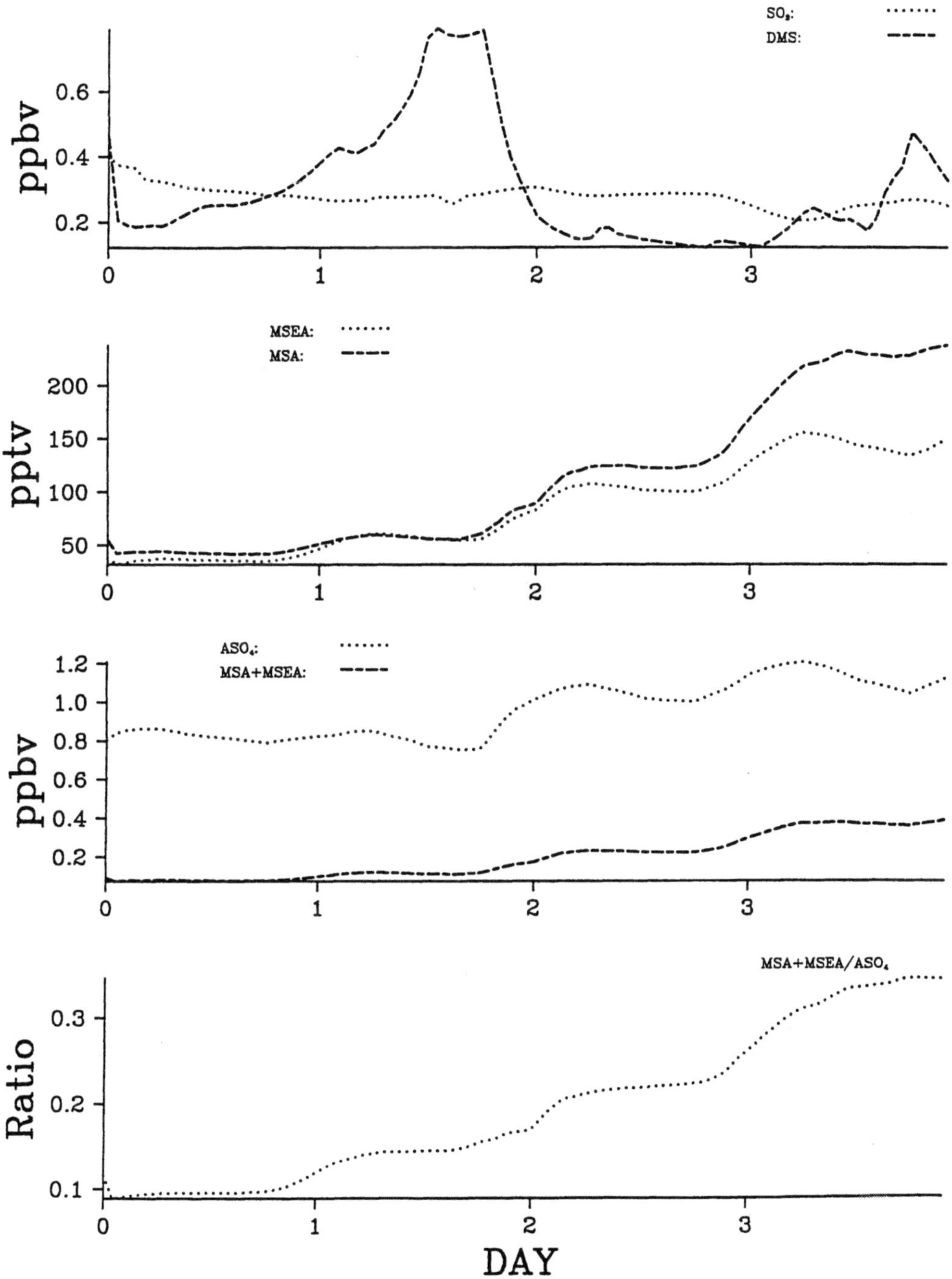

Fig. 4 Mixing ratios at 2 m height along the 925 hPa backtra-jectory arriving at 1200 h 20 June 1989. SO_2 emissions were set to zero.

A PHYSICAL RECEPTOR MODEL APPLIED TO AEROSOL DATA FROM NORTHEASTERN GREENLAND

PETER WÅHLIN
National Environmental Research Institute
Division of Emissions and Air Pollution
Frederiksborgvej 399, DK-4000 Roskilde, Denmark.

SUMMARY

A receptor model has been developed that is based on the Chemical Mass Balance model, but incorporates multiple regression to fit the components of the source profiles. In an iterative method, which minimizes the total chi-square, constraints are used to exclude non-physical or mixed solutions. The model was tested on aerosol samples from Northeastern Greenland that were collected on a weekly basis in the period from August 1990 to April 1991, and analyzed for elemental contents by Particle Induced X-ray Emission. Involving a soil, a sea, and two anthropogenic source profiles, extremely good fits to measured concentrations were obtained when sulphur, chlorine, and bromine were excluded. The model could give important input to a discussion about the reasons to the deviant behaviour of these elements. Thus, the poor fit of sulphur could be well explained in the winter by a high fraction of non-oxidized sulphur dioxide, while a still unexplained excess of sulphur in August and September might be of marine biogenic origin.

1. INTRODUCTION

The two main branches of receptor models are Chemical Mass Balance (CMB) models and multivariate factor analysis models (1). CMB models need only one sample and assume knowledge of the sources and their composition, while factor analysis, by artful mathematics, attempts to apportion the sources and determine their composition on the basis of a series of observations at the receptor alone. That factor analysis without physical presumptions attempts to get more information out of the data than is really there, has been emphasized by HENRY (2). The hybrid model used in this paper on aerosol data from Northeastern Greenland unifies qualities from both branches of receptor models.

2. THE PHYSICAL RECEPTOR MODEL

The fundamental hypothesis in mathematical receptor models is that the measured concentrations x_{ij} of n different chemical components in N samples can be satisfactorily expressed as a sum of contributions from a small number p ($p<n$) of sources with a constant composition,

$$x_{ij} \cong \sum_k a_{ik} f_{kj} \ . \qquad \text{(eq. 1)}$$

In the equation (eq. 1) a_{ik} are constants representing the relative composition of the sources (the source profiles), and f_{kj} are the source strengths found in the different samples. Index i refers to the n chemical variables, index j refers to the N samples, and index k refers to the p sources. A simple, illustrative example is shown in fig. 1 with three chemical variables and two sources.

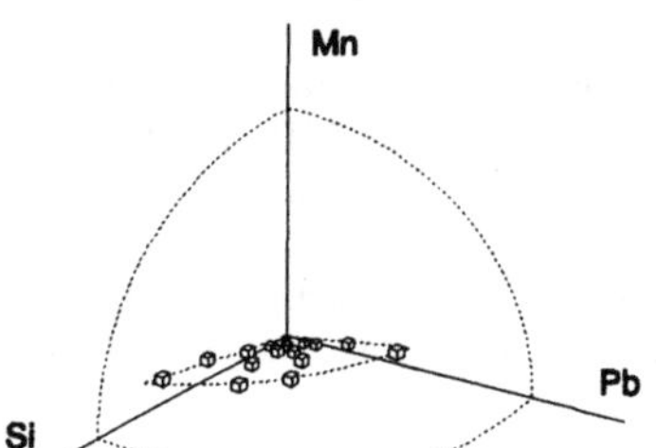

Fig. 1. Illustrative plot of measurements on an aerosol originating from two sources.

Solving (eq. 1) means to find any source profiles a_{ik} and source strengths f_{kj} that, by insertion in the right side of (eq. 1), give results close to the measured x_{ij}. From a physical point of view, the best solution will minimize χ^2, given by

G. Restelli and G. Angeletti (eds.), Dimethylsulphide: Oceans, Atmosphere, and Climate, 385–391.

$$\chi^2 = \sum_j \sum_i \frac{(x_{ij} - \sum_k a_{ik} f_{kj})^2}{\sigma_{ij}^2}, \quad \text{(eq. 2)}$$

where σ_{ij} are the uncertainties of x_{ij}. In general, an infinity of solutions are equally good in a purely mathematical sense, but the arbitrariness tends to disappear when physical restrictions are considered. To a simple linear model like (eq. 1), without mutual influence by chemical components from different sources, a fundamental physical demand is that both a_{ik} and f_{kj} must be non-negative. This demand can be complied with, if constraints are introduced in the minimizing process every time a_{ik} or f_{kj} hit the non-physical barrier. In addition, general constraints can be set up for the source profiles by the research worker using knowledge of real source composition. An iterative process is used, in which the minimizing of χ^2 is performed solving the systems of linear equations derived from

$$\frac{\partial \chi^2}{\partial f_{\ell' j}} = 0 \quad \text{(eq. 3)}$$

and

$$\frac{\partial \chi^2}{\partial a_{i\ell'}} = 0 \quad \text{(eq. 4)}$$

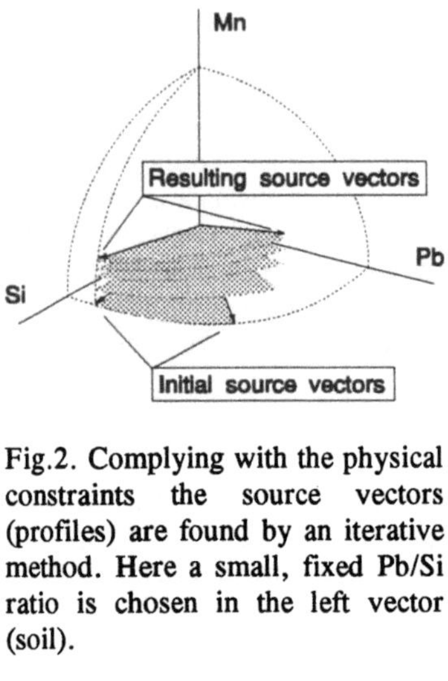

Fig.2. Complying with the physical constraints the source vectors (profiles) are found by an iterative method. Here a small, fixed Pb/Si ratio is chosen in the left vector (soil).

alternately. Solving the equations derived from (eq. 3) corresponds to the approach in the CMB model, which presupposes known source profiles, while (eq. 4) leads to a multiple regression analysis searching for better fitting source profiles. The source index label (') indicates that some sources may be omitted in order to prevent violation of constraints. An initial profile matrix a_{ik} is set up in which the column vectors are chosen freely. The idea in the iteration is illustrated in fig. 2. The iteration is continued until χ^2 reaches a stagnant minimum value. The degrees of freedom, ν, depends upon the number of negative values of a_{ik} and f_{kj} found in the last iteration loop, therefore both χ^2 and ν are determined by the iterative process. The research worker can use background knowledge to direct the iteration to a rational result by choosing, for example, vectors that are proportional to known source profiles, or vectors with a single non-zero component used as tracer, and by setting up constraints to prevent mixing of the source vectors. As more constraints are added to the hybrid model, it can be gradually changed from a pure multivariate model to a pure CMB model.

A code implemented on IBM PC has been written to perform the iteration. As inputs are used three matrices: An $n \times p$ profile matrix containing the initial source vectors, an $n \times p$ form matrix containing information about the constraints on the profile matrix elements in 0=fixed/1=free format, and a $2 \times n \times N$ data matrix containing the measured values and their absolute uncertainties.

3. METHODS AND RESULTS

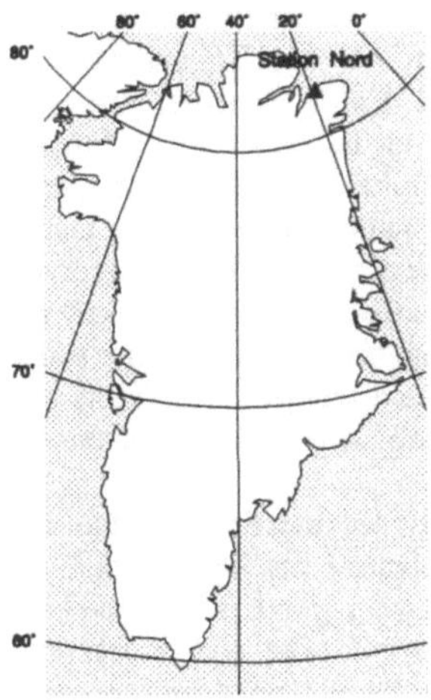

Fig. 3. Greenland map.

Samples were collected at a site 2 km north of Station Nord in Northeastern Greenland (fig. 3) as small spots on one single grease-coated impaction foil and one single 0.4 μm Nuclepore filter on a weekly basis from August 1990 to April 1991 with an automatic two-stage discrete streaker designed for PIXE (Particle Induced X-ray Emission) analysis (3). The 35 coarse fraction samples (aerodynamic cut-off diameter $\cong$ 2.0 μm) and 35 fine fraction samples were analyzed for elemental contents by PIXE, and weekly mean concentrations were calculated using measured air volumes. Background corrections were performed using measured values on non-exposed areas of the same impaction foil and filter. Twenty elements (Si, S, Cl, K, Ca, Ti, V, Cr, Mn, Fe, Ni, Cu, Zn, As, Se, Br, Rb, Sr, Zr, and Pb) were detected with uncertainties down to 5%. Some elements in some samples were below detection limits, and background corrections produced some negative concentrations, but no special handling of this was needed in the fitting process thanks to the consistent use of absolute uncertainties.

The weekly mean concentrations were fitted with an initial profile matrix containing four source vectors: A soil profile, a sea salt profile, a general anthropogenic profile (in this paper called the urban profile), and an additional anthropogenic profile containing only Cu and Ni (in this paper called the metal profile). As soil and sea profiles were used the proportion by weight in ppm of the elements occurring in igneous rocks of the

	Fine fraction					Coarse fraction				
	Soil	Sea	Urban	Metal	Chi2	Soil	Sea	Urban	Metal	Chi2
Si	20.6746	0.00028	10.9338	0	56	20.9741	0.00026	4.0912	0	39
S	0.05075	12.4188	497.814	0	0	0.07458	11.6312	88.3206	0	0
Cl	0.04684	266.638	0	0	0	0.06884	249.729	0	0	0
K	2.49422	5.33838	5.46904	0	67	3.25616	4.99985	1.60903	0	32
Ca	4.65952	5.61935	2.27365	0	34	5.66968	5.263	1.51443	0	19
Ti	0.42942	0	0.15977	0	63	0.63108	0	0.13024	0	95
V	0.02022	4.21E-06	0.25924	0	183	0.03739	3.95E-06	0.03830	0	9
Cr	0.00976	0	0.13182	0	46	0.01434	0	0.07166	0	33
Mn	0.06759	1.40E-05	0.2669	0	70	0.09365	1.32E-05	0.12250	0	68
Fe	5.56765	2.81E-05	4.33886	0	24	7.23842	2.63E-05	2.76546	0	44
Ni	0.00625	1.40E-06	0.12042	0.03092	140	0.00917	1.32E-06	0.03474	0.01397	61
Cu	0.00683	1.40E-05	0.13689	0.11869	28	0.01004	1.32E-05	0.05578	0.08248	54
Zn	0.00780	7.02E-05	1.63484	0	60	0.01147	6.58E-05	0.62891	0	13
As	0.00048	0.00014	0.14297	0	57	0.00071	0.00013	0.08143	0	29
Se	8.78E-06	5.62E-05	0.02538	0	42	1.29E-05	5.26E-05	0.01016	0	35
Br	0.00024	0.91314	0	0	0	0.00035	0.85523	0	0	0
Rb	0.01491	0.00281	0.02522	0	26	0.02225	0.00263	0.01648	0	75
Sr	0.01299	0.18262	0.03061	0	46	0.02992	0.17104	0.03284	0	37
Zr	0.01997	0	0.00644	0	10	0.03011	0	0.00832	0	11
Pb	0.00156	5.62E-05	1.24454	0	71	0.00229	5.26E-05	0.46484	0	48
Sum					1022					701
ν					545					552

Table I. Resulting source profiles normalized to arithmetic mean concentrations in the sampling period (in ng/m³). Also summarized are the chi-square values and the degrees of freedom.

continents and the average composition of sea water in ppm, both quoted in (4). Using zeros in the soil column of the corresponding form matrix, Ti was defined as soil tracer by fixing the mass ratios of the essential non-crustal elements S, Cl, Cr, Ni, Cu, Zn, As, Se, Br, and Pb to this crustal element, while the ratios of the remaining crustal elements K, Ca, V, Mn, Fe, Rb, Sr, and Zr were allowed to vary freely. The elements S, Cl, and Br could not be satisfactorily fitted into the four-vector space, and they were excluded by introducing infinite uncertainties in the data matrix. With S, Cl, and Br missing only K, Ca, and Sr were left as important sea source markers, and therefore the sea profile was fixed by a null sea column in the form matrix to preserve the initial average sea water composition. Pb was used as tracer in the urban profile with fixed initial ratios of S, Cl,, and Br to Pb. Because S, Cl, and Br were of no importance in the fitting process, the mass ratios of these elements were arbitrarily fixed to values to be discussed later in this paper. The other elements in the urban profile were allowed to vary freely with over-estimated initial values of the crustal elements.

The resulting profile matrices were calculated independently for the fine and the coarse samples and are summarized in table I. The goodness of the single element fits and the total fit are expressed by the chi-square values. It can be seen that the fits are very good, which indicates a remarkable constancy of the anthropogenic profiles during the period. The coarse fraction fit (reduced chi-square, $\chi^2/\nu = 1.27$) is so close that the deviations can be explained by the statistics alone. The fine fraction fit is slightly more deviant ($\chi^2/\nu = 1.87$) mostly due to (correlated) divergences of V and Ni.

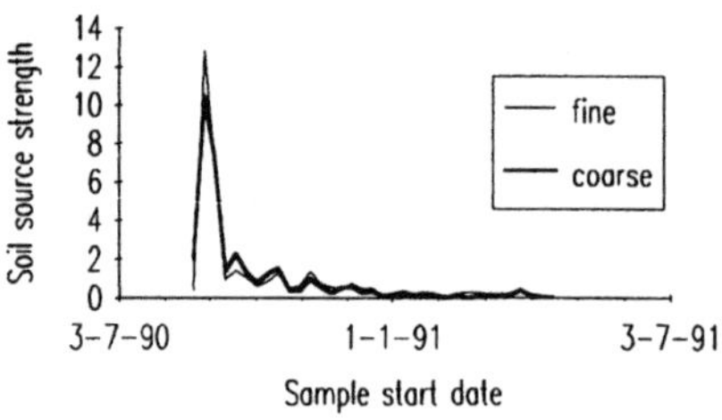

Fig. 4. Soil source strength signal.

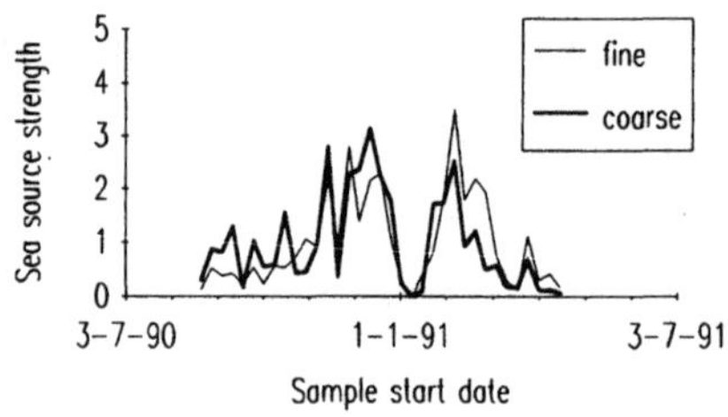

Fig. 5. Sea source strength signal.

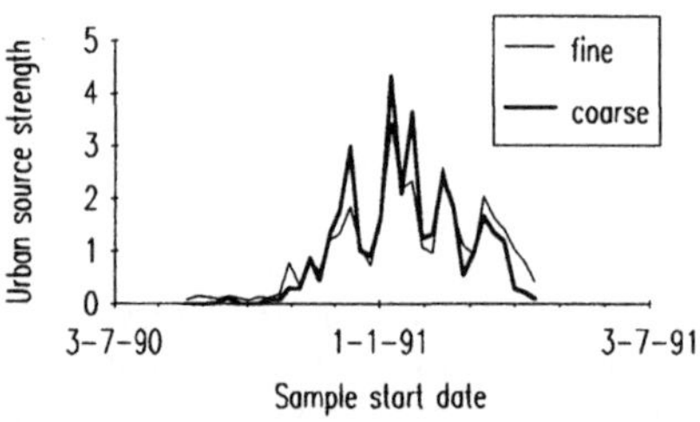

Fig. 6. Urban source strength signal.

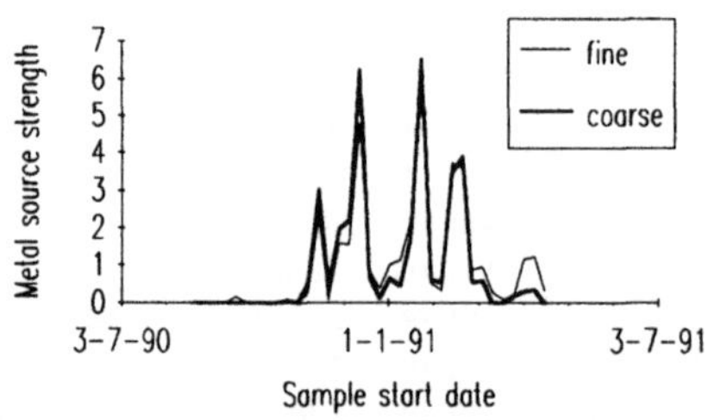

Fig. 7. Metal source strength signal.

4. DISCUSSION

The source strength time series are plotted in figs. 4 – 7. As far as can be deduced from the fine and the coarse soil signals the soil aerosol seems to have a rather constant size distribution in the period. The strengths decrease in the winter to almost zero, which is natural because of the snow cover. The sea aerosol size distribution is much more varying, and the signals do not show a clear seasonal tendency. The urban and the metal signals seem to have nearly constant size distibutions, and they both show the typical winter appearance of the long range anthropogenic pollution in the Arctic (arctic haze). The rapid variation of the metal source strength signals suggests that Cu and Ni are emitted from a strong point source positioned in a larger urban source

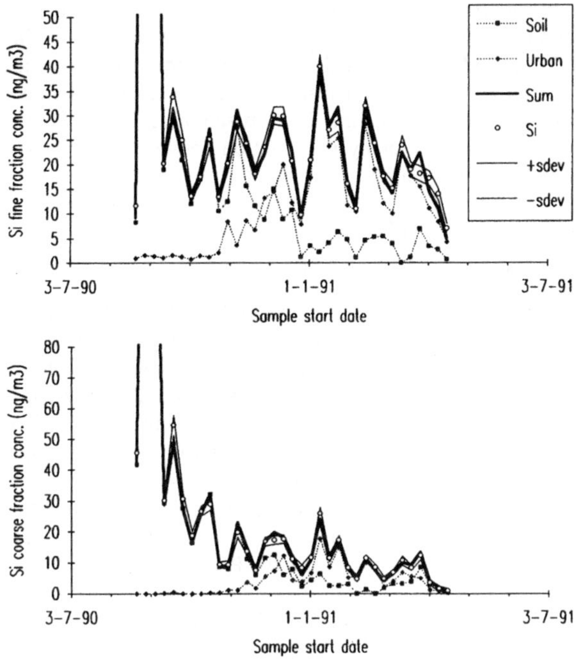

Fig.8. Fine and coarse fraction measurements of Si (with uncertainties) plotted together with the fits.

area. The "ravines" between the peaks in the urban signal also appear in the metal signal but not in the sea signal, so they probably reflect injections of non-polluted air from south.

The fitted concentrations are obtained from the calculated source profiles and source strengths. I fig. 8 this is done for Si and the result is plotted together with the measurements. As can be seen from the chi-square values in tabel I the goodness of the silicon fits is representative of all the fitted variables apart from fine fraction V and Ni, which both display a small growth during the winter compared with the fits (fig. 9).

The misfits Sulphur, Chlorine, and Bromine

It was a necessity to exclude S, Cl, and Br from the analysis to get good fits, but calculated concentrations of these elements could be obtained by giving them reasonable values in the source profiles. The average composition of sea water was used throughout as the sea source profile, and the contributions from the sea source to S, Cl, and Br were on this basis obtained as a direct result of the calculations. The mass ratios of S to Pb in the urban profiles were arbitrarily fixed by inspection of the plots to values that looked reasonable (fig. 10), when it was taken into account that a large fraction of total sulphur was in a gaseous state in midwinter. Simultaneous, but two days displaced, filterpack measurements of sulphur dioxide and sulphate at Station Nord demonstrate this, and the total-sulphur plot (fig. 11) of the filterpack measurements

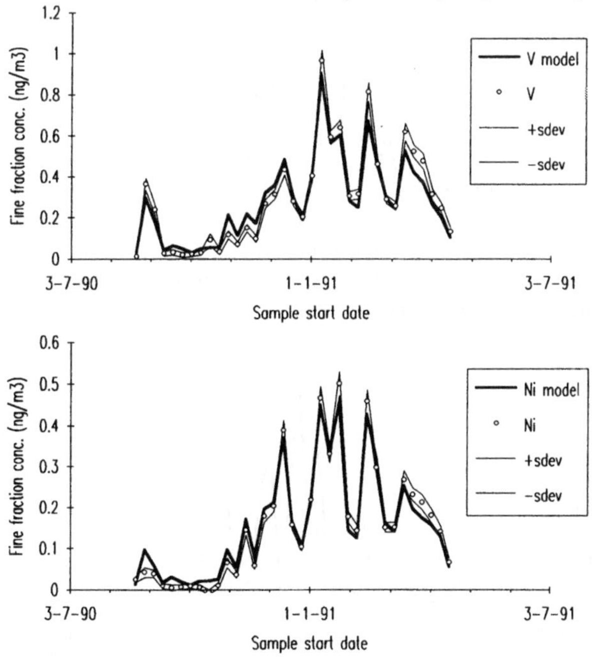

Fig. 9. Poor fine fraction fits for vanadium and nickel. Both elements exhibits a growth during the winter compared with the fits.

has indeed some resemblance to the "fits" in fig. 10, but a still unexplained occurrence of sulphur, especially in the coarse fraction, in August and September might be of marine biogenic origin. The chemical elements Cl and Br were modeled with zero contributions from the anthropogenic sources, but even in this case, regarding sea salt as the only important source, the calculated values were generally too large. In many samples, especially in the fine fraction samples, Cl was totally absent, so it may be assumed that it was lost from the filter by the reaction between chloride and sulphuric acid (and, probably, ammonium hydrogen sulphate):

$$H_2SO_4 + 2Cl^- \rightarrow SO_4^{2-} + 2HCl(g) ,$$

$$NH_4HSO_4 + 2Cl^- \rightarrow SO_4^{2-} + NH_3(g) + 2HCl(g) .$$

This hypothesis is supported by fig. 12, in which measured and calculated Cl concentrations (in nmol/m³) are compared with the concentration of S multiplied by 2. The stoichiometrical correspondence between S and missing Cl is good, especially on the filter stage. Br may be lost by similar reactions, and it was confirmed that Br to some extent disappeared during the PIXE analysis itself.

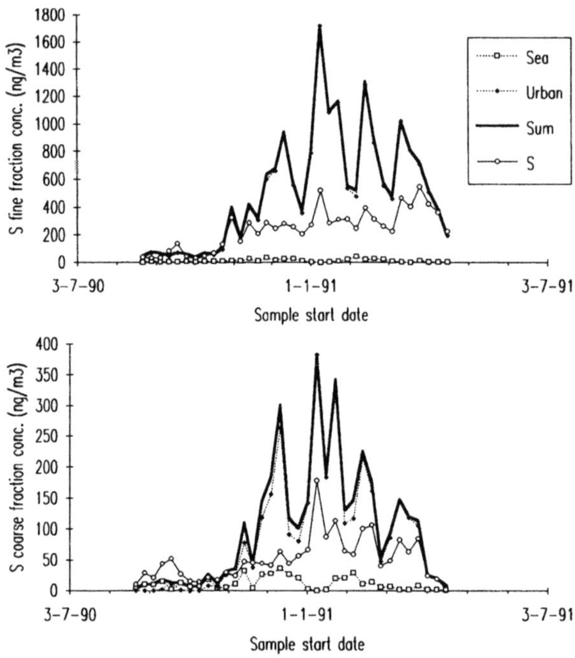

Fig. 10. Fine and coarse fraction measurements of sulphur compared with model calculations.

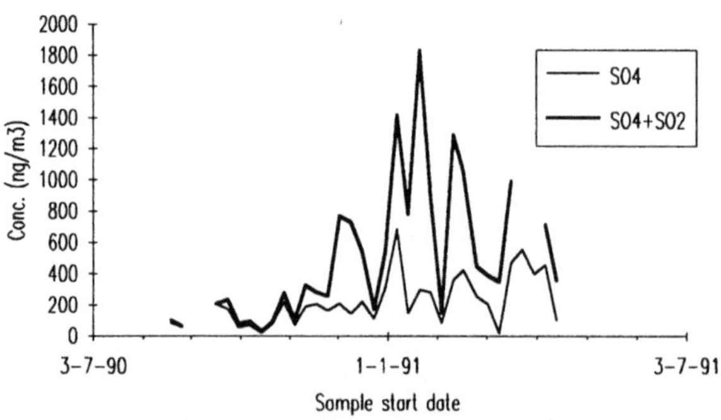

Fig. 11. Filterpack measurements of sulphur in sulphurdioxide and sulphate at Station Nord.

5. CONCLUSION

It is clear from the chi-square values that good fits could be obtained for most chemical elements when the physical receptor model was applied. A soil source profile, a sea source profile, and an urban source profile were upon the whole sufficient for the description. Episodes of enhanced Cu/Ni contents could be described by an additional metal source profile, and the residue could be characterized by a slow increase in the fine fraction V/Ni content during the winter.

The elements S, Cl, and Br presented a singular behaviour and could not be fitted, but it was possible by application of the model to calculate expected S, Cl, and Br concentrations. In this way the model could give input to a discussion about the reasons for the deviant behaviour.

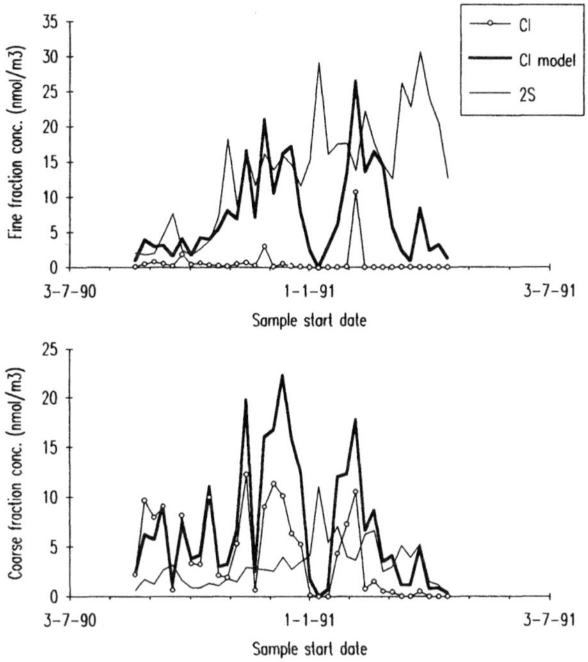

Fig. 12. Fine and coarse fraction measurements of chlorine (in nmol/m3) compared with model calculations and the concentration of sulphur multiplied by 2.

REFERENCES

(1) HENRY R.C., LEWIS C.W., HOPKE P.K. and WILLIAMSON H.J. (1984) Review of receptor model fundamentals. Atmospheric Environment 18, 1507-1515.

(2) HENRY R.C. (1987) Current factor analysis receptor models are ill-posed. Atmospheric Environment 21, 1815-1820.

(3) KEMP. K. and WÅHLIN P. (1992) Aerosol sampler for all-year sampling at remote sites. Proceedings of 5th International Symposium on Arctic Air Chemistry.

(4) KAYE G.W.C and LABY T.H. (1959) Abundances of the elements. Physical and chemical constants, pp. 108-110. Longmans, London.

RECOGNITION AND INVENTORY OF OCEANIC CLOUDS FROM SATELLITE DATA USING AN ARTIFICIAL NEURAL NETWORK TECHNIQUE

G.G. WILKINSON, C. KONTOES and C.N. MURRAY

Institute for Remote Sensing Applications
Joint Research Centre
Commission of the European Communities
21020 Ispra, Varese, Italy.

SUMMARY

The use of satellite remote sensing provides the possibility for monitoring the genesis and evolution of oceanic clouds. In order to link cloud genesis and dimethylsulphide production it is necessary to develop a method for accurately classifying and extracting the properties of cloud fields in imagery. An experiment is reported in which an artificial neural network has been used to identify the main types of clouds found in NOAA/AVHRR imagery of the northern Atlantic ocean. The method is based on the use of an unsupervised pattern recognition approach: the topological (or self-organising) map neural network. Such a network has been trained with 362 examples of 40x40 cloud fields extracted from imagery and has been used to provide 25 different cloud classes.

1. INTRODUCTION

The Institute for Remote Sensing Applications of the Joint Research Centre possesses a large dataset of daily raw satellite images from the NOAA Advanced Very High Resolution Radiometer (AVHRR) spanning more than a decade with coverage over a significant portion of the eastern Atlantic Ocean [6°-30°E, 0°-40°N]. These images with 4km. 'Global Area Coverage' resolution in 5 spectral channels (2 visible, 1 near infra-red and 2 thermal infra-red) have been used in the past to produce an archive of sea surface temperature data for coastal upwelling studies. However in principle this same raw dataset could be used also to categorise and make inventories of clouds over the ocean. This could lead to studies of time trends and possible linkages to cloud generation effects such as the phytoplankton-created Dimethylsulphide (DMS).

In order to study possible links between DMS production and cloud generation or modification, it will be useful to make accurate maps of cloud types and amounts over the very long time period permitted by the JRC's AVHRR dataset. However the recognition and categorisation of clouds in satellite images is a non-trivial problem. The visible channels of the AVHRR provide cloud reflectance information related to optical thickness and the infra-red channels provide brightness temperature information which is related to cloud height. However the recognition of cloud types also depends

G. Restelli and G. Angeletti (eds.), Dimethylsulphide: Oceans, Atmosphere, and Climate, 393–399.

on their spatial textures which can also be extracted from imagery. The cloud recognition problem in satellite images is thus a difficult discrimination problem involving multi-parameter pattern recognition. In our recent work at the JRC we have demonstrated the potential value of artificial neural networks in recognition tasks of this complexity e.g. for classifying complex satellite image data of land surfaces (1). Besides their capability for solving complex pattern recognition problems, neural networks offer the additional possibility of direct encoding in parallel computer hardware allowing very fast analysis of large data archives.

The application of neural networks to cloud classification is not new to this study and has already been reported in (2) and (3). In both of these applications, neural networks have been used to carry out supervised classification. Examples of particular image classes have been used to train an artificial neural network to encode the cloud class characteristics. The network can then be used for classification. In the applications reported multi-layer perceptron networks were used based on the backpropagation training algorithm (4).

For oceanic cloud recognition however, for DMS studies, we have chosen a different approach. The NOAA/AVHRR dataset provides a very extensive volume of images from which training examples could be extracted. However instead of manually extracting and naming class examples, we have chosen an unsupervised neural network approach to learn the characteristics of the main classes which can in principle be extracted from the dataset. This is effectively a data clustering procedure -i.e. we allow the neural network algorithm to identify the most meaningful types of cloud fields present in typical AVHRR imagery of the north east Atlantic Ocean. We do not wish to make an arbitrary manual classification of clouds in the imagery at the beginning in order to train a supervised classifier. If the unsupervised approach performs well, it will be possible at a subsequent stage to label the prototype cloud classes derived by the neural network approach and to use those prototypes as a basis for classification of new cloud field data.

2. THE TOPOLOGICAL MAP NEURAL NETWORK

The T-Map neural network algorithm (also called the self-organising map or Kohonen map) is an unsupervised pattern recognition technique (5), (6). For our application it provides a relatively easy way of using both radiance and texture information present in cloud field images.

The T-map consists of two layers of neural nodes. The first layer has one neuron node per component of input feature vectors. The second layer is a two-dimensional array of competitive neural nodes each of which will represent a prototype pattern learned from a training data set. This second layer effectively becomes a 'map' showing the natural relationships between the patterns that are given to the network.

All interconnections in the network go from the first layer to the second. The two layers are fully interconnected, thus each node of the input layer is connected to all of the nodes in the second (competitive) layer. Each interconnection has an associated weight value -see figure 1.

The nodes in the second layer each form weighted combinations of the outputs from the first (input) layer. Initially the weights between the nodes in the two layers are random but they are progressively modified during an iterative training procedure to encode a mapping from the input feature space to the classification space. The weight values are updated during the training procedure as follows:

Suppose an n-dimensional input feature vector **E** is fed to the n-node input layer of the network and is denoted as:

$$\mathbf{E} = \{e_1, e_2, e_3, \ldots, e_n\}$$

Let the set of weights **W** associated with the interconnections be denoted as:

$$\mathbf{W} = \{w_{i1}, w_{i2}, \ldots w_{in}\}$$

where 'i' denotes the ith. node in the competitive layer.

When pattern **E** is presented to the network the extent to which it matches the weights associated with each node in the competitive layer is computed. The match extent for node 'i' is $\| \mathbf{E} - \mathbf{W_i} \|$ which is the difference between the vectors **E** and $\mathbf{W_i}$. The node in the competitive layer for which this value is the smallest is the winner.

After the winning node has been identified a procedure is invoked to find the neighbourhood around it denoted as N_c. This consists of nodes close to the winning node 'c'. The size and shape of the neighbourhood can be selected by the user. Its size is normally reduced as training proceeds. Weights are then updated for all neurons that are in the neighbourhood of the winning node according to the equation:

$$\Delta w_{ij} = \alpha(e_j - w_{ij}) \text{ if node 'i' is in the neighbouhood, otherwise } \Delta w_{ij} = 0, \text{ and}$$

$$w_{ij\,new} = w_{ij} + \Delta w_{ij}.$$

This adjustment results in the winning node and its neighbours becoming more like the input pattern. The winner then becomes more likely to win if the same or a similar input pattern is presented at a later stage. The parameter 'α' is the learning rate which is also normally decreased during the training process.

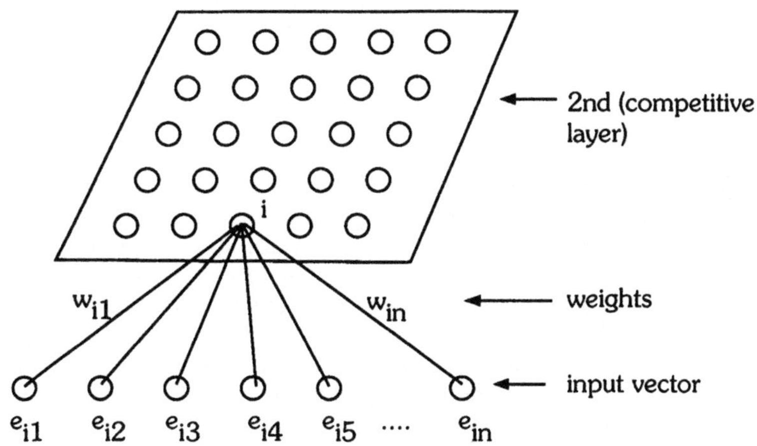

Figure 1. Architecture of the topological map network.

As the training procedure runs, the weights associated with the output layer begin to take on the characteristics of the main clusters of input feature vectors. Hence the main types of cloud fields present in the dataset begin to emerge as prototypes.

For identifying cloud field prototypes we have used a 3200-component feature vector (i.e. all pixel values in a 40x40 pixel window in both a visible and an infra-red channel) and have used a 5x5 array of nodes in the second layer to extract 25 prototype cloud fields. Both spectral and spatial information are present in the input feature vectors.

3. CLUSTERING CLOUD FIELDS USING THE T-MAP APPROACH

In order to extract the significant types of oceanic clouds which can be identified in NOAA AVHRR imagery, a total of 362 sample cloud fields were extracted from the JRC's Atlantic Ocean dataset and used to train a T-map (7). The image data were initially radiometrically calibrated and adjusted for solar zenith angle variations etc. The T-map algorithm used in these experiments was provided by a commercial neural network package 'MIMENICE' (Mimetics corporation) on a Sun SparcStation-2 computer.

Figures 2 - 4 show the 5x5 array of prototype cloud fields at the second (output) layer of the T-map. Figure 2 shows the output of the neural network before training -i.e. all cloud prototypes are blank. Figure 3 shows the output of the neural network after 30 training iterations. Already at this stage the network has been able to extract quite well the characteristics of 25 prototype cloud fields. Figure 4 shows the final cloud field prototypes at the end of training after 1530 iterations through the training set. The network has now learned the characteristics of 25 prototypes and very little change was noted between the last few iterations prior to this. Each prototype is shown as a pair of images- the one on the right is in a visible channel and the one on the left is in a thermal infra-red channel. Figure 5 gives an interpretation of the network output. These prototypes indicate the 25 most significant types of cloud fields which can be extracted from a typical AVHRR dataset. The prototypes include cumuliform, stratiform and cirriform clouds.

4. CONCLUSIONS

Satellite remote sensing has the potential to allow us to monitor the genesis and evolution of cloud patterns over large oceanic areas and through long periods of time. At the present time, one of the greatest challenges in linking DMS production and clouds is the problem of adequately identifying cloud types in satellite imagery and measuring their properties and then establishing links with plankton data available from other satellite datasets. Alongside the recognition problem there is the difficulty of handling the large volumes of satellite data involved. Here we have shown that artificial neural networks, of the topological map type, can be used as a recognition tool besides allowing us to process large datasets quickly. This research is still at an early stage, but if the T-map can be used to pick out more classes of clouds with a greater variety of properties, and if it can be encoded for parallel computing, then operational use for application to the DMS problem will be possible.

REFERENCES

(1) KANELLOPOULOS, I., VARFIS, A., WILKINSON, G. G. and MÉGIER, J. (1992): "Land cover discrimination in SPOT imagery by artificial neural network -a twenty class experiment", International J. Remote Sensing, 13, 5, 917-924.

(2) LEE, J., WEGER, R. C., SENGUPTA, S. K. and WELCH, R. M. (1990): "A neural network approach to cloud classification", IEEE Trans. Geoscience and Remote Sensing, 28, 5, 846-855.

(3) KEY, J., MASALANIK, A. and SCHWEIGER, A. J. (1989): "Classification of merged AVHRR and SMMR Arctic data with neural networks", Photogrammetric Engineering and Remote Sensing, 55, 1331-1338.

(4) RUMELHART, D. E., HINTON, G. E. and R. J. WILLIAMS (1986): "Learning internal representations by error propagation", in: Parallel Distributed Processing: Explorations in the Microstructure of Cognition -Vol. 1. Foundations, Cambridge, Massachusetts: MIT Press, pp. 318-362.

(5) KOHONEN, T. (1984): Self organization and associative memory, Springer series in Information Sciences, Vol. 8, Springer-Verlag (also 2nd. edition 1988).

(6) DAYHOFF, J. (1990): "Neural network architectures", New York: Van Nostrand Reinhold, 259pp.

(7) KONTOES, C. and WILKINSON, G. G. (1992): "Study on investigation of cloud class detectability in AVHRR GAC data by artificial neural network for studies of atmosphere-upper ocean interaction processes", Internal Report, June, Joint Research Centre, Commission of the European Communities, Ispra Establishment, Italy.

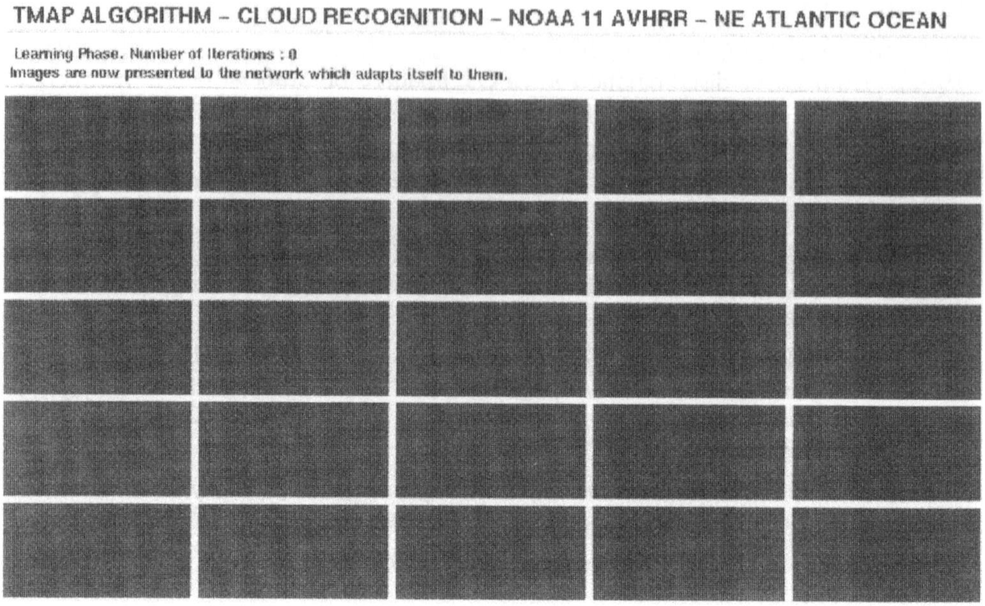

Figure 2. Prototype cloud fields from the T-map at iteration 0.

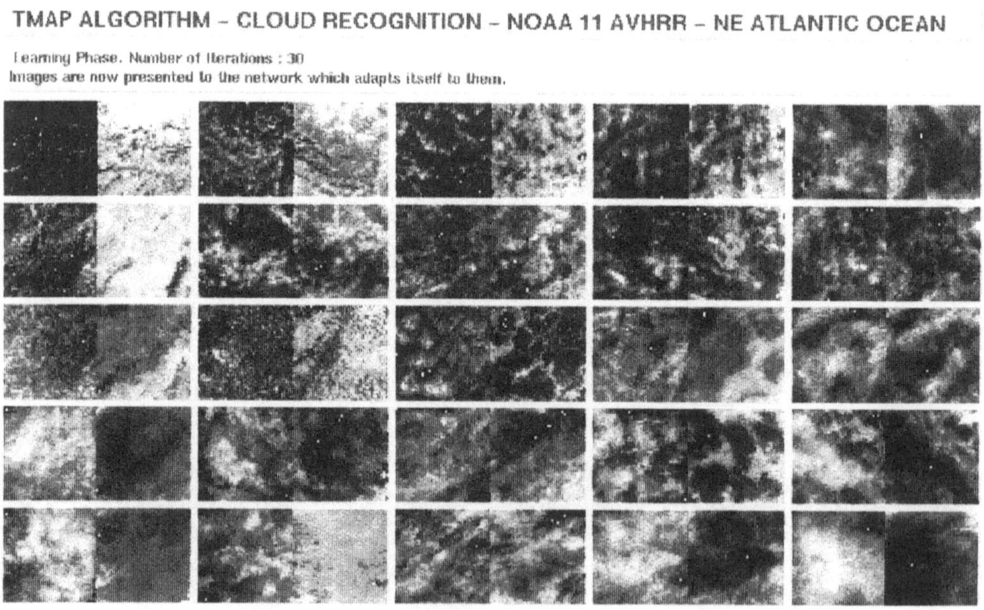

Figure 3. Prototype cloud fields from the T-map at iteration 30.

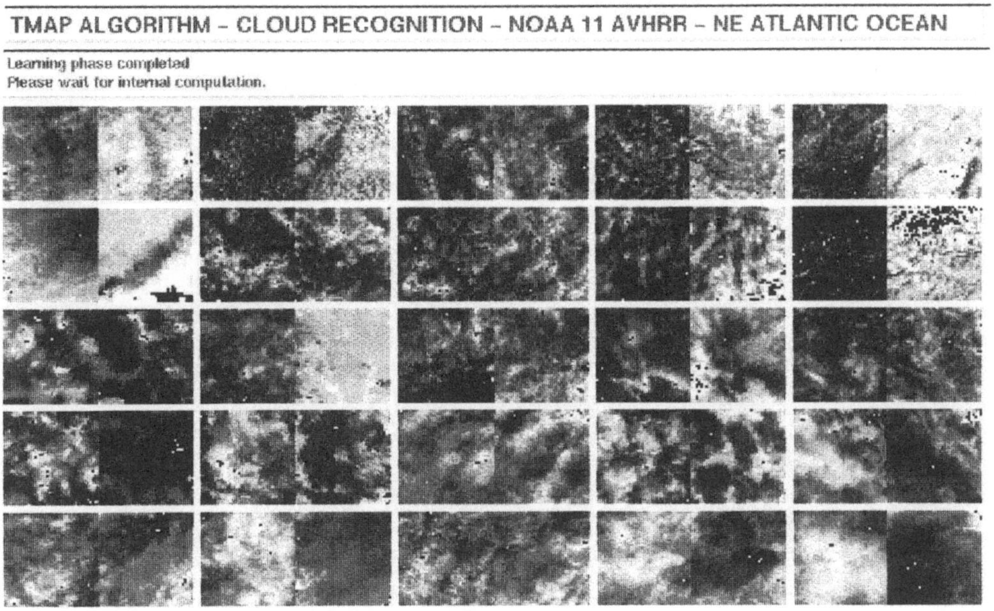

Figure 4. Prototype cloud fields from the T-map at iteration 1530 (final iteration).

F	E	C	A	A
F	E	C	B	A
J	I	M	B	G
J	K	M	H	D
K	K	L	D	D

A: Cumuliform, 4-10km. cells. Cumulus streets/Stratocumulus
B: Non-cumuliform, bands up to 100km. Low brightness temp. Altocumulus.
C: Cumuliform, 20-50km. cells. Wide albedo range. Thick stratocumulus.
D: Uniform texture, extensive (~100km.), low brightness temperature. Altostratus.
E: Cumuliform, 4-20km. cells. Wide albedo range. Stratocumulus.
F: Uniform texture, low optical thickness. Stratus/fog/cirrus.
G: Cumuliform, vermiculated, 100-150km. bands. Cumulus congestus.
H: Cumuliform, cell size ~50km. Cumulus congestus/Cumulonimbus.
I: Uniform textuire, high brightness temperature. Stratus and overlaying Cumulus.
J: Cumuliform, cell size ~20-50km. Altocumulus/Cumulus congestus/ Small Cumulonimbus.
K: Non-cumuliform, high optical thickness. Nimbostratus/Altostratus/Altocumulus.
L: Uniform texture, wide albedo range. Stratus/Stratocumulus.
M: Uniform texture, medium reflectance and brightness temperature. Thick Stratus.

Figure 5. Key to figure 4 - labels for prototypes extracted from T-map network.

[illegible]

A. [illegible]
B. [illegible]
C. Cumuliform. Small cells. Wide albedo range. Thin stratocumulus
D. Uniform texture, [illegible], low brightness temperature. Altostratus
E. Cumuliform. [illegible] cells. Wide albedo range. Stratocumulus
F. Uniform texture, low optical thickness. [illegible]
G. Cumuliform, [illegible] 100-150km [illegible] congestus
H. Cumuliform, cell size [illegible] Cumulonimbus
I. Uniform texture, [illegible] Stratus [illegible] Cumulus
J. Cumuliform. Cell size ~20-50km. Altocumulus, Cumulus congestus/
   Small Cumulonimbus.
K. Non-cumuliform. High optical thickness. Nimbostratus/[illegible]
L. Uniform texture, wide albedo range. Stratus/Stratocumulus
M. Uniform texture, medium reflectance and brightness temperature. Thick Stratus

Figure 5. Key to figure 4 - labels for prototypes extracted from [illegible]

MIX
Papier aus verantwortungsvollen Quellen
Paper from responsible sources
FSC® C105338

If you have any concerns about our products,
you can contact us on
ProductSafety@springernature.com

In case Publisher is established outside the EU,
the EU authorized representative is:
Springer Nature Customer Service Center GmbH
Europaplatz 3, 69115 Heidelberg, Germany

Printed by Libri Plureos GmbH
in Hamburg, Germany